AF517410

Theory and Practice of Integrated Pest Management

ABOUT THE AUTHORS

Dr Ramesh Arora is Senior Entomologist in the Department of Entomology, Punjab Agricultural University, Ludhiana and obtained his Ph.D. in Entomology from the same institute in 1990. He has been working on development and implementation of IPM programmes for the last more than three decades. He has more than 200 publications including 65 research articles, 10 books and 28 book chapters to his credit. He has been Editor-in- Chief (1995-2003) of the Indian Journal of Ecology and is associated with a number of professional societies.

Dr. Balwinder Singh is working as Professor-cum-Head, Department of Entomology, Punjab Agricultural University, Ludhiana. He did his Ph.D in Entomology with specialization in Insect Toxicology in 1984 from PAU, Ludhiana. He was Commonwealth Fellow at Rothamsted Research Harpenden, UK for six months. He has more than 28 years experience in teaching and research in the field of pesticide residues. He has handled many research projects funded by various international and national agencies. He has more than 200 publications to his credit. He has also edited three books and contributed 11 chapters in different books. He is President of the Indian Society for the Advancement of Insect Science and also the Vice-President, Society of Pesticide Science, India. He is acting as a reviewer for number of international journals related to field pesticide residues. He is associated with a number of professional societies at the National level.

Dr. A.K. Dhawan, Former Professor-cum-Head, Department of Entomology, Punjab Agricultural University, Ludhiana started his career an Assistant Entomologist in 1973. He has more than 302 research papers, 10 books, 22 chapters in books, 31 review articles and 121 research papers in national and international conference to his credit. He is a fellow of 4 professional societies, life members of 11 professional societies, referee of 7 scientific journals of entomology and acted as Vice President, Joint Secretary and Managing Editors in different professional societies. He is at present General Secretary of Indian Ecological Society and Society for Sustainable Cotton Production and is also Managing Editor of Indian Journal of Ecology. He participated in 8 international and 63 national workshops/conferences. He has won several awards including the ICAR Team award in "Insecticide Resistance Management in Cotton" and National Award for "Dissemination of IPM Technology Through Street Play" in 2004. He was awarded Role of Honour and Best Extension Worker Award in 2008 and 2010 by PAU, Ludhiana. He did a commendable work in transfer of technology on IPM of cotton recommended by Punjab Agricultural University, Ludhiana among the cotton growers of Punjab.

Theory and Practice of Integrated Pest Management

Ramesh Arora
Balwinder Singh
A.K. Dhawan
Department of Entomology
Punjab Agricultural University, Ludhiana

Published by

SCIENTIFIC PUBLISHERS (INDIA)
5 A, New Pali Road, P.O. Box 91
Jodhpur 342 001 (INDIA)
E-mail: info@scientificpub.com
Website : http://www.scientificpubonline.com

Print: 2026

ISBN: 978-81-72337-60-5
eISBN: 978-93-86347-82-4

Printed in India

PREFACE

Insect pests have been associated with agricultural crops since the dawn of civilization. In traditional agriculture, the yields obtained were low and farmers accepted some losses due to insect pests as inevitable. Slowly, the farmers developed a number cultural and mechanical control of practices to minimize these losses. The increasing food and fibre requirements of expanding population necessitated intensification of agriculture through introduction of high yielding varieties, expansion in irrigation facilities and application of increased amounts of agro-chemicals. The advent of synthetic organic insecticides during 1940s and 1950s enabled us for the first time to gain an upper hand in the battle against insects.

However, there is evidence that insect pest problems have escalated with an increasing cropping intensity and with the use of agrochemicals inherent in modern agriculture. Consequently, Indian plant protection scientists have intensified research on the development of pest management tactics and effective pest management systems have been designed for all the important crops in the country.

This book, consisting of 29 chapters, draws together the diverse literature on the subject of insect pest management in agriculture and contains contributions written by scientists having extensive experience with insect pest problems in Indian agriculture.

The first half of the book is devoted to the principles and components of pest management including factors affecting pest populations, construction of life tables, coevolution of insects and plants, pest forecasting, pesticides, IGRs, botanicals, entomopathogenic nematodes and molecular approaches, etc. A separate chapter has been devoted to the role of information technology in IPM. The different tactics for the management of major insect pests of principal agricultural crops of India, viz. rice, maize, wheat, forage crops, cotton, sugarcane, vegetables, fruits, oilseeds, pulse crops, jute and mesta and tobacco have been discussed in the second half of the book. The book is largely based on chapters contributed by invited experts for a 'Winter School' on Advances in Agricultural Entomology held during Dec., 2010 to Jan., 2011 at the Department of Entomology, PAU, Ludhiana. Some chapters have, however, been added to strengthen the subject matter.

We are thankful to all the contributors for the meticulous job they have done in preparing their respective chapters in accordance with the theme of the book and within the stipulated period. We are also thankful to Mr. Sandeep Singh and Mr. Sudhendu Sharma, both from the Department of Entomology, PAU, Ludhiana and Mr. H.M. Yeshwanth Department of Entomology, UAS, Bangalore for permission to use the photographs reproduced on the cover page of the book.

The book contains a wealth of information on all aspects of insect pest management in agriculture under Indian conditions and would prove indispensable for not only students, teachers and researchers in agricultural entomology but also for administrators, planners and field level extension functionaries implementing agricultural development programmes in India and other developing countries.

Editors

15th August, 2011
Ludhiana

CONTRIBUTORS

Naveen Aggarwal
Department of Entomology
Punjab Agricultural University
Ludhiana-141 004

P.K. Arora
Punjab Agricultural University
Regional Research Station
Abohar-152 116

Ramesh Arora
Department of Entomology
Punjab Agricultural University
Ludhiana-141 004

D.S. Brar
Department of Entomology
Punjab Agricultural University
Ludhiana - 141 004

H.K. Cheema
Department of Plant Breeding and Genetics
Punjab Agricultural University
Ludhiana-141 004

A.K. Dhawan
Department of Entomology
Punjab Agricultural University
Ludhiana - 141 004

L.V. Ghetiya
Bidi Tobacco Research Station
Anand Agricultural University
Anand-388 110

R.S. Gill
Department of Entomology
Punjab Agricultural University
Ludhiana- 141 004

V.K. Gupta
Department of Entomology
Punjab Agricultural University
Ludhiana-141 004

Vikas Jindal
Department of Entomology
Punjab Agricultural University
Ludhiana-141 004

B.K. Kang
Department of Entomology
Punjab Agricultural University
Ludhiana-141 004

Uma Kanta
Department of Entomology
Punjab Agricultural University
Ludhiana-141 004

Sandeep Kaur
Department of Vegetable Crops
Punjab Agricultural University
Ludhiana-141 004

Sarwan Kumar
Department of Plant Breeding and Genetics
Punjab Agricultural University
Ludhiana-141 004

Vijay Kumar
Department of Entomology
Punjab Agricultural University
Ludhiana-141 004

M.S. Mahal
Department of Entomology
Punjab Agricultural University
Ludhiana-141 004

D.M. Mehta
Bidi Tobacco Research Station
Anand Agricultural University
Anand-388 110

A.D. Patel
Bidi Tobacco Research Station
Anand Agricultural University
Anand-388 110

T. Ramasubramaniana
Division of Crop Protection
Central Research Institute for
Jute and Allied Fibres (ICAR)
Barrackpore, Kolkata-700 120

I.P.S. Sandhu
Department of Entomology
Punjab Agricultural University
Ludhiana-141 004

P.S. Sarao
Department of Plant Breeding
and Genetics
Punjab Agricultural University
Ludhiana- 141 004

S. Satpathy
Division of Crop Protection
Central Research Institute for
Jute and Allied Fibres (ICAR),
Barrackpore
Kolkata-700 120

D.R. Sharma
Department of Horticulture
Punjab Agricultural University
Ludhiana-141 004

P.S. Shera
Department of Plant Breeding
and Genetics
Punjab Agricultural University
Ludhiana-141 004

Balwinder Singh
Department of Entomology
Punjab Agricultural University
Ludhiana-141 004

Beant Singh
Department of Plant Breeding
and Genetics
Punjab Agricultural University
Ludhiana-141 004

Gursharan Singh
Department of Entomology
Punjab Agricultural University
Ludhiana-141 004

Ravinder Singh
Department of Plant Breeding
and Genetics
Punjab Agricultural University
Ludhiana-141 004

Sandeep Singh
Department of Plant Breeding
and Genetics
Punjab Agricultural University
Ludhiana-141 004

K.S. Suri
Department of Entomology
Punjab Agricultural University
Ludhiana-141 004

G.K. Taggar
Department of Plant Breeding
and Genetics
Punjab Agricultural University
Ludhiana-141 004

CONTENTS

Theory and Practice of Integrated Pest Management
R. Arora, B. Singh and A.K. Dhawan (Eds)
Scientific Publishers (India), Jodhpur 2012, pp. 1 - 17.

1

INTEGRATED PEST MANAGEMENT: CONCEPT AND APPROACHES

Ramesh Arora and A.K. Dhawan
Department of Entomology,
Punjab Agricultural University, Ludhiana - 141 004, India

1.1. HISTORICAL PERSPECTIVE

During ancient times, humans had to live with and tolerate the ravages of crop pests, but they gradually learned to improve their condition through trial-and-error experiences. Over the centuries, farmers developed a number of mechanical, cultural, physical and biological control measures to minimize the damage caused by phytophagous insects. Synthetic organic insecticides developed during the mid-20th century provided spectacular control of these pests to begin with but resulted in the abandonment of traditional pest-control practices. Many components of pest management were developed in the late 19th and early 20th centuries. Thus, the evolution of the concept of pest management spans a period of more than a century (Table 1.1). Rapidly developing technologies and changing societal values had their impact on pest control tactics also. The history of agricultural pest control thus has three distinct phases, namely, the era of traditional approaches, the era of pesticides and the era of integrated pest management (IPM) (Dhaliwal and Arora, 2006).

Table 1.1. Landmarks in the history of agricultural pest management

Period	Landmark(s)
(1)	(2)
Ancient	The Chinese used wood-ash for the control of insect pests in enclosed spaces and botanical insecticides for seed treatment. They also used ants for biological control of stored grain as well as foliage insects. In India, neem leaves were placed in grain bins to keep away troublesome pests. In the Middle and Near East, powder of chrysanthemum flowers was used as an insecticide
1782	'Underhill' variety of wheat reported resistant to Hessian fly in the USA
1831	'Winter Majetin' variety of apple reported resistant to woolly apple in the USA
1855	Fitch reported the role of ladybird beetles, green lacewings and other predacious insects in the control of insect pests of crops
1858	Pyrethrum first used for insect control in the USA
1889	Biological control of cottony cushion scale on citrus on the USA by use of the vedalia beetle imported from Australia
1890	Control of grape phylloxera in Europe by grafting of European grapevine scions to resistant North American rootstocks
1923	Multiple-component suppression techniques involving the use of resistant varieties, sanitation practices and need-based application of insecticides developed for the control of boll-weevil in the USA
1939	• Insecticidal properties of DDT reported by Paul Muller in Switzerland • *Bacillus thuringiensis* Berliner first used as a microbial insecticide
1941	Insecticidal activity of hexachlorocyclohexane (HCH) discovered in France
1946	Parathion, the first organophosphatic insecticide developed
1948	'Doom' based on *Bacillus popilliae* Dutky and *Bacillus lentimorbus* registered in the USA of the control of Japanese beetle larvae on turf.
1951	• Painter published his classic book *Insect Resistance in Crop Plants* • Introduction of first carbamate insecticide, isolan
1959	• Concept of integrated control involving integration of chemical and biological control introduced • Concept of economic injury level and economic threshold developed by Stem and Co-workers
1962	Publication of the book *Silent Spring*, by Rachel Carson, which dramatized the impact of the misuse and overuse of pesticides on the environment
1964	Publication of the book *Biological Control of Insect Pests and Weeds*, by Paul DeBach, which established biological control as a separate discipline in entomology

(1)	(2)
1975	• Elcar (*Helicoverpa nucleopolyhedrovirus* (NPV)) registered for the control of bollworm and tobacco budworm on cotton • First insect growth regulator (Methoprene) registered for commercial use in USA • Publication of the book *Introduction to Insect Pest Management* by Metcalf and Luckmann, which was the first comprehensive treatise on IPM and established the concept on a firm footing.
1980	The interest in botanical pesticides revived and the First International Conference on Neem was held at Rottach-Egern, Germany
1987	Development of first transgenic plant, reported by Vaeck and co-workers of Belgian biotechnology company, Plant Genetic System, by transferring *B. thuringiensis* δ-endotoxin gene to tobacco for the control of *Manduca sexta* (Johannsen)
1989	An IPM Task Force was established to garner international support for the development and implementation of IPM programmes. A team of consultants appointed by the Task Force reviewed the status of IPM and made recommendations. The Task Force was later reconstituted as the Integrated Pest Management Working Group (IPWG) in 1990
1990	• Concept of environmental economic injury levels proposed by Pedigo and Higley • Knipling and Bushland were awarded the World Food Prize for developing sterile-insect technique • United Nations Conference on Environment and Development (Rio de Janeiro, Argentina) assigned a pivotal role to IPM in the agricultural programmes and policies envisaged as part of its Agenda 21
1994	A Task Force consisting of FAO, the World Bank, UNDP and UNEP co-sponsored the establishment of the Global IPM Facility with the Secretariat located at FAO, Rome
1995	Herren was awarded the World Food Prize for developing and implementing the world's largest biological control project for cassava mealy bug, which had almost destroyed the entire cassava crop of Africa
1996	Insect-resistant transgenic (*Bt*) cotton, maize and potato were commercialized in the USA
1997	Smith and Adkisson were awarded the World Food Prize for their pioneering work in the development and implementation of the IPM concept
2002	*Bt* cotton approved for commercialization in India

Source: (Koul *et al.,* 2004; Dhaliwal and Arora, 2006).

DDT, dichlorodiphenyltrichloroethane; FAO, Food and Agriculture Organization; UNDP, United Nations Development Programme; UNEP, United Nations Environment Programme; *Bt, Bacillus thuringiensis*

Era of Traditional Approaches (Ancient-1938)

Cultural and mechanical practices, such as crop rotation, field sanitation, deep ploughing, flooding, collection and destruction of damaging insects/insect-infested plants, etc., developed by farmers through experience were among the oldest methods developed by humans to minimize the damage caused by insect pests. These were followed by the use of plant products from neem, chrysanthemum, rotenone, tobacco and several other lesser-known plants in different parts of the world. The Chinese were probably the pioneers in the use of botanical pesticides as well as biological control methods for the management of insect pests of stored grains and field crops (Dhaliwal and Arora, 1994).

However, systematized work on many important tactics of pest control, including the use of resistant varieties, biological control agents and botanical and inorganic insecticides, was done in the USA from the end of the 18th to the end of the 19th century. Remarkable success was achieved in the management of grape phylloxera, caused by *Viteus vitifoliae* (Fitch), by the grafting of *European grapevine* scions to resistant North American rootstocks during the 1880s. At around the same time, cottony cushion scale, *Icerya purchasi*, which was causing havoc in the citrus industry in California, USA, was successfully controlled by release of the vedalia beetle, *Rodolia cardinalis*, imported from Australia (Dent, 1995).

A number of synthetic inorganic insecticides containing arsenic, mercury, tin and copper were also developed towards the end of the 19th and the beginning of the 20th century. With the developing of these insecticides, the focus of research in pest control slowly shifted from ecological and cultural control of chemical control, even before the development of synthetic organic insecticides (Luckman and Metcalf, 1994).

Era of Pesticides (1939-1975)

The synthetic inorganic insecticides were broad-spectrum biocides and were highly toxic to all living organisms. These were followed in due course by the synthetic organic insecticides, such as alkyl thiocyanates, lethane, etc. The era of pesticides, however, began with the discovery of the insecticidal properties of 2,2-*bis* (p-chlorophenyl)-1,1,1-trichloroethane (DDT) by Paul Muller in 1939. The impact of DDT on pest control is perhaps unmatched by any other synthetic substance and Muller was awarded a Nobel Prize for this work in 1948.

The DDT was soon followed by a number of other insecticides, such as hexachlorocyclohexane (HCH), chlordane, aldrin, dieldrin, heptachlor (organochlorine group), parathion, toxaphene, schradan, O-ethyl O-4 (nitrophenyl) phenyl phosphonothionate (EPN) (organophosphorus group) and allethrin (synthetic pyrethroids), during the 1950s and a large number of other popularly used organophosphates and carbamates in the ensuing decade (Radcliffe *et al.*, 2011).

Due to their efficacy, convenience, flexibility and economy, these pesticides played a major role in increasing crop production. The success of high-yielding varieties of wheat and rice, which ushered in the 'green revolution', was partially due to the protective umbrella of pesticides. The spectacular success of these pesticides masked their limitations. The intensive and extensive use, misuse and abuse of pesticides during the ensuing decades caused wide spread damage to the environment. The most serious effects of pesticides are those on human health and life. It has been reported that at least 3 million, and perhaps as many as 25 million, agricultural workers are poisoned each year by pesticides, and some 20,000 deaths can be directly attributed to agrochemical use. Nearly three-quarters of the documented deaths take place in the Third World, even though it consumes only 15% of the global pesticide output. In addition, long-term teratogenic, carcinogenic and mutagenic effects of pesticides are well documented. In a landmark report, the National Resource Defence Council (NDRC) of the USA reported that one out of every 3400 children between 1 and 5 years of age could one day get cancer because of the pesticides they consumed as young children.

Only a small amount (< 1%) of the pesticide applied to a crop reaches target pests and the remaining (>99%) enters different components of the environment to contaminate soil, water, air, food, feed, forage and other commodities. Nearly 100% of the human population has been found to contain some residues of pesticides, such as DDT and HCH. In developing countries, such as India, a large proportion of market samples of nearly all types of food commodities have been found to contain pesticide residues above the legal maximum residue limits (MRLs). In addition to human beings, the whole range of living organisms, including natural enemies, pollinators, domestic and wild animals, birds, fish and other aquatic organisms and even soil fauna are affected by the use of insecticides in agriculture. In addition, pest problems in some crops increased following the continuous application of pesticides. This, is turn, further increased the consumption of pesticides, resulting in the phenomenon of the pesticide treadmill. The combined impact of all these problems, together with the rising cost of pesticides, provided the necessary feedback for limiting the use of the chemical control strategy and led to the development of the IPM concept.

Era of IPM (1976 onwards)

Although many IPM programmes were initiated in the late 1960s and early 1970s in several parts of the world, it was only in the late 1970s that IPM gained momentum. The first major IPM project in the USA, commonly called the Huffaker project, spanned 1972-1978 and covered six crops, i.e. alfalfa, citrus, cotton, pines, pome and stone fruits, and soybean. This was followed by another large-scale IPM project called the Consortium for Integrated Pest Management (CIPM) (1979-1985), which focused on alfalfa, apple, cotton and soybean. The average adoption of IPM for four crops was claimed to be about 66% over 5.76 million ha. In 1993, the US government set up the National IPM Initiative and submitted that implementing IPM practices on 75% of the nation's crop area by 2000 was a national goal. In a recent accounting of the progress by the US Department of Agriculture (USDA) in achieving the goal, it has been estimated that some level of IPM has been implemented on about 70% of the US crop acreage.

National IPM programmes were launched in the late 1980s and early 1990s in several developing countries. The most outstanding success has been the Food and Agriculture Organization (FAO) IPM programme for rice in South-east Asia. By the end of 1995, 35,000 trainers and 1.2 million farmers had been exposed to IPM through this programme (FAO, 1995). A recent development at FAO in support of IPM is the establishment of the Global IPM Facility, co-sponsored by the United Nations Development Programme (UNDP), the United Nations Environment Programme (UNEP) and the World Bank. The concept is in response to the United Nations Conference on Environment and Development, held in Rio de Janeiro, Brazil, in 1992, which assigned a central role for IPM in agriculture as part of Agenda 21. The Facility will serve as coordination, consulting, advising and promoting agency for the advancement of IPM worldwide (Kogan, 1998).

1.2. DEFINITIONS OF IPM

The integrated control was first defined during 1950s as applied pest control which combines and integrates biological and chemical control. Since then more than 65 definitions of integrated control, pest management or integrated pest management (IPM) have been proposed (Bajwa and Kogan, 2002). A broader definition was adopted by FAO Panel of Experts: "Integrated pest control is a pest management system that, in the context of associated environment and population dynamics of the pest species, utilizes all suitable techniques and methods in as compatible a manner as possible and maintains pest populations at levels below those causing economic injury". It is not simply the juxtaposition or superimposition of two control techniques but the integration of all suitable management techniques with the natural regulating and limiting elements of the environment. According to National Academy of Sciences, IPM refers to an ecological approach in pest management in which all available necessary techniques are consolidated in a unified programme, so that pest populations can be managed in such a manner that economic damage is avoided and adverse side effects are minimized.

According to a recent definition "IPM is a decision support system for the selection and use of pest control tactics, singly or harmoniously coordinated into a management strategy, based on cost/benefit analyses that take into account the interests of and impacts on producers, society and the environment" (Kogan, 1998).

1.3. ESSENTIAL PRE-REQUISITES FOR IPM

The IPM is a knowledge intensive system and a lot of background information regarding the pest, abiotic and biotic factors, agroecosystem and management tactics is required for taking decisions regarding execution of IPM programmes (Dhaliwal and Arora, 2006).

Identification of insect pest(s). Proper identification of the insect pest causing damage in a given situation is essential for collecting further information about the pest as well as for undertaking necessary control measures. Correct identification will lead to published information on the insects' life history, behavior and ecology which are important in the development of suitable control measures.

Correct identification may help to direct biological control workers to know that area of origin of the pest. On the other hand, wrong identification of the target organism in the native habitat may lead to the introduction of the wrong natural enemy, which may also fail to get established in the new area. Such mis-identification was one of the reasons for repeated failures in the biological control of California red scale. *Aonidiella aurantii* (Maskell) which was misidentified as a species of *Chrysomphalus*, a genus believed to be of South American origin.

Life history and behavior of the pest. In addition to the information obtained from literature, it is essential to study the life history of the pest including duration of different developmental stages, fecundity and number of generations per year, in the target area. Host range of the insect pests including the sequence of crops, non-cultivated plants and weeds should also be known.

Behavior of the pest including mode of feeding, plant parts (roots, foliage, reproductive parts, etc.) attacked, external or internal feeding behavior and mobility of the pest are also important considerations for undertaking control measures.

Natural regulating factors. Most of the insects occurring on a crop are held in check by natural regulating factors. It is essential to know the biotic (parasitoids, predators, pathogens, etc.) and abiotic (temperature, moisture, etc.) factors which regulate the population of important insect pests on a particular crop. Correct identification of natural enemies is also important in order to manipulate them for their utilization in IPM.

Need for control measures. Insects are ubiquitous in agricultural as well as natural ecosystems. The need for taking control measures against a particular insect pest depends on a large number of interacting factors including number of insects, crop value, type of damage, idiosyncrasies of the consumer and economic and environmental costs of the control measures. In IPM, location specific environmentally based economic injury levels may be used for taking decisions regarding the need for control measures.

Timing of control measures. Timing of application of control measures is a critical factor in IPM decision making. The application must be made at a time at which the pest is present at a susceptible stage. For instance, it is important to apply insecticides against early instars of *H. armigera* before they bore into the pod of fruit where they are difficult to control. Timing may also be important in relation to augmentative releases of parasitoids. Parasitoids should not be released before a pest population is large enough to allow natural enemy establishment but not too late that they cannot exert sufficient levels of control before damage is caused. Where a combination of measures is used, for instance, pesticides and natural enemies, then the timing of pesticide application needs to be adjusted to minimize the harm to populations of natural enemies and other beneficial organisms.

Selection of suitable control measures. An array of control measures is usually available for use against a particular pest. The prophylactic measures have to be taken well in advance of the appearance of the pest or even before the time of sowing of the crop. These include resistant varieties, cultural control measures and environmental manipulations for strengthening of natural control. The curative control measures including the use of chemical and biologically based pesticides may be undertaken based on economic injury levels. It is important that appropriate control measures are selected that are compatible and their combined use is practicable and effective. The selection of chemical or biological pesticide depends on its efficacy against the pest, formulation and method of application, cost of application, residual effectiveness, effect on non-target organisms, environmental cost, etc.

1.4. FRAMEWORK OF IPM

Every IPM programme involves application of pest management tactics based on pest surveillance and forecasting. The decision making system is utilized for choosing the most suitable pest control option which is then applied or implemented.

Pest Management tactics. The pest management tactics are either preventive or therapeutic. *Preventive practices* utilizes tactics to lower environmental carrying capacity (reduce the general equilibrium position) or increase tolerance of the host to pest injury. Prevention relies on an intimate understanding of the pest's life cycle, behavior and ecology. The preventive tactics involve natural enemies, host resistance and cultural practices. In addition, quarantines are also an important component of preventive tactics. *Therapeutic tactics* are applied as a correction to the system when necessary. The objective of therapy is to dampen pest population below the economic injury level. The only widely used therapeutic tactic is the use of conventional insecticides but other approaches like microbial agents, augmentation of natural enemies, use of growth regulators, etc. may also play a vital role (Pedigo and Rice, 2008).

Actual integration involves proper choice of compatible tactics and blending them so that each component potentiates or complements the other. Probably, the earliest example of integration of techniques was the use of a combination of resistant varieties and sanitation practices as prophylactic measures combined with application of calcium arsenate at high population level in case of boll weevil on cotton in USA during first quarter of the twentieth century. Similar programmes were being developed for other pests also but the advent of synthetic organic insecticides intervened and these techniques were relegated to the background. The misuses and abuses of insecticides have again focused attention on integrated control measures.

Pest Surveillance and Forecasting. Pest surveillance is the watch kept on a pest for the purpose of decision making. The ultimate goal of pest control programmes is the accurate forecasting of pest attacks before they actually take place, so that control measures can be planned with maximum efficiency. The forecasting may be in the form of emergence or occurrence warnings or it may be quantitative based on sampling. The person employed for sampling growing areas regularly for the purpose of making pest management decisions are called pest management scouts and their activity is called scouting.

The forecasting may also be done by prediction. Temperature is the single most important factor controlling insect development and hence population numbers. Most prevalent method of measuring the effect of temperature on insect development is the degree-day method. Degree-days represent the accumulation of heat units above some temperature for a 24-hr period. Below this minimum, no development takes place, but above it, heat units are accumulated toward development. The latest method of pest forecasting is by modeling. A model is an abstract representation of a real-world situation that behaves like the real world system in certain respects. Sampling models, phenology models and economic-injury level models are commonly used in pest forecasting.

1.5. DECISION MAKING SYSTEMS

Pest management is a combination of processes that include decision making, taking action against a pest, and obtaining the information to be used in reaching these decisions. In assessing, evaluating and choosing a particular pest control option, farmers are likely to take three major factors into account.

- Farmer's perception of the problem and of potential solutions is the most important factor. Here, the farmer's ability to identify pests, his assessment of likely and potential pest losses, and his opinion regarding the efficacy of different control options will affect the decision process.
- The way in which control options are assessed will depend on the farmer's objectives. Subsistence farmers may opt for a guaranteed food supply while commercial farmers are more concerned with profit.
- The number of options that a farmer can feasibly use, would depend on the constraints set by the resources available.

Various alternative pest control options could be evaluated for their cost effectiveness. The practice that achieves effective results with least cost is the most cost-effective practice and is adopted.

1.6. IMPLEMENTATION

Although IPM has been accepted in principle as the most attractive option for the protection of agricultural crops from the ravages of insect and non-insect pest, implementation at the farmer's level has as yet been rather limited. Some of the important constraints to wider adoption of IPM and suggested measures to overcome them are discussed in this section (NRI, 1992).

Constraints in IPM Implementation

The Consultant Group of the IPM Task Force has conducted an in-depth study of the constraints on the implementation of IPM in developing countries, which can be categorized into the following five main groups.

Institutional constraints. IPM requires an interdisciplinary, multifunctional approach to solving pest problems. Fragmentation between disciplines, between research, extension and implementation and between institutes leads to a lack of institutional integration. The traditional top-down research in many cases does not address the real needs of farmers, who eventually are the end-users and who elect to adopt or reject the technology based on its appropriateness.

Informational constraints. The lack of IPM information that could be used by farmers and extension workers is a major constraint in implementation. While the individual control techniques are well known, little knowledge is available on using them in an integrated fashion under farm conditions. The lack of training materials, curricula and experienced teachers on the principles and practice of IPM is another major constraint. In many cases, the field level extension workers are not sufficiently trained in IPM to instill confidence in the farmers.

Sociological constraints. The conditioning of most farmers and farm level extension workers by the pesticide industry has created a situation where chemicals are presented as highly effective and simple to apply. This acts as a major constraint in IPM implementation. There appears to be a direct conflict between the industry's objective of more sales and the IPM message of rational pesticide use in the eyes of farmers. There is a need for private industry and public-sector extension agencies to work in a more complementary manner.

Economic constraints. A major constraint, even if IPM is adopted in principle, is the funding for research, extension and farmer training needed for an accelerated programme. IPM must be viewed as an investment and, as with other forms of investment, requires an outlay. In the long run, IPM programmes may become self-generating due to savings on resource inputs for production. A majority of the farmers purchase pesticides on credit and depend on shopkeepers and pesticide dealers for information about pest-control methods.

Political constraints. The relatively low status of plant-protection workers in the administrative hierarchy is a constraint to general improvement in plant protection. Associated with the above is the moral and financial standing of these workers. The continuance of pesticide subsidies by the government for political reasons and their tie-up with government-provided credit for crop production act as a major constraint to farmer's acceptance of IPM. Various vested interests associated with the pesticide trade also act as a political constraint on the implementation of IPM.

Strategies for IPM Implementation

Acceleration of IPM implementation in developing countries requires farmers' participation, increased government support, an improved institutional infrastructure and a favourable environment.

Farmer's participation. It would not be an exaggeration to say that the dawn of civilization started with farmer innovation. Ever since that day, farmers have improved ways of growing crops through successive innovations. Farmer's innovations were the only source of improvements in crop production and protection technology until formal research by on-station scientists started complementing it during the late 18th and 19th centuries.

Unfortunately, with the advent of modern high-tech agriculture, farmers have been completely displaced from the research and development process. Instead, this role has been usurped by private industry and government agencies. The alternative path of sustainable agriculture requires farmer's participation at every step of the research and development process in order to draw on their understanding of the local conditions and constraints, their innovativeness and their skills at making the best possible use of limited resources.

Placing the farmer at the centre of the development process is wholly consistent with the IPM goal of making the farmer a confident manager and decision maker, free from dependence on a constant stream of pest-control instructions from outside. The role of researchers, extension workers and non-governmental organizations (NGOs) is to act as consultants, facilitators and collaborators, stimulating and empowering the farmers to analyse their own situation, to experiment and to make constructive choices (Heinrichs, 1998).

Government support. Both the national programmes of developing countries and the donor agencies must have a policy commitment to IPM in the context of national economic planning and agricultural development. The costs to developing countries of not bringing their policies in line with the objectives of IPM are relatively greater than the costs to developed countries. National policies to promote IPM requires close regulation at all stages related to the importation and/or manufacture, distribution, use and disposal of pesticides. In the case of pesticides that do not meet prescribed standards for safety, persistence, etc., important and manufacturing bans should be enacted. At a minimum, the conditions laid out by the FAO *Code of Conduct on the Regulation, Distribution and Use of Pesticides* should be adopted. Pesticide subsidies need to be eliminated in order to make IPM an attractive alternative. The funds thus saved may be utilized for the implementation of IPM. Funds may also be diverted from some of the current research programmes to IPM-oriented plant-protection programmes. Additional monetary resources may be generated through cooperation with bilateral/multilateral agencies willing to support such programmes (NRI, 1992).

Legislative measures. IPM is an information system and its adoption reduces pest-control costs. The alternative to IPM is the indiscriminate use of broad-spectrum synthetic organic pesticides. Unfortunately, while pesticide manufacturers and users (farmers) derive the full benefits from the use of these chemicals, they pass on the environmental and ecological costs of their use to the society as a whole. If they are made to bear the full cost of the use of these toxicants, they may find IPM a more economical and attractive alternative. This could be achieved by enforcing suitable legislative measures.

Secondly, the success of an IPM programme in any geographical region depends upon its implementation by all the farmers in the area. Ideally, farmers may voluntarily adopt an IPM programme but some farmers may hold out. Such farmers, called 'spoiler holdouts', may impair the success of a programme by failing to adopt a necessary practice, thus causing damage to adjacent areas. This is especially important in the case of mobile pests. In addition, some farmers may free-ride and thus shift the costs of implementing and managing a programme to a group of participating farmers. To overcome 'spoiler holdouts' and 'free riders', it may be necessary to impose a programme upon an unwilling minority through suitable legislative measures.

Improved institutional infrastructure. IPM cannot be implemented unless there is a basic infrastructure for plant protection in a country. There is a need to develop and support national programme capabilities for on-farm testing and technology extrapolation. At the international level, the establishment of an IPM working group to coordinate and monitor the funding of IPM projects is bound to provide an impetus to the implementation of IPM.

IPM is predominantly a knowledge technology, the use of which requires training of the many groups involved. There is currently little training material for most of these groups, including farmers, extension personnel and researches. If IPM is to become the major approach for pest management in the developing world, this deficiency must be remedied urgently.

Lack of a reliable database has also hampered the progress of IPM programmes. A reliable source of accurate information on the status of crops and pests in farmers' fields is necessary for many IPM activities. Most of the successful IPM programmes in both developed and developing countries have a reasonably accurate system of monitoring and evaluating various biological and environmental parameters in the agro-ecosystem. A reliable database on crop yield and pest losses is required for planning and resource allocation at the national and international level. Systems analysis has been used as a problem-diagnosis tool for IPM in developed-country cropping systems and may be used in developing countries as well.

Improved awareness. Increased education and awareness regarding the objectives techniques and impact of IPM programmes are required at all levels including policy makers, planners, farmers, consumers and the general public. The importance and benefits of pesticides are being overemphasized by a multibillion dollar industry utilizing the services of not only their salesman but also agricultural scientists, administrations and planners. There is not yet a strong market in IPM information. Policy makers and planners need to be convinced that without IPM current agricultural production systems are not sustainable. Similarly, much important information that might induce a farmer to adopt IPM is not immediately observable and is therefore not sought by the farmer. A manufacturer has no incentive to recommend a programme that uses less pesticide or even selective pesticides that kill a limited range of pests.

Consumer groups and the general public may also be able to support the implementation of IPM programme by demanding residue-free commodities. There is now a distinct market for organically produced food and other products and consumer groups need to be strengthened in developing countries, so that there is a public-oriented movement for implementation of IPM.

1.7. PERSPECTIVES IN IPM

Major efforts in implementing IPM in irrigated rice have been carried out in Asia by the FAO through the Inter-Country Programme for the Development and Application of Integrated Pest Control in Rice in South-east Asia. This programme remains one of the best examples of IPM implementation in the tropical region. It involves purposeful, direct efforts to change farmers' practices, in contrast to some more indirect routes of IPM technology diffusion in many industrialized, temperate environments. The programme itself has evolved into its present transnational form from a relatively small project supported by Australia in the late 1970s, following the large-scale pest outbreaks in several South-east Asian countries (NRI, 1992).

The first phase of the FAO programme (1980-1986) focused on developing and testing the technical aspects of the IPM concept in its seven participating countries, namely, Bangladesh, India, Malaysia, the Philippines, Sri Lanka and Thailand. More recently, the project has been directed towards enhancing farmers' adoption of IPM. Australia, The Netherlands and the Arab Gulf fund have supported the programme. One significant accomplishment of the programme has been to cause policy changes within several governments, in the form of official support of IPM as the means for national plant protection in the Philippines, Indonesia, India, Sri Lanka and Malaysia (APO, 1996; Ooi, 2000).

A case study of the National IPM Programme in Indonesia as a part of the regional programme during 1989-1991 provides an interesting scenario. Following research findings showing the relation between brown planthopper outbreaks and high pesticide use, the Indonesian government banned the use of 57 broad-spectrum pesticides and declared IPM as the national pest-control strategy. These measures, together with the gradual abolition of the high subsidies on pesticides, created a favourable climate for the large-scale implementation of IPM.

To assess the impact of IPM on pest control and yield, the latter were compared with those of non-IPM farmers. IPM farmers really decreased their frequency of pesticide spray during and after the training to a level consistently lower than that of non-IPM farmers' expenditure on insect chemical control decreased. The number of farmers not applying pesticides was also significantly higher among IPM farmers. In spite of this, IPM farmers obtained higher yields than non-IPM farmers, which, in addition to lower pest-control expenditure, resulted in higher returns from the crop. Additionally, there were fewer variations in fields among IPM farmers than among non-IPM farmers, indicating less risky farm management under the IPM regime. Timeliness and adequacy of various cultivation and pest-control practices, resulting from better monitoring and decision making skills, seem to be most important in this respect (Van de Fliert, 1998).

During the second phase of the National IPM Programme (1994/95), 310,550 farmers participated in rice IPM field schools. IPM farmers applied pesticides, on average, 1.25 times while non-IPM farmers made an average of 3.62 applications during the season. Similarly, IPM farmers used fewer formulations (cocktails) per spray event, with an average of 2.6 for IPM farmers versus 7.2 for non-IPM farmers.

In India, FFSs were organized at locations covering all the major rice growing areas of the country during 1994/95. Pesticide use in IPM fields decreased by 50-100% as compared with non-IPM areas. There was an increase in yield between 6.2 to 42.1% in IPM fields as compared with non-IPM areas.

The success of the rice programme persuaded FAO to launch similar IPM Intercountry Programmes for vegetables (1996) and cotton (1999). Moreover, in the individual countries, FFS activities were started in a range of other crops, sometimes without. The Asian model of FFS IPM training has been applied in other continents as well, albeit with adjustments for different cropping and socio-economic conditions (Bartlett, 2006).

1.8. POTENTIAL OF IPM

Initially, IPM programmes evolved as a result of the pest problems caused by repeated and excessive use of pesticides and increasing cases of pest resistance to these chemicals. It is only during the past few years that economic and social aspects of IPM have also received increasing attention. Some of the important advantages offered by IPM over the pesticide-based plant protection programmes are listed here (NRI, 1992).

(i) **Sustainability.** It is now being increasingly recognized that modern agriculture cannot sustain the present productivity levels with the exclusive use of pesticides. Increasing pest problems and disruptions in agroecosystem can only be corrected by use of holistic pest management programmes.

(ii) **Economics.** If the environmental and social costs of pesticide use are taken into account, IPM appears to be a more attractive with lower economic costs.

(iii) **Health.** Production, storage, transport, distribution and application of pesticides involves greater health hazards than the safe inputs used in IPM. In developing countries, it is almost impossible to implement residue limits or waiting periods for pesticides on food products and other commodities. This endangers the safety of the entire population of these countries.

(iv) **Environmental quality.** The IPM programmes, do not endanger non-target organisms, nor do they pollute the soil, water and air. The clean air, water and soil are now being recognized as non-renewable resources which once polluted are almost impossible to purify.

(v) **Social and political stability.** The pesticides used by the farmers are obtained from the corporate houses and even from other countries. The inputs used in IPM are usually based on local resources and outside dependence is minimized. This helps in maintaining social and political stability.

(vi) **Local knowledge.** IPM builds upon indigenous farming knowledge, treating traditional cultivation practices as components of location-specific IPM practices. This is especially important for the farmers in developing countries where traditional agricultural systems are based on indigenous farming practices. The incorporation of IPM into these practices helps the farmers to modernize while maintaining their cultural roots.

(vii) **Export of agricultural commodities.** The presence of pesticide residues is affecting our exports of agricultural and horticultural commodities. There is a growing demand for organically cultivated, fresh and processed fruits and vegetables. Thus, implementation of IPM in these crops will give boost to export of fresh and processed agricultural commodities from India and other Asian countries.

REFERENCES

APO (1996) *Integrated Pest Management in Asia and the Pacific.* Asian Productivity Organization, Tokyo, 170 pp.

Bajwa, W.I. and Kogan, M. (2002) *Compendium of IPM Definitions (CID) – What is IPM and how is it defined in Worldwide literature?* IPPC Publication No. 998, Integrated Plant Protection Centre (IPPC), Oregen State University, Corvallis, OR 97331, USA.

Bartlett, A. (2006) *FAO Community-IPM.* www.communityipm.org

Dent, D. (1995) *Integrated Pest Management.* Chapman & Hall, London, 356 pp.

Dhaliwal, G.S. and Arora, R. (eds) (1994) *Trends in Agricultural Insect Pest Management.* Commonwealth Publishers, New Delhi

Dhaliwal, G.S. and Arora, R. (2006) *Integrated Pest Management: Concepts and Approaches.* 2nd Revised edition, Kalyani Publishers, New Delhi.

FAO (1995) *Intercountry Programme for the Development and Application of Integrated Pest Control in Rice in South and South-East Asia.* FAO Plant Protection Service, Rome, Italy.

Heinrichs, E.A. (1998) *IPM in the 21st century: challenges and opportunities.* In: Dhaliwal, G.S. and Heinrichs, E.A. (ed) Critical Issues in Insect Pest Management. Commonwealth Publishers, New Delhi, pp. 267-276.

Koul, D., Dhaliwal, G.S. and Cuperus, W. (eds) (2004). *Integrated Pest Management: Potential, Constraints and Challenges.* CABI, Wallingford.

Kogan, M. (1998) Integrated pest management: historical perspectives and contemporary developments. *A. Rev. Ent.* 43: 243-270.

Luckman, W.H. and Metcalf, R.L. (1994) The pest management concept. In: Metcalf, R.L. and Luckman, W.H. (eds) *Introduction to Insect Pest Management.* John Wiley & Sons, New York, pp. 1-34.

NRI (1992) *Integrated Pest Management in Developing Countries: Experience and Prospects.* National Resources Institute, Chatham, UK.

Ooi, P.A.C. (2000) Present status of IPM in the Asian region. In: *Farmer-led Integrated Pest Management*, Asian Productivity Organization, Tokyo, pp. 21-29.

Pedigo, L.P. and Rice, M.E. (2008) E*ntomology and Pest Management.* 6th Edition, Macmillan, New York.

Radcliffe, E.B., Hutchison, W.D. and Canalado, R.E. (eds) (2011) *Radliffe's IPM World Textbook.* url:http//ipmworld.umn.edu, University of Minnesota, St Paul, MN.

Van de Fliert, E. (1993) *Integrated Pest Management: Farmers' Field Schools Generate Sustainable Practices.* Wageningen Agricultural University, Wageningen, The Netherlands.

Theory and Practice of Integrated Pest Management
R. Arora, B. Singh and A.K. Dhawan (Eds)
Scientific Publishers (India), Jodhpur 2012, pp. 18 - 37.

2

FACTORS AFFECTING INSECT-PEST POPULATIONS

P.S. Sarao and M.S. Mahal[1]
Department of Plant Breeding & Genetics, [1]Department of Entomology
Punjab Agricultural University, Ludhiana - 141 004, India

Population is a group of organisms of the same species occupying a particular area at a specific time. The magnitude of decrease and increase of population is governed by the degree of influence of various physical and biotic factors of the environment as well as the genetic make-up of the species. The rate of change in a pest population is determined by the fecundity, speed of development and survival among its members. The environmental factor that favour fecundity and speed of development are not inimical to survival but promote increase numbers but those having reverse influence would cause decline in numbers. The same factor may be favourable in case of one population but may become unfavourable of another. It is thus necessary to consider the influence of various factors with respect to a particular pest population (Atwal and Singh, 1990).

The environment of an animal is everything that may influence its chances to survive and multiply (Andrewartha and Birch, 1954). The Environmental factors are abiotic and biotic. Among abiotic factors, it is temperature, moisture and light that have a direct influence on the population of insect-pests. These factors also influence the pest population indirectly by modifying the biotic factors. The biotic factors include food and other populations, primarily the natural enemies of the insect-pests.

2.1. ABIOTIC FACTORS

Abiotic factors also refers to climatic factors. The climate is a broader term and explains the, "the prevailing atmospheric phenomena and conditions of temperature, humidity, wind etc. of country and region", whereas, weather is, "the condition of the atmosphere at a given place and time with respect to heat, cold, sunshine, rain, cloud, wind etc." Abiotic factors of environment have a direct influence on animal distribution and abundance. These factors interact with one another to influence the ecology of insects, and are often correlated with one another. The number of insects caught per week in a malaise (flight intercept) traps were positively related to air temperature, but negatively related to wind speed (Brier *et al.,* 2003). But another approach is to study the effect of each component separately on the development and physiology of individuals. Such studies can be made keeping one component constant under laboratory conditions. After knowing the influence of one component, other components can be varied. Temperature and moisture below and above the favourable zone has lethal affect for a species. However, light do not kill animals but influence their behaviour (both photo tactic and physiology). Affect of different abiotic factors can be studied as given below:

2.1.1. Temperature

Temperature influences every thing that an organism does (Clarke, 2003). Depending on temperature there are two type of animals:

(a) Poikilothermic or cold blooded animals. Insects are poikilothermic, i.e. their body temperature depends upon the environmental conditions. Different metabolic processes in the body occur at a rate depending upon the environmental temperature. So, temperature plays an important role in the life of insects, since it affects either directly or indirectly their development, survival, abundance and reproduction. Densities of pest populations tend to fluctuate with the prevailing temperature conditions. Some insects like bees or moths can elevate the temperature of their flight muscles by rapid contractions before take off. The body temperature can be regulated by moving to the place of favourable range of temperature. Insect body temperature rises and falls with increasing rates of respiration and water evaporation.

(b) Homeothermic or warm blooded animals. These animals maintain a constant temperature, irrespective of changes in environment, e.g. Mammals, birds etc.

There is a gradient of temperature on land, water and soil. Land surface is directly exposed to changes in atmosphere so it shows great variations in the diurnal and seasonal temperatures. Such variations in water and soil are relatively low. Most of individuals live under moderate conditions as no individual can thrive on such a wide range of 0 to 60°C. In general terrestrial animals (20-40°C) have a wider range of favourable temperature than the aquatic ones (5-10°C). The growth of poikilothermic animals arrested at about 0°C. This temperature is called Development Zero or threshold of development. In high temperature, the metabolism is affected adversely and also there is desiccation. These two factors cause death in animals.

Effect of temperature can be discussed in the three ranges i.e., favourable range, lethal high range and lethal low range.

Effect of favourable temperature. Most of animals tend to congregate between quite a narrow limit of temperature which is preferred. This narrow range of temperature is called Temperature preferendum or preferred temperature. Temperature in favourable zone influences the behaviour, development, fecundity, sex ratio, size and colour of insects.

Temperature. Influences behaviour of insects e.g. locust have a tendency to congregate in sunny places to get warmth and when the ground temperature approaches 17-22°C, locust crowd become gregarious and then take to flight.

Calandra oryzae (stired grain pest) generally feeds near the surface of bins. With the development of colony, temperature rises due to metabolism. As temperature reaches 32°C, adults move away to cooler places. In case of rose thrips (*Thrips imaginis*), also average number of thrips on any flower on any day increases by 25% with every 5°C increase in the temperature.

A temperature between 20-30°C may not have the same effect as the constant temperature of 25°C. It has been observed that organism accustomed to fluctuating temperature cannot go on well with the constant temperature Results obtained at constant temprature can not be applied to field conditions and vice-versa because of different considerations like:

1. Life cycle of insects may include diapause stage which requires low temperature than that favourable for morphogenesis.
2. Life cycle of insects may include exposure to low temperature Short exposure of this may hamper the development.
3. Different stages in life cycle may have different response to temperature within the favourable zone. Within the same stage, this response may vary with age or sex, e.g. In holometabolus insects, different stages live at different times of the year and at different places and experience different range of temperature
4. Short exposure to extreme temperature may follow healthy development, but if this extreme temperature is experienced continuously it may be lethal.

5. Relationship between temperature and rate of development is not linear but it is curvilinear.

Short exposure of insects to extremes of temperature may even impair/injure its competency to develop at a favourable temperature e.g. *Locusta migratoria* is a pest of tropical and sub-tropical climate. When eggs of this insect are incubated at 5°C for 16 days, half of the eggs died and when these were exposed for 32 days at 5°C all the eggs died. Even the eggs incubated at this temperature did not develop normally when returned to favourable temperature. Whereas, in some healthy development may follow short exposure of insects to extremes of temperature which would otherwise be harmful or even lethal if experienced continuously e.g. post diapause eggs of *Austroicetes cruciata* (Australian grasshopper) will develop and hatch between 16 and 33°C. But at 33°C, harmful effects of continuous high temperature is manifested in eggs. When eggs were incubated on alternate days at 33°C and 19°C, it was found that rate of development was greater than when eggs were exposed to 33°C continuously. Development rate of the predatory bug, *Joppeicus paradoxus* increases with temperature but shows a decline in the nymphal stage after around 32°C (Morimoto *et al.,* 2007).

The total heat energy required for completion of a certain stage of development in the life history of a species or in the completion of a physiological process is constant and this is called Thermal constant. Within the favourable range, thermal constant is not affected by the level of temperature It can be measured as Day Degree (one degree of mean temperature lasting for one day). Development in insects is often predicted and modeled by the accumulation of day-degrees. E.g. if an insect requires 100 day-degrees to develop from egg ro adult, then in theory this could be consista of 10 days at 10°C or 5 days at 20°C. This is not that simple, very high and very low temperature is deleterious. There is normally a base temperature below which the accumulation of day-degrees does not take place (Hunter, 1993). The scale insect, *Hemiberlesia rapax,* a pest of kiwi fruit in New Zealand, has a total development time of 1056 day degree above the base temperature of 9.3°C (Blank *et al.,* 1995). For shorter life cycles, it is more convenient to use the measure Hour-Degree. The total heat energy (heat summation) can be expressed as Development unit, "It is the amount of development produced in a given unit time by the increase of one degree centigrade".

$$\text{Development units (U)} = \frac{\text{Total duration of development period (T) / No. of days or hours}}{\text{(nt) required for its completion}}$$

The development units are the reciprocal of fraction of amount of development, which takes place in a day or an hour. This information can be used to predict the time of appearance of a given pest. Mean development time of the pine false webworm, *Acantholyda erythrocephala*, a sawfly development rate of larvae increases as development time decreases, but, in case of eggs development stops altogether at around 30°C and the eggs fail to hatch at temperature below 6°C (Lyons, 1994). Clearly there is a fairly restricted temperature range over which insect eggs can survive and mature. However, this prediction is difficult as the temperature in the field is not constant and other factors also interfere.

Generally, fecundity of insects is more influenced by moisture, food, number of other animals and age of animals, but temperature also affects the fecundity. In some insects, no eggs are laid below and above a certain temperature However, exposure to low temperature preceding the optimum temperature may increase the rate of egg production. The gravid females od diapausing curculionids require an exposure to a low temperature before they can lay eggs. Fecundity of insects tends towards maximum at moderately high temps. Which may be near the upper limit of the favourable range. As the upper reaches the fecundity again decreases. In *Plutella maculipenis*, when larvae and adults were reared at 18°C more eggs were produced than when it was reared at 22°C. In most insect species, rate of egg production at temperature below a certain well defined lower limit is virtually zero. Development rate of predatory ladybeetle, *Propylea dissecta* increases linerly with temperature while survival from egg to adult peaks at around 27°C (Omkar & Pervez, 2004).

In case of plague flea, *Xylopsylla clavipes*, males outnumbered females on the rats when temperature was between 21-25°C, but the position wsa reversed on more cooler days. In case of *Macrocyclops albidu*, as the temperature rises there is a significant increase in the number of males.

In warm and humid climates, many animals like mammals, birds and insects bear darker pigmentation. This phenomenon is called Gloger's Rule. Walking stick (*Carausius* sp.) is known to become black at 15°C and brown at 25°C.

Lethal influence of temperature. In nature, temperature is not always in the favourable range. It tends to fluctuate.

Low temperature slows down the growth of insects. At still lower temperature (just below 0°C), the growth or activity is totally stopped. In this case, the activity is resumed again if the insects are subjected to warm temperature Crozier (2003) found that the minimum winter temperature constrained the range of the skipper butterfly, *Atalopedes campestris* and concluded that winter warming is required for the expansion of the species. Insects that livein temperate or artic regions must be able to survive extremes of cold and they adopt a combination of tactics such as diapause or cold-hardiness to ensure that atleast some of their population survives to the following spring (Andreadis *et al.*, 2005). However, prolonged exposure to low temperature may produce ill-effects like deformities of antennae, wing pads and legs etc. If cold hardy insects is chilled (below 0°C) the body tissue do not freeze immediately but are super cooled beyond the freezing point. Generally, body fluid remains unfrozen in a super cooling state when chilled at below 0°C. Super cooling takes place, because water is present in the body in a finally divided state in association with colloids. When chilling takes place, small isolated ice-crystals are formed here and there throughout the animal body. The colloids in the body act as insulator and do not allow the unfrozen super cooled water to freeze (Sinclair *et al.,* 2003; Powell and Bale, 2004). When accident occurs, isolates break down and wave of freezing sweeps through most parts of the body.

Insects could be divided into three groups according to their exposure to low temperatures:

(a). The ones that can not survive for any considerable period if temperature falls below the favourable range for normal development. They can not become dormant at low temperature They must either develop or die. Species originating or living in tropical or subtropical regions e.g. Cimex, Calendra and Locust, no stage in the life cycle is capable of becoming dormant.

(b). Species, originating from temperate climates, which have a stage in the life-cycle adapted to survive exposure to low temperature and other stages resembles the tropical or subtropical forms in lacking the capacity to become dormant at low temperature e.g. *Helicoverpa* and *Agrotis* sp.

(c). The third group has a special over wintering stage of species adapted to cold temperate climate. Cold hardiness and dormancy is nearly always associated with a condition known as diapause. e.g. Sugarcane borer, Red hairy caterpillar

The species adapted to live in temperate climates, hibernate in a particular developmental stage and frequently this very stage and none other going diapause. Some species has insulating protection of hibernacula such as debris, soil, plant remnants, snow and ice and also their ability to withstand exposure to low temperature through under cooling. The combination of these two mechanisms prevents mortality, but in case of incomplete protection exposure to severe cold period may lead to eradication of the insect population. This is more relevant in case of introduced natural enemies. In extreme cases, aquatic insects such as various dipterans larvae may have to diapause inside ice forming on their ponds in winter (Bouchard *et al.,* 2006). This occurs by way of a process known as super-cooling. The super-cooling point (SCP) is considered to be as absolute lower limit for survival. Worland *et al.* (2004), reported that SCP in case of cockroach, *Celatoblatte quinquemaculata* was -3.2°C, whereas, rebound temperature (where the body actually appears to resist freezing) was -1.5°C and minimum temperature was -12°C. The mechanistic basis for insect super-cooling seems to be enhanced concentration of glycerol in the body fluids, which act as anti freeze. Spruce budworm, *Choristoneura fumiferana* produces higher levels of glycerol when exposed to lower temperature so that as winter sets in the insect is able to respond by setting up its supercooling abilities in advance of serious cold (Han and Bauce, 1995). However, if these cold spells arrive too early, and the larvae are not in the physiological state to synthesize glycerol, they may be frozen to death. Extreme prolong periods of relatively extreme cold have shaped the life cycle of some modern-day insects. Periodical cicadas in the genus *Magicicada* have unusually long life cycles for insects, with periodicities of 13 to 17 years. Nymphs develop very slowly in the soil, feeding on the xylem sap of tree roots, and emerge synchronously as adults to mate and return to soil for further decade and half. This life history may have evolved in response to perpetually low temperature and hence extremely scarce resources availability during the last few ice ages (Grant, 2005).

Death at high temperature may be caused by desiccation and increased rate of metabolism, with consequent wastage of energy. There ia an optimum range for every vital activity. When the temperature exceeds this optimum level chemical reactions are speeded up and death occurs. The highest temperature at which survival is possible is the maximum survival temperature. According to Vant's Hoff's rule also, the rate of chemical reaction are doubled (1.5 to 3 times) for every 10°C rise in temperature at high temperature the animals dies quickly or they may survive for sometime and fail to grow or reproduce. In case of the parasitic wasp, *Meteorus trachynotus* which attack larvae of the spruce budworm in N. America, adult females live for a much shorter time when exposed to higher temperature A 40 day adult lifespan at 15°C is reduced to mere 10 days or so at 30°C (Thireau and Regniere, 1995). Similarly, in case of European butterflies, the grayling, *Hipparchia semele* and small heath, *Coenonympha pamphilus* (species from open and dry habitat) and the ringlet, *Aphantopus hyperantus* and speckled wood, *Pararge aegeria* (species from shady habitat) the adult lifespan declines linearly as temperature increases with shade species suffering greater decline in longevity than open landscape species (Karisson and Wiklund, 2005). The high temperature at which the animals die immediately will be the upper limit of the lethal high temperature range. The adverse effects of upper lethal temperature is not reversible. When in nature temperature becomes too high, insects seek shelter and move to more favourable sites. If they are exposed to such temperature beyond the limits of tolerance, they die. This reaction would vary from individual to individual as the population is never homogenous. This mortality would be independent of density / unit area and thus climate is not considered a density dependent factor. The insects which are generally exposes to unfavourable high temperature in nature becomes dormant in summer. This state of inactivity is called aestivation and insects resume activity when the temperature becomes congenial.

2.1.2. Moisture

In nature insects are found under a wide range of moisture conditions from fresh water to the desert sand-dunes. In aquatic animals, the dryness is expressed in terms of osmotic pressure of the water. The driest place for them would be the place where sea-water osmotic pressure may exceed that of body. In fresh water, where osmotic pressure is less than body fluid, the animals are so adapted that they do not accumulate excess water. The terrestrial insects can live in dry places under varying conditions of moisture. *Collembola* live in very humid places where air is almost saturated with moisture. Whereas, desert insects can survive in air containing less than 10 per cent moisture. Since food is the source of moisture for practically all insects, their feeding habits reflect adaptations to cope with conditions of excessive moisture or shortage of water. e.g. aphids and other bugs ingest large quantities of plant sap and get rid of excessive water through excretion. Water from various sources is utilized by ecosystem in protoplasmic synthesis and is returned to the atmosphere through transpiration. Of various forms of water, the vapour form plays the most important role in the life of terrestrial organisms. A certain amount of humidity is essential for controlling the loss of water from the body.

For a given species, however, living within that range, they maintain their body water at a constant level. The amount of water in the body of animals generally exceeds 50 per cent. In the larvae of some Scarabaeidae it amounts to 78 to 67% in adults. There is a correlation between the water content of an insect's food and its own body. Larvae of *Sitophilus granarius* feeding on grain with 9-10% water content, contain only 46-47% of water; whereas the larvae of gelechid (*Teleoides* sp.) feeding on hazel leaves containing 71-73% of water, have a water content as high as 90-92%. This shows that food is the main source of water. Insects can also directly absorb the water from the environment. Eggs of *Melolontha* absorb water from the soil in amounting to three times the original volume. Eggs of *Acantholyda erythrocephala* laid in incisions made in pine needles absorbs water from the plant sap. Grasshoppers' absorb water through their body cover directly from the humid air. Excess water in the body is eliminated with faeces, by evaporation and in some cases by glands. The loss of water from the body is prevented by insect cuticle having waxy layer. The humidity preference is influenced by the prevailing temperature.

The lethal influence of dry air can be expressed in terms of longevity or survival rate of a population. Both are influenced by:

(i) rate at which insects loses water into the unsaturated air,

(ii) the extent to which insects can withstand desiccation of the tissues.

The lethal influence of dry air on insects depends on the rate at which water loss is regulated by atmospheric temperature, wind velocity and body movements. Insects in most stages of their development cannot survive a loss of even a small per cent age of water for a long time. However, during aestivation or diapause insects lose a high proportion of body water without any ill-effect. e.g. adult beetle, *Leptinotarsa decemilineata* lose large quantities of water before aestivation and its internal organ shrivel up and alimentary canal lies along the ventral body wall as a thin ribbon. This adaptation allows these beetles to survive exceptionally dry conditions. The eggs of *Austroicetes* and larvae of *Tenebrio* sp. Can survive for several months if they loose sufficient quantity of water from their bodies.

The longevity of insects will be inversely proportional to amount of water lost in unit time and also to saturation deficit. The eggs of *Rhizopertha dominica* are more resistant to dryness than eggs of *Sitophilus oryzae*. Some insects are drought hardy through out their lives, while some become only during certain life stages. *Austroicetes cruciata* have adapted to live in hot and dry summers as only drought hardy diapausing eggs shall be present during summers and less hardy stages will be present in other seasons.

The excessive moisture can prove lethal to insect population in number of ways

1. It adversely affects the normal development and feeding activity of insects e.g. silkworm do not pupate in moist air. The tsetse fly, *Glossina tachinoides* does not feed at a relative humidity higher than 88%. Life cycle is also short at that humidity.
2. It aversely affects the death rate in insects by favouring the spread of epizootics of virus, fungal and bacterial diseases. This also causes fungal infections in congregating insects like Bihar hairy caterpillar, red hairy caterpillar and *Spodoptera*.
3. Excessive moisture in insect body during winter reduces its capacity to withstand exposure to low temperature and thus leads to an adverse effect on cold hardiness.

Moisture also affect the speed of development and fecundity of insects as follows

1. In nymphs of *Locusta* and *Schistocerca* speed of development is retarded at high humidity.

2. In eggs of *Locusta* and *Musca* speed of development is enhanced at high humidity.
3. in eggs of *Bruchus* and *Tineola* rate of development is independent of humidity.

The fecundity of weevil, *Sitophilus* increases steadily upto 70% RH but there is no decline in the number of eggs laid, if the humidity is increased to 90%. Sugarcane black bug, *Cavelerius excavatus* multiplied more rapidly at high humidity (90% RH), whereas, RH above 70% was harmful for cotton jassid, *Amrasca* sp. For eggs and pupae of spotted bollworm, *Earias insulana* both very high (100%) and low (<40%) humidity were not conducive for development of these stages (Atwal and Singh, 1990).

Majority of insects lose water almost continuously through respiration from their tracheae. Insects in general have some control over the openings of their spiracles and under conditions of low humidity; those are partially closed to ensure the minimum loss of moisture. One of the usual methods of giving out water is as a part of excreta. However, the species that live under dry conditions or on dry food cannot afford to excrete too much water; therefore they have specially adjusted rectal caecae that reabsorb water from the rectum, so their droppings become dry. Water loss through cuticle is variable in different species. *Collembola* lives in damp places have cuticle permeable to water, so water is absorbed or lost readily from the body, whereas cockroaches have cuticle coated with waxes of low melting point and lose water even at moderately low temperatures. Insects from hot and arid zones have a coating of hard waxes and their cuticle remains impermeable to water at high temperature for long times.

2.1.3. Rainfall

The effect of rainfall can be direct or indirect. Heavy rain can take aphids off their host plant and both beetles and bugs may be killed by violent thunderstorms. Absence of rain may cause desiccation and death. Seasonal rains influence the ways in which host plants flush and grow and provide food, in particular, the quality of plant phloem (Hale *et al.*, 2003) for herbivores. Droughts can stress trees, rendering them susceptible to insect attack. Rainfall also affects humidity, which combine with temperature and wind to effect local microclimate conditions. Some insect species are benefited from wet conditions while others are benefited from dry conditions. Mosquito population increases in wet weather because rain fills all available containers in which larvae can live (Vezzani *et al.*, 2004). However, the bird cherry-oat aphid, *Rhopalosiphum padi* has a higher intrinsic rate of increase when feeding water stressed on host plants having water stress (Hale *et al.*, 2003), may be due to drought conditions increase the quality of phloem sap in terms of nitrogen content. Association between insect distribution and rainfall patterns are frequently linked to the availability of host plant, particularly in semi desert conditions. The solitary phase of the locust, *Schistocerca gregaria* in Sudan become concentrated into areas of sufficient rainfall and humidity as dry conditions proliferate at the end of the wet season (Woldewahid *et al.,* 2004).

2.1.4. Light

Sun is the main source of radiant energy on the earth. Light come towards earth in the form of electromagnetic waves like cosmic rays, X-rays, UV rays, visible rays, infra red light. Human eye can see in the range of wavelengths of 3900-7600 Å in the visible spectrum. The basic process of food production i.e. photosynthesis is carried out under this visible portion of light spectrum, in this blue and red light are more strongly absorbed by the chlorophyll in the leaves. Green part of the spectrum is less strongly absorbed, infra red band of light has bulk of Sun heat and leaves of plants reject it and save themselves from lethal temperature Light affects the organism through different components, like, intensity of light (brightness of light), quality of light (wave length of visible light), photoperiod (duration of day and night), direction of rays (over head or slanting rays) and polarized light (diffused light).

The effect of light on insect life is different from that of temperature There is no lethal effect of either high or low level of light. Light acts as a token stimulus by enabling insects to regulate and synchronize their life cycles with the change in season. The properties of light that influence insect life are illumination or intensity, wave length or quality and photoperiod or duration of light hours. The insects orientate to the source of light and thus reach the right place at right time. Such phototactic behaviour of insects is altered by temp., humidity or moisture and food. Several species of moths, leaf and planthoppers and beetles are attracted to light at dusk or during the night and this behaviour has been extensively used through light traps for observing brood emergence or fluctuations in their population (Atwal and Singh, 1990).

The light orientation (phototaxix) of insects is associated with their activity and photokinesis (motion induced by light). Phototaxis is a reaction to the intensiveness and quality of light as well as as its direction. Some species exibit positive phototaxis (moving towards the source of light), whereas others react negatively (moving away from the source of light). For some species only light waves of a particular length, mostly ultraviolet and yellow-green rays of the spectrum, are of significance; for other species the light intensity also plays an important role in phototaxis. Some insects are able to detect polarized light (diffused). Honey bee can perceive the degree of polarization of light in different regions of sky and enable them to determine the proper hive of flight even on a cloudy day to locate their hive.

2.1.5. Photoperiodicity

The length of daylight during a 24 h day-night cycle, known as photoperiod, has a fundamental influence on the development and ecology of insects that live in seasonal climates. Photoperiodism influe-nces the motor activity rhythms of insects such as locomotion, feeding, adult emergence, mating and oviposition of some insects. The daylength experienced by an insect provides information about the progression of the seasons, and the ability to vary growth and development rates enab-les the insect to achieve efficient timing relative to favourable conditions (Leimar, 1996). The induction of diapause,a genetically determined state of suppressed development and manifestation of which is induced by environmental factors (Beck, 1968), is influenced by photoperiod in most of lepidopterous insects. Ishihara and Shimada (1995) studied duration of various life stages in the life cycle of bean weevil, *Kytorhinus sharpianus* from egg to adult according to vary photoperiod at constant temperature. The whole cycle can be accomplished between 50 and 80 days when the insect receives 15 or 16 h of daylight per day-night cycle, but this period increases dramatically as the hours of daylight diminish to 14 h and 12 h. with only 12 h of daylight the pupal stage is never reached until longer hours of light return.

Day length has a fundamental effect on diapause induction and cessation e.g. when larvae of parasitic wasp, *Cotesia malanoscela* were exposed to long daylengths (greater than 18 h), they developed contin-uously all the way to adulthood. However, when larvae were exposed to short daylengths (less than 16 h) entered diapause, which halted development in the cocooned prepupal stage (Nealis *et al.,* 1996). In ecological terms this system makes sure that adult parasitoid appear only during the long days of summer when their host insects, gypsy moth larvae are most likely to be abundant. The damson-hop aphid, *Phorodon humuli* produces overwintering eggs that are laid on sloe, *Prunus spinosa*; damson (*P. insititia*) or plum (*P. domestica*) and when eggs are hatch in spring the resulting nymphs undergo several wingless genera-tions on these host species before producing winged adults (emigrants) that fly to hops (*Humulus lupulus*) the only summer host (Campbell and Muir, 2005). No individual climatic effect is easily separable from another. Photoperiod combined with temperature determine the propo-rtion of adult female *Empoasca fabae* ovipositing (Taylor *et al.,* 1995) and female monarch butterflies entering diapause (Goehring and Oberhauser, 2002). Under both decreasing and long photoperiod regimes, constant rather than fluctuating temperature reduce incidence of diapause. Han *et al.,* 2005 reported that diapuse induction in the pine caterpillar, *Dendrolimus tabulaeformis* was in 100% larvae after 9 or 10 h in dark period and at 24 and 28°C temp.

2.1.6. Wind

Wind is a vital component of more general weather patterns, giving rise to fronts and convergence zones. It has a variety of influences on the ecology of insects. It can carry insects many miles to new habitats and regions (Coulson *et al.,* 2002, Otuka *et al.,* 2005 Harrison and Rasplus, 2006), it can destroy trees to provide new breeding sites (Eriksson *et al.,* 2005), it brings rain to produce new food (Ji *et al.,* 2006). *Plutella xylostella* is a pest of *Brassica* crops within the plant species Cruciferae. In the parts of the worlds where winters are too extreme for overwinter survival, new diamondback moth outbreaks are entirely the result of regular, long-range immigrations of adult moths, some of these windborne immigrations can cover more than 2000 km (Chapman *et al.,* 2002, Coulson *et al.,* 2002).

The influence of wind patterns on the movement of insects in the Far East is complex. The trajectory analysis of winds at both 10 m and 1.5 km above ground has been used to determine the direction and extent of windborne movements of insect pests of rice (Mills *et al.,* 1996). Planthoppers are thought to be transported by prevailing winds and these can be used to develop weather forecasting models (Otuka *et al.,* 2005). Aerial traps caught two species of planthoppers, viz., brown planthopper, *Nilaparvata lugens* and whitebacked planthopper, *Sogatella furcifera.* Northward migrations of these and other species were detected along a broad front in prevailing summer monsoon and trade winds. Feng *et al.*(2006) reported the nocturnal migrations of *Pantala flavescens* (Odonata) in china and observed that these insects flew at heights up to and beyond 1000 m, traveling between 150 and 400 km in a single flight. Complex interaction of temperature, height above ground, wind direction, wind speed help in prediction of the population.

2.2. BIOTIC FACTORS

The biotic components such as food and natural enemies, are important components of the life-system besides the abiotic components of the environment. It is not only the individuals of the species that are influenced by the abiotic environment but also the biotic factors. There exists a strong relationship between weather and food.

2.2.1. Food

Food is an important component of the environment. It influences the survival and multiplication in animals by modifying its fecundity, longevity and speed of development. Both quality and quantity of food are important. Quality influences the physiological aspect, whereas quantity influences the total density of a species.

Quantity of food. It is unusual for extensive population to consume all food in a particular area. Only few species do it regularly or often and there are 3 possibilities in nature:

(i) Most of species consume only a small portion of their food. Even in the presence of abundant food, some other components of the environment might be pressing on them so hard that they do not become numerous to consume all the food. E.g. *Aphilinus mali* does not exceed a limit to control *Eriosoma* sp. (Wooly apple aphid) an some population remains in nature even after predation.

(ii) Certain species have the capacity to multiply right up to the level where food supply is almost exhausted. E.g. *Cactoblastis cactorum*, which controls *Opuntia* weed, multiplies to an extent that it eats up nearby all the food. Absolute shortage of experienced by some predators which press heavily on their preys.

(iii) In certain cases, food may be plentiful but it may be inaccessible to the organism concerned. This may be due to insufficient synchronization between the host plant and the preying organisms. This situation has same effect as the lack of food. This is known as relative shortage of food.

Food and behaviour of insect. Number of insects is determined by their behaviour in relation to resources of the area. There is an innate behaviour of every spp. In which they select the locations having adequate food for themselves as well as for their offsprings before the reproductive period starts. The reproductive period generally coinci-des with the natural increase of food. This is mainly to avoid overcrow-ding under limited food conditions. Most of the natural populations consume only a small portion of the total food and its accessibility to population also depends upon the capacity of dispersal to search new area of food if in the earlier one it is in short supply. E.g. in case of Thrips imagines thrips develop innate urge to disperse when flowers in which these are live, die. The thrips will fly to new uncrowded flower food. So chances of thrip population to survive and multiply will depend upon the abundance of flowers in the area.

Weather and food. The distribution of an organism depends upon distribution of certain plants. E.g. *Pieris rapae* occur only in places where cruciferous plants are found, The distribution of Parasitoid apanteles, which preys upon the *Pieris,* is also restricted by distribution of these plants. The distribution of plants is normally determined by the climate. So kind of insect found in an area depends largely upon climate or weather conditions of that area. Population of *Thrips imaginis* in Australia depends upon the amount of food present (flowers) and this in turn not only depends upon the amount of rain fallen during the proceeding season but the dates on which it was actually fallen.

Other insects of the same kind and food. In situation when only a small portion of the total food available is eaten, the shortage (relative) of food affects the rate of increase of insects. The chance of survive and multiply depends upon the distribution of food and on its own behaviour but not on the number of other insects seeking the same food. However, when all the food in the area is eaten (absolute shortage) then the numbers of animals of same kind may be of major importance. This type of situation may occur occasionally in the nature. Every individual needs a sufficient amount of food for its survival and to reach to maturity. So if more number of individuals is feeding the same food but only a few will gain maturity and reproduce. Food taken by rest of these is considered as waste food. The food that is eaten by the reproducing insects is the effective food. Thus distribution and abundance of individuals depends upon the effective food present.

Insect food relationships. An insects chance to get food also depends upon the presence of organisms of different kind as under:

(i) The two species may consume the same food and thereby reduce the total stock

(ii) They may spoil or destroy some or all of the stock of food incidently without themselves consuming it. Some more complex relationship are also possible. An epizootic of fungus *Entomophthora* reduces the stock of available food (*Plutella*) for the predator species (*Angitia* spp). The predator species suffered severely from shortage of host food. The activity of disease causing pathogens is determined largely by the weather and this adds to the complexity of the relationships.

(iii) Sometimes insects chance of getting enough food may be enhanced by the presence of insects of the other kind in its environment. Such insects do not serve as food e.g. Dung beetle live in a field of clover which serve as its food. But unless the cows, which eat the clover and the drop the dung, are present, the food is of no use and is not utilized by the beetles. Similar relationships are found in case of protozoans in the gut of termites. Protozoans help to digest cellulose, which form a large part of the termites food. Similarly, many species of small insects (*Diptera* and *Hymenoptera*), which live on honey dew secreated by aphids, coccids and other Homoptera.

Quality of food. The quality of food may influence the innate capacity of increase of an organism by modifying the longevity, speed of development and fecundity. There are some essential food constituents which an animal must get from outside and there are others it can synthesize.

Longevity. Insects which are mainly monophagous restrict themselves to particular type of food for the survival and multiplication e.g. outbreaks of Spruce budworm, *Christoneura fumiferana* occur in forests where balsam firs were available in full bloom. The larvae prefer staminate flowers as they feed on pollen and also young foliage. The survival rate of young larvae feeding on pollen and young foliage are equally high. *Thrips imaginis* breeds only in flowers. In the laboratory, these can be reared on entire stamens but when anthers with the pollen were removed, none survived to become adult. Species *Ephestia* and *Oryzaephilus* commonly found in store houses food on flour, whole meal and bran. These restrict to such foods as these cannot survive except on diet rich in carbohydrates. These species require Vit-B in the food. *Ephestia elutella* no growth in the absence of riboflavin, *Ephestia kukniella* survived on diet which lacked riboflavin. *Lasioderma, Stegobium* and *Oryzaephilus* donot require vit-B in their food as they get it from symbiotic m.o. which live in special cells in their bodies.

Fecundity: many species which have a complete metamorphosis in their life cycle usually store enough food during larval stages so that adults may produce eggs without taking any food. *Thrips imagines* require pollen for normal egg laying. Blow flies *Lucilia* and *Calliphora* produced eggs normally when their diet included extract of liver or muscle or some other animal protein. They did not produce any eggs when these were made to feed only on cane sugar and water. *Musca domestica* also require protein in its diet to produce eggs. Mosquitoes get their protein from blood. Aphids, *Brevicoryne brassicae* produced more young ones when the diet was rich in protein.

Speed of development. On unsuitable diet, insects require longer to complete their development e.g. larvae of Tribolium confusum required 28 days to complete the larval stage when the food was high extraction flour. It takes 32-50 days with yeast or meat meal as food, which were poor in carbohydrate. However, larvae of *Ptinus tectus* developed slowly on diet with more carbohydrates. In wire worms, search for food is through random movements. These are stimulated to eat the root by the presence of certain sugars, fats and polypeptides. Number of caterpillars is known to select their food by recognizing a specific chemical compound. Larvae of butterfly, *Papilio anax* were attracted to filter paper soaked in methylonyl ketone, which is present in species of Umbelliferae and Rutaceae on which these larvae are known to feed in nature. Most larvae are required to be very close to their food before they can smell, but adults have highly sensitive olfactory organs.

In general, the search for food start with movements that is random with respect to food. Insects, however, tend to move in gradient of temperature, moisture and light. These may bring the insects to place where food is present. When insect comes close enough to see or smell its food, it moves directly towards it. The influence of quality of food on rate of increase of insect spp. has been used to develop resistant varieties. Rice variety Mudgo has been resistant to *Nilaparvata lugens* as it lacked proper proportion of amino acid (asparagine) in it which affects planthopper. Generally speaking, resistant varieties exhibit antibiosis which may be due to lack of essential nutrient or the presence of inhibitors for the pest feeding on them.

2.2.2. Natural enemies

The natural enemies of insect pest populations include predators, parasitoids and disease-causing micro-organisms such as bacteria, fungi, viruses and rickettsiae. Since the natural enemies are living entities, they are also influenced by various environmental factors such as weather and hyper-parasitism. The degree of influence of various natural enemies on the pest population would thus vary. A number of models concerning the predator-prey and host-parasite relationships are available. The competitive displacement of predators has been discussed by DeBach (1964) and the functional response of invertebrate predators to the density of their prey has been studied by Holling (1965, 1966). In the study of a pest population, its natural enemies have to be identified and their role in causing numerical fluctuations needs to be quantified (Atwal and Singh, 1990).

Weather is known to determine the effectiveness of a natural enemy. For example, in the control of woolly apple aphid with *Aphelinus mali*, the prevalent temperature in spring would determine the effectiveness of the parasitoid. Both the host and the parasitoid remain dormant during winter but the activity of the host is resumed at a slightly lower temperature, and thus it is able to breed in the absence of the parasitoid in the early spring. In such a situation the predator may fail to press heavily on the host population as compared with the one when relatively higher temperature prevailed during early spring. The larval parasitoid (*Apanteles flavipes*) of maize borer in the Punjab fails to exercise a regulatory effect on the host population. The parasite after successful overwintering in the host larvae resumes activity in the spring earlier than its host but its unable to parasitize this particular host during dry and hot summer (May-June). This is beacause of lethal effects of high temperature (above 35°C) and low humidity (below 60% R.H.) on the parasitoid. In the absence of this particular natural enemy the pest population reaches its peak during this period. It is, thus, imperative that while studying the population of an insect pest the influence of changing environmental conditions on its natural enemies may also be taken in to consideration (Atwal and Singh, 1990).

2.3. CONCLUSION

The above studies showed that how different factors like abiotic (temperature, moisture light, and wind) and biotic factors (quality and quantity of food and natural enemies) interact to determine insect-pest population dynamics. These studies provide ample evidence that to understand the fluctuations of insect populations all available tools are to be used to get information about the ecology of insect pests.

REFERENCES

Andreadis, S.S., Milonas, P.G. and Savopoulou-Soultani, M. (2005) Cold hardiness of diapausing and non-diapausing pupae of the European grapevine moth, *Lobesia botrana, Entomol. Exp. Appl.* 117: 113-118.

Andrewartha, H.G. and Birch, L.C. (1954) *The Distribution and Abundance of Animals,* p. 782. University of Chicago Press, Chicago, IL.

Atwal, A.S. and Singh, B. (1990) *Pest Population and Assessment of Crop Losses.* Indian Council of Agricultural Research, New Delhi, pp. 131.

Beck, S.D. (1968) *Insect Photoperiodism.* Academic Press, Inc. New York.

Blank, R.H., Gill, G.S.C., Oslon, M.H. and Upsdell, M.P. (1995) Greedy scale (Homoptera: Diaspididae) phenology on taraire based on Julian day and degree-day accumulations. *Environ. Ent.* 24: 1569-1575.

Bouchard, R.W., Carrillo, M.A. and Ferrington, L.C. (2006) Lower lethal temperature for adult male *Diamesa mendotae Muttkowski* (Diptera: Chironomidae), a winter-emerging aquatic insect. *Aquatic Insects* 28: 57-66.

Brier, R.A., Cariss, H.M. and Gee, J.H.R. (2003) Flight activity of adult stoneflies in relation to weather. *Ecol. Ent.* 28: 31-40.

Campbell, C.A.M. and Muir, R.C. (2005) Flight activity of the damson-hop aphid, *Phorodon humuli. A. Appl. Biol.* 147: 109-118.

Chapman, J.W., Reynolds, D.R., Smith, A.D., Riley, J.R., Pedgley, D.E. and Woiwood, I.P. (2002) High-altitude migration of the diamondback moth *Plutella xylostella* to the UK: a study using radar, aerial netting and ground trapping. *Ecol. Ent.* 27: 641-650.

Clarke, A. (2003) Costs and consequences of evolutionary temperature adaptation. *Trends in Ecol. Evol.* 18: 573-581.

Coulson, S.J., Hodkinson, I.D., Webb, N.R., Mikkola, K., Harrison, J.A. and Pedgeley, D.E. (2002) Aerial colonization of high Arctic islands by inverteb-rates: the diamondback moth *Plutella xylostella* (Lepidoptera: Yponomeu-tidae) as a potential indicator species. *Diver. and Distribut.* 8: 327-334.

Crozier, L. (2003) Winter warming facilitates range expansion: cold tolerance of the butterfly Atalopedes campestris. *Oecologia* 135: 648-656.

Erikson, M., Pouttu, A. and Roininen, H. (2005) The influence of windthrow area and timber characteristics on colonization of wind-felled spruces by *Ips typographus* (L.). *Forest Ecol. Magmt.* 216: 105-116.

Feng, H.Q., Wu, K.M., Ni, Y.X., Cheng, D.F. and Guo, Y.Y. (2006) Nocturnal migration of dragonflies over the Bohai Sea in northern China. *Ecol. Ent.* 31: 511-520.

Goehring, L. and Oberhauser, K.S. (2002) Effects of photoperiod, temperature and host plant age on induction of reproductive dispause and development time in Danaus plexippus. *Ecol. Ent.* 27: 674-685.

Grant, P.R. (2005) The priming of periodical cicada life cycles. *Trends in Ecol. Evol.* 20: 169-174.

Hale, B.K., Bale, J.S., Pritchard, J., Masters, G.J. and Brown, V.K. (2003) Effects of host plant drought stress on the performance of the bird cherry-oat aphid, *Rhopalosiphum padi* (L.): a mechanistic analysis. *Ecol. Ent.* 28: 666-677.

Han, E.N. and Bauce, E. (1995), Glycerol synthesis by diapausing larvae in response to the timing of low temperature exposure, and implications for overwintering survival of the spruce budworm, *Choristoneura fumiferana. J. Insect Physiol.* 41: 981-985.

Han, R., Ge, F., Yardim, E. N. and Hc, Z. (2005) The effect of low temperatures on diapause and non-dispause larvae of the pine-caterpillar, *Dendrolimus tabulaeformis* Tsai et Liu (Lepidoptera: Lasiocampidae). *Appl. Ent. Zool.* 40: 429-435.

Harrison, R.D. and Rasplus, J.Y. (2006) Dispersal of fig pollinators in Asian tropical rain forests. *J. Tro. Ecol.* 22: 631-639.

Hunter, A.F. (1993) Gypsy moth population sizes and the window of opportunity in spring. *Oikos* 68: 531-538.

Ishihara, M. and Shimada, M. (1995) Photoperiodic induction of larval dispause and temperature-dependent termination in a wild multivoltine bruchid, Kytorhinus sharpianus. *Entomol. Exp. Appl.* 75: 127-134.

Ji, R., Li, D.M., Xie, B.Y., Li, Z. and Meng, D.L. (2006) Spatial distribution of oriental migratory locust (Orthoptera : Acrididae) egg pod populations: implications for site -specific pest management. *Environ. Ent.* 35: 1244-1248.

Karisson, B. and Wiklund, C. (2005) Butterfly life history and temperature adaptations; dry open habitats select for increased fecundity and longevity. *J. Evol. Biol.* 74: 99-104.

Leimar, O. (1996) Life history plasticity: influence of photo period on growth and development in the common blue butterfly. *Oikos* 76: 228-234.

Lyons, D.B, (1994) Development of the arboreal stages of the pine false webworm (Hymenoptera: Pamphiliidae). *Environ. Ent.* 23: 846-854.

Mills, A.P., Rutter, J.F. and Rosenberg, L.J. (1996) Weather associated with spring and summer migrations of rice pests and other insects in south-eastern and eastern Asia. *Bull. Ent. Res.* 86: 683-694.

Morimoto, S., Imamura, T., Visarathanonth, P. and Miyanoshita, A. (2007) Effects of temperature on the development and reproduction of the predatory bug *Joppeicus paradoxus* Puton (Hemiptera: Joppeicidae) reared on *Tribolium confusum* eggs. *Biol. Cont.* 40: 136-141.

Nealis, V.G., Oliver, D. and Tchir, D. (1996) The diapause response to photoperiod in Ontario populations of *Cotesia melanoscela* (Ratzeburg) (Hymenoptera: Braconidae). *Can. Entomol.* 128: 41-46.

Omkar, and Parvez, A. (2004) Temperature-dependent development and immature survival of an aphidophagous ladybeetle, *Propylea dissecta* (Mulsant). *J. Appl. Ent.* 128: 510-514.

Otuka, A., Dudhia, J., Watanabe, T. and Furuno, A. (2005) A new trajectory analysis method for migratory planthoppers. *Sogatella furcifera* (Horvath) (Homoptera: Delphacidae) and *Nilaparvata lugens* (Stal), using an advanced weather forecast model. *Agricul. Forest Ent.* 7: 1-9.

Otuka, A., Watanabe, T., Suzuki, Y., Matsumura, M., Furuno, A. and Chino, M. (2005) A migration analysis of the rice planthopper *Nilaparvata lugens* from the Philippines to East Asia with three-dimensional computer simulations. *Popl. Ecol.* 47: 143-150.

Powell, S.J. and Bale, J.S. (2004) Cold shock injury and ecological costs of rapid cold hardening in the grain aphid *Sitabion avenae* (Hemiptera: Aphididae). *J. Insect Physiol.* 50: 277-284.

Sinclair, B.J., Vernon, P., Klok, C.J. and Chown, S.L. (2003) Insects at low temperatures: an ecological perspective. *Trends Ecol. and Evol.* 18: 257-262.

Taylor, P.S., Shields, L.J., Tauber, M.J. and Tauber, C.A. (1995) Induction of reproductive dispause in *Empoasca fabae* (Homoptera: Cicadellidae) and its implications regarding southward migration. *Environ. Ent.* 24: 1086-1095.

Thireau, J.C. and Regniere, J. (1995) Development, reproduction, voltinism and host synchrony of *Meteorus trachynotus* with its hosts *Choristoneura fumiferana* and *C. rosaceana. Ent. Experi. Appl.* 76: 67-82.

Vezzani, D., Velazquez, S.M. and Schweigmann, N. (2004) Seasonal pattern of abundance of *Aedes aegypti* (Diptera: Culicidae) in Buenos Aires city, Argentina. *Memorias do Instituto Oswaldo Cruz* 99: 351-355.

Woldewahid, G., Van der Werf, W., Van Huis, A. and Stein, A. (2004) Spatial distribution of populations of solitarious adult desert locust (*Schistocerca gregaria* Forsk.) on the coastal plain of Sudan. *Agricul. Forest Ent.* 6: 181-191.

Worland, M.R., Wharton, D.A. and Byars, S.G. (2004) Intracellular freezing and survival in the freeze tolerant alpine cockroach *Celatoblatta quinquemaculata. J. Insect Physiol.* 50: 225-232.

Theory and Practice of Integrated Pest Management
R. Arora, B. Singh and A.K. Dhawan (Eds)
Scientific Publishers (India), Jodhpur 2012, pp. 38 - 48.

3

CONSTRUCTION OF LIFE-TABLES FOR INSECT SPECIES

M.S. Mahal

Department of Entomology
Punjab Agricultural University, Ludhiana - 141 004, India

3.1. INTRODUCTION

Populations of insect species in various ecosystems fluctuate under the influence of both abiotic and biotic components of the environment. In nature, numerical changes can be described by knowing the birth rate, death rate and rate of to and fro movements of a population. These factors are function of additive (which contribute towards increase in numbers) and subtractive (which contribute towards decrease in numbers) ecological processes. Main aim of studying natural populations is to know how many insects are being added to a population or taken away and what type of ecological processes are responsible for these additions or subtractions. This can be done by understanding functioning of Life system.

In nature, usually a large number of interactions of complex nature are going on between insect species and environment, which may be rather difficult to study. Under the circumstances, approach of studying life system of a species is more realistic. In this case, a part of the ecosystem is studied which explains the existence, abundance and evolution of a particular population (Clark *et al.,* 1967). The word population refers to the number of individuals of a species existing in a particular place, while the effective environment of an individual in a population is everything else in the universe that may affect its chance to survive and multiply. Accordingly, population is itself an environmental factor which may affect the individual depending upon its density.

Life-system is composed of a single subject species possessing intrinsic characteristics and its effective environment. The latter consists of resources (food & space) required for population maintenance and the inimical agencies constituting both abiotic and biotic which oppose the survival and reproduction of individuals of subject species. These interacting co-determinants allow the individuals to exist with certain species characteristics and decide population size. The co-determinants perform a duel role of controlling the life functions of individuals which lead to population increase by reproduction and immigration and the circumstances which lead to limitation of numbers by premature mortality, restricted natality and emigration. In field populations, a few key influences operate usually in one or two critical age intervals in the life-cycle and this may determine the population trend to a large extent.

Life system of a species can be studied by following the technique of **life-table**. It is a concise summary of certain vital statistics of a population whose members start life together. A life-table describes the mortality occurring in successive age intervals during the life-cycle of an insect species. The differences in the numbers at successive developmental stages are assumed to result from mortality or dispersal/emigration or both. Life-table provides an important tool in understanding the changes in population of insect pests during different developmental stages throughout the life-cycle.

Key mortality factor(s). The developmental stage in which mortality is maximum is the critical age interval and the environmental factor(s) causing that mortality is the 'Key mortality factor(s)'.

Generation survival. The proportion of individuals in the first developmental stage and the last in the generation explains the 'Generation survival ($SG=N_3/N_1$)'. This gives an index of population trend in a generation without the effects of fecundity and adult mortality.

Trend index. The proportion of the same developmental stage in two successive generations indicates the 'Trend index ($I = N_2/N_1$)'.

Budget. Absolute insect populations in successive developmental stages recorded in the field are plotted in a table along with information on numbers dying in different stages owing to action of various mortality factors and such tables are called 'Budget' (Richards, 1961).

Initially, life-tables were used by actuaries for working out the expectancy of human life at various ages for the purpose of life insurance. It was modified for insect populations by making it age specific instead of time specific. In insects, survivorship is measured by periodic sampling of individuals of the same population. The life-tables record a series of sequential measurements that indicate population changes throughout the life-cycle of species under natural conditions. Morris and Miller (1954) were the first to apply this approach for studying the population dynamics of spruce budworm under natural conditions. Later, this technique was discussed comprehensively by Morris (1963), Harcourt (1969) and Southwood (1978) for studying the natural populations of insect pests.

Life-table technique is used to determine (i) the mathematical relationship of various environmental factors with population changes and (ii) the key factors that account for a large part of the change in population size.

According to Southwood (1978), life-tables are of two types:

(i) **Age-specific.** These tables are also known as horizontal life-tables and are based on the fate of real cohort, the individuals belonging to single generation. The population may be stationary or fluctuating.

(ii) **Time specific.** These tables are also known as vertical life-tables and are based on the fate of an imaginary cohort formed by determining the age structure, where there is considerable overlapping of generations. Age determination is a pre-requisite for time specific life-tables.

3.2. CONSTRUCTION OF LIFE-TABLE

Basic requirements. Life tables have to be constructed for different populations of the same species throughout its range of occurrence and from different habitats.

(i) Intensive studies are required for recording age specific survival of species. The survivorship is measured by periodic sampling of individuals of the same population from the same habitat.

(ii) Measurements of independent variables, including weather factors, parasitism and predation.

Steps. Consult literature on biology and behaviors of the insect, otherwise generate this information through studies.

(i) To have more realistic study of LT, go for area where insect spp is inhabiting in abundance or where its most favoured crop is grown in abundance, e.g. for cotton pests in cotton belt area. Area should representative of natural habitat of insect spp.

(ii) The life-table for a species can be prepared only after an intensive study of population is carried out at same place in 3-4 different habitats.

(iii) For precise estimates of absolute population, the first step is to choose basic population unit or sample size, then decide the sampling technique especially the number of samples to be taken keeping in view the pattern of insect distribution.

(iv) Care should be taken to sample all the stages of life-history. If not possible, then as many stages as possible should be observed.

(v) Decide the time of the day for sampling, it will depend upon behaviour of different stages. Time should be such that when maximum population can be easily sampled.

(vi) Time interval (after how many days) for recording observations is decided keeping in view duration of life stages.

(vii) Time interval for recording observations is decided keeping in view the duration of life stages. It must be decided in such a way that new stage must be sampled on subsequent date(s) of sampling.

(viii) Periodic record of incidence of parasitoids, predators and disease should be made in the field itself but these observations also need to be supplemented through laboratory studies for assessing the role of various mortality factors.

(ix) Effect of various weather factors should be studied experimentally under controlled conditions.

Determination of absolute populations. In case of insect species having discrete generations, only a single developmental stage would be available at any particular time. So, multi-stage strategies are needed for determining the influence of various environmental factors in bringing about change in numbers. It is desirable to estimate the total population of as many stages as possible through the appropriate sampling procedures. Number of individuals that pass through a particular stage in one generation are considered as absolute numbers for use in the life-table. At a point of time, when all the individuals of a generation are in a given stage, sampling population at that time will provide peak estimates. It is also easier to determine deaths occurring during the stage through direct observations. The variations in the size of population are recorded for a number of successive generations and the role of various environmental factors in bringing about numerical changes in successive developmental stages within a generation and also between generations are then determined. For example, sugarcane top borer has 4 distinct generations occurring in different seasons and there was not even the slightest overlapping. Roy and Bains (1983) used absolute population (total) of larvae and pupae of top borer for construction of life-tables to determine the key mortality factors.

In case the generations overlap, a number of estimates of population have to be integrated for obtaining the total population entering a particular developmental stage. Selection of appropriate integration procedure will depend upon the degree of overlapping of stages. Most of the methods are complex and require even computer facilities. When there is complete overlapping of all the stages, integration methods based on age structure of population (Southwood, 1978) are most appropriate.

There are many graphical and statistical techniques which can used to estimate the numbers entering a particular stage or at mid-point of the stage from a series of population samples. Graphical method is the simplest integration method and gives reasonable estimates of numbers entering mid-point of a stage. It is applicable when either a heavy mortality occurs at the end of stage or there is light mortality at constant rate throughout the stage. In second method, total time for a given stage is divided into periods corresponding to developmental time of an individual. The mean population is calculated for each of these periods and then these means are summed up to work out the numbers entering that stage.

Third method is applicable when (i) a developmental stage has well marked peak and (ii) has a steady mortality rate. A single impulse of recruitment is considered and the number of estimates from successive estimates is plotted. The regression line of numbers (after peak) on time is drawn and extended back. The population corresponding to zero time will be the number alive just before that stage is seen for the first time. This method overestimates the recruitment as it overlooks the hatching failure and mortality in egg stage. The fourth method is applicable when there is no distinct peak and recruitment and mortality are overlapping. This method is more precise as it also takes into account the survival rate of individuals of a particular stage. However, it requires accurate estimate of initial population and duration of various stages. Some 4-5 other methods of integration are also available but all these are quite complex and require even computer facilities.

However, some simple methods can also be adopted to work our absolute populations in different developmental stages from periodic population estimates, like (i) mean population, (ii) total population or (iii) peak population. Selection of any method will depend upon (i) duration of life stages and (ii) interval of sampling or both. Singh *et al.* (1977) used a simple technique to determine the numbers entering a particular stage in case of maize borer, *Chilo partrellus* having overlapping generations on maize crop. They recorded population of various stages at 7-10 days interval and took the mean of total population of a particular stage from successive sampling estimates on which that stage was noticed during a particular season.

3.3. FIELD LIFE-TABLE

Morris and Miller (1954) and Harcourt (1969) suggested the following columns for constructing a life-table for natural populations. All the lx and dx values in the life-table indicate density of individuals per unit of habitat/host.

x = age interval (egg, larva, pupa, adult)

lx = number surviving at the beginning of stage noted in the x column

dx = number dying within the age interval stated in x column

100 qx = percentage mortality

Sx = survival rate within the stage mentioned in x column

Richards (1961) suggested the following three columns in the preparation of Budget i.e. numbers/ha, log numbers/ha and "k" values.

Case Study. Key mortality factors in population dynamics of sugarcane top borer, *Tryporza (Scirpopgaga) nivella* (Fab.) in Punjab (Roy and Bains, 1983)

The population counts of various stages were made at 10-15 day interval. The estimates of maximum absolute population of larvae and pupae of sugarcane top borer per 30 quadrat (1 x 1 m each) have been used in the construction of life-tables for determining the key mortality factor(s) in different generations occurring in different seasons from March 1980 to 1981. The average population of four fields located in the four different locations in Ludhiana and Jalandhar districts (Punjab) were taken for preparing life-tables during different seasons. Life-tables have been prepared separately for four generations during 1980-1981. The four generations covered the same period as given below:

I generation = IV week March to II week May (Table 3.1)

II generation = III week May to I week July (Table 3.2)

III generation = II week July to III week August (Table 3.3)

IV generation = IV week August to III week March (Table 3.4)

The contribution of successive developmental stages to the overall population trend and key factors influencing the build-up of sugarcane top borer population in different generations was determined following Southwood (1978).

Table 3.1. Life table of sugarcane top borer for the first generation during 1980 (IV week March to II week May)

x	lx	dxf	dx	100 qx	Log (lx)	"k"
Expected eggs	640	Parasitoids	73	11.41	2.8062	0.0961
		Sterility	54	8.45		
Expected larvae	513	Failure to enter mid-ribs & unknown causes	399	77.78	2.7101	0.6535
Larvae - L1 (N1)	114	Unsuccessful boring through mid-ribs/leaf whorls & unknown causes	73	64.03	2.0569	0.4441
Larvae - L2	41	Parasitoids	6	14.63	1.6128	0.4087

		Unsuccessful boring through growing points & unknown causes	19	46.34		
Pupae	16	Parasitoids	2	12.50	1.2041	0.0580
Moths	14	Sex ratio (1:1)	-	-	-	-
Females x 2(N3)	14	-	-	-	-	-
Normal ?s x 2					1.1461	
		Total	626	97.81	-	1.6601

Expected eggs = 7 x 35.3	=	247
Number of parasitised and sterile eggs	=	90
Expected number of viable eggs	=	157
Expected number of early instar larvae	=	157
Actual number of early instar larvae (L_1) in the second generation during 1980 (N_2)	=	74
Trend index (I) = N_2/N_1 = 74/114	=	0.65
Generation survival (SG) = N_3/N_1 = 14/114	=	0.12

Table 3.2. Life table of sugarcane top borer for the second generation during 1980 (III week to I week July)

x	lx	dxf	dx	100 qx	Log (lx)	"k"
Expected eggs	247	Parasitoids	70	28.34	2.3927	0.1968
		Sterility	20	8.10		
Expected larvae	157	Failure to enter mid-ribs & unknown causes	83	52.86	2.1959	0.3267
Larvae -L1(N1)	74	Unsuccessful boring thro-ugh mid-ribs/leaf whorls & unknown causes	49	66.21	1.8692	0.4713
Larvae -L2	25	Parasitoids	6	24.00	1.3979	0.3565
		Unsuccessful boring through growing points & unknown causes	8	32.00		
Pupae	11	Parasitoids	1	9.09	1.0414	0.0872
Unknown causes	1	9.09				
Moths	9	Sex ratio (1:1)	-	-	-	-
Females x 2(N3)	9	-	-	-	-	-
Normal ?s x 2	-	-			0.9543	-
		Total	238	96.35		1.4385

Expected eggs = 4.5 x 84.5	=	380
Number of parasitised and sterile eggs	=	122
Expected number of viable eggs	=	258
Expected number of early instar larvae	=	258
Actual number of early instar larvae (L_1) in the second generation during 1980 (N_2)	=	205
Trend index (I) = N_2/N_1 = 205/74	=	2.77
Generation survival (SG) = N_3/N_1 = 9/74	=	0.12

Table 3.3. Life table of sugarcane top borer for the third generation during 1980 (II week July to II week August)

x	lx	dxf	dx	100 qx	Log (lx)	"k"
Expected eggs	380	Parasitoids	90	23.68	2.5798	0.1662
		Sterility	32	8.42		
Expected larvae	258	Failure to enter mid-ribs & unknown causes	53	20.54	2.4116	0.0999
Larvae -L1(N1)	205	Unsuccessful boring through mid-ribs/leaf whorls & unknown causes	135	65.85	2.3117	0.4666
Larvae -L2	70	Parasitoids	19	27.14	1.8451	0.4834
		Unsuccessful boring through growing points & unknown causes	28	40.00		
Pupae	28	Parasitoids	3	13.04	1.3617	0.1064
		Unknown causes	2	8.69		
Moths	18	Sex ratio (1:1)	-	-		-
Females x 2(N3)	18	-	-	-	-	-
Normal ?s x 2		-	-	-	1.2553	-
		Total	362	95.26	-	1.3245

Expected eggs = 9 x 83.2	=	749
Number of parasitised and sterile eggs	=	172
Expected number of viable eggs	=	577
Expected number of early instar larvae	=	577
Actual number of early instar larvae (L_1) in the second generation during 1980 (N_2)	=	295
Trend index (I) = N_2/N_1 = 295/205	=	1.44
Generation survival (SG) = N_3/N_1 = 18/205	=	0.09

Table 3.4. Life table of sugarcane top borer for fourth generation during 1980-81 (IV week August to III week March)

x	lx	dxf	dx	100 qx	Log (lx)	"k"
Expected eggs	749	Parasitoids	124	16.55	2.8745	0.1133
		Sterility	48	6.41		
Expected larvae	577	Failure to enter mid-ribs & unknown causes	282	48.87	2.7612	0.2914
Larvae -L1(N1)	295	Unsuccessful boring through mid-ribs/leaf whorls & unknown causes	141	47.79	2.4698	0.2823
Larvae -L2	154	Parasitoids	13	8.44	2.1875	0.4471

		Unsuccessful boring through growing points & unknown causes	86	55.84		
Pupae	55	Parasitoids	5	9.09	1.7404	0.2490
Unknown causes	19	34.54				
Moths	31	Sex ratio (1:1)	-	-	-	
Females x 2(N3)	31	-	-	-	-	-
Normal ?s x 2		-	-	-	1.4914	-
		Total	718	95.86	-	1.3831

Expected eggs = 15.5 x 69.1 = 1071

Number of parasitised and sterile eggs = 77

Expected number of viable eggs = 994

Expected number of early instar larvae = 994

Actual number of early instar larvae (L_1) in the second generation during 1980 (N_2) = 157

Trend index (I) = N_2/N_1 = 157/295 = 0.53

Generation survival (SG) = N_3/N_1 = 31/295 = 0.11

At the end of the fourth generation during third week of March 1980, 7 females were observed in the laboratory culture, with average fecundity of 35.3 eggs. In the first generation during 1980, negative trend index showed that mortality factors were operative effectively and population showed a decline in next generation. The newly hatched failed to enter leaf mid-ribs of first open leaf from the top. In second and third generations, the trend index was positive, showing that mortality factors were not able to reduce the population in subsequent generations. Here also, larvae failed to enter the leaf mid-ribs and parasitization by egg parasitoid *Telenomus* sp in second generation and parasitoids Isotima javensis, *Rhaconotus scirpophage* and *Stenobracon nicevillei* during third generation was high. In the fourth generation again, the trend index was negative showing that the population will reduce in the next generation. There was high mortality in the late instars when the larvae were boring through cane tops and water was collected in their feeding tunnels and also due to unsuccessful entry of early instar larvae in leaf mid-ribs.

3.4. LABORATORY LIFE-TABLE

Insects living in a habitat under certain environmental conditions are expected to grow at a certain rate, to live for a certain period, and to produce certain number of offspring over a span of its life. Each individual in a population will have its own speed of development, longevity and fecundity at different ages of its life. But for a population, normally mean values are considered. These mean values are usually influenced by

(i) the environment and (ii) certain innate quality of the animal, i.e innate capacity for increase in numbers. If the environment is favourable for growth then individuals in population will show maximum capacity to increase, i.e. biotic potential. However, various factors in the environment act as a check on this maximum potential, which is termed as environmental resistance. It includes the physical factors like temperature, moisture and light and the biotic factors which include food, parasitoids, predators and competition. Environment is rarely constant in nature and whatsoever rate of growth of a population is observed is the actual rate of increase.

Innate capacity for increase in numbers (rm) can be measured experimentally under controlled environment conditions with the help of laboratory life-tables. Some components of the environment influence individuals regarding their birth rate and death rate, which ultimately influence the innate capacity for increase in numbers. A population is balance between the birth and death rates. When environment is defined experimentally, then 'rm' can be measured by measuring the rate of increase. Here,

(i) certain components of the environment are excluded, i.e. other animals of different kind (predators, parasitoids, pathogens)

(ii) some components are kept at an optimal level, i.e. quantity of food, space, other animals of same kind

(iii) some of the remaining components are kept as constant, i.e. quality of food

Innate capacity for increase in numbers. It can ne defined as maximal rate of increase of a species attained at any particular combination of temperature, moisture, quality of food and so on, when quantity of food, space and other animals of the same kind are kept as optimum and other animals of different kind are excluded from the environment.

Determination of 'rm'. The actual calculation of 'rm' involves knowledge about both birth and death rates in terms of age groups. For this, start the experiment with a known cohort (individuals of same age). The females developing from eggs laid on the same day are observed for age specific fecundity (mx). These observations are continued till the death of females. The survival of females at different age intervals are expressed as 'lx '. The value x" refers to mid-point of each age group. Observations on immature stages are not taken into account for calculation of 'rm'.

Following columns are suggested for construction of life-table for laboratory populations:

x = Pivotal age (days/weeks)

lx = Numbers surviving at particular age

mx = Number of female eggs laid

lx.mx = Total number of female births

x.lx.mx = Age schedule of total female births

From this table, the following parameters are worked out:

(i) Net reproductive rate (∑lx.mx): The summation of product of 'lx ' and 'mx ' values for entire duration of female life will give net reproductive rate, i.e. increase in numbers per generation.

(ii) Mean duration of generation (t) = ∑x.lx.mx / ∑lx.mx

(iii) Innate capacity for increase in numbers (rm)

rm = log eRo / t

REFERENCES

Clark, L.R., Geier, P.W., Hughes, R.D. and Morris, R.F. (1967) *The Ecology of Insect Populations in Theory and Practice*. Methuen and Co. Ltd., London.

Harcourt, D.G. (1969) The development and use of life tables in the study of natural insect populations. *A. Rev. Ent.* 14: 175-96.

Morris, R.F. (ed) (1963) The dynamics of epidemic spruce budworm populations. *Mem. ent. Soc. Can.* 31: 1-332.

Morris, R.F. and Miller, C.A. (1954) The development of life tables for the spruce budworm. *Can. J. Zool.* 32: 283-301.

Richards, O.W. (1961) The theoretical and practical study of natural insect populations. *A. Rev. Ent.* 6: 147-62.

Roy, T.C.D. and Bains, S.S. (1983) Role of key factors in population dynamics of sugarcane top borer, *Tryporza nivella* (Fab.) in Punjab. *Proc. Symp. Insect Ecol. and Resource Mgmt.,* pp 71-79.

Singh, B., Dhaliwal, J.S. and Atwal, A.S. (1977) Population studies on maize borer, *Chilo partellus* (Swinhoe) in the Punjab. IV. Life tables for determining key mortality factors. *Indian J. Ecol.* 4(1): 107-17.

Southwood, T.R.E. (1978) *Ecological Methods with Particular Reference to the Study of Insect Populations.* The English Language Book Society and Chapman and Hall.

Theory and Practice of Integrated Pest Management
R. Arora, B. Singh and A.K. Dhawan (Eds)
Scientific Publishers (India), Jodhpur 2012, pp. 49 - 75.

4

COEVOLUTION OF INSECTS AND PLANTS

Ramesh Arora
Department of Entomology
Punjab Agricultural University, Ludhiana - 141 004, India

4.1. INTRODUCTION

The insects are among the most fascinating organisms on earth. The story of insects is a tremendous success story. Insects appeared on earth at least four hundred million years ago. The earliest known fossil insect is *Rhyniella praecursor*, a Collembolan from the Devonian period of Scotland. Available evidence indicates that it might have fed on plant juice through punctured wounds. Down through the ages, the insects have struggled endlessly to survive. During this struggle, insects have gradually developed an incredible variety of body forms and ways of life (Dhaliwal and Arora, 2006). The Carboniferous saw the rise of all modern plant divisions except the angiosperms, and without doubt their main herbivores were insects. By the end of Carboniferous period, several modern groups of insects including Ephemeroptera, Orthoptera, Hemiptera and Mecoptera were well developed. Many of the insect orders (Dictyoneurida, Micropterida, Diaphanopterida) that flourished in the Carboniferous period are now extinct. These insects had beak-like, apparently piercing mouthparts, assumed to be reminiscent of the mouthparts of modern Hemiptera. These insects were believed to be phytophagous, feeding on the reproductive parts of fruits of Carboniferous and Permian plants (Panda and Khush, 1995).

In the Permian period, nineteen orders of insects were in existence. The endopterygota (Neuroptera, Mecoptera, Trichoptera, Coleoptera) were well developed by then. By the early Jurassic period, with the addition of the Demaptera, Phasmatidea and Thysanoptera, and the loss of (now) extinct orders, the fauna was essentially a modern one. The most prominent groups that emerged during the Mesozoic were the Hymenoptera, Cecidomyiidae (Diptera) and Lepidoptera (Strong *et al.,* 1984). Ditrysian (Lepidoptera) presence during the mid-Cretaceous, in addition to Lepidopteran body-fossil evidence from Early Cretaceous and Late Jurassic deposits, suggests that the radiation of major Lepidopteran lineages probably occurred during the late Jurassic on a gymnosperm-dominated flora (Labandeira *et al.,* 1994).

Much of the flowering plants and modern insect life was established during the Cretaceous period. The primitive angiosperms were woody and the evolution of herbaceous angiosperms was a later phenomenon. Many of these trees started producing nectar and colourful scented petals, for pollination. The dominance of angiosperms, with their diversified plant surface characteristics, architecture and chemical factors, provided the three factors needed for insect proliferation i.e. food, shelter and oviposition sites. These factors offered an ideal habitat for insects, which, consequently diversified in the ecosystem. Reciprocal adaptation and counter adaptation between flowering plants and insects has been an important mechanism driving a steady increase in plant and insect biodiversity over the last more than 100 million years (Panda and Khush, 1995, Futuyma and Agrawal, 2009).

Presently, Plant Kingdom and the Class Insecta represent two very extensive taxa of living organisms, both in abundance of species and in biomass. Green plants, by far the most voluminous compartment of living matter, whereas the insects are the leaders in number of species.

The two empires, herbivorous insects and flowering plants are united by intricate relationships. Animal life including that of insects cannot exist in the absence of green plants, which serve as the primary source of food for heterotrophic organisms. On the other hand, insects with their overwhelming variation in form and life history, have in all likelihood been one of the major forces in shaping the plant world. The study of these inter-relationships is great practical importance. The productivity of a large number of crops is dependant on insect pollination. The development of resistant varieties of crop plants as well as progress in integrated pest management both require an understanding of insect-plant interrelationships (Schoonhoven *et al.,* 2005)

4.2. CONCEPT OF COEVOLUTION

As early as 1859, Darwin in his magnum opus *'Origin of Species'* wrote of "Coadaptations of organic beings to each other..." Virtually all organisms interact and presumably influence their evolution in some way. Coevolution is a change in the genetic composition of one species (or group) in response to a genetic change in another species. In other words, co-evolution is reciprocal evolutionary change in interacting species. The term was originally used by C.J. Mode in 1958 for the 'Co-evolution of obligate parasites and their hosts'. Ehrlich and Raven (1964) were the first to apply the term in the context of insect-host plant co-evolution based on their study of Monarch butterfly-milkweed plant interactions.

A plant is neither susceptible to all the phytophagous insects nor any insect species is a pest on all the species of plants it encounters in nature. One of the most striking features of insect-plant relationships is the high degree of food specialization among herbivorous insects. Most insects are either monophagous or oligophagous. Less than 10 per cent of herbivorous species feed on plants in more than three plant families. All plants have developed a variety of morphological and biochemical adaptations to defend themselves against herbivores; only those species which are able to breach these defenses are able to feed on that plant (Arora and Dhaliwal, 1999). The plants keep developing new phytochemicals or other strategies to protect themselves from insect attack and insects keep on developing pathways for overcoming these defenses. These antagonistic interactions are akin to an escalating arms' race.

4.3. HOST PLANT DEFENSE STRATEGIES

Defense is a common strategy in plants which, being immobile, have difficulty in avoiding their enemies. There are two basic approaches to defense: passive or constitutive defense and dynamic or induced defense (Singh and Dhaliwal, 2004).

(i) Constitutive defense. Plants have evolved a multitude of structures and chemicals that are incorporated into their tissues for the purpose of defense. These constitutive defenses can deter, repel, intoxicate or interfere with the development or reproduction of herbivores. It may depend on the texture and composition of the plant surface, the presence of anatomical structures such as thin veins, thorns, silica, trichomes or resin ducts, the absence of nutrients required by the pest, the presence of hormone like substances that affect the development of insects, unsuitable pH or osmotic pressure or the accumulation of secondary metabolites. These products are diverse, ranging from amino acids to alkaloids, terpenes, phenolics, steroidal, cyanogenic and mustard oil glycosides (Table 4.1). In addition healthy plants may tie-up nitrogen in forms unavailable to herbivores. The advantage of some constitutive defenses is that they can be laid down during periods of energy or nutrient surplus and may then serve over an extended period of time.

Although passive constitutive defenses may be effective against an array of generalist herbivores, they encourage the evolution of specialists because they exert strong selective pressure continuously on the herbivore population. Thus, even the best defended species are attacked by a few specialist herbivores.

Table 1. Major classes of secondary plant metabolites acting as defense herbivorous insect

Class	Sub-class	Approximate number of structures	Distribution	Physiolo-gical activity
(1)	(2)	(3)	(4)	(5)
Nitrogen Compounds Alkaloids	Benzylisoquinoline Monoterpene indole Simple indole Pyrrolizidine Quinolizidine Polyhdroxy	6500	Widely distributed in angiosperms, in roots, leaves and fruits	Many toxic and bitter Tasting
Amines	L-Canavanine	100	Widespread, often in flowers	Many repellent smelling
Non-protein amino acids	L-Arginine	400	Especially seeds of legumes but relatively widespread	Many toxic
Cyanogenic glycosides	Dhurrin, amygdalin	60	Sporadic, especially in fruit and leaf	Poisonous
Glucosin-olates	Isothiocyanate, Sinigrin	100	Cruciferous and 10 other Acrid families	Acrid and bitter
Phenolics	Simple phenols	200	Universal in leaf, often in other issues as well	Antimicrobial
	Coumarins	300	From 70 families of dicot	
Flavonoids	Anthocyanins	4000	Universal in angiosperms, gymnosperms and ferns	Often coloured for attracting pollinators
	Flavonols Insoflavonoids			
Quinones	Benzoquinone Naphthaquinone Anthraqujinone Extended quinone	800	Widely distributed in all plant families except ferns and mossese specially Rhamnaceae	Coloured

(1)	(2)	(3)	(4)	(5)
Polyacet-ylenes		650	Mainly in compositae and Umbelliereae	Some toxic
Terpenoids	**Monoterpenoids**			
	Iridoids	600	Restricted to 57 families in Dicotyledoneae	
	Others	700	Well known in conifers	Some bitter and toxic, other allergenic
	Sesquiterpenoids		Widespread	
	Sesquiterpenoids	3500	Mostly in Compositae, localized in galndular hairs	
	Lactones			
	Others	3000	Widespread	Some toxic
	Diterpenoids		Widely distributed, especially in latex and plant resins of conifers	
	Clerodanes	400	Found in Lamiaceae, Asteraceae, and others	
	Others	2500	Widespread	
	Triterpenoids			Many bitter and toxic
	Cucurbitacins	20	Cofined mainly to Cucubitaceae	
	Cardenolides	150	Asclepiadaceae and others	
	Limonoids	300	Mainly in Rutaceae, Meliaceae	
	Quassinolds	200	Simaroubaceae	
	Saponins	1200	Widespread	Haemolyze blood cells
	Phytoecdysones	70	Found in more than 100 plant families	Haemolyze blood cells
	Others	1500	Widespread	
Polyacet-ylenes		650	Mainly in composital some toxic and umbelliferae	

Source: Arora and Dhaliwal (2004)

(ii) Induced defense. Induced defense involves the production of chemicals or physical structures, or the removal of nutrients essential to the herbivore, in response to attack. Induced defenses may take several forms.

(a) Nutrient removal. It involves the tie-up of nutrients (such as nitrogen) that are essential for herbivore growth in a form that is unavailable to the herbivore.

(b) Cell lignification. In which tougher (heavily lignified), less digestible food is produced after attack.

(c) Controlled chemical biosynthesis. In which relatively small quantities of defensive chemicals are produced by living cells close to or even distant from the point of injury. Controlled biosynthesis does not result in cell death or necrosis because the toxic chemicals are contained within membrane limited structures like vacuoles, intercellular spaces or are impregnated into cell walls. Maize (*Zea mays* L.) lines resistant to feeding by a number of lipidopteran species, rapidly mobilize a unique 33-k Da Cysteine protease in response caterpillar feeding. The accumulation this protease in the maize mid-whorl was correlated with a significant reduction in caterpillar growth that resulted from impaired nutrient utilization (Pechan *et al.,* 2002).

(d) Uncontrolled chemical biosynthesis. In which very large quantities of defensive chemicals are produced by living cells close to the site of injury, sometimes called as a hypersensitive reaction. During this response, sugars and starches are rapidly converted into indigestible or toxic compounds (terpenes, phenol, etc.) as a result of apparently uncontrolled chemical reactions.

(e) Green leafy volatiles. Green leafy volatiles (GLVs), six-carbon aldehydes, alcohols and esters are involved in priming the corn plant defenses against herbivorous insects. These GLVs induced intact undamaged corn seedlings to rapidly produce jasmonic acid and emit sesquiterpenes. More importantly, corn seedlings previously exposed to GLV from neighbouring plants produced significantly more jasmonic acid and volatile sesquiterpenes when mechanically damaged and induced with caterpillar regurgitant than seedlings not exposed to GLV. In contrast, GLV priming did not affect jasmonic acid production induced by mechanical wounding alone. Thus, GLVs enhance inducible defense responses triggered during attack and may play a key role in plant-plant signaling and plant-insect interactions (Engelberth *et al.,* 2004).

(iii) Cross-talk between plant defense signaling pathways. Plants are exposed to very different attackers, including herbivorous insects and microbial pathogens. To counteract potential invaders, plants have evolved defensive strategies as mentioned above. But recent advances in plant defense signaling research have revealed that plants are capable of differentially activating inducible, broad-spectrum defense mechanisms, depending upon the type of invader encountered. The plant hormones salicylic acid (SA), jasmonic acid (JA) and ethylene (ET) are major player in the network of defense signaling pathways. Cross-talk between SA-JA-and ET-dependent signaling pathways is thought to be involved in fine-tuning the defense reaction, eventually leading to the activation of an optimal mix of defense responses to resist the intruder. Genetic engineering of the biosynthetic pathways of these signaling compounds and the development of new chemicals mimicking their mode of action provide useful tools for the development of new strategies for crop protection. However, there is evidence for antagonism between SA-dependent resistance to microbial pathogens and JA-dependent resistance to herbivorous insects. But through genetic engineering, application of defense signal-mimicking plant protectants will boost the plants' immunity to potential invaders (Pieterse *et al.,* 2001; Kessler and Baldwin, 2002).

(iv) Plant Defense through Association of Organisms. Another mode of investing in defensive chemicals through association with other organisms is a fascinating relationship. Some of them are exemplified as under:

Endophyte enhanced plant defense. Several endophytic fungi have been associated with plant defense against insect herbivores. The relationship between the endophytic fungus and its host plant may be mutualistic or pathogenic depending upon the specific fungal isolates and on environmental conditions. However, many endophytes do not produce pathogenic effects but enhance resistance to insects. The effect of the endophyte *Rhabdocline parkeri* Sherwood on a complex of midges, *Contarinia* spp. (Diptera : Cecidomyiidae) is one of the most studied examples. Fungal toxins are responsible for the larval kill of insects. The *Acremonium* endophytes are obligate seed-borne fungi that form inapparent infections in leaves, stems and inflorescences of wide range of turf, conservation and pasture grasses. Resistance to at least 23 species of insects in 10 families and 5 orders has been found associated with endophyte infection of perennial ryegrass and tall fescue.

Ant-Acacia relationship. Most acacias have thorns, bitter leaves, and other protection against a hungry world. One, the "swollen thorn acacia," learned to encourage a species of ant *Pseudomyrmex* sp. to monopolize it as a food source and kill or run off all other predators. Enticements gradually included nifty water-proof swollen thorns to live in, handy nectar fountains, and special ant-food buds at the leaf tips. The ants, whose interests increasingly coincided with the acacia's, learned to inhabit the thorns, patrol the acacia day and night, attack every acacia-hungry organism, and even prune away invading plants such as vines and tree seedlings that might shade Mother *Acacia*. The acacia gave up its bitter leaves, sharp thorns, and other devices and now requires the acaia-ant for survival. Likewise the ant colonies can no longer live without the acacia. Together they're unbeatable.

4.4. STRATEGIES OF INSECTS TO OVERCOME PLANT DEFENSE

Several strategies have been employed by insects to overcome plant defenses.

(i) Sequestration of Defensive Substances from Plants

Instead of detoxifying plant poisons, several species of insects sequester and deploy the poisons for their own pheromone system and defense. These herbivores are so intimately adapted to the host plants that they can use these compounds to their own advantage. A number of aposematic butterfly and diurnal moth species sequester unpalatable or toxic substances (secondary metabolites) from their host plants to achieve effective protection from predators. It is evident from Table 4.2 that large number of secondary metabolites are sequestered by Lepidoptera.

Table 4.2. Plant allelochemicals sequestered by Lepidoptera

Chemical class of sequestrates	Butterflies	Moths
Terpenic metabolites		
Cardenolldes	Danalnae	Arctiidae, Ctenuchidae
Grayanolds	-	Geometridae
Podocarpaceae norditerpenolds	-	Geometridae
Iridold glycosides	Nymphalinae	Geometridae, Sphingidae
Phenolic metabolites		
Lichen phenolics	-	Arctiidae
Nitrogen-containing metabolites		
Pyrrolizidine alkaloids	Ithomiinae, Danainae	Arctiidae
Tropane alkalolds	-	Pyralidae
β-Carboline alkaloids	Heliconlinae	-
Amaryllidaceae alkaloids	-	Noctuidae
Phenanthroindolizine alkaloids	Danainae	-
Azafuranose/azapyranose	Nymphalinae	Uraniiae
Aristolochic acids	Paplionidae	-
Cyanogenic glycosides	Heliconiinae, Acaeinae	Zygaenidae
Samentosin	Papiiionidae	Geometridae, Zygaenidae
Cycasin	Lycaenidae	Arctiidae
Glucosinolates	Pieridae	-

Source: Nishida (2002)

Several pharmacophagous insects have been shown to sequester specific substances or their derivatives in their body tissues. Adults of *Athalia rosae* (Linnaeus) visit a plant *Clerodendrum trichotomum* Thunberg and feed voraciously on the leaf surface. Clerodendrins were characterized as the potent phagostimulants of *A. rosae* adults. The insect sequesters some of the analogues and becomes extremely bitter on its body surface. Some chrysomelids associated with cucurbits were found to store high concentrations of bitter cucurbitacins in their body. Four species of Asian pumpkin leaf beetles belonging to the genus Aulacophora were shown to sequester 23, 24-dihydrocucurbitacin D in body tissue as the major bitter principle. Three phenylpropanoids closely related to methyl eugenol were found to accumulate in the rectal glands of the male of *Bactrocera dorsalis* (Hendel). One of the rectal gland components, 2-ally1-4, S-dimethyloxyphenol, was shown to be released in the air during courtship. In all of these cases, selectively sequestered compounds strongly deterred feeding by some predators, thus serving as allomones in this context.

Pyrrolizidine alkaloids are defensive secondary metabolites found in numerous plant groups. Various insects belonging to different orders have special requirements for these compounds and sequester them from such plants for their own defense and often as pheromone precursors.

The sexual behaviour of phytophagous insects is often integrated in a variety of ways with their host plants. This integration may be manifested as effects or influences of host plants on insect physiology and behaviour, including sex pheromone communication. Some insects produce or release sex pheromones in response to particular host plant cues. Chemicals from host plants enhance insect responses to sex pheromones. By these means, insects use host plants to mediate sexual communication and regulate the reproduction to coincide with the availability of food and oviposition sites.

(ii) Detoxification of Defense Chemicals

Insects possess a range of detoxifying enzymes which constitute a powerful defense against plant toxicants. The involvement of the cytochrome P-450-dependent microsomal monooxygenase system in interactions between plant terpenes and insect herbivores is well established Cytochrome P-450-metabolises many terpenes. Potato aphid *Macrosiphum euphorbiae* (Thomas) is well adapted to feed on secondary metabolites of potato such as glycoalkaloids. The foliage of sweetbay magnolia (*Magnolia virginana*) contains at least two biologically active phenylpropanoid compounds (magnolol and a biphenylether) that are toxic to number of generalist insect herbivores. But Sweetbay silk moth, *Callosamia securifera* (Maassen) is not affected by such compounds. The degradation of these compounds by midgut homogenate suggests the involvement of cytochrome P-450 detoxification enzymes.

(iii) Detoxification of inducible defense chemicals

Behavioural adaptations in herbivores that reduce or negate induced plant defenses have been recognized. They suggest that the characteristic feeding behaviour of certain species of *Epilachna* (Coleoptera: Coccinellidae) represents an adaptation to the induction of cucurbitacin defenses in their cucurbit hosts. *Epilachna tredecimnotata* (Latreille) and *Epilachna borealis* Fabricius chew a circular trench through all but the lower epidermal leaf tissue, effectively isolating the leaf area on which they are about to feed. Such behaviour is thought to prevent the rapid translocation (within 40 min.) of cucurbitacins and other antifeedants to the damaged site, thus preserving the palatability and quality of the leaf resource while it is consumed in the ensuing hours.

The Gypsy moth, *Lymantria dispar* (Linnaeus) is known as a polyphenol specialist which is relatively tolerant of the phenolic compounds by detoxifying them during passage of food through gut. Larvae of the tortoise beetle, *Eurypedus nigrosignata* Boh. construct fecal shields containing the same terpene pattern as that of leaves of larval host, *Cordia curassavica* (Jacq.) Roem and Schult. These terpenes act as repellent for ants and other predators.

(iv) Use of Allomones as Kairomones or Nutrients

Some specialist insects are able to utilize not only primary nutrients for growth but allelochemicals as well. For example, canavanine in seeds of *Dioclea megacarpa* Rolfe is a harmful non-protein amino acid. The bruchid beetle, *Caryedes brasiliensis* Thunberg is a highly adapted predator of this leguminous vine. Thus, the poison canavanine becomes an important nutrient for this particular insect. Several insect herbivores have evolved a wide variety of behavioural responses and physiological mechanisms to tolerate host plant toxins. It is also suggested that induction of detoxification enzymes by plant chemicals may be involved in insect's adaptation to its unsuitable host plant. For example, the limited adaptation of the desert locust, *Schistocerca gregaria* (Forskal) to plant glucosides is compensated by an ability to tolerate high concentrations of allelochemicals (thioglucosides) for a short period. Glucosinolates induced the production of myrosinase in the mid gut of locust.

Some Hemipteran insects, when feed in large numbers, result in enha-nced nutrient flow in infested leaves. The relationship between the gall-forming insects and their host plant is remarkable example of insect-plant interactions. Flower galls of the cruciferous plant *Diplotaxis muralis* are formed from the saliva of the gall midge, *Paragephyraulus diplotaxis* Solinas. The flower galls are nutritive tissues and provide food to the larvae of gall midge. Some phytochemicals such as quinines, phenolics, aldehdes, sesquiterpene lactones, isothiocyanates and pyrrolizidine alkaloids present in the plants when ingested with protein may negatively impact on protein utilization. But some insects like *Heicoverpa zea* (Boddie) has the physiological adaptation to low protein quality.

(v) Avoiding Pheromonal Communication between Plants

It is postulated that plants could receive pheromonal signals emitted by nearby attacked leaves. Such signals are highly advantageous to plants as a form of advance warning of herbivore attack. Some herbivores could evolve adaptations to suppress pheromonal communication between damaged plants. This communication suppression could occur as an indirect result of suppression of the recognition of attack or through herbivore secretions that block the release of communication substances from the wound. Some herbivores may themselves emit pheromonal countersignals to confuse, mimic or otherwise interfere with the signals generated by plants. Red alder trees *Alnus rubra* Bong attacked by the western tent caterpillar, *Salix sitchensis* Sanson exhibit a change in foliage quality. Tent caterpillars that were fed with unattacked leaves from attacked trees grew more slowly, died at a faster rate than those fed on leaves from unattacked trees.

(vi) Avoiding Plant Allelochemicals

The mode of insect feeding can greatly reduce the exposure of insects to plant allelochemicals. The furanocoumarins in the plant family Apiaceae are localized in the seeds and companion canals adjacent to the veins. The insects avoid feeding on them. The neonate larvae of *Heliothis virescens* (Fabricius) avoid feeding on gossypol glands during first 48 h after hatching. When neonate larvae were fed squares of lines with many glands, growth was decreased by 25 to 75 per cent.

(vii) Avoiding Physical Defense

Insects have evolved many ways of avoiding physical defenses such as leaf spines, hairs and trichomes. The caterpillar of the noctuid moth, *Pardasena* sp. could mow the trichomes on the leaves of *Solanum coccineum* Jacq. before eating the mown strip. Small herbivores like *Tupiocoris notatus* (Distant) is found predominantly on *Datura wrightii* Regel plants with glandular trichomes. This sap sucking herbivore had significantly higher survival rates on such plants than those raised on non-glandular plants.

(viii) Utilizing Alternate Diets/Host Plants

Insects are able to discriminate between good and less suitable host plants from nutrition point of view. The generalist insects avoid young leaves that are rich in toxins and deterrents while specialist herbivores prefer them because the specialists have tolerance for the toxins produced by their specific hosts. *Pristiphora abietina* Christ avoids feeding on Norway spruce trees containing more defensive compounds like fibre, quinic acid and tannins. The larvae of *Ceratitis capitata* (Wiedemann) can discriminate between nutritionally superior and inferior diets.

(ix) Increase in Consumption Rate

Many herbivores increase their consumption rate as dietary nutrient concentration declines. This compensatory response can mitigate the fitness lowering impact of reduced food quality.

(x) Nutritional Enhancement through Association with Microorganisms

Utilization of microorganisms by insects helps in overcoming their nutritional constraints and improves the insect-plant nutritional relationships. Termites are capable of producing enzymes supplied by protozoan flagellates or bacterial symbionts. Aphids can obtain many amino acids via their gut flora. The yeast of *Lasioderma serricorne* (Fabricius) can produce detoxifying enzymes that can metabolize plant allelochemicals, meal toxins, mycotoxins, insecticides and even herbicides which improve fitness of the insect drastically.

The insect-plant interactions summarized above reveal the adaptation of plants in variety of ways to reduce the impact of herbivory. Counter adaptation of insects to overcome the hurdles put forth by plants seems to be a never ending race. In the process huge data have been generated for theoretical considerations of the evolutionary processes thereby leading to formulation of various theories of evolution.

4.5. THEORIES OF INSECT-PLANT CO-EVOLUTION

Reciprocal adaptations and counter-adaptations between plants and phytophagous insects have been an important mechanism leading to a steady increase in plant and insect biodiversity over the last 400 million years. Insects have acquired a particularly well developed and diversified chemical communication system; successful host finding and acceptance are primarily controlled by chemical cues. Concomitantly, the plants have also evolved a rather bewildering array of chemicals for protection against insects and other herbivores. A number of theories have been propounded to explain this evolutionary arms' race between two interdependent groups of organisms.

Theory of co-evolution

This theory was elaborated by Ehrlich and Raven (1964) and later supported by Berenbaum (1983). The main points of the theory are listed below.

(i) Many plant taxa manufacture a prototypical photochemical that is mildly noxious to phytophages and that may have an autecological or physiological function in the plant.

(ii) Some insect taxa feed upon plants with only this and other, similarly mild, phytochemicals, thus reducing plant fitness.

(iii) Plant mutation and recombination cause novel, more noxious phytochemicals to appear in the plants. The same chemical can appear independently in distantly related plant groups.

(iv) Insect feeding is reduced because of toxic or repellent properties of the novel phytochemicals; thus plants with increasingly noxious chemicals are selected for by the pressure of insect herbivory.

(v) The plant, 'protected from the attacks of phytophagous animals, would in a sense have entered a new adaptive zone. Evolutionary radiation of the plants might follow...'

(vi) Insects evolve tolerance of, or even attraction to and utilization of, the novel compound and the plant producing it. An insect can specialize in feeding upon plants with the novel compound 'here it would be free to diversify largely in the absence of competition from other phytophagous animals'.

(vii) The cycle may be repeated, resulting in more phytochemicals and further specialization of insects.

Theory of Sequential Evolution

The theory of sequential coevolution (Jermy, 1976, 1984) proposes that the evolution of herbivorous insects follows the evolution of plants without, however, significantly affecting plant evolution.

The sequential or asymmetrical nature of the evolution of insect-plant relationships is indicated by the following.

(i) Reciprocal selective interactions between plants and herbivorous insects have not been proved so far. Plants exert strong selection pressure on insects, whereas the latter exert selective pressure on the plants only in rare cases and even in these only weakly.

(ii) As a consequence, speciation in herbivorous insects may be mediated by plants but speciation in plants has not been proved to occur as a consequence of interaction with herbivorous insects. Compared with alternative resources, plants are characterized by both high availability and high diversity. Insect diversification rates could conceivably be influenced by both resources abundance (decreased competition) and diversity (large number of potential niches)

Recently, Janz *et al.* (2006) applying a variant of independent contrast analysis, specially tailored for use on questions of species richness showed that species richness in butterfly family Nymphalidae was strongly correlated with diversity of hose use. Furthermore, by comparing the results from reciprocal sister group selection it was observed that diversity of host use was driving insect species richness, rather than vice versa.

Theory of Diffuse Coevolution or Community Co-evolution

The theory of diffuse co-evolution proposes that, instead of the pair-wise reciprocal evolutionary interactions, coevolution must be considered in a community context and not simply as an isolated two species interaction. The community affecting any one plant species includes several type of herbivores, alternative host plants, potentially competing plants and species in high trophic levels that may respond directly to the herbivores or to the plants, changes induced by one member of the community may affect many other species.

Similarly, the theory of community co-evolution proposes that interactions between the members of a community are so tight that the evolution of any member affects the evolution of all other members.

The Geographic Mosaic Theory of Co-evolution

The mosaic theory of coevolution has been defined as the reciprocal evolutionary change in interacting species and assumes that all interacting species, at least in some locations, necessarily coevolve. Thompson (1994, 2005) summarized this theory as follows.

(i) Interspecific interactions commonly differ in outcome among populations. These differences result from the combined effects of differences in the physical environment, the local genetic and demographic structure of populations and the community context in which the interaction occurs.

(ii) As a result of these differences in outcomes, an interaction may coevolve some populations (coevolutionary hot spot), affect the evolution of only one of the participants in other populations (coevolutionary cold spot) and have no effect on evolution in yet other local populations (again coevolutionary cold spot).

(iii) In addition, populations differ in the extent to which they show extreme specialization to one or more species. Some populations may specialize on and sometimes coevolve locally with only one other species, other populations may specialize on and perhaps coevolve with different species, and yet others may coevolve simultaneously with multiple species.

(iv) These interpopulational differences in outcome and specialization create a geographic mosaic in interactions.

(v) Gene flow among populations, random genetic drift, selection for novel traits and extinction of some demes reshape the geographic mosaic of coevolution as the adaptations and patterns of specialization developed locally spread to other population or are lost.

(vi) The result is a continually shifting geographic pattern of coevolution between any two or more species.

Recently Toju and Sota (2006) presented fresh evidence that climatic conditions could affect the local and overall trajectory of co-evolutionary arms race. These authors investigated the geographic variation in physical defensive/offensive traits among 17 populations of the Japanese camellia (*Camellia japonica*) and its obligate seed predator, the camellia weevil (*Curculio camelliae*) over a 700-Km-wide area in Japan. The sizes of the plant defensive apparatus (pericarp thickness) and the weevil offensive apparatus (rostrum length) clearly correlated with each other across populations. Nevertheless, the balance in armaments between the two species are geodgraphically structured. The Camellia pericarp thickness varied among populations from 9 to 20 mm. Both pericarp thickness and rostrum length increased clinically from north to south. In the south, the balance was relatively advantageous for the plants' defense and natural selection on the trait was stronger because in the other populations (towards north), most plant individuals were too vulmerable to resist the attack of weevils, and their seeds were infested independent of pericap thickness.

4.6. EXAMPLES OF CO-EVOLUTION

The coevolution between insects and plants could be of two types: 'arms race' coevolution and the mutualistic coevolution.

4.6.1. 'Arms Race' co-evolution

In this type of coevolution, one organism (usually the plant) evolves so as to reduce or eliminate the attack of the second organism (usually the insect). The insect must then counteract these changes in the plant, so that it can more effectively attack it. This type of coevolution is common between plants and herbivorous insects.

A possible example is the evolution of milkweeds (*Asclepias* spp.) and their insect predators. Some individuals develop foul-tasting and toxic chemicals (cardenolide glycosides) which deter plant feeders. In turn, some insects develop ability to detoxify these chemicals either getting rid of them or storing them in their bodies. These include the Monarch butterfly, two species of milkweed beetles and two species of milkweed bugs.

Another example of human-driven co-evolution is between Hessian fly and wheat. The Hessian fly, *Mayetiola destructor* (Say) has a long history of pestilence in the United States. The insect was first found infesting wheat on Long Island, New York in 1779 and was believed to have been introduced by Hessian soldiers in straw bedding for horses during the American Revolutionary War. The Hessian fly now occurs in all major wheat-growing areas of the United States. The fly is capable of infesting and injuring all class of wheat growing in the United States. Although wheat is the preferred host of the Hessian fly, barley, rye and triticale are other cultivated cereal crops occasionally infested. Several wild grasses may serve as hosts when these crops are not available.

Feeding injury by the Hessian fly is caused entirely by the larva. On young plants, the larva feeds beneath the leaf sheath at the base of the plant. The feeding mechanism of the larva and how it obtains food from the plant are not fully understood. A study of the larval mouthparts has revealed highly specialized mandibles that are probably used to inject salivary fluids into plants. These secretions are believed to contain enzymatic substances that inhibit plant growth and increase cell wall permeability that allows the larva to suck the juices from the plant. No injury to plant tissues has been observed at feeding sites, although infested plants show a characteristics stunted appearance. The number of stems and leaves and foliar and root weight are reduced in wheat plants infested with Hessian fly larvae. Leaves of infested plants also appear more erect and are shorter and darker green than those of uninfested plants. When infestations are severe, stunting of seedlings occurs earlier and many of the young plants die after the larvae have matured (Buntin and Chapin, 1990).

Larval feeding also reduces the winter hardiness of wheat plants that survive the fall infestation. This contributes to further loss during the winter. Heavily infested fields show areas of dead plants and thin stands. At lower infestation levels, seedlings may survive and develop new tillers that enable the plant to compensate for stunting of the main stem. In the spring when plants are in the jointing stage, the larva feeds just above the node between the leaf sheath and culm. Feeding causes injury that prevents normal elongation of internodes and transport of nutrients to the developing spike. This injury reduces both quantity and quality of the grain. The most obvious damage occurs when infested culms break over at the weakened nodal areas. Much of the grain of broken culms is lost when the crop is harvested.

In 1915, Hessian flies caused more than $ 100 million damage to wheat in the United States. Current estimates of yield reduction caused by Hessian fly in susceptible wheat varieties range from 5 to 75 per cent. Major programs to develop and release Hessian fly-resistant wheat varieties adapted to the Great Plains hard wheat and eastern U.S. soft winter wheat regions began in the 1920s and 1940s, respectively. The documented release of sixty Hessian fly-resistant varieties during the period from 1950 to 1983 is evidence of the success of these programs. In areas where resistant varieties have been grown for several years, losses by the Hessian fly have been reduced to less than one per cent. Numerous sources of resistance have been identified and incorporated into resistant wheat varieties; varieties with new sources of resistance are continually developed and released when older varieties lose effectiveness. Resistance in these sources is dominant, partially dominant, or recessive, and conditioned by single, duplicate or multiple genetic factors derived from common wheat *Triticum aestivum* L., durum wheat, *T. turgidum* L., the wild wheat, *T. tauschii* (Coss) Schmal. and rye, *Secale cereale* L. Thirty two major genes designated H1 through H32 have been described (Ratcliffe, 2006).

There are two types of Hessian fly/wheat interactions. First compatible interactions allow first-instar larvae to survive on susceptible wheat plants. In these interactions, larvae establish a sustained feeding site, develop normally, and complete their life cycle. However, the susceptible wheat seedlings are severally affected. Second, incompatible interactions inhibit survival of first-instar larvae on resistant wheat plants. These interactions are characterized by larvae that fail to establish a sustained feeding site or develop normally and usually die within a period of 5-6 days after hatching. Resistant wheat undergoes little or no physiological stress during Hessian fly attack and yields normally (Mittapalli *et al.,* 2007).

Recent work has indicated that resistant plants produce a specific protein, called HFR-3, one of a group of substances called lectins. When the larvae attack a resistant plant, the plant's lectin production quickly increases by as much as 300 times. The larvae then ingest the lectin. This interaction probably damages the larva's chitin-rich mid-gut lining so that it can not absorb nutrients from the plant, causing the insects to starve to death. Virulent biotypes of Hessian fly are capable of ridding their bodies of lectins and, thus, survive. The researchers now believe that plants resistant to Hessian fly invasions may be making several types of lectins in response to virulent larvae (Giovanini *et al.,* 2006). According to Mittapalli *et al.* (2007) the incompatible interaction is characterized by higher mRNA levels of antoxidant genes in larvae on resistant plants. The larvae in incompatible interaction experience significantly greater oxidative stress. This may be due to increased reactive oxygen species (ROS) in the resistant wheat plants attacked by Hessian fly larvae.

The development of virulent biotypes of the Hessian fly poses the greatest threat to the permanence of resistance in wheat varieties. The growing of wheat varieties with high levels of resistance exerts a strong selection pressure on Hessian fly populations that favors races, or biotypes, capable of surviving and reproducing on resistant wheat. Therefore, the biotype composition of Hessian fly populations within wheat growing regions changes with exposure to wheat varieties carrying specific resistance genes, and over time the effectiveness of varieties in preventing infestation by specific biotyps is reduced.

In this extensively studied system a gene-for-gene association has been well demonstrated. For every resistance gene in wheat that can stop the fly, there is a complimentary avirulence gene in the fly that can overcome the resistance. Sixteen Hessian fly biotypes, classified as 'GP' (Great Plains) and “A” through 'O' differ in the number of resistance genes to which they express virulence. Biotype 'L' is the most virulent and can attack wheat varieties with any of these genes (Ratcliffe, 2006). Now the scientists are trying to sequence all the Hessian fly genes which help the fly in successfully attacking resistant wheat plants in order to develop genotypes with durable resistance. A team of scientists from Agricultural Research Service, USA has identified at least 2000 genes that play some role in churning out the toxic salivary brew that the fly injects into wheat plant when taking a bite. Molecular marker-assisted selection (MAS) methods have also been developed for some of the resistance genes in wheat and are being increasingly adopted in resistance breeding programmes. Similar development of biotypes has now been reported in a number of Homopterous, Dipterous, Coleopterous and Lepidopterous insects (Table 4.3).

Table 4.3. Important insect-pests of agricultural crops for which biotypes have been reported

S.No.	Species	Common name	Crop	No. of entries
1	2	3	4	5
		COLEOPTERA **Bruchidae**		
1	*Callosobruchus maculates* (Fabricius)	Cowpea weevil	Cowpea	1
	Curculioidae			
2	*Phaedoia inclusa* Stal	Soybean weevil	Soybean	1

1	2	3	4	5
3.	*Sitophilius oryzae* (Linnaeus)	Rice weevil	Com wheat	2
		DIPTERA		
		Curculionidae		
4.	*Agromyza oryzae* (Munakata)	Rice leaf miner	Rice	2
5	*Mayetiola destructor* (Say)	Hessian fly	Wheat, barley	14
6	*Orseolia orzea* (Wood-Mason)	Aisan rice gall midge	Rice	10
	Chloropidae			
7	*Chlorops oryzae* (Metsumara)	Rice stem maggot	Rice	2
	Tephritidae			
8.	*Rhagoletis cerasi* Linnaeus	Cherry fruit fly	Cherry	2
		HOMOPTERA		
		Aleyrodidae		
9.	*Bemisia tabacci* (Gennadius)	White fly	Cotton, Okra, Cassava, Squash, etc.	9
		Aphididae		
10	*Acyrthosiphon plsum*(Harris)	Pea aphid	Alfalfa, etc.	9
11.	*Aphis kondoi* Shinji	?	Alfalfa	1
12	*Amphorophora runi* (Kaltenbach) *A.idaei* (Born)	Raspberry aphid	Berries	6
13	*Aphis craccivora* Koch	Cowpea aphid	Groundnut, Bush sitao	2 5
14	*Aphis fabae* Scopoli	Bean aphid	Broad bean	2
15	*Aphis gossypii* Glover	Cotton aphid	Pepper	2
16	*Aphis nasturtii* (Kaltenbach)	Buckhorn aphid	Potato	2
17	*Aulacorthum solani* (Kaltenbach)	Greenspoted potato aphid	Potato	2
18	*Brevicoryne brassicae* (Linnaeus)	Cabbage aphid	Cabbage	2
19	*Chaetosiphon fragaefolii* (Cockerell)	Strawberry aphid	Cabbage	2
20	*Dysaphis directa* (Walker)	Apple aphid	Apple	3
21	*Eriosoma lanigerum* (Hausmann)	Wooly apple aphid	Apple	3
22	*Macrosiphum euphorbiae* (Thomas)	Potato aphid	Potato	2
23	*Myzus persicae* (Sulzer)	Green peach aphid	Potato	3

1	2	3	4	5
24	*Rhopalosiphum maidis* (Fitch)	Corn leaf aphid	Barley	5
			Corn	5
			Sorghum	2
25	*Rhopalosiphum padi* (Linnaeus)	Oat-bird cherry aphid	Cherry	2
26	*Schizaphis graminum* (Rondani)	Greenbug	Barley, Oat, Sorghum, Wheat	11
27	*Therioaphid maculata* (Buckton)	Spotted alfalfa aphid	Alfalfa	6
28	*Therioaphis trifolii* (Monell)	Yellow clover aphid	Alfalfa	8
		Cicadellidae		
29	*Muelerianella farmairei* (Perris)	Leafhopper	-	1
30	*Nephotettix virescens* (Distant)	Green leafhopper	Rice	3
		Coccidae		
31	*Saissetia oleae* (Olivier)	Black scale	Citron melon	1
		Delphacidae		
32	*Nilaparvata lugens* (Stal)	Brown planthopper	Rice	5
		Diaspididae		
33	*Lepidosaphes ulmi* (Linnaeus)	Oystershell scale	Apple	2
		Phylloxeridae		
34	*Daktulosphaira vitrifoliae* (Fitch)	Grape phylloxera	Grapevine	2
		LEPIDOPTERA		
		Pyralidae		
35	*Ostrinia nubilalis* (Hubner)	European corn borer	Corn	4
		Tortriidae		
36	*Laspeyresia pomonella* (Linnaeus)	Walnut coddling moth	Apple, plum, walnut	3
		Yponomeutidae		
37	*Yponomeuta padells* (Linnaeus)	Small ermine moth	Apple hawthorn	1
		THYSANOPTERA		
		Thripidae		
38	*Thrips tabaci* (Lindeman)	Onion thrips	Tobacco deadnettle	2

Source: Bentur (2004)

4.6.2. Mutualistic Co-evolution

In mutualistic coevolution, the two organisms change (evolve) in such a way as to enhance the effectiveness of the interaction. This type of interaction is common between plants and their insect pollinators, as well as some other specialized cases. If the coevolution is extreme enough, the two partners may become completely dependent on each other. Pollination is beneficial to both the insect and the plant. Insect gets food (pollen and/or nectar) while plant increases probability of successful reproduction. Flowers have evolved as an indicator of food (a reward) to their pollinators. Flowers anatomy ensures that, while feeding, the pollinator also picks up pollen.

Early gymnosperms and angiosperms were wind-pollinated. Carter (2005) presented an interesting account of the origin and evolution of pollination in plants. Like modern gymnosperms the ovule exuded droplets of sap to catch pollen grains. Insects (beetles) on the plant found this protein/sugar mix and used it as food. Insects became dependent on this food source and started carrying pollen from plant to plant. Beetle-pollination must have been more efficient than wind for some species, so there was natural selection for plants that attracted insects.

Next to occur would have been the evolution of nectaries, nectar-secreting structures, to lure the pollinators. Development of white or brightly-colored, conspicuous flowers to draw attention to the nectar and/or other food sources would also have occurred. The carpel (female reproductive structure) was originally leaf-shaped. It became folded on itself to enclose and protect the ovule from being eaten by the pollinators (hence Angiosperms). Plants with protected ovules would have been selected over ones with ovules that got eaten.

By the beginning of the Cenozoic Era, the first bees, wasps, butterflies, and moths had evolved. From this point on, certain plant and insect species have had a profound influence on one another's evolution. A flower that attracted specific pollinators on a regular basis had an advantage (less wasted pollen) over flowers that attracted "promiscuous" pollinators. It is also an advantage for the pollinator to have its own "private" food source because there is, thus, less competition. The varied shapes, colors, and odors of flowers allowed sensory recognition by pollinators and excluded unwanted, indiscriminate pollinators. Today, over 65 per cent of Angiosperms are insect-pollinated and 20 per cent of insects, at least at some stage, depend on flowers for their food (Carter, 2005). Different pollinators are attracted to different colours of flowers. Blue flowers tend to be bee-pollinated while butterflies prefer white and yellow flowers. Some many bees see in UV, bee-pollinated flowers often have markings in UV light.

There are some species of insects in most of the twenty-nine insect orders that are involved in pollination. Solitary bees, followed by bumble bees and honey bees are the most important groups of insect pollinators (Table 4.4). The other important groups include some flies, butterflies and moths and beetles. Among crops, a large number of fruit trees, vegetables, oilseeds and several seed crops are insect pollinated (entomophilous) (Table 4.5).

Trait complimentary and exploitation barriers are the two important factors determing which insect species pollinates a particular crop. Each flower type conceals its rewards behind barriers and only those pollinators whose trait allow them to overcome the barriers have access to the reward. An insect will be unable to reach nector in floral tubes longer than its proboscis; the tube length sets up a barrier to some species, but not to others. Each plant species also has a given flowering period. The specific activity period of each insect species will complement the flowering of some plant species more than others. Other barriers and other complementary traits have been described for a variety of plant-pollinator pairs. In general, plant species evolve to be pollinator specialists on the most effective and common pollinators when their abundance is low relative to other plant species in the community (i.e. conspecific pollen is relatively rare) and evolve to be pollinator generalists when they are numerically dominant (i.e. conspecific pollen in abundant) (Sargent and Otto, 2006).

Table 4.4. Important groups of insect pollinators

(a) Hymenoptera

Colletidae (Plasterer Bees; Membrane Bees)

Colletes spp. } Holarctic originally; now widespread

Hylaeus spp. }

Halictidae (Mining Bees; Sweet Bees)

Halictus spp. temperate

Nomia spp. more tropical, especially in New World

Andrenidae (Burrowing Bees; Digger Bees)

Anthidiumspp. Holarctic, etc.

Megachilidae (Leaf-cutting Bees and Mason Bees)

Anthidium spp. tropics

Chalicodomaspp. (Mason Bees, etc.)

Megachile spp. (Leaf-cutting Bees) Worldwide

Anthophoridae (Cuckoo Bees; Mining Bees; etc.)

Anthophora spp. Most in Europe and Asia

Hemesia spp. Important in USA

Xylocopidae (Carpenter Bees)

Allodape spp. Small; nest in plant stems in tropics

Ceratina spp. Small; nest in stem pith; Africa, Europe, etc.

Xylocopa spp. Large; tropical, timber-tunnellers

Apidae (Social Bees)

Bombinae (Bumble Bees)

Bombus spp. Holarctic

Meliponinae (Mosquito Bees; Stingless Bees)

Melipona spp. South America; largish in size

Trigona spp. Afria, Asia; tiny

Apinae (Honey Bees; Social Bees)

Apis mellifera (Honey Bee) Worldwide

***Apis* dorsata** (Giant/Rock Honey Bee) India; wild

Apis florae (Little Honey Bee) India

(b) Diptera

Members of most families are recorded as anthophilous, but the more important groups include:

Midges	Pollinate cocoa in South America
Syrphidae	(Hover Flies)
Muscidae	(House Flies, etc.)
Calliphoridae	(Blow Flies; Bluebottles, etc.)

(c) Lepidoptera

Butterflies are diurnal; most moths nocturnal; nearly 100 families known and most adults are anthophilous and involved in plant pollination

(d) Coleoptera

Some 95 families and more than 330000 species are named Many adults anthophilus and some are regular pollinators of certain plants.

(e) Thysanoptera

Thripidae (Thrips)

Thrips spp.

Frankliniella spp. (100+ spp.) (Flower Thrips)

Source: Hill (1997)

A long-recognized association of this kind, between yucca moths (Prodoxidae) and yucca plants (Agavaceae), has become an important model in understanding how obligate mutualisms coevolve. In this association, established at least 40 million years ago, yuccas are pollinated exclusively by yucca moths, whose larvae in turn consume some of the developing yucca seeds. This has been an evolutionarily and ecologically highly successful association, with some 30-45 yucca species being important vegetation components throughout much of the North American deserts and semiarid regions (Pellmyr and Krenn, 2002).

Table 4.5. Economically important entomophilous plants

Trees	Acacia	Ficus	Peach
	Almond	Hawthorn	Pear
	Apple	Lime	Plum
	Avocado	Litchi	Rubber
	Cashew	Mango	Spindle
	Coconut	Maple	Sycamore
	Date Palm	Nutmeg	Tung
	Eucalyptus	Oil Palm	some Willows
Shrubs	Cassava	Coffee	Guava
	Cinnamon	Cotton[a]	Lantana
	Citrus[a]	Currants (*Ribes*)	Tea
	Cocoa	Gardenia	
Vegetables	Asparagus	Cucurbits	Sweet Potato
	Beans	Onions	Tomato[a]
	Beets	Peas	
	Cabbage	some Solanaceae[a]	
Oilseeds	Rapeseed	Mustard	Sunflower
Miscellaneous	Clovers	Ivy	Poppy
	Flax	Lavender	
	many Flowers	Mint	Ranunculus
	Geranium	some Orchids	
	Heathers	*Oxalis*	Tobacco[a]
	Honeysuckle	Passion fruit	*Viola*

Source: Hill (1997)

The pollinating yucca moth genera *Tegeticula* and *Parategeticula* constitute a monophyletic group within the Prodoxidae. Jointly they contain at least 25 extant species, two of which are derived nonpollinating *Tegeticula* species that oviposit into yucca fruit created by coexisting pollinator species. The sister group *Prodoxus* coexist with the pollinators on yuccas but feed as larvae on plant parts other than the seeds. Their radiation was thus directly facilitated by the pollinator radiation.

Together these genera constitute a major adaptive radiation on yuccas, with a species diversity more than 20-fold that of their sister group, the nonpollinating seed-parasitic *Mesepiola*, whose larvae feed on plants in the Nolinaceae.

Female yucca moths possess unique tentacles on their mouthparts that are used to actively pollinate host flowers where they oviposit. The female moth gathers the glutinous pollen of yucca flowers by scraping if off the anthers with her tentacles. The pollen is immediately compacted by using tentacles and sometimes the forelegs as well, and placed as a solid batch on the concave posterioventral surface of the head. The pollen mass may approach 10,000 grains and weigh up to 10% of the moth body mass. Prolific pollen coating maintains batch cohesion, and the tentacles are not involved in its retention. After pollen gathering, the moth seeks out flowering yucca plants where she oviposits into (*Tegeticula*) or near (*Parategeticula*) pistils. As oviposition is completed, the female flexes her tentacles and uses the apical portion to remove a small pollen load from her batch. She walks to the floral stigma and very deliverately places the pollen on it. In all but one host species, the stigmatic papillae line the interior of the hollow style, and the moth packs in the pollen with 10-20 repeated bobbing motions in the course of 3-10 seconds.

The tentacular appendages of the maxillary palp in the female yucca moths are the rare example of a complex key innovation that emerged in oe clade without any homologous structure. However, the morphological data strongly indicate that the unique tentacles of female yucca moths originated through expression in a novel site of the genetic template for the elongated galeae of Lepidoptera. The tentacles are the only truly novel morphological trait in these moths, and the acquisition of this complex trait and its mechanism of movement may have been evolutionarily simple. The behavioral component of pollination is likely derived from a common probing behavior for nectar in floral tubes in more basal prodoxids and on the wet stigmas of yuccas, thus a passive, less efficient pollination mechanism may have preceded the origin of the tentacles and active pollination in the yucca moths (Pellmyr and Krenn, 2002).

4.7. CONCLUSIONS

Terrestrial biodiversity is dominated by plants and the herbivores that consume them, and they are one the major conduits of energy flow upto higher trophic levels. Reciprocal adaptations and counter-adaptations between these two groups has been an important mechanism leading to a steady increase in the plant and insect biodiversity over the last 400 million years. On the one hand the flowering plants have developed colorful, scanted flowers largely to attract pollinators. On the other hand, they have also developed a rather bewildering array of exploitation-barriers to save themselves from insects and other herbivores. The insects, which are able to overcome these barriers, can access an abundant supply of food with very little competition from other herbivores. The study of these inter-relationships is of great practical importance for the future agricultural production. We are only now beginning to understand the intricacies of these interrelationships. The new techniques of molecular genetics, biotechnology and genomics offer exciting opportunities for an intricate understanding of insect-plant relations as well as for exploiting them in the development of pest resistant varieties.

REFERENCES

Arora, R. and Dhaliwal, G.S. 1999. *The Insects: Diversity, Habits and Management*. Kalyani Publishers, New Delhi.

Arora, R. and Dhaliwal, G.S. 2004. Biochemical bases of resistance in plants to insects. In: G.S. Dhaliwal and R. Singh (eds). *Host Plant Resistance to Insects*. Panima Publ. Corp., New Delhi, pp. 84-125.

Bentur, J.S. 2004. Genetics of insect resistance in crop plants. In: G.S. Dhaliwal and R. Singh (eds). *Host Plant Resistance to Insects*. Panima Publ. Corp., New Delhi, pp. 126-158.

Berenbaum, M. 1983. Coumarins and caterpillars : A case for coevolution. *Evolution* 37: 163-179.

Buntin, D.G. and Chapin, J.W. 1990. Biology of Hessian fly (Diptera: Cecidomyiidae) in the Southeastern United States: Geographic variation and temperature-dependent phenology. *J. econ. Ent.* 83:1015-1024.

Carter, J.S. 2005. Coevolution and Pollination. Available: http://biology.clc.uc.edu/courses/bio 303/coevolution.htm.

Dhaliwal, G.S. and Arora, R. 2006. *Integrated Pest Management : Concepts and Approaches*. Kalyani Publishers, New Delhi.

Ehrlich, P.R. and Raven, P.H. 1964. Butterflies and plants:a study in coevolution. *Evolution* 18: 586-608.

Engelberth, J., alborn, H.T., Schmelz, E.A. and Tumlinson, J.H. 2004. Airborne signals prime plants against insect herbivore attack. *Proc. natn. Acad. Sci.*, USA 101:1781-1785.

Futuyma, D.J. and Agrawal, A.A. 2009, macroevolution and the biological diversity of plants and herbivores. *Proc. natn. Acad. Sci.* USA 106: 18054-18061.

Giovanini, M.P., Saltzmann, K.D., Putoff, D., Gonzalo, M., Ohm, H.W. and Williams, C.E. 2006. A novel wheat gene encoding a putative chitin-binding lectin is associated with resistance against Hessian fly. *Molec. Pl. Path.* Available: http://doi: 1364-3703.2006.00371.x.

Hill, D.L. 1997. *The Economic Importance of Insects*. Chapman & Hall, London.

Janz, N., Nylin, S. and Wahlberg, N. 2006. *Diversity begets diversity:host expansions and the diversification of plant-feeding insects. BMC Evol. Biol.* Available: http://doi:10.1186/1471-2148-6-4.

Jermy, T. 1976. Insect-host plant relationship-coevolution or sequential evolution? *Symp. Biol. Hung.* 16: 109-113.

Jermy, T. 1984. Evolution of insect/plant relationships. *Am.Nat.* 124: 609-630.

Kessler, A. and Baldwin, I.T. 2002. Plant response to insect herbivory: The emerging molecular analysis. *A. Rev. Pl. Biol.* 53: 299-328.

Labandeira, C.C., Dilcher, D.L., Davis, D.R. and Wagner, D.L. 1994. Ninety-seven million years of Angiosperm-insect association: Paleobiological insights into the meaning of coevoluton. *Proc. natn. Acad. Sci.,* USA 91: 12278-12282.

Mittapalli, O., Neal, J.J. and Shukle, R.H. 2007. *Antioxidant response in a galling insect. Proc. nat. Acad. Sci.*, USA. 104: 1889-1894.

Nishida, R. 2002. Sequestration of defensive substances from plants by Lepidoptera. *A. Rev. Ent.* 47: 57-92.

Panda, N. and Khush, G.S. 1995. *Host Plant Resistance to Insects.* CAB International, Wallingford, UK.

Pechan, T., Cohen, A., Williams, W.P. and Luthe, D.S. 2002. Insect feeding mobilizes a unique plant defense protease that disrupts the peritrophic matrix of caterpillars. *Proc. nat. Acad. Sci.,* USA 99: 13319-13329.

Pellmyr, O. and Krenn, H.W. 2002. Origin of a complex key innovation in an obligate insect-plant mutualism. *Proc. nat. Acad. Sci.*, USA 99: 5498-5502.

Pieterse, C.M.J., Ton, J. and van Loon, L.C. 2001. Cross-talk between plant defense signaling pathways : boost or burden? *Agbiotechnet* 3: 1-8.

Ratcliffe, R.H. 2006. Breeding for Hessian fly resistance in wheat In: Radcliffe, E.B. and Hutchison, W.D. (eds). *IPM World Textbook.* URL:http://ipmworld.umn.edu/ chapters/ratcliffe.htm.

Robinson, R. 2007. Both barriers and trait complementarity govern pollination network structure. *Plos Biol.* 5(2): e54doi:10.1371/journal/pbio.0050054.

Sargent, R.D. and Otto, S.P. 2006. Role of local species abundance in the evolution of pollinator attraction in flowering plants. *Am. Nat.* 167: 67-80.

Schoonhoven, L.M., VanLoon, J.J.A. and Dicke, M. 2005. *Insect Plant Biology.* Oxford University Press, Oxford.

Singh, R. and Dhaliwal, G.S. 2004. Insect-plant interactions and host plant resistance In: G.S. Dhaliwal and R. Singh (eds). *Host Plant Resistance to Insects: Concepts and Applications.* Panima Publ. Corp., New Delhi, pp. 3-41.

Strong, D.R., Lawton, J.H. and Southwood, S.R. 1984. *Insects on Plants: Community Patterns and Mechanisms.* Blackwell Scientific Publ., Oxford, London.

Thompson, J.N. 1994. *The Coevolutionary Process.* University of Chicago Press, Chicago.

Thompson, J.N. 2005. Coevolution: The geographic mosaic of coevolutionary arms race. *Curr. Biol.* 15(24): R 992-R 994.

Toju, H. and Sota, T. 2006. Imbalance of predator and prey armament; geographic clines in phenotypic interface and natural selection. *Am. Nat.* 167: 105-117.

Theory and Practice of Integrated Pest Management
R. Arora, B. Singh and A.K. Dhawan (Eds)
Scientific Publishers (India), Jodhpur 2012, pp. 76 - 83.

5

ROLE OF PESTICIDES IN MANAGEMENT OF CROP PESTS

Balwinder Singh
Department of Entomology
Punjab Agricultural University, Ludhiana - 141 004, India

Pesticides have made a great impact on human health, production and preservation of foods, fiber and other cash crops by controlling disease vectors and by keeping in check many species of unwanted insects and plants. With the introduction of pesticides, farm practices have undergone revolutionary changes leading to incredible possibility that hunger can be vanished from the earth. Although various methods are used to control pests in different pest management systems, pesticides continue to be the major component of most of the pest control programmes and will probably remain so during the near future. More than 55 per cent of the land used for agricultural production in developing countries uses about 26 per cent of the total pesticides produced in the world (Dollacker, 1991). However, the rate of increase in the use of pesticides in developing countries is considerably higher than that of the developed countries. Pesticides are necessary to protect crops from losses that may amount to about 45 per cent of total food production world-wide (Pimentel and Levitan, 1986)

Pesticide means any substance intended for preventing, destroying, attracting, repelling or controlling any pest including unwanted species of plants or animals during the production, storage, transport, distribution and processing of food, agricultural commodities or animal feeds or which may be administered to animals for the control of ectoparasites.

5.1. LOSSES DUE TO PESTS

Assessment of crop losses due to pests has been a difficult and often controversial subject. There are many example of total crop failures or losses of more than 50 per cent of their potential production but to cite 40-50 per cent of average yield loss per year appears to be somewhat difficult to believe. The average crop loss resulting from animals (mostly insects), diseases, viruses and weeds has been calculated as high as 42 per cent of the potential production of all crops with a higher percentage of 40 per cent potato for and the lowest being 28 per cent for wheat (Table 5.1).

Table 5.1. Crop Losses due to various pests

Crop	Loss (%)				
	Weeds	Animal pests	Pathogens	Viruses	Total
Wheat	7.7	7.9	10.2	2.4	28.2
Rice	10.2	15.1	10.8	1.4	37.4
Maize	10.5	9.6	8.5	2.7	31.2
Potato	8.3	10.9	14.5	6.6	40.3
Soybean	7. 5	8.8	8.9	1.2	26.3
Cotton	8.6	12.3	7.2	0.7	28.8
Average	8.8	10.8	10.0	2.5	32.1

Source: Oerke (2006)

These losses differ depending on the geographical regions and on the causal agent. The higher losses occur in Africa and Asia, reaching 48.9 and 47.1 per cent, respectively or their relative potential crop value.

5.2. PESTICIDE DEVELOPMENT : HISTORICAL PERSPECTIVE

The use of chemicals to control insect pests possibly dates back to classical Greece and Rome, for Homer mentioned the fumigant value of burning sulphur in about 1000 B.C. Pliny the Elder advocated the insecticidal use of arsenic and referred to the use of soda and olive oil for the treatment of legumes. Pliny reported the use of sulphur in 77 A.D. The Chinese were employing moderate amounts of arsenicals as insecticides in 1900 A.D. and not long afterwards nicotine, in the form of tobacco extracts, was used to control the plum curculio (Hassal, 1990). The first stomach poison was probably arsenic, which was mixed with honey in the mid-1600 on an ant bait. In the late eighties, Paris Green, which consisted of copper salts of arsenic, was used to control Colorado potato beetles, codling moth and other leaf-eating insects when dusted on plant foliage. In 1892 lead arsenate was introduced as one of the most effective inorganic insecticides for control of pests such as gypsy moth, apple maggot and various soil insects.

In 1874, Zeidler synthesized the DDT and its insecticidal properties were discovered by Paul Muller in 1939. In 1944, parathion (Organophosphates) was discovered by Gerhard Schrader in Germany. Carbamates constitute the third major group of synthetic insecticides. Discovered in early 1950s, the first commercial product was carbaryl marketed by Union Carbide Corporation. In 1976, the first synthetic pyrethroid was introduced. These compounds are effective at extremely low rates and are among the least toxic to mammals and humans. Another advance has been the introduction of several cuticle synthesis inhibitors, juvenile hormone mimics and antijuvenile hormones. After 1980, the new molecules viz avermectins, neonicotinoids, spinosyns and thiamethoxam which are effective at very low dosage, have quick knockdown effect, less persistence in the environment and safe to human beings and non targeted species were discovered (Table 5.2).

Table 5.2. History of Chemical Control

Year	Development of pesticides
900	Arsenite in China (Inorganic compounds)
1690	Tobacco used in Europe (Plant and natural products)
1787	Soaps used in Europe
1867	Paris Green in US
1874	DDT synthesized by Zeidler
1939	DDT insecticidal property discovered by Paul Muller of Switzerland
1941	BHC in France and UK (in 1942)
1944	Parathion (Organo phosphates) discovered by Gerhard Schrader in Germany
1947	Carbamate insecticides in Switzerland
1956	Introduction of carbaryl, the first successful carbamate insecticide
1962	Rachel Carson's Silent Spring appears (US)
1967	First JH mimic
1970	Development of synthetic pyrethroids
1980	Discovery of avermectins
1990	Discovery of newer groups like Neonicotinoids and Spinosyns
2001	The neonicotinoid, thiamethoxam, was commercially developed
2005	Flonicamid, a novel insecticide with a rapid inhibitory effect on aphid feeding was launched in the world market
2008	The Pesticides Management Bill was introduced in the Indian Parliament

5.3. GLOBAL PESTICIDE CONSUMPTION PATTERNS

Approximately 70 per cent of the pesticides used in the world is applied in developed countries, the remaining 30 per cent in the developing countries. In all nations, pesticide use is concentrated on a relatively small portion of the total crops. The consumption of pesticide in world wide account for 46 per cent as herbicides followed by 26 per cent insecticide, 23 per cent fungicides and others 5 per cent. In India consumption of pesticide account for 61 per cent as insecticide followed by 19 per cent fungicides, 17 per cent herbicides and others 3 percent (Fig. 1).

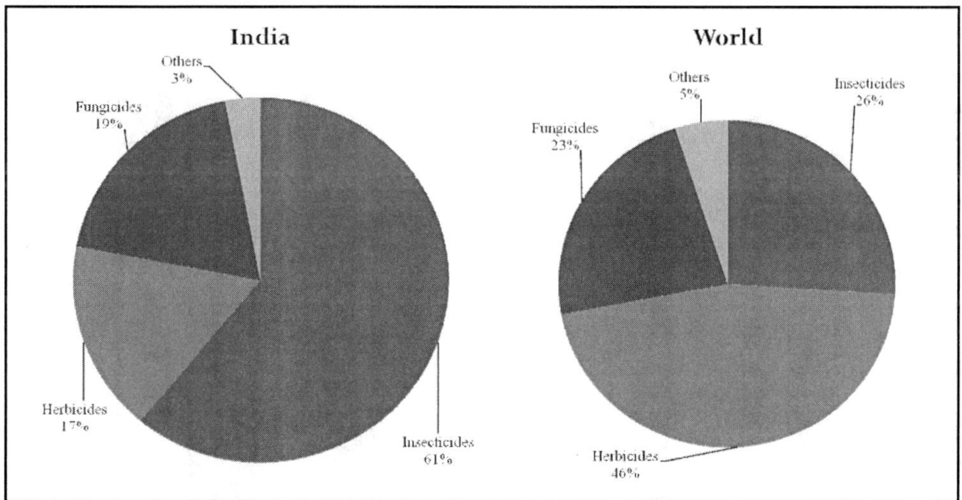

Fig. 1. Pattern of pesticide consumption in India and World

5.4. CONSUMPTION OF PESTICIDES IN PUNJAB AND INDIA

In India, during 1996-1997 consumption of pesticide was 56,114 metric ton but during 2008-2009 consumption of pesticide was 43,860 metric ton. This reduction could be attributed to the availability of newer pesticides which are effective at very low dosages, introduction of GM crops and adoption of IPM strategies for the control of insect pests. In Punjab, consumption of pesticides has gone down from 7300 MT in 1996-97 to 6900 MT during 2004-05. Because Punjab is a largest consumer of herbicides in India (Table 5.3).

5.5. PROBLEMS ASSOCIATED WITH PESTICIDES

(a) Pesticide poisoning. Human pesticide poisoning are a major concern in producing and using pesticides in agricultural and public health programmes. It has been estimated that annually in the world, there are about 7,50,000 reported pesticide poisonings with about 13,800 deaths. More than half of these poisoning cases and three-forth of the documented deaths take place in the third world even though these consume only 15 per cent of the total world pesticide output. In the United States, the best estimate is 45,000 human pesticide poisonings annually, with about 300 of these serious enough to require hospitalization. Indiscriminate use of pesticides apart from being occupationally hazardous in developing world, is today posing a serious threat to human health. International Development Research Centre Ottawae, has claimed that every year about 10,000 people die and 4,00,000 people suffer from various effects of pesticide poisoning in developing countries. Despite the restrictions and regulations on pesticide use, India accounts for 1/3 of pesticide poisoning cases in the world. Cases of blindness, cancer, diseases of liver and nervous system from pesticide poisoning have been identified in cotton growing areas of Maharashtra and Andhra Pradesh in India (Singh and Dhaliwal, 2000).

Table 5.3. Consumption of Pesticides in Punjab and India

Year	M.T. technical grade	
	Punjab	India
1996-1997	7,300	56,114
1997-1998	7,150	52,239
1998-1999	6,760	49,157
1999-2000	6,972	46,195
2000-2001	7,005	43,584
2001-2002	7,200	47,020
2002-2003	7,200	48,350
2003-2004	6,780	41,020
2004-2005	6,900	40,672
2005-2006	5,610	39,773
2006-2007	5,975	41,510
2007-2008	6,080	44,770
2008-2009	5,760	43,860

Source: www.indiastat.com

(b) Resistance. Large scale use of pesticides to control pests has resulted in the development of resistance which is most serious in the successful use of pesticides. When Rachel Carson wrote her famous book silent spring about 40 years ago, 137 species of insects and mites had already become resistant to chemical pesticides. To-day the number of resistant pests has risen to above 700. This phenomenon has also appeared in 100 species of plant pathogens and 55 species of weeds, as well as in some nematodes and rodents. In Asia-Pacific region, available records indicate that at least 13 species of insect have developed resistance to insecticides to insecticides. Rice pests such as brown planthopper and the green leafhopper have developed high levels of resistance to organophosphates, carbamates and pyrethroids. The most serious pesticide resistance problem in this region is that of diamondback moth. This has become a limiting factor in the production of vegetables in many Asian countries.

(c) Resurgence of pests. Use of broad spectrum and persistent insecticides kill the natural enemies which exert natural control under field conditions. In the absence of natural enemies and other insects, competition is reduced as a result there is resurgence of insect pests due to availability of good quality and quantity of food. Sometimes, unjudicious use of pesticides kills the key pests and under such conditions, secondary pests become the primary pests. The problem of resurgence is assuming serious proportion in rice and cotton involving many insecticides belonging to predominate groups like organophosphates, carbamates and pyrethroids. The major insect pests responding to resurgence include brown planthopper *Nilaparvata lugens* (Stal), green leafhopper *Nephotettix virescens* (Distant), whitebacked planthopper *Sogatella furcifera* (Horvath), whitefly *Bemisia tabaci* (Gennadius) and aphid *Aphis gossypii* Glover (Jayaraj, 1987).

(d) Effect on non-target organisms. Repeated use of pesticides on cotton, fruits, tobacco and other crops has destructive effects on beneficial insects like pollinators, biocontrol agents, soil, wild and aquatic life. The residues of pesticides in soil may create a variety of hazards. Many invertebrates take up pesticides from soil into their bodies and may concentrate pesticides several times greater in their tissues than those in the surrounding soil. The animals that feed upon these invertebrates may in turn concentrate these residues to levels that may kill them or affect their normal activities. Soil microorganisms which cause breakdown of cellulose, nitrification, turn over of organic matter and other biological materials may also be adversely influenced by pesticides. The application of carbofuran in rice leads to microbial population favoring accumulation of NO_2 from NH_4. This converted form of soil nitrogen is not preferred by paddy plants and thus the ultimate yield could be low even if the insecticide protected the crop against insect pests (Dhaliwal and Singh, 2000).

(e) Pesticide Residue Problems. Even though the use of pesticides in developing countries is low, residue problems have become quite alarming which are due to the continuous and uncontrolled use of highly persistent pesticides in agriculture and public health programmes. The occurrence of pesticide residues in food is thus viewed with great concern. The studies conducted under the All India Coordinated Research Project on Pesticide Residues have revealed that 60 per cent of food commodities were found tobe contaminated with pesticide residues, out of which 14 per cent contained residues above the maximum residue limit (MRL) as compared to 21 per cent contamination with only 2 per cent above MRL on world-wide basis (Dhaliwal and Singh, 2000).

Pesticides residues or their toxic metabolites could affect man' health. The occurrence of appreciable amounts of certain pesticides in human tissues of general population gives sufficient reason for concern. The population of developing countries is known to carry heavy burden of pesticides in their bodies (WHO, 1990). The principal source of these residues is believed to be the diet which contains significant quantities of the persistent chemicals.

5.6. CONCLUSIONS

Pesticides are essential tools of modern agriculture and form an integral part of overall strategy because production has hardly any meaning without protection. Pesticides are also a help to the environment when they are used carefully and wisely. The use of pesticides is necessary today, especially in developing countries, where population growth is very high. The use of pesticides is energy-efficient, because it safeguards high productivity per unit of area.

Pesticide use in developing countries is indispensable in the struggle against hunger and diseases. Emphasis should, therefore, be given to judicious and safe use rather than the banning or restriction of pesticides. Nevertheless, in view of the adverse effects resulting from pollution of the environment by these chemicals, e.g. damaged to, or mass killings of non target organisms, there is a great need to monitor pesticide usage. The effectiveness of pesticide is at present determined only by product yield. There is no monitoring of the effect on users and the non- target flora and fauna, nor is there any analysis of the final product that goes to the markets. Applicators wear little or no protective clothing due to ignorance and discomfort in the hot and sometimes humid climate.

Innovative strategies for crop protection, the development of new active ingredients and on-going programmes to develop resistant varieties will help to maintain and, where necessary increase the effectiveness of the available crop protection measures. Finally, it is important to ensure better protection for workers in the third world and the environment in the tropics. This can be achieved by (i) better information and training of applicators and an overall reduction of pesticides to the bare minimum using different techniques; (ii) adopting plant species resistant to insects and diseases; (iii) increasing biological control through integrated management for a gradual reduction of pesticides; and (iv) techniques relating to usage in the field and protection of workers and the environment cannot be automatically transferred from temperate regions without being adjusted to the tropics taking into account the substantial differences in the environment.

Thus, the need and use of pesticides are certain to grow, despite the adverse publicity to which these chemicals have been subjected to in recent years. The mechanization of agricultural production to compensate for labour shortages, and the monitoring food requirements of an expanding world population, make this necessary. The hope of mankind to escape hunger and diseases is closer to reality today than it was ever before. This hope rests, to a large degree, on continued research and developmental efforts directed towards the production of safer and more effective pesticides. More efficient use of present pesticides in well organized pest management programmes will greatly increase our worldwide capacity to produce more food and feed for expanding population.

Therefore, for the foreseable future, pesticides will continue to be an indispensible weapon in the battle to protect crops. Society expects manufacturers to develop safe and environmentally friendly active ingredients and it expects farmers to use these products properly and in a responsible manner so as to get wholesome food of high quality.

REFERENCES

Dhaliwal, G.S. and Singh, B.S. (eds) (2000) *Pesticides and Environment.* Commonwealth Publishers, New Delhi. pp. 1-439.

Hassal, K.A. (1990) *The Biochemistry and Uses of Pesticides.* VCH, Weinheim.

Jayaraj, S. (ed) (1987) *Resurgence of Sucking Pests.* Tamil Nadu Agricultural University, Coimbatore.

Oerke, E.C. (2006) Crop losses to pests. *J. Agric. Sci.* 144 : 31-43

Pimentel, D. and Levitan, L. (1986) Pesticides: Amounts applied and amounts reaching pests. *Bioscience* 36 : 86-91.

WHO (1990) *Public Health Impact of Pesticide Use in Agriculture.* World Health Organization, Geneva.

Theory and Practice of Integrated Pest Management
R. Arora, B. Singh and A.K. Dhawan (Eds)
Scientific Publishers (India), Jodhpur 2012, pp. 84 - 109.

6

MOLECULAR APPROACHES IN ENTOMOLOGICAL RESEARCH

V.K. Gupta and Vikas Jindal
Department of Entomology
Punjab Agricultural University, Ludhiana - 141 004, India

6.1. INTRODUCTION

Rapid advances in molecular biology (originally developed for plant molecular biology) resulted in availability of extremely powerful tools and techniques for in-depth analysis of various biological structures and functions. These techniques having universal application, opened new platforms for identifying various biological problems, establishing principals and providing solutions to the same with high level of precision. Although application of molecular techniques for studies in insect science was earlier considered expensive and sophisticated, but with the considerable reduction in cost of same and of most informative DNA sequencing, its exploitation to insect science has gained increasing significance. In entomology, molecular approaches find application in principally three important types of areas: molecular diagnostics or molecular markers, molecular analysis and understanding gene functions

6.2. MOLECULAR DIAGNOSTICS/ MARKERS

Faced with the difficulties of taxonomic assignation and establishment of systematic / evolutionary relationships amongst various insect species, research in past few years have been intensified on molecular approaches in an attempt to clarify the taxonomic status, systematic relationships, evolution of important insect taxa and emergence of insect species with new phenotypes. The exploitation of these versatile molecular tools in entomology though had been initially slow is now gaining wider application. This has primarily contributed to understanding the phylogeny of important insect species and precise identification of insect pests and their dispersal (spread).

Various molecular biology procedures primarily help to identify those biological molecules (DNA, RNA, proteins) / molecular regions that are equally shared by different but related species and others which are specific to species of the organism under investigation. As these regions serve as specific markers these are termed as 'molecular markers'. Precisely, molecular marker is a piece of the DNA molecule that is associated with a specific trait of an organism. This identified pieces of DNA by acting as molecular flag, helps us to identify specific DNA segments/ regions/ sequences/genes that are either associated with or control important traits in specific insect species. Thus, it is the identification and the availability of a variety of molecular markers in different insect species, the application of which has led to precise understanding on the ecology and identity of different insect species. Earlier used biochemical markers (allozymes) in insect science have generated a wealth of information on identification, taxonomy and phylogenetic relationship amongst different insect species. Though, allozyme analysis proved as valuable informative techniques, but it suffered from serious limitations of sample handling and number of loci (isozymes) that can be studied. Now it has more or less has been taken over by DNA based approaches.

In insect science, DNA markers have been used to provide raw information based upon which an ecologist make estimates on genetic diversity and gene flow between species. Molecular data provides means to differentiate sympatric species from allopatric and parapatric species by providing mode of evolution and evolutionary trajectories. Also, diagnostic molecular markers based upon linkage to certain traits or genes are used for diagnostic purposes of individual insect species and strains (Antonini *et al.,* 2009). The DNA marker technology also finds equally useful applications of these markers especially in molecular ecology research in insects. Over the last 15 years or so, DNA markers have made a significant contribution to rise of molecular studies of genetic relatedness, phylogeny and population dynamics and gene and gene mapping in insects. In fact molecular markers are increasingly being exploited in all the different aspects of entomological research including insect plant interactions, insect pathogen interactions, insecticide research, genetic diversity, ecological entomology, genetic linkage maps, insect behaviour and forensic studies.

6.3. MOLECULAR ANALYSIS

DNA based molecular analysis has revolutionized insect science particularly for distinction amongst biological entities and the assessment of their systematic and evolutionary relationships. This has primarily contributed to the understanding of the phylogeny of important insect species, precise identification of insect pests and their spread and movement to newer areas. Different DNA-regions (genomic regions) have been widely analyzed using different molecular markers for various entomological applications. The analyses of these regions with different types of markers have provided stable and precise information for understanding important aspects of insect sciences and hold tremendous potential for exploitation in specific insect pest diagnostics. The so called 'molecular biology tools' are applied direct to the isolated (purified) genomic DNA. As these genomic regions serve as specific markers these are so termed as 'molecular markers'. A number of techniques have come up and utilized for direct analysis of genetic material in the study of populations, species, and deeper phylogenetic clades within the particular insect specie and have yielded a greater wealth of information and produced key knowledge for the design of rational disease control strategies.

The methodology to develop these markers is essentially derived from those earlier developed for plants but has been modified for application to small amount of DNA that can be obtained from insect individuals. In general, the molecular analysis or screening involves investigating whole of the genomic DNA (nuclear- nDNA and mitochondrial- mtDNA). The purified DNA is analyzed after it is digested prior to being probed to detect differences (as in RFLP) or used as such to amplify regions of interest using sequence specific oligonucleotide primers (as in PCR based techniques). Both nuclear and mt- gene sequence polymorphisms have been surveyed. Use of DNA based molecular studies have provided new in sights into the systematics and evolutionary trends, with surveys ranging from intertribe to intraspecific levels. Many of the results have been used to refine and improve insect control strategies in particular geographic areas. Vector control programs are already benefiting from knowledge derived from molecular research. Several DNA-based techniques have been hitherto applied, which include randomly amplified polymorphic DNA (RAPD), species-specific length variation of rDNA amplicons, single strand conformational polymorphism analysis, sequencing of selected nuclear and mitochondrial gene fragments and characterization of polymorphic microsatellite markers.

The molecular tools used to analyze these regions primarily are based upon different variations of polymerase chain reaction (PCR) to amplify specific DNA regions of interest, though some utilize the earlier popular restriction fragment length polymorphism (RFLP) and sometime coupled with DNA sequencing for obtaining more specific analysis at base sequence level. Both coding (mitochondrial, nuclear) and noncoding (whole genomic DNA irrespective of coding properties) DNA have been the equal focus of attention for these molecular analysis. Different segments of the mitochondrial (mt) and nuclear genome are known to evolve at different rates. Whereas fast-evolving regions are suitable for the study of closely related organisms, the conserved sections are more appropriate for comparisons among more diverged taxa.

6.3.1. DNA sequencing for nucleotide specific information

DNA sequencing has become the dominant technique for generating comparative molecular data. Besides many other desirable properties of sequence data, two of these properties (a) the inherent comparability of sequence data across studies- the "connectibility" of Avise, and (b) a unique insight into the evolutionary processes driving the diversification of DNA itself, have revolutionized various ecological and evolutionary aspects of insect sciences. Automated sequencing facilities are accessible to most biologists within a reasonable budget, and appropriate markers have been developed for most evolutionary applications. Sequencing of selected genomic fragments (both mt and nuclear DNA after PCR amplification using specific primers) allows for the direct assessment of DNA polymorphisms, providing researchers with the ultimate information for phylogenetic inference and evaluation of kinship among organisms and populations. With drastic reduction in cost of custom DNA sequencing, the DNA sequencing has become the method of choice for most molecular characterization studies, but many alternatives may still be more appropriate or more practical for certain applications. Sequencing has been considered as the most appropriate for studies at inter-specific levels and higher and is the most used procedure for all types of molecular analysis of individuals as well as populations.

6.3.2. Coding DNA regions and markers

Intensive research efforts in the recent past have resulted in identification of specific coding regions in the insect genome that hold specific significance for molecular diagnostics as well as molecular analysis of insects and insect populations. These primarily include mitochondrial genes although some nuclear genes have also been found useful for the purpose.

Mitochondrial genes. To date the complete or near complete mitochondrial genomes have been sequenced from 175 insect species in 22 orders. Most species come from the order Diptera, Hemiptera, Hymenoptera, Lepidoptera and Orthoptera. Mitochondria (~102-104 per cell) as cytoplasmic organelles of eukaryotic cells are maternally inherited. Each mitochondrion with a single circular mtDNA molecule

(15-20 kbp) encodes 37 genes including 13 protein-coding subunits *(COI-III, CytB, ND1-6, ND4L, ATP6* and *ATP8*) from three of the oxidative phosphorylation complexes, plus the 2 ribosomal RNA *(rrnS and rrnL)* and 22 transfer RNA *(trnI, trnQ, trnM, trnW, trnC, trnY, trnL1, trnK, trnD, trnG, trnA, trnR, trnN, trnS1, trnE, trnF, trnH, trnT, trnP, trnS2, trnL2, trnV)* genes necessary to translate the protein-coding genes. The mtDNA molecule replicates with no recombination and presents faster evolution rates than the nuclear genome. High substitution rates that occur due to commonly occurring defective repair mechanisms during replication affect the degenerate third codon positions as compared to functional and structural constraints, which lower the substitution rates in second codon positions. In various arthropod taxa, at low levels of divergence (up to ~ 3 million years), the mtDNA substitutions accumulate at a constant rate of ~2.3 per cent as compared to average of ~2 per cent for animal mtDNA sequence divergence per million years.

Thus molecular markers based upon mitochondrial genes are widely used in population genetics and in evolutionary studies at different levels of divergence for example, in assessment of relationships among recently diverged taxa (Abad-Franch *et al.,* 2004). Direct sequencing of mtDNA regions is easier than other nuclear genes (fragments) due to provision of straight forward sequence alignment. For mtDNA, the most frequently sequenced genes are cytochrome oxidase I (COI), COII, 16S rDNA and 12S rDNA. Of these, COII has been sequenced over the widest variety of taxa, with homologous sequences available for nearly all orders. COI has also been sequenced in many different groups, but due to its greater size, the specific region studied varies from study to study. A region of the 16S rRNA gene (corresponding to positions 12900-13400 of *Drosophila yakuba*) had been consistently sequenced across most taxa. A higher level of sequence diversity can be found using two mtDNA fragments combined (12S+16S rDNA, totaling 878 bp). The phylogeny of 11 species of the triatomines complex has been explored using mt12S and 16S rRNA; and cytochrome oxidase I (COI) genes. mtCOI sequences and corresponding amino acid data have been used to calibrate a mitochondrial molecular clock for various insect orders. mtDNA has also been used widely in taxonomic and population studies. mtDNA data have played a major role in studies of molecular evolution that have helped to refine models of evolution and understand how rates and patterns of substitution vary across sequences and over time. Information is also available on universal primers, which amplify across the entire insect mtDNA for molecular eveolution and phylogenetic analysis. Several studies are available laying emphasis on use of mtDNA for taxonomic and population genetic studies on biotypes of cereal aphids, fruit flies and hymenopterous parasitoids, phylogenetic analysis of leafhoppers, mimicry in Lepidoptera, genetic variation in grasshoppers, grouping the honey bees into major clusters, maternity studies in ants and their movement between nests and the tracking of founder effects. mtDNA is by far the most widely used of DNA regions for insects as well as for animals in general. The popularity of mtDNA markers is due to its relative ease of isolation and amplification, even from marginally preserved specimens; and amenability to straight orward analyses. Complete mitochondrial genome sequences are being determined at an ever-increasing rate, providing many new opportunities for comparing genome structure. With the completion of whole genome sequencing projects accelerating, this will serve as a model for interpreting broader aspects of genome evolution and insect phylogeny. In addition to expanding this data set with broader taxonomic representation, it is important that studies address the molecular mechanism leading to rearrangement and develop better theoretical models and computer reconstructions.

Nuclear genes. Nuclear genomic fragments are generally more conserved than mt-genes. They are therefore better suited for analyses of diversity and relationships above the species level. Hence a number of nuclear genes had been investigated for arthropod phylogenetics. Nuclear rRNA (ribosomal DNA) genes had been especially used due to their abundance within a genome and concomitant ease of amplification and sequencing of the same. The nuclear ribosomal repeats (an array containing the 18S, 5.8S, and 28S genes, separated by transcribed and non-transcribed spacers, ITSI & ITSII), which are mosaics of highly conserved and variable regions are informative for resolving relationships at higher taxonomic levels, but main concerns with the use of ITS regions in insect systematics refer to problematic sequence alignment and intragenomic variability (Ramsey and Schofield, 2003). Nonetheless, rDNA sequences have proven highly informative for phylogenetics. However, ribosomal RNA spacers such as the internal transcribed spacers (ITSs) have proved informative for population analyses, molecular taxonomic and evolutionary studies. Apart from the combined analysis of mt gene fragments with the D2 variable region of the 28S rRNA gene, the second internal transcribed spacer (ITS-II) of the nuclear rDNA has been studied as a molecular marker for populations, species, and phylogenetic relationships and phylogenetic analyses of ITS-2 sequence data corroborated important findings based on mtDNA.

Studies based upon nuclear rDNA show a greater consistency. Hence, sequencing either partial or the entire 18S rRNA gene has become common for most high-level studies. However, the internal transcribed spacer (ITS) and 28S regions predominate in studies of Diptera and Hymenoptera. Generally, complete 18S analyses support relationships that are largely in agreement with previous hypotheses based on morphology (Saccone *et al.,* 2003). Compared to this, the 28S data sets generated are typically much smaller (less than 500 bp) and comparison of results for this gene is difficult. A number of loci representing few nuclear protein coding genes have shown great promise for resolving relationships at levels too conserved to be easily examined with mitochondrial proteins. Many of these loci may be effective for more recent divergences as well and can be useful in identification of molecular variations say in an insect pest due to environmental pressure or adaptation to new environment. These provide unbiased nucleotide composition of most nuclear loci compared to mtDNA and thus provide a higher saturation threshold. However, working with nuclear genes poses potential difficulties with heterozygosity, and low copy number may hinder amplification; in many cases reverse transcriptase PCR (RT-PCR) may be required.

Phylogenetic studies utilizing nuclear protein-coding genes are far fewer in num-ber than studies of either mtDNA or nuclear rDNA. However, a few loci have come into wider use, of which EF I-alpha is the most popular. Its sequences had proved very useful for studies among species groups and genera within subfamilies. These support relationships in conformity with mitochondrial COI-II in *Papilio*. However, amino acids in EF I-alpha are highly conserved and third codon positions provide essentially all the phylogenetic information and amino acid variation may be informative at the interordinal level and above. Besides EF I-alpha, no other single copy nuclear genes have been widely applied, though several show significant promise (particularly *ddc, g6pdh, hb, per, pepck* and *wg*). The alcohol dehydrogenase (*adh*) locus had been useful in Drosophilidae, but not applied outside of this group. A number of other nuclear genes that have been found applications include, alanyl-tRNA synthetase (*AATS*), carbamoylphosphate synthase (*CAD*), 6-phosphogluconate dehydrogenase (*PGD*), triosephosphate isomerase (TPI) and RNA polymerase II (*RNA POL II*), that have been used in molecular phylogeny and evolution of insect species belonging to 29 taxa representing the 11 holometabolous orders and two hemimetabolous out-groups.

There have been many investigations that have used nDNA regions to distinguish between members of species populations and complexes across a number of insects belonging to different orders. Several agricultural and forestry pests and medically-important vectors of disease have shown variations between populations and biotypes using this method. Such studies have quickly identified the usefulness of this molecular marker for differentiating at the level of sub-species, races and strains, etc (Ellis *et al.,* 209). However, some times continuous variation of the intergenic spacer (IGS) and multiple copies of the gene within a single individual can confound analysis based upon PCR amplification and sequencing results analysis. This is well demonstrated by IGS variability within an individual which has been found to be as great as that across a sample population of the aphid, *Schizaphis graminum* Rondani (Hemiptera: Aphididae) at various spatial scales. In general, problems with multiple copies mean that this technique, unlike others, should be used with careful consideration.

6.3.3. Non- coding DNA regions and markers

Seeking variation, which may often be elusive, systematists and population biologists are delving increasingly into the vast noncoding portions of the genome. Noncoding markers are identified irrespective of their position on genome and can be randomly distributed or specifically positioned on genomic DNA regions that may be nuclear or mitochondrial or coding or noncoding (Cartetto *et al.,* 2009). Vast noncoding portions of the genome comprise repetitive, or satellite DNA to as much as 45 per cent of the entire genome in some cases. Because much of this satellite DNA may exhibit highly conserved higher order structure, it is apparent that this DNA plays an important role in chromosome structure and function. Hence, the analysis of such regions provides direct interpretations at species level and higher taxa. These are particularly useful for examining relationships among individuals and breeding groups within insect populations. As noncoding DNA regions are simple to analyze a number of techniques/ markers have come up for molecular analysis of individual insect species as well as populations as described below:

(i) Restriction Fragment Length Polymorphism - RFLP. Among DNA markers, restriction fragment length polymorphism (RFLP) has proved to be a powerful tool in germplasm classification and genome mapping of higher plants and animals. The potential of RFLP markers has been massively exploited in various aspects of rice breeding. However, as RFLP analysis requires large quantities of high quality DNA and involves detection systems that use either radioisotopes or complex biochemistry, data generated via RFLP restriction enzymes has largely disappeared from phylogenetic studies of insects, although the technique is still commonly applied in plant studies. RFLP analysis still offers a quick alternative for studying sequence variation across broad cross sections of the genome. Whole mitochondrial digestions had proved very useful and reasonably accurate for phylogeny reconstruction. The most common use for restriction enzymes in insect systematics is for the diagnosis of otherwise cryptic taxa. RFLP have been applied for analysis of genetic linkage (chromosomal mapping) in *Heliothis* species, population genetics and parental analysis of honeybees and developing detailed genetic and physical maps for *Ae. aegypti* and *Drosophila*, which has resulted in considerable insight into genome organization. Provided that the restriction sites are mapped, RFLP analysis still offer a quick alternative for studying sequence variation across broad cross sections of the genome, and in this respect whole mitochondrial digestions have proven very useful and reasonably accurate for phylogeny reconstruction. The most common use for restriction enzymes in systematics is for the diagnosis of otherwise cryptic taxa. In this application, diagnostic digestion profiles are sought within amplified DNA fragments (most frequently mtDNA) either with sequencing or without sequencing first to identify diagnostic sites directly. This technique has found ample acceptance for molecular identification and sustematics in various insect species and is most commonly employed e.g., Strain Identification of Spodoptera frugiperda insects and cell lines, genetic variations in populations of Oriental fruit fly, *Bactrocera dorsalis*, RFLP analysis of *Bactrocera dorsalis* complex species, species determination of Malaysian *Bactrocera* pests using PCR-RFLP analyses, Diagnostic molecular markers of six lepidopteran insect pests infesting apples and molecular identification of Japanese Tetranychus spider mites (Acari: Tetranychidae) and thrip species (Brunner *et al.,* 2002).

(ii) Single stranded conformational polymorphism (SSCP) and double stranded conformational polymorphism (DSCP). Single-strand conformation polymorphism (SSCP), or single-strand chain polymorphism, is defined as conformational difference of single-stranded nucleotide sequences of identical length as induced by differences in the sequences under certain experimental conditions. This property allows distinguishing the sequences by means of gel electrophoresis, which separates the different conformations. A single nucleotide change could dramatically affect the strand's mobility through a gel by altering the intrastrand base pairing and its resulting 3D conformation. After denaturation, single-stranded DNA undergoes a 3-dimensional folding and may assume a unique conformational state based on its DNA sequence. The difference in shape between two single-stranded DNA strands with different sequences can cause them to migrate differently on an electrophoresis gel, even though the number of nucleotides is the same. Thus, SSCP is the electrophoretic separation of single-stranded nucleic acids based on subtle differences in sequence (often a single base pair) which results in a different secondary structure and a measurable difference in mobility through a gel. The mobility of single strands, however, is noticeably affected by very small changes in sequence, possibly one changed nucleotide out of several hundred. Small changes are noticeable because of the relatively unstable nature of single-stranded DNA (in the absence of a complementary strand), the single strand may experience intrastrand base pairing, resulting in loops and folds that give the single strand a unique 3D structure, regardless of its length.

SSCP polymorphism analysis takes advantage of this quality of single-stranded DNA. First announced in 1989 as a new means of detecting DNA polymorphisms, or sequence variations, SSCP analysis offers an inexpensive, convenient, and sensitive method for determining genetic variation. Like restriction fragment length polymorphisms (RFLPs), SSCPs are allelic variants of inherited, genetic traits that can be used as genetic markers. However, unlike RFLP analysis, SSCP analysis can detect DNA polymorphisms and mutations at multiple places in DNA fragments. As a mutation scanning technique, SSCP is more often used to analyze the polymorphisms at single loci, especially when used for medical diagnoses.

The DSCP- A single nucleotide change in a particular sequence, as seen in a double-stranded DNA, cannot be distinguished by electrophoresis, because the physical properties of the double strands are almost identical for both alleles. However, the mobility of double-stranded DNA in gel electrophoresis is dependent on strand size and length but is relatively independent of the particular nucleotide sequence. Whereas, the conform-ational polymorphism holds valuable potential for genetic and ecological studies of Arthropods, DSCP typing has shown its potential in identifying phytophagous insect genetic entities based on direct-PCR of fecal DNA to allow the differentiation of morphocryptic entities within the species of *Ceutorhynchus assimilis* (Coleoptera : Curculionidae), a candidate biocontrol agent of a noxious weed.

Among the techniques that identify sequence polymorphism without sequencing, both SSCP and DSCP are becoming increasingly popular. Single-stranded DNA mobilities are dependent on temperature and pH. Thus, gel electrophoresis must be run in a constant temperature and pH. Besides, for optimal results, DNA fragment size should fall within the range of 150 to 300 bp, although SSCP analysis of RNA allows for a larger fragment size. Under optimal conditions, approximately 80 to 90% of the potential base exchanges are detectable by SSCP. If the specific nucleotide responsible for the mobility difference is known, a similar technique called 'Single Nucleotide Polymorphism' (SNP) may be applied. In entomological applications, SSCP is typically used in conjunction with PCR amplification of a particular fragment for the purpose of species diagnosis. SSCP also provides a quick method of screening for polymorphisms among populations and species that can then be sequenced for phylogenetic purposes. A particularly interesting application employs SSCP as an alternative to cloning for the isolation of nuclear alleles. Atkinson and Adams, (19970 utilized DSCP aanalysis mtDNA control region in termite *Nasutitermes corniger* to reveal existence of 21 haplotypes in 44 colonies as compared to only nine haplotypes detected by RFLP.

(iii) Random amplified polymorphic DNA (RAPD). RAPD also known as arbitrarily primed PCR (or AP-PCR) burst onto the systematics in 1991 and immediately achieved wide use. These provided sudden access to variability in otherwise monomorphic or in previously unstudied taxa; and are inexpensive methods of generating comparative genetic data that is valuable in diagnosis of pest host races, biotypes and species. Data needs to be generated using a fairly large number of RAPD primers for ensuring that sufficiently large portion of genome has been taken into account. Amplification of polymorphic DNA with random primers has been used to investigate genetic variability in several economically important insects such as triatomines, whitefly and aphids and more. A serious though solvable problem with RAPDs is that, because primers anneal randomly, there is little guarantee of homology between co-migrating bands. RAPD markers being dominant, there is no way to determine whether a single band on the gel corresponds to a homozygous or heterozygous individual. However, many weaknesses of the RAPD technique quickly became apparent. Most obviously problematic is the assumption that comigrating PCR products represent homologous pieces of DNA, particularly given that RAPD markers are expected largely to come from repetitive regions of the genome. A more insidious problem is the difficulty of replicating results under even slightly different PCR conditions. Unpredictable variation in specimen preservation will also affect results, as will endosymbionts, parasites and phoretics. The confident use of RAPDs requires, as an absolute minimum, scrupulous standardization of all conditions of collection, extraction, and amplification; sample sizes sufficient to confirm consistency; and some check of band homology, whether through sequencing, SSCP or hybridization experiments. Given these cautions, however, RAPDs providing a quick and inexpensive method of generating comparative genetic data have proved valuable in the diagnosis of pest host races, biotypes, and species.

Due to same reason, RAPD-PCR is a still a one of the most common practice in insect identification and differentiation. In molecular markers RAPD-PCR is a conceptually simple technique for estimation of genetic diversity of organisms. Due to same reason RAPD markers have found application in all the various fields of entomology e.g., insect plant interactions, insect pathogen interactions, insecticide research, insecticide resistance, resistance monitoring, population restructuring, genetic diversity, developing genetic maps and insect behaviour studies (Sharma *et al.*, 2008; Jain *et al.*, 2010).

(iv) Sequence characterized amplified regions (SCARs). SCARs take advantage of a band generated through a RAPD experiment. They use 16-24 bp primers designed from the ends of cloned RAPD markers. This technique converts a RAPD marker band that was prone to difficulties in interpretation and/or reproducibility into being a very reliable marker. For obtaining SCAR polymorphisms, a potentially interesting band is identified in a RAPD gel and DNA is purified from the cut band out of the gel. The DNA fragment is cloned in a vector and sequenced. Specific primers (16-24 bp long) for the sequenced fragment are designed. Re-amplification of the template DNA with the new primers will show a new and simpler PCR. Because the primers used are longer than is usual for RAPDs, SCARs are typically more reproducible than the RAPDs from which they were derived. SCARs are usually codominant, although not if one or both primers overlap the site of sequence variation pattern and show Mendelian inheritance. SCARs generate simpler patterns than RAPDs in a robust assay due to the design of specific long primers. However, their generation requires at least a small degree of sequence knowledge and effort and expense in designing specific primers for each locus. Nevertheless, though SCAR markers had found adequate application in MAS in Plants, these have been largely exploited for molecular differentiation in various insect species, identification of necrophagous fly species in forensic samples, establishing host plant specificity in whitefly *Bemisia tabaci* and molecular identification of B-biotype of *B. tabaci* (Gupta *et al.,* 2008, 2010).

(v) Cleaved amplified polymorphic sequence (CAPS). Cleaved amplified polymorphic sequences (CAPS) are like SCARs, but with an additional step of a restriction digest of amplified product to help identify polymorphisms that may not be identifiable from whole PCR products. Both SCARs and CAPS are based on the presence of nucleotide changes or insertions and/or deletions causing differences between the test sequences. One drawback of both is that they detect polymorphism only over a small range of the genome, the area between the primers being typically less than 5 kb. This method is based on the design of specific primers, amplification of DNA fragments, and generation of smaller, possibly variable, fragments by means of a restriction enzyme. This technique aims to convert an amplified band that does not show variation into a polymorphic one. For generation of CAPS, a band, DNA, gene sequence or other type of marker is identified as important either through PCR (and cut out of the gel, and the fragment cloned and sequenced) or the fragment sequence is already available Specific primers are designed from the fragment sequence. The newly designed primers are used to amplify the template DNA. The PCR product is subjected to digestion by a panel of restriction enzymes. Once a polymorphism is identified with a particular restriction enzyme, the primers may be redesigned, based on the newly generated fragments, to optimize the detection and visualization of the polymorphism. Primers, when possible, should be chosen so that the PCR products are likely to include introns. This increases the chances of obtaining polymorphisms. The CAPS are codominant markers that are produced from robust assays using specific long primers based upon previously mapped sequences and identify polymorphisms in markers that were previously not informative. However, their initial development requires at least a small amount of sequence and effort and expense required to design specific primers for each locus. CAPS have been useful in identifying resistance genes to green rice leafhopper resistance gene, *Grh3(t)* in rice to and rootknot nematode resistance in sugar beet and demonstration of genetic exchange during cyclic development of *Leishmania* in sand fly vector. SSRs both individually as well as in combination with CAPS have been used to analyze the genetic diversity as well as to develop a consensus linkage map of *Bombyx mori.*

(vi) Amplified fragment length polymorphism- AFLP. This is a PCR based technique technique, which like RAPDs quickly locates widely dispersed genetic polymorphisms among organisms. AFLP-amplified-fragment length polymorphism analysis developed by Keygene BV, Wageningen, The Netherlands. This method combines universal applicability with high powers of discrimination and reproducibility. AFLP analysis belongs to the category of selective restriction fragment amplification techniques, which are based on the ligation of adapters (i.e., linkers and indexers) to genomic restriction fragments followed by a PCR-based amplification with adapter-specific primers. For AFLP analysis, only a small amount of purified genomic DNA is needed; this is digested with two restriction enzymes, one with an average cutting frequency (like *Eco*RI) and a second one with a higher cutting frequency (like *Mse*I or *Taq*I). Double-stranded oligonucleotide adapters are designed in such a way that the initial restriction site is not restored after ligation, which allows simultaneous restriction and ligation, while religated fragments are cleaved again. An aliquot is then subjected to two subsequent PCR amplifications under highly stringent conditions with adapter-specific primers that have at their 3′ ends an extension of one to three nucleotides running into the unknown chromosomal restriction fragment. An extension of one selective nucleotide amplifies 1 of 4 of the ligated fragments, whereas three selective nucleotides in both primers amplify 1 of 4,096 of the fragments. The PCR primer which spans the average-frequency restriction site is labeled. After polyacrylamide gel electrophoresis a highly informative pattern of 40 to 200 bands is obtained. The patterns obtained from different strains are polymorphic due to (i) mutations in the restriction sites, (ii) mutations in the sequences adjacent to the restriction sites and complementary to the selective primer extensions, and (iii) insertions or deletions within the amplified fragments. Several enzyme combinations have been used, such as *Eco*RI, *Pst*I, *Hind*III, or *Apa*I combined with *Mse*I or *Taq*I. For animal genomes *Eco*RI and *Taq*I appear to be the most suitable. Alternative AFLP typing procedures are based on one enzyme with a single adapter and analysis by agarose gel electrophoresis. A major improvement has been obtained by switching from radioactive to fluorescently labeled primers for detection of fragments in an automatic sequence apparatus. An increasing number of reports describe the use of AFLP analysis for different insect species. AFLP have been variousely applied for different insect species such as, assessment of geographic and host-associated population variations of the carob moth, *Ectomyelois ceratoniae*, on pomegranate, fig, pistachio and walnut, genetic divergence among tomato leafminer populations, and analyis of wild and captive testse fly. Other examples of use of AFLP in entomologica research include, genotyping diptera for species in the families *Calliphoridae* and *Sarcophagidae* insects are *Cochliomyia macellaria, Phormia regina* and *Sarcophaga bullata*, identification of Species boundaries and genetic diversity among Hawaiian crickets of the genus Laupala, genetic differentiation in the winter pine processionary moth, *Thaumetopoea pityocampa*-wilkinsoni complex, population genetics of (Diptera : Culicidae), development of a diagnostic DNA probe for the fruit flies and (Diptera : Tephritidae) and revealing cryptic populations structures in European red Admirals (Brattstrom *et al.,* 2010). AFLP analysis has established itself as a broadly applicable genotyping method with high degrees of reproducibility and discriminatory power. AFLP are codominants by targets same locus and needs fewer primers to screen all possible sites. However, if the polymorphism is due to presence/absence of a priming site, the relationship is dominance. The non-priming allele will not be detected as a band. The AFLP holds advantage as this procedure typically detects more polymorphisms per reaction than RFLP or RAPD analysis. AFLP analysis is especially useful in screening backcross individuals and protocols for AFLP analysis are available. Several applications AFLP in taxonomy, diagnostics, and epidemiology have already been realized. Its reproducibility may allow compilation of a database of genotypes and the exchange of data between laboratories. Despite apparent promise for entomological applications, AFLP has not been widely applied in studies of characterization of insect biodiversity.

(vii) Sequence-tagged sites (STS). Unlike PCR with arbitrary primers, sequence-tagged sites (STS) are primers that are based on some degree of sequence knowledge. These unique, sequence-specific primers detect variation in allelic, genomic DNA. STS have a particular advantage over RAPDs in that they are codominant, that is, they can distinguish between homozygotes and heterozygotes. They also tend to be more reproducible, because they use longer primer sequences. However, they have the disadvantage of requiring some pre-existing knowledge of the DNA sequence of the region, even if only for a small amount. The investment in effort and cost needed to develop the specific primer pairs for each locus is their primary drawback. As observed with RAPDs, using PCR produces a quick generation of data while requiring little DNA. All STS methods use the same basic protocols as RAPDs (DNA extraction and PCR) and require the same equipment. As an alternative, sequence tagged sites (STS) generated through the polymerase chain reaction (PCR) offer a simple, high-speed, non-radioisotopic approach to the detection of DNA polymorphism. In this procedure, PCR amplified specific DNA regions using region specific primers are exposed to restriction digestion which reveals some DNA polymorphisms as PCR-based RFLP. Whereas Southern-based RFLP detects DNA variation present within as much as 30 kb of the marker locus, PCR-based RFLP can detect polymorphism occurring only within the DNA segment delimited by the primers.

However, PCR-based RFLP offers higher resolution in the detection of variation and can be applied to both nuclear and mitochondrial DNA.

(viii) Microsatellites. Microsatellites are also called simple sequence repeats (SSRs), sequence-tagged sites (STS), sequence-tagged microsatellite sites (STMS) or simple sequence repeat polymorphisms (SSRPs). Microsatellites are series of short repetitive motifs [e.g., (GT)n or (AT)n] within nuclear DNA. This type of repeated DNA is common in eukaryotes, their number of repeated units varying widely among organisms to as high as 50 copies of the repeated unit. These polymorphisms are identified by constructing PCR primers for the DNA flanking the microsatellite region. The flanking regions tend to be conserved within the species, although sometimes they may also be conserved in higher taxonomic levels. The loci identified are usually multi-allelic and codominant. Bands can be scored either in a codominant manner, or as present or absent. Because flanking DNA is more likely to be conserved, the microsatellite-derived primers can often be used with many varieties and even other species. These markers are easily automated, highly polymorphic, and have good analytical resolution, thus making them a preferred choice of markers. Microsatellites hold several advantages such as, they require very little and not necessarily high quality DNA, are highly polymorphic and evenly distributed throughout the genome, simple in interpretation of results, are easily automated, allowing multiplexing and provide good analytical resolution and high reproducibility. They are highly polymorphic, neutral, and exhibit Mendelian inheritance and codominance. They are highly variable and evenly distributed throughout the genome that are in widespread use for genetic analysis at or below the species level and prove significantly useful in molecular diagnostics of insect pests. Their generally high mutation rate ensures high levels of polymorphism and they have therefore, been particularly useful for examining relationships among individuals and breeding groups within insect populations. They are by far the most widely used type of markers. Microsatellites are applied variousely for individual genotyping, germplasm evaluation, genome mapping, assessing the genetic diversity and in phylogenetic and evolutionary studies. Microsatellites are simple sequence repeats (SSRs) that exhibit complex patterns in their frequency of occurrence, genomic distribution, mutability, function and evolution. Apart from being the source of informative genetic markers, microsatellites per se have attracted a lot of attention with respect to their origin, distribution, expansion, mutation and disintegration and regulation of gene expression.

Microsatellites have become the tools of choice for population genetics when other markers do not show an adequate degree of polymorphism. Particularly exciting applications of microsatellites have focused on testing hypotheses of kin selection and behavior as they covary with individual relatedness in social insects (Tay *et al.,* 2010). A surprising result of these studies is that our ability to discern relationships in many cases exceeds that of the insects themselves and, therefore, behaviors predicted by kin selection are not realized. Microsatellites have also provided fascinating insights into sperm competition and paternity, the origins of sociality, and population structure. A large number of microsatellites have been identified and characterized in *T. dimidiate* and *T. infestans* populations from Mexico, Guatemala and Honduras (Anderson *et al.,* 2002). Like earlier used allozymes, microsatellites have proved valuable in insect population genetic studies, e.g. Hymenoptera, aphids, mosquitoes, moths, butterflies, whiteflies and African fruit fly species and in disease vectors *Triatoma infestans* and paternity studies in Hymenoptera.

Microsatellite based whole genome sequences in insects. Insects exhibit the greatest genetic diversity and biologists have relied on insects to unravel many fundamental tenets of biology. Whole genome sequences of insects have lived up to the reputation of diversity and have thrown immense variability in size and organization of their genomes. Among others, there are five fully sequenced insect genomes: *Drosophila melanogaster* (model organism providing maximum annotated data), *Anopheles gambiae* (another Dipteran-economically important as malarial vector), *Tribolium castaneum* (early insect order of Coleoptera), *Apis mellifera* (Hymenoptera, a recent insect order) and *Bombyx mori* (economically important as silk-producing member of Lepidoptera as model organism for crop pests) that are available with InSatDb (www.cdfd.org.in /insatdb). InSatDb, besides providing information on microsatellites as markers, allows users to address various questions by accessing qualitative and quantitative genome level microsatellites profile of a single insect or to carry out comparative genomic analysis using all the five genomes. It presents an interactive interface to query information regarding microsatellite characteristics per se of five fully sequenced insect genomes. InSatDb allows users to obtain microsatellites annotated with size (in base pairs and repeat units); genomic location (exon, intron, up-stream or transposon); nature (perfect or imperfect); and sequence composition (repeat motif and GC per cent). One can access microsatellite cluster (compound repeats) information and a list of microsatellites with conserved flanking sequences (microsatellite family or paralogs). InSatDb is complete with the insect information, web links to find details, methodology and a tutorial. A separate 'Analysis' section illustrates the comparative genomic analysis that can be carried out using the output. InSatDb is available at www.cdfd.org.in/insatdb.

(ix) Inter-simple sequence repeats (ISSRs). These are regions found between microsatellite repeats. Because of the known abundance of repeat sequences spread all over the genome, ISSRs target multiple loci.

The technique is based on PCR amplification of intermicrosatellite sequences. For the identification of ISSR polymorphisms a typical PCR is performed in which primers have been designed, based on a microsatellite repeat sequence, and extended one to several bases into the flanking sequence as anchor points. Different alternatives are possible: use of only one primer, two primers of similar characteristics or combinations of a microsatellite-sequence anchored primer with a random primer (i.e. those used for RAPD). ISSRs hold advantage in not requiring prior sequence information, in detecting variation within unique regions of the genome at several loci simultaneously, in identifying significant levels of variation, specificity to corresponding microsatellite sequence and being very useful for DNA profiling, especially for closely related species. However, these represent only dominant markers and their observation require polyacrylamide gel electrophoresis and detection with silver staining or radioisotopes. The so called nonanchored inter simple sequence repeats (ISSRs) are arbitrary multiloci markers produced by PCR amplification with a microsatellite primer and hence serve as good choice for DNA fingerprinting. They are applied to different plants as an easy and informative genetic marker system in cotton for revealing both inter- and intraspecific variations.

In contrast to other dominant markers, such as RAPDs, the ISSR technique uses longer primers, thus allowing higher annealing temperature and greater reproducibility of the DNA fragments. Inter-simple sequence repeat (ISSR) markers being cost-effective, rapid and efficiently sensitive are extremely useful for assessing genetic variability in some species.

The high degree of polymorphism and good repeatability of Dominant ISSRs have allowed the successful detection of intra-specific polymorphisms and characterization of genetic diversity in various plant species and are being increasingly used in insects, such as cyclically parthenogenetic aphids, *Acyrthosiphon pisum* and Pemphigus obesinymphae, mosquito *Aedes aegypti*, planthopper, *Sogatella furcifera*, spiders and other arthropod invertebrates with very high (97.5 per cent) percentage of polymorphism (Grasela & McIntosh, 2003) ISSRs have also been utilized to distinguish amongst 16 established insect cell lines derived from an assortment of lepidopteran, dipteran, and coleopteran species and which generated several unique markers distinguishing between the parental cell lines and their clones as well as amongst cell lines deived dfrom specific tissues from the same species/different species.

6.4. UNDERSTANDING GENE FUNCTIONS - RNA INTERFERENCE TECHNOLOGY

The recognition of RNA interference (RNAi) as a method for analyzing the functions of individual genes paved an easier way to analyze the functions of genes in insects by the use of loss-of-function analyses in many animals, including insects. Before the discovery of RNAi, it was almost impossible to analyze the functions of genes in insects for which genetic analyses are difficult. RNA interference (RNAi) is a cellular process by which a specific mRNA is targeted for degradation by introduced double-stranded RNA (small interfering RNA) that contains a strand complementary to a fragment of the target mRNA, resulting in sequence specific inhibition of gene expression. The double-stranded RNA is broken down into small, ~21-mer RNA by the RNase III-like protein Dicer. Thus, RNAi has become a very widely used tool for functional genomics after synthetic small interfering RNA (siRNA) was shown to be able to induce RNA degradation also in mammalian cells. The small RNA is incorporated into the RNA-induced silencing complex (RISC) - a multi-turnover enzyme. RISC consists of several components, some of which are still uncharacterized, and genetic screens in lower organisms have unearthed a long list of candidate RNAi genes. Target degradation seems to be initiated by mRNA cleavage, both in Drosophila and in mammalian cells. A central candidate for being the mRNA endonuclease "Slicer" is the Argonaute protein Ago-2. In general, when a double-stranded RNA (dsRNA) with a strand complementary to a fragment of the target mRNA is injected into insects, phenotypes appear, depending on the concentration of dsRNAs, the types of targeted genes and some other still unknown factors (Mito *et al.*, 2010).

The discovery of RNAi enabled the use of loss-of-function analyses in model (Drosophila) and many non-model insects to elucidate the roles of specific genes besides many other fields, such as embryogenesis, pattern formation, reproduction, biosynthesis and behavior. The increasing availability of insect genome data has made the RNAi technique an indispensable technique for characterizing gene functions in insects. Now, RNAi allows the rapid and straightforward analysis of gene function in non-model insects. Currently, many RNAi-based studies with various insects are in progress, and several reviews on their results have already been published. However, this approach is still relatively new, and this field will continue to expand as the number of insect genome projects increases and as RNAi methods are used to analyze the functions of new genes. Even in *Drosophila melanogaster*, after the completion of its genomic sequencing, a large number of genes whose functions are unknown were identified, and their functions have been investigated with RNAi methods. Thus, the combination of genomic sequencing projects with RNAi methods will lead to an increase in our understanding of gene functions.

6.4.1. Current status of RNAi-based research in insects

The advent of RNAi represents a new experimental paradigm beyond Drosophila, as it opens the door to the study of gene functions in other species. RNAi has been used to phenocopy the mutations of the deformed orthologue in the flour beetle, *Tribolium castaneum* and to dissect gene function in the development of the milkweed bug, Oncopeltus fasciatus, which undergoes hemimetabolic maturation. RNAi-based experiments reported in non-drosophilid insects can be classified into four types: parental, embryonic, larval/nymphal/pupal, and regeneration-dependent RNAi.

Parental RNAi. The observations on injection of dsRNA into the body cavity or the application of dsRNA via ingestion leading to gene inactivation in offspring embryos in *Caenorhabditis elegans* has enabled efficient, genome-wide functional screens in this organism. Similar injection of dsRNA into the mother's hemocoel results in a knockdown of the zygotic genes in the offspring embryos of *Tribilium castaneum* ("parental RNAi", paRNAi) to result in 100% RNAI phenotypes. Since then RNAis have been used to know the role of gap genes hunchback and Krüppel, in segmentation in milkweed bugs, Oncopeltus fasciatus (Hemiptera: Lygaeidae). Ronco *et al.* (2008) and Mito *et al.* (2008) found that paRNAi have also been found to in the cricket, *Gryllus bimaculatus*. Effect of paRNAi on caudal development in cricket, *Gryllus bimaculatus*, as observed by injection of dsRNA targeting Gb'caudal in female crickets results in formation of only the head region in eggs, paRNAi against Distal-less in the two-spotted spider mite, *Tetranychus urticae*, resulting in phenotypes with canonical limb truncation as well as the fusion of leg segments. These results established that paRNAi functions in many metazoan taxa, and the transmission of dsRNA across oocyte membranes is a conserved feature of the RNAi response.

Since large numbers of RNAi embryos can be readily obtained with paRNAi and analyzed using standard histochemical and in situ hybridization procedures, the functional characterization of genes has been greatly facilitated. Thus, the use of paRNAi is an important method for analyzing early embryogenesis, even for species where embryonic RNAi is already available, and it is crucial for many insects whose eggs are not accessible or do not survive with microinjection. Thus, paRNAi should allow large-scale RNAi screens in many species and will greatly facilitate functional approaches in the burgeoning field of evolutionary developmental biology.

Embryonic RNAi. When dsRNA for a target gene is injected into developing eggs, the RNAi phenotypes appears in embryos larvae/ nymphs, or even adults. Such type of RNAi- called embryonic RNAi (emRNAi) has been used to analyze gene function during embryogenesis under circumstances when target a gene is indespensible for egg formation for example, role of wingless (wg) gene in the leg development of *Tribolium castaneum* using stage-specific staggered emRNAi because the paRNAi experiments lead to either empty eggshells or wild-type cuticles demonstrating the separate functions of *Tribolium* wg in distal and ventral leg development.

Larval/nymphal/pupal RNAi. RNAi has been used to create pupal and adult loss-of-function phenotypes in *T. castaneum* by injecting dsRNA into late instar larvae (larval RNAi, laRNAi). The laRNAi technique has been used to analyze gene functions in postembryonic development to study the molecular basis of adult morphological diversity in various insect species for example, function of laRNAi in *T. castaneum* to determine the function of *Ubx/Utx* during hindwing/elytron development, which induced a complete transformation of hindwing to elytron. Larval RNAi have proved useful for analyzing the functions of various enzymes in biosynthesis pathways for example, laccase 2 in the formation of white and soft bodies *T. castaneum* and the production of the pheromone bombyko through suppressing effect on accumulation of cytoplasmic lipid droplets by preventing the synthesis of triacylglycerols.

Similarly, a number of nymphal RNAis have been used to analyze various gene functions in different insect species, such as *'vermilion eye color gene'* of first instar nymphs, which triggered a suppression of ommochrome formation in the eye and early retinal genes *eyes absent (eya)* or *sine oculis (so)* suppressing eye development in juvenile grasshopper, *Schistocerca americana*, function of RXR/ USP (in *Blattella germanica*), which as a one component, along with EcR, of the heterodimeric nuclear receptor of 20-hydroxyecdysone (20E) allows normal development of nymphs through the instar only but making them unable to molt into adults, identifying genes for leg regeneration of the cricket, *Gryllus bimaculatus*, and those involved in circadian rhythm in Gryllus (Dong *et al.,* 2010). To explore the possibility that *G. bimaculatus* could be a useful model for characterizing the molecular mechanisms of human diseases, loss-of-function analyses have been performed on *G. bimaculatus* genes homologous to the human genes that are responsible for certain human disorders: *fragile X mental retardation 1 (fmr1)* and *Dopamine receptor (DopR).* For Gb'fmr1, they observed three major nyRNAi phenotypes: (i) abnormal wing postures; (ii) abnormal calling song; and (iii) loss of the circadian locomotor rhythm; indicating that the cricket has the potential to become a novel model system to explore human neuronal pathogenic mechanisms and to screen therapeutic drugs with RNAi (Hamada *et al.,* 2009). The conservation of systemic RNAi in these insects suggests that this pathway can be exploited for the gene-specific manipulation of larvae/nymphs and pupae/adults in a wide range of insects.

Regeneration-dependent RNAi. Nymphs of hemimetabolous insects, such as cockroaches and crickets, exhibit a remarkable capacity for regenerating complex structures from damaged legs. Using the cricket, *G. bimaculatus*, as a model, RNAi phenotypes could be observed during regeneration following the amputation of legs. Because no phenotype is induced by nyRNAi in an intact cricket leg, this effect was designated as regeneration-dependent RNAi (rdRNAi). Using specific rdRNAis team of Nakamura at *University of Tokushima, Tokushima City, Japan*, established the role and function of various genes encoding signaling factors and cellular adhesion proteins, such as *Fat* and *Dachsous, Gb'armadillo* (the orthologue of b-catenin); and *Gryllus* orthologues of *Drosophila hedgehog (Gb'hh), wingless (Gb'wg)* and *decapentaplegic (Gb'dpp), Gb'ft, Gb'ds, Gb'expanded* and *Gb'Merlin* during leg regeneration. Their results indicated that the canonical Wnt/wg signaling pathway is involved in the process of leg regeneration and determination of the positional information in the leg segment. The study also provided an important insight about how regenerating blastemal cells are aware of both their position and the normal size of the leg.

6.4.2. Applications of RNAi for development of species specific dsRNA insecticides

Possibility of using RNAi to kill only the target insect-pests by down-regulating their essential gene functions have been recognized recently. For the time being the major problems lies with efficient methods for mass introduction of dsRNA into insects for making the technology feasible for practical application. Although dsRNA can prove useful in a bait to cause target mortality, the problem remains with continuous feeding of dsRNA as an insecticide (Whyard *et al.*, 2009). On the contrary, success of transgenic corn plants engineered to express dsRNAs for the western corn rootworm (WCR) in controlling WCR suggesting that the RNAi pathway can be exploited to control insect pests via the in planta expression of a dsRNA. Similar effect has been seen in *Arabidopsis thaliana* and *Nicotiana tobacum* transgenic plants expressing dsRNA specific to a cytochrome P450 gene (CYP6AE14) of the cotton bollworm (*Helicoverpa armigera*), which permits this herbivore to tolerate the cotton metabolite, gossypol. Thus, feeding insects with plant material expressing dsRNA may be a general strategy for the delivery of RNAi and could find applications in entomological research and field control of insect pests. Sequence specificity of RNAi has been harnessed to design orally-delivered dsRNAs that selectively killed target insects. Thus, *D. melanogaster*, *T. castaneum*, pea aphids (*Acyrthosiphon pisum*), and tobacco hornworms (*Manduca sexta*) could be selectively killed when fed species-specific dsRNA targeting vacuolar-type ATPase transcripts.

For the aphid nymphs and beetle and moth larvae, dsRNA could simply be dissolved into their diets. However, to induce RNAi in the drosophilid species, the dsRNAs needed to be encapsulated in liposomes to help facilitate the uptake of the dsRNA. This method may lead to higher throughput RNAi screens and the development of a new generation of species specific insecticides. In addition to liposomes, polycation polymers, such as cyclodextrin or dynamic polyconjugates, may be another promising delivery method for dsRNAs into cells. Overcoming the specific delivery of dsRNA or siRNA into the cytoplasm of target cells is still an important issue for the use of RNAi for insecticides.

6.4.3. Efficiency of RNAi depends on Developmental stages and species

The efficiency of systemic RNAi in vivo depends on the species; less-derived species are generally more sensitive than more-derived species. Genome-wide surveys to compare genes involved in the machinery of RNAi between *Tribolium* and *C. elegans* have shown a robust systemic RNAi response. Significant differences have been found between the genes involved in the RNAi machinery of these organisms suggesting that insects might use an alternative mechanism for the systemic RNAi response. Because over-expression of dsRNAs within cells using hairpin RNAs readily triggers RNAi in *D. melanogaster* tissues resistant to systemic RNAi, it can be speculated that the poor sensitivity is not related to the RNAi core machinery, but rather to the penetration and transmission of the interfering signal through cells and tissues and to the occurrence of degradation mechanisms that are able to remove alien RNA. Two mechanisms of dsRNA uptake include (i) the transmembrane channel mediated uptake mechanism, in which orthologs of *C. elegans systemic RNAi defective mutant gene (sid)* may be involved; and (ii) the endocytosis-mediated uptake mechanism, in which vacuolar H^+-ATPase, clathrin, scavenger receptors, etc. may be involved. This uptake mechanism may be linked to innate antiviral immunity responses. Thus, the dependence of RNAi efficiency on developmental stages or species is attributed to differences in the expression patterns of the *sid* orthologs and/or differences in the sensitivity of dsRNA detection systems for endocytosis in insect cells. If this is the case, one of the ways to circumvent the problem of low RNAi efficiency is to use chemically modified siRNAs or carriers which may stabilize siRNAs and transport them into cells, as investigated for human RNAi therapy. Using such methods in insects could help to render other insects amenable to systemic RNAi and may influence pest control approaches.

6.4.4. Future Prospects for RNAi-Based Experiments in Insects

RNAi-based experiments have provided several interesting results and shed light on gene functions in non-drosophilid insects. However, the mechanisms underlying RNAi phenomena, such as species or tissue dependent changes in the sensitivity to dsRNAs, are not fully understood. Thus, a more precise elucidation of RNAi mechanisms is an important issue in Entomology.

The potential application of RNAi techniques to any gene and any species could lead to comparative studies for the function of a gene, or gene network, in species covering a large spectrum of insect orders. Moreover, RNAi can facilitate comparative studies and evolutionary insight into other processes, such as social behaviors, reproductive strategies, and host-parasite interactions (Belles, 2010). Systemic RNAi-based genome-wide screening is particularly useful to identify genes involved at the whole animal level, e.g. in determination of life span and size, metabolic controls, circadian clock systems, ecdyses, etc.

With the increase in the number of genome projects, the RNAi method will be more useful for genome-wide analyses of gene functions. This will provide an access to a large pool of information on various insect genes to clarify what insects are and how to treat insects in the real world. However, the facilities and manpower to analyze such a large body of information are currently unavailable. In the future, a worldwide forum about RNAi should be established within the Entomological Society.

6.5. CONCLUSIONS

Different molecular approaches, combining sound sampling strategies with an intelligent choice of markers, can be used to address diverse questions on the biogeography, behavior, taxonomy, evolution, and population structure of insects. The database of insect sequences is growing rapidly. GenBank now boasts nearly 100,000 insect sequences. Although 80,000 of these are *Drosophila* sequences, a substantial fraction of the same have been used in phylogenetic studies of other insects. More than 40 protein-coding genes, all of the major ribosomal RNA genes (both mitochondrial and nuclear) as well as numerous non-coding regions have now been examined for systematic studies of insects. This number continues to grow as more number of additional loci (most of which are nuclear protein-coding regions derived from *Drososphila* studies) is increasingly being exploited for general applications to other insects. The diversity of available markers has unquestionably assisted the cause of insect molecular systematics and related biodiversity studies. However, the numerous markers also bring with it the risk of pluralism, making comparison and synthesis difficult or impossible. The most commonly sequenced regions in insects are mtDNA and nuclear rDNA attributed to their ease of development and comparison. The other class is that of nuclear protein coding genes, which are also more or less similar to mtDNA and rDNA. However, the available information will reveal that still there is substantial incompatibility among studies; there are few commonly used regions for both mitochondrial and ribosomal genes, but in no cases are these consistent across all taxa; in many cases the same gene has been sequenced across several studies but these regions often do not overlap with regions sequenced for related taxa e.g., 28S for Diptera. This stresses upon need for extensive explorations to identify consistent markers which in large can be directly linked to established phylogenetic relations.

The use of STS as markers relies on some sequence knowledge. They are codominant and highly reproducible. The SSRs are the type of STS most widely used. Others are SCARs, CAPS and ISSRs. SSRs are short tandem repeats, highly variable and evenly distributed in the genome. These features make SSRs a good marker of choice for genetic diversity analyses. Polymorphisms in SSRs are caused by differences in the number of repeat units. In comparison, questions of intraspecific population structure, species limits, and species diagnosis often can be effectively addressed using time tested techniques like allozyme electrophoresis, restriction fragment length polymorphisms (RFLP), or any of a variety of newer DNA-based techniques such as single-stranded conformational polymorphisms (SSCP), amplified fragment length polymorphism (AFLP), or random amplified polymorphic DNA (RAPD). The best strategy for certain problems often utilizes one or more of these in combination with sequencing. Choosing similar loci, primers, and analytical methods can helps to ensure that a study adds to a body of accumulating knowledge, thus accomplishing more than merely solving the problem at hand.

Advent of siRNA technology and its increasing exploitation in insect sciences have opened a new field in assigning functions to various genes and their role in insect ohysiology and development. Such an analysis is urgent and immediate need as the derived information can be utilized for manipulating both the beneficial insects as well as effective control of harmful insect pests in different agroecological environments. In fact the new and upcoming approaches in molecular biology have started generating new biological information that is perhaps more useful to insect sciences than any other field of biology.

REFERENCES

Abad-Franch, F., Monteiro, F.A., Aguilar, V.H.M. and Miles. M.A. 2004. Population-level mitochondrial DNA sequence diversity in *Rhodnius ecuadoriensis*. *Programme and abstracts, IX European Multicolloquium of Parasitology,* Valencia, Spain, p 201.

Anderson, J.M., Lai, J.E., Dotson, E.M., Cordón-Rosales, C., Ponce, C., Norris, D.E. and Beard, C.B., 2002. Identification and characterization of microsatellite markers in the Chagas disease vector *Triatoma dimidiata. Infect. Genet. Evol.* 1: 243-48.

Antonini, G., Coletti, G., Serrani, L., Tronci, C., Cristofaro, M. and Smith, L. 2009. Using molecular genetics to identify immature specimens of the weevil *Ceratapion basicorne* (Coleoptera: Apionidae). *Biol. Cont.* 51: 152-57.

Bargues, M.D., Marcilla, J.A., Dujardin, J.P. and Mas-Coma, S. 2002. Triatomine vectors of Trypanosomacruzi: a molecular perspective based on nuclear ribosomal DNA markers. *Trans R Soc Trop Med Hyg* 96 (Suppl. 1): 159-64.

Bellés, X. 2010. Beyond Drosophila: RNAi in vivo and functional genomics in insects. *A. Rev. Ent.* 55: 111-28.

Brattström, O., Akesson, S. and Bensch, S. 2010. AFLP reveals cryptic population structure in migratory European red admirals (*Vanessa atalanta*). *Ecol. Ent.* 35: 248-52.

Brunner, P.C., Fleming, C. and Frey J.E. 2002. A molecular identification key for economically important thrips species (Thysanoptera : Thripidae) using direct sequencing and a PCR-RFLP-based approach. *Agric. For. Entomol.* 4: 127-36.

Carletto, J., Blin, A. and Vanlerberghe-Masutti, F. 2009. DNA-based discrimination between the sibling species *Aphis gossypii* Glover and Aphis frangulae Kaltenbach. *Syst. Ent.* 34: 307-14.

Dong, Y. and Friedrich, M. 2010. Enforcing biphasic eye development in a directly developing insect by transient knockdown of single eye selector genes. *J. Exp. Zool. Part B, Mol. Devel. Evol.* 314: 104-14.

Ellis, J.S., Blackshaw, R., Parker, W., Hicks, H. and Knight, M.E. 2009. Genetic identification of morphologically cryptic agricultural pests. *Agric. For. Entomol.* 11: 115-21.

Grasela, J.J. and Mcintosh, A.H. 2003. Application of inter-simple sequence repeats to insect cell lines: Identification at the clonal and tissue-specific level, *In Vitro Cell. & Devel. Biol. – Animal* 39: 353-63.

Gupta, V.K., Sharma, P.K., Sharma, R.K., Jindal, J. and Dilawari, V.K. 2008. Development of SCAR markers for specific identification of B-biotype of *Bemisia tabaci* (Genn). *Ann. Pl. Prot. Sci.* 16: 274-81.

Gupta, V.K., Sharma, R.K., Jindal, V. and Dilawari, V.K. 2010. SCAR markers for identification of host plant specificity in whitefly, *Bemisia tabaci* (Genn.). *Indian J. Biotech.* 9: 360-66.

Hamada, A., Miyawaki, K., Honda-Sumi, E. *et al.* 2009. Loss-offunction analyses of the fragile X-related and dopamine receptor genes by RNA interference in the cricket *Gryllus bimaculatus*. Develop. Dynam. 238: 2025-33.

Jain, S.K., Neekhra, B., Pandey, D. and Jain, N. 2010. RAPD marker system in insect study: A review. *Indian J. Biotechnol.* 9: 7-12.

Mito, T., Nakamura, T., Bando T., Ohuchi, H. and Noji, S. 2010. The advent of RNA interference in Entomology. *Entomol. Sci.* 1-8 (doi: 10.1111/ j.1479-8298.2010.00408.x)

Mito, T., Ronco, M., Uda, T., Nakamura, T., Ohuchi, H. and Noji, S. 2008. Divergent and conserved roles of extradenticle in body segmentation and appendage formation, respectively, in the cricket *Gryllus bimaculatus*. *Develop. Biol.* 313: 67-79.

Ronco, M., Uda, T., Mito, T., Minelli, A., Noji, S. and Klingler, M. 2008. Antenna and all gnathal appendages are similarly transformed by homothorax knock-down in the cricket *Gryllus bimaculatus*. *Develop. Biol.* 313: 80-92.

Saccone, C., Gissi, C., Lanave, C., Larissa, A. and Pesole, G.R. 2000. Evolution of the mitochondrial genetic system: an overview. *Gene* 261: 153-59.

Sharma, R.K., Gupta, V.K., Jindal, J. and Dilawari, V.K. 2008. Host associated genetic variations in whitefly *Bemisia tabaci* (Genn). *Indian J. Biotech.* 7: 366-70.

Tay, W.T., Behere, G.T., Batterham, P. and Heckel, D.G. 2010. Generation of microsatellite repeat families by RTE retrotransposons in lepidopteran genomes. *BMC Evol. Biol.* 10: 144-48.

Whyard, S., Singh, A.D. and Wong, S. 2009. Ingested double-stranded RNAs can act as species-specific insecticides. *Insect Biochem. Mol. Biol.* 39: 824-832.

Theory and Practice of Integrated Pest Management
R. Arora, B. Singh and A.K. Dhawan (Eds)
Scientific Publishers (India), Jodhpur 2012, pp. 110 - 119.

7

POTENTIAL OF INSECT GROWTH REGULATORS IN INSECT PEST MANAGEMENT

Gursharan Singh and K.S. Suri
Department of Entomology
Punjab Agricultural University, Ludhiana - 141 004 India

The threats arising due to indiscriminate use of conventional insecticides are really very challenging and demand serious attention of all concerned. Development of insecticide resistance and resurgence has made the management of insect pests extremely difficult apart from other environmental concerns such as residue in food and feed and influence on humans and other nontarget organisms besides resulting in environmental pollution and contamination. Since decades, entomologists world over had been attempting to find eco-friendly and sustainable alternatives for management of insect pests with least disturbances to the environment. Out of various strategies tried, one is the application of Insect Growth Regulators (IGRs), which are known to be extremely safe due to their biodegradable nature, selective in action and have mode of action different from the conventional insecticides which make these compounds less prone to development of cross-resistance. The IGRs comprises two sub-groups i.e. insect hormones and their analogues and chitin synthesis inhibitors (CSIs).

7.1. INSECT HORMONES

Neurohormones. Though there are more than 40 neurosecretions on record in the insects, but brain hormone more appropriately called prothoracicotropic hormone (PTTH) or ecdysiotropin, secreted from the groups of neurosecretory cells (median and lateral) in the brain of insects, is the best studied. The PTTH and other neuropeptides could be applied to induce various metabolic disturbances in the target insect species, leading to derangements through excessive production of neurohormones which could be exploited for insect pest management. However, no breakthrough could be achieved, as yet, regarding their use in insect pest management. The major limitation being their costly synthesis in addition to other problems such as their inability to penetrate through cuticle of insects because of their proteinaceous nature, apart from being heat-labile and highly unstable under field conditions.

Juvenile Hormones (JH) and their analogues (JHAs). Juvenile hormones (JHs) are secreted by the cells of *corpora allata*, which are bilaterally paired structures in Lepidoptera, but often fused into one mass of tissue in other groups of insects. Chemically JHs are sesquiterpenoids. Six JHs (JH I, JH II, JH III, JH 0, *iso* JH 0 (also called 4-methyl JH I) and JH bisepoxide, have been found to be present in insects; the first to be characterized was JH I (C-18 analogue) from *cecropia* oil i.e. Methyl-10, 11-epoxy-7-ethyl-3, 11–dimethyl-2-*trans*-6-*trans*-tridecadienoate. JH II is the most widely found in class Insecta. The concentration of JHs in the haemolymph is determined by the rate of its biosynthesis in the *corpora allata* and its degradation in the tissues. JHs perform differential roles in immature and mature insects. While in the larval/nymphal stages, the higher concentrations of JHAs inhibit metamorphosis, thus affecting morphogenesis, but in adults they perform an entirely different function, i.e. synthesis and uptake of yolk; the effects known as gonadotropic actions of JHAs. Thus, any exogenous application of JHs will derange the process of metamorphosis producing adultoids, larval-pupal intermediates or supernumerary instars, which rarely moult into proper adults. On the basis of such observations, Williams (1967) was the first to raise the possibility of use of JHs as the 'Third Generation Pesticides' which led to the synthesis of a large number of JHAs. The natural JHs could not be of much use in IPM programmes given to their instability in UV light, rapid metabolim in insects coupled with difficulties in their synthesis.

Farnesol was synthesized as the first synthetic hormone for use in pest control as a JHA. Unfortunately, its use in pest control has been limited since it does not penetrate insect cuticle well, maintains insect in the juvenile (and possibly injurious) stage and does not have a great deal of residual activity following application (Thacker, 2002). Methoprene (ALTOSID®) was the first JHA, commercially introduced in the market by Zoecon Corporation, California and approved by Environmental Protection Agency, USA for use against the California flood-water mosquito, *Aedes nigromaculus* (Linnaeus) USA. It was effective in micro-encapsulated form at <1.0 ppb concentration in water, where these mosquitoes breed. Subsequently, it was also registered in UK in 1980 to control pharaoh's ants under the trade name PHARORID®. Methoprene has also been used in combination with the organophosphate insecticide methamiphos to control grain pests *Sitophilus oryzae* (L.) and *S. granaries* L. (Thacker, 2002).

Hydroprene (GenTrol®/ GENCOR®/ GENCORPLUS®/ MATRON®/ MATOR®/TURBOCIDE GOLD®) was another JHA marketed for the control of cockroaches that acts by inducing sterility in treated insects (Sehnal, 1985). Hydroprene resulted in arrested larval growth, morphological deformities, incomplete emergence and twisted wings in *T. castaneum* and *T. confusum* Duval (Arthur, 2003). Hydroprene and another JHA (R-20458) demonstrated appreciable morphogenetic activity by producing adultoids, nymphal-adult intermediates and supernumerary nymphs upon their topical application to mustard aphid, *Lipaphis erysimi* (Kalt.), apart from affecting reproduction of the aphid, apterising and fumigant effects of the JHAs (Singh & Sidhu, 1992, 1991, 1990a & 1988).

Fenoxycarb (INSEGAR®/PRECLUDE®/ PRECISION®) was registered in Switzerland and Italy for the control of several fruit pests such as pear psylla, *Cacopsylla pyricola*, grape moths, *Lobesia botrana* Schiff. and *Eupoecilia embiguella*, plum moth, *Grapholita funebrana* Treitschke; codling moth, *Cydia pomonella* L.; summer fruit tortix, *Adoxophyes orana* (Fischer von Rösslerstamm); fruit tree tortrix, *Archips podana* (Scopoli), and several other tortricid species (Solomon & Fitzgerald, 1990; Charmillot *et al.,* 1987; Pluciennik *et al.,* 1999; Charmillot *et al.,* 1989; Charmillot & Brunner, 1989; De Reed *et al.,* 1985). Application of fenoxycarb on lime in Florida gave a satisfactory control of citrus leaf miner, *Phyllocnistis citrella* Stainton.

Pyriproxyfen (ADMIRAL®/DISTANCE®/NYLAR®/NEMESIS®/ TIGER® 10 EC), a phenylether juvenile hormone mimic, marketed for the control of public health pests (Estrada & Mulla, 1986; Okazawa *et al.,* 1991), field pests and as a feed-through compound in poultry, cattle, and swine for the control of house fly, *Musca domestica* L. and the face fly, *M. autumnalis* De Geer (Miller, 1989), was also shown to suppress egg hatch in pear psylla, *C. pyricola*, egg hatch and adult formation in cotton whitefly, *Bemisia* tabaci (Gennadius) and greenhouse whiteflies, *Trialeurodes vaprariorum* (Westwood) and *Haematobia irritans* (L.) in Israel (Higbee *et al.,* 1995; Ishaaya & Horowitz, 1992; Bull & Meola, 1993; Ishaaya *et al.,* 1994). It also caused sterility and reduction in adult longevity in cotton aphid, *Aphis gossypii* Glover on cotton (Kerns & Stewart, 2000). Owing to its environmental instability, pyriproxifen can be effectively used in greenhouse conditions.

Epofenonane, another JHA, was reported to be effective against last larval instars of *S. litura* on cauliflower (Singh & Singh, 1988). Toxic effects along with abnormal morphogenesis were observed after foliar treatment with a JHA, R-20458 and hydroprene against mustard aphid, *Lipaphis erysimi* (Kaltenbach) on cabbage (Singh & Sidhu, 1990b). JH analogue, ZR-777 drastically reduced the fecundity of red rust flour beetle, *T. castaneum* (Singh, 1994).

Plants with JH factors. In addition to the above discussed JHAs, a large number of JHAs of plant origin have also been identified from a number of plant species. The compound juvocimene I and II, isolated from sweet basil, *Ocimun basilicum* showed JH activity against milkweed bug, *Oncopeltus fasciatus* Dallas. Roots of the American coneflower, *Echinacea angustifolia* has a compound called echinolone, which showed JH activity against *O. fasciatus* and *T. molitor* (Jacobson *et al.,* 1975). Biological activity of some plant oils, especially the terpenoid lactones has also been noted against the mustard aphid, *L. erysimi* (Arora *et al.,* 1982; Singh & Kalsi, 1996).

Anti-Juvenile Hormones (AJHAs). The knowledge that some plants produce juvenile hormone mimics has stimulated the search for antagonists of JHs within plants. The action of anti-JHs is accomplished by competing with JH in binding to the JH receptors or to the JH-carrier proteins, killing the *corpora allata* cells, or interfering with JH biosynthesis (Leighton *et al.,* 1981). There are Hübner JHAs that compete with JH at the receptor site and become feedback inhibitors of JH biosynthesis. An example is ETB [ethyl 4-(2-pivaloyloxybutyloxy)-benzoate], which showed JH agonist and antagonist activities in *M. sexta* larvae (Staal, 1986).

Moulting Hormone Analogues (MHAs). Ecdysones or moulting hormones are secretions of prothoracic glands of insects. The prothoracic glands sequester cholesterol from the circulating haemolymph and convert it into ecdysone, which is considered a prohormone and is secreted into the haemolymph as it is produced, which is soon converted into the active form i.e. 20-hydroxyecdysone (-ecdysone) by 20-monooxygenase enzyme by adding a hydroxyl group at C-20 position. Several other ecdysteroids i.e. 26-hydroxyecdysone and 20, 26-dihydroxyecdysone with hormonal activity are known from different insects. Exogenous application of moulting hormone analogues (MHAs) or ecdysone agonists result in hormonal imbalance since these cannot be metabolized or excreted rapidly and consequently precocious metamorphosis is ensued.

Rohm and Haas Company in 1983 serendipitously discovered the first bisacylhydrazine ecdysteroid agonist. Soon after, a simple and more potent MH analogue, RH-5849 was synthesized that targeted many lepidopteron, coleopteran and dipteran insect pests (Aller & Ramsay, 1988). It indicated high effectiveness on *Leptinocarsa decemlineata* (Say), *Mamestra brassicae* (L.), *Pieris brassicae* (L.) and *O. fasciatus*. RH-5849 also induced premature initiation of moulting at all stages of larval development of the tobacco hornworm, *M. sexta* (Wing *et al.,* 1988). It was superseded by a group of some cost effective MHAs. The best known representatives of this group comprise the chemical compounds tebufenozide, methoxyfenozide and halofenozide. Both methoxyfenozide and halofenozide were developed by the joint venture of Rohm and Haas and American Cynamid Company. These are similar to tebufenozide, but differ in respect that both are plant systemic, thus also effective against soil-dwelling insects. These compounds appear to work by binding to the ecdysone receptor protein, causing a precocious and unsuccessful moult to take place.

Multiple number of commercial formulations of ecdysteroids are available in the market for use against several insect pests belonging to Lepidoptera, Coleoptera, Diptera and insects belonging to several other orders, but analogues of moulting hormones (MHAs) do not appear to have a bright scope in IPM programmes due to certain serious limitations. One of the very serious limitations is their steroidal nature which makes these compounds more hydrophilic than lipophilic, and hence restricts their penetration through the waxy layer of the insect cuticle making them ineffective when applied topically and must be fed orally to the target species. Moreover, their complex steroidal structure also makes their synthesis costly and laborious. The amount required to affect precocious moulting is generally quite high which often deter feeding of insects on the treated material. Morethan all this, their regulatory role in the development of crustaceans and their close proximity to steroid hormones (found in man and other higher animals), is a matter of concern and hence we need to adopt a very cautious approach for their use in pest control programmes.

Anti-Moulting Hormone Analogues (AMHAs). The AMHAs are the compounds antagonizing the action of moulting hormones. Chemically, the AMHAs are azosterols and non-steroidal compounds which act by antagonizing the conversion of phytosterols to cholesterol, which is further required as a base for the synthesis of ecdysone in insects. As a result, the moulting hormone titer is affected and consequently, the moulting process is delayed (Walker & Svoboda, 1973). Certain plants producing AMHAs have been identified as *Ajuga decumbens, Ajuga remota, Azadirachta indica* and *Plumbago capensis,* which can be used in the near future for their synthesis in pest control programmes.

7.2. CHITIN SYNTHESIS INHIBITORS (CSIs)

The second sub-category of IGRs i.e. chitin synthesis inhibitors (CSIs) are chemically benzoylphenylureas, which act by inhibiting the enzyme chitin synthetase required for polymerization of N-acetylglucosamine in the bio-synthesis of chitin during the process of moulting of insects. The insecticidal activity of theses compounds was first discovered by Philips-Duphar Company around 1970 for the first time and now multiple numbers of commercial formulations are available for the control of insect pests of field crops.

The first chitin synthesis inhibitor introduced into the market as a novel insecticide was a benzoylphenylurea named diflubenzuron (Miyamoto *et al.,* 1993). Diflubenzuron (marketed under the trade names DIMILIN / ADEPT® 25 WP/ MICRO®) is toxic to a wide range of pests, has a low (negligible) mammalian toxicity, a low toxicity to most beneficial species, possesses a high degree of environmental persistence and has been registered in India (CIBRC, 2005). It is an effective tool for controlling fungus gnats, shore flies, and foliar feeding pests (armyworms, leaf miners) on ornamental crops grown in greenhouses, shadehouses, and interiorscapes. Lufenuron (MATCH® 5 EC/PROGRAM®), another CSI has been found quite effective against several insect pests of field crops. Sublethal concentrations of lufenuron affected various morphological, biological and physiological parameters of *P. xylostella* (Josan & Singh, 2000). Lufenuron was found to be very effective against several leaf-rollers such as *Ctenopseustis obliquana* Walker, *C. herana* Walker, *Planotortrix octo* Dugdale, *P. excessana* (Walker) and *Epiphyas postvittana* (Walker) infesting apples and kiwi fruit in New Zealand (Follas *et al.,* 1994). Butter *et al.* (2003) reported a reduction in pupal weight, adult emergence and fecundity of *Helicoverpa armigera* on cotton treated with lufenuron. It has also been registered in India against phytophagous mites. Another CSI, hexaflumuron (SENTRICON®) has been used as bait against subterranean termite species.

Triflumuron and teflubenzuron have been found to be promising against *C. molesta* on apple and pear and against soft scale *C. floridensis* (Pollini & Bariselli, 1993; Eisa *et al.,* 1991). Triflumuron, when used against red rust flour beetle, *T. castaneum*, resulted in arrested spermatogenesis and production of abnormally large sperm bundles (Parween, 2001).

Novaluron, another very potent CSI, was evaluated against *S. exigua, S. littoralis* Boisduval and *B. tabaci* on cotton under field conditions and was very effective against these pests, besides being safe for non-targets and exhibiting no cross resistance to other groups (Ishaaya *et al.,* 2003).

Teflubenzuron, another CSI, significantly reduced the population of black scale, *Saissetia oleae* (Olivier). Teflubenzuron resulted in ovicidal action, interference with all developmental stages and mortality of eggs at blackhead stage in rice grain moth, *Corcyra cephalonica* Stainton (Chakraborti & Chaterjee, 2001). Teflubenzuron and hexaflumuron were also found to reduce adult emergence in pulse beetle, *C. maculates* (Su *et al.,* 2003).

Buprofezin (APPLAUD®/ACCOLADE®/ KNACK®/TALUS®), another chitin synthesis inhibitor, is a thiadiazine insect growth regulator. It acts by inhibiting cuticle deposition. It also suppresses egg laying in female adults with inhibition of prostoglandin synthesis and has effects on levels of hormones associated with moulting in nymphs. APPLAUD® has been registered in over 40 countries around the world. Buprofezin was found to be potent against homopteran pests including nymphs of brown plant-hoppers, *N. lugens*, leafhoppers, *Nephotettix cincticeps* Uhler, whiteflies *B. tabaci*, *T. vaporariorum*, and scale insects attacking fruit crops and certain species of Colepotera and Acarina (Asai *et al.*, 1985; Ellsworthip & Martinez, 2001; Palumbo *et al.*, 2001).

7.3. FUTURE PROSPECTS

Insect growth regulators can become a viable component of IPM programme if used judiciously, since many commercial formulations of these are now available. Low mammalian toxicity, biodegradability and specific nature of these compounds make them eco-friendly and add to the diversity spectrum of chemical control agents. This is further supplemented by their novel modes of action, making them less prone to development of cross-resistance. Among the IGRs, juvenile hormone analogues have better prospects due to their easy synthesis and availability of multiple formulations especially developed for field stability. Apart from being selective in action, these act at extremely low concentrations and are biodegradable and eco-friendly. The second group of IGRs, i.e. Chitin synthesis inhibitors, has proven field efficacy against a variety of insect pests. On the whole, JHAs and CSIs have demonstrated their potential to become a viable component of IPM programmes.

REFERENCES

Aller, H.E. and Ramsay, J.R. (1988) RH-5849- A novel insect growth regulator with a new mode of action. *Brighton Crop Prot. Conf. Pests Dis.* 2: 511-18.

Arora, R., Singh, G., Kaily, J.S. and Kalsi, P.S. (1982) Biological activity of terpenoid lactones as juvenile hormone analogues against the mustard aphid, *Lipaphis erysimi. Phytoparasitica* 10(1): 57-60.

Arthur, F.H. (2003) Efficacy of a volatile formulation of hydroprene (Pointsource) to the control of *Tribolium castaneum* and *Tribolium confusum* (Coleoptera: Tenebrionidae). *J. Stored Prod. Res.* 39(2): 205-12.

Asai, T.A., Kajihara, M., Fukada, F. and Makekawa, S. (1985) Studies on the mode of action of buprofezin II. Effects on reproduction of the brown plant-hopper, *Nilaparvata lugens* Stal. (Homoptera: Delphacidae). *Appl. Ent. Zool.* 20: 111-17.

Bull, D.L. and Meola, R.W. (1993) Effect and fate of the insect growth regulator pyriproxyfen after application to the horn fly (Diptera: Muscidae). *J. econ. Ent.* 86: 1754-60.

Butter, N.S., Gurmeet, S., Dhawan, A.K. and Singh, G. (2003) Laboratory evaluation of insect growth regulator lufenuron against *Helicoverpa armigera* on cotton. *Phytoparasitica* 31(2): 200-03.

Chakraborti, S. and Chaterjee, C. (2001) Ovicidal effects of some chitin inhibitors on the eggs of *Corcyra cephalonica. Pestology* 25(6): 41.

Charmillot, P.J., Baillod, M., Bloesch, B., Guignard, E., Antonin, P., Frischknecht, M.L., Hoehn, H. and Schmid, A. (1987) An insect growth regulator used for its ovicidal action in the control of the grape moths, *Lobesia botrana De. & Schiff, and Eupoecilia ambiguella Hb. Revue Suise de Viticulture d'Arboriculture et d. Hortic.* 19(3) : 183-91.

Charmillot, P.J., Bloesch, B. and Benz, M. (1989) Control of the codling moth, *Cydia pomonella* L. with fenoxycarb and teflubenzuron. *Revue Suise de Viticulture d'Arboriculture et d. Horticulture* 21(3) : 187-93.

Charmillot, P.J. and Brunner, J.F. (1989) Summerfruit tortix, *Adoxophyes orana*: Life cycle, warning system and control. *Ent. Hell.* 8: 17-26.

CIBRC (2005) Pesticides registered under Insecticide Act, 1968 of India. *http://cibrc.nic.in/reg_products.htm*

De Reed, R.H., Alkema, P. and Blommers, L.H.M. (1985) The use of the insect growth regulators fenoxycarb and epofenonane against leafrollers in integrated pest management in apple orchards. *Ent. Exp. Appl.* 39: 265-72.

Eisa, A.A., El-Fatah, M.A., El-Nabawi, A. and El-Dash, A.A. (1991) Inhibitory effects of some insect growth regulators on developmental stages, fecundity and fertility of the Florida wax scale, *Ceroplastis floridensis. Phytoparasitica* 19(1): 49-55.

Ellsworthip, P. and Martinez, C.J.L. (2001) IPM for Bemisia tabaci: A case study from North America. *Crop Prot.* 20(9): 853-69.

Estrada, J.G. and Mulla, M.S. (1986) Evaluation of two new insect growth regulators against insects in the laboratory. *J. Am. Mosq. Contr. Assoc.* 2: 57-60.

Follas, G., Welsh, R., Bland, S., Patterson, T.M. and Popay, A.J. (1994) Lufenuron- an insect growth inhibitor for the control of leaf-rollers in apples and kiwi fruit. Proc. *47th New Zealand Plant Prot. Conf. New Zealand,* 341-45.

Higbee, B.S., Horton, D.R. and Krysan, J.L. (1995) Reduction of egg hatch in pear psylla (Homoptera: Psyllidae) after contact by adults with insect growth regulators. *J. econ. Ent.* 88: 1420-24.

Ishaaya, I., De Cock, A. and Degheele, D. (1994) Pyriproxyfen, a potent suppressor of egg hatch and adult formation of the greenhouse whitefly (Homoptera: Aleyrodidae). *J. econ. Ent.* 87(5): 1185-89.

Ishaaya, I. and Horowitz, A.R. (1992) Novel phenoxy juvenile hormone analog (pyriproxyfen) suppresses embryogenesis and adult emergence of sweet potato whitefly (Homoptera: Aleyrodidae). *J. econ. Ent.* 85: 2113-17.

Ishaaya, I., Kontsedalov, S. and Horowitz, A.R. (2003) Novaluron (Rimon), a novel IGR: Potency and cross resistance. *Arch. Insect Biochem. Physiol.* 54(4): 157-64.

Jacobson, M., Redfern, R.E. and Mills, G.D. Jr. (1975) Naturally occuring insect growth regulators III. Echinolone, a highly active juvenile hormone mimic from *Echinacea angustifolia roots*. *Lloydia* 36(6): 473-76.

Josan, A. and Singh, G. (2000) Sublethal effects of lufenuron on the diamondback moth, *Plutella xylostella* (Linnaeus). *Insect Sci. Applic.* 20(4): 303-08.

Kerns, D.L. and Stewart, S.D. (2000) Sublethal effects of insecticides on the intrinsic rate of increase of cotton aphid. *Ent. Exp. App.* 94(1): 41-49.

Leighton, T.E., Marks, D. and Leighton, F. (1981) Pesticides: Insecticides and fungicides are chitin synthesis inhibitors. *Sci.* 213: 905-7.

Miller, E.W. (1989) Evaluation of S-31183 for fly (Diptera: Muscidae) control as a feed-through compound for poultry, cattle, and swine. *J. Agri. Ent.* 6: 77-81.

Miyamoto, J., Hirano, M., Takimoto, Y. and Hatakoshi, M. (1993) Insect growth regulators for pest control, with emphasis on juvenile hormone analogs: Present status and future prospects. *ACS Symp. Ser.*, ACS, Washington, DC. 524: 144-68.

Okazawa, T., Bakotee, B., Suzuki, H., Kawada, H. and Kere, N. (1991) Field evaluation of a new insect growth regulator, pyriproxyfen, against *Anopheles punctulatus* on North Guadalcanal, Solomon Islands. *J. Am. Mosq. Contr. Assoc.* 7: 604-7.

Palumbo, J.C., Horowitz, A.R. and Prabhakar, N. (2001) Insecticidal control and resistance management for *Bemisia tabaci. Crop Prot.* 20(9): 739-65.

Parween, S. (2001) Effect of triflumuron on the male reproductive system of *Tribolium castaneum* (Herbst.) (Coleoptera: Tenebrionidae). *Bangladesh J Ent.* 11: 127-31.

Pluciennik, Z., Olszak, R.W. and Tworkowska, U. (1999) Modern insecticides in controlling plum fruit moth (*Grapholita funebrana* Tr.). *Prog. Pl. Prot.* 39: 448-51.

Pollini, A. and Bariselli, M. (1993) *Cydia molesta*: Pest on the increase and defense of pome fruits. *Informatora Agrario Supplemento* 49(14): 19-21.

Sehnal, F. (1985) Use of insect hormones, their analogues and anti-hormonal compounds in pest control. In: *Metodi alternative alla lotta chimica nella difesa delle colture agrarie* (eds. Goidanich & Baronio), 1° Convegno Internazionale, Cesena Agricultura, Cesena, pp. 223-36.

Singh, G. (1994) Effect of juvenile hormone analogues on the reproduction and longevity of rust red flour beetle, *Tribolium castaneum* (Herbst.). *J Insect Sci.* 7(1): 91-92.

Singh, G. and Kalsi, P.S. (1996) Juvenile hormone activity of some plant oils against mustard aphid, *Lipaphis erysimi* (Kalt.). In: *Allelopathy in Pest Management for Sustainable Agriculture* (eds. Narwal & Tauro), Scientific Publishers (India), Jodhpur, India, 175-78.

Singh, G. and Sidhu, H.S. (1988) Demorphogenesis of mustard aphid, *Lipaphis erysimi* (Kalt.) by vapours of juvenile hormone analogues. *Curr. Sci.* 58: 270-72.

Singh, G. and Sidhu, H.S. (1990a) Effect of juvenile hormone analogues on the reproduction and longevity of mustard aphid, *Lipaphis erysimi* (Kalt.). *Insect Sci. Applic.* 11(1): 69-72.

Singh, G. and Sidhu, H.S. (1990b) Toxic effects of juvenile hormone analogues on mustard aphid, *Lipaphis erysimi* (Kalt.). *J Insect Sci.* 3: 18-22.

Singh, G. and Sidhu, H.S. (1991) Apterizing effect of juvenile hormone analogues on mustard aphid, *Lipaphis erysimi* (Kalt.). *J. Insect Sci.* 4(1): 71-72.

Singh, G. and Sidhu, H.S. (1992) Morphogenetic abnormalities of alatiform mustard aphid, *Lipaphis erysimi* following topical application of juvenile hormone analogues. *Insect Sci. Applic.* 13(1): 27-33.

Singh, G. and Singh, D. (1988) Bioassay of a juvenile hormone analogue against tobacco caterpillar, *Spodoptera litura* (Fab.). *J Insect Sci.* 1: 12-16.

Solomon, M.G. and Fitzgerald, J.D. (1990) Fenoxycarb, a selective insecticide for inclusion in integrated pest management systems for pear in the UK. *J. Hortic. Sci.* 65: 535-39.

Staal, G.B. (1986) Anti-juvenile hormone agents. *A. Rev. Ent.* 31: 391-429.

Su, T.Y., Mulla, M.S. and Zaim, M. (2003) Laboratory and field evaluation of a new insect growth regulator (IGR) against *Culex mosquitoes. J. Am. Mosq. Control Assoc.* 19(4): 408-18.

Thacker, J.R.M. (2002) An Introduction to Arthropod Pest Control. Cambridge Univ. Press, UK. pp: 343.

Sehnal F (1985) Use of insect hormones, their analogues and anti-hormonal compounds in pest control. In: *Metodi alternative alla lotta chimica nella difesa delle colture agrarie* (eds. Goidanich & Baronio), 1° Convegno Internazionale, Cesena Agricultura, Cesena, pp. 223-36.

Walker, W.F. and Svoboda, J.A. (1973) Suppression of Mexican bean beetle development by foliar applications of azosterols and reversal by cholesterol. *Ent. Exp. Appl.* 16: 422-26.

Williams, C.M. (1967) Third generation pesticides. *Sci. Amer.* 217: 13-15.

Wing, K.D., Slawecki and Carlson, G.R. (1988) RH 5849, a nonsteroidal ecdysone agonist: Effects on larval Lepidoptera. *Sci.* 241: 470-72.

Theory and Practice of Integrated Pest Management
R. Arora, B. Singh and A.K. Dhawan (Eds)
Scientific Publishers (India), Jodhpur 2012, pp. 120 - 134.

8

BOTANICAL PESTICIDES IN INSECT PEST MANAGEMENT

Naveen Aggarwal
Department of Entomology
Punjab Agricultural University, Ludhiana - 141 004, India

8.1. INTRODUCTION

Agriculture in a tropical country like India, owing to its climatic conditions and its particular environment, suffers severe losses of agricultural produce due to attack by various insect pests. The Indian farmers are in need of effective tools to fight against these pests and rely largely on chemical pesticides to save their crops from insect pests. Twenty years after synthetic insecticides were firmly entrenched in 'modern' agricultural production, overzealous use of these synthetic insecticides led to numerous problems unforeseen at the time of their introduction like acute and chronic poisoning of applicators, farm workers, and even consumers; destruction of fish, birds, and other wildlife; disruption of natural biological control and pollination; extensive groundwater contamination, potentially threatening human and environmental health; and the evolution of resistance to pesticides in pest populations (Perry *et al.,* 1998). After severe setback arising from the use of chemical pesticides on living systems and the environment, therefore, it has now become necessary to search for the alternative means of pest control, which can minimize the use of synthetic pesticides and it led to interest in 'natural' means of pest control (Roger *et al.,* 2005). Now these days, the use of eco-friendly bio pesticides is gaining momentum. The use of plants, plant material or crude plant extracts for the protection of crops and stored products from insect pests is probably as old as crop protection itself (Thacker, 2002). The practice of using plant derivatives, or botanical insecticides as we now know them, in agriculture dates back at least two millennia in ancient China, Egypt, Greece, and India. Indeed, prior to the development and commercial success of synthetic insecticides beginning in the 1940s, botanical insecticides were major weapons in the farmer's arsenal against crop pests. According to Purohit and Vyas (2004) as many as 2121 plant species are reported to be useful in pest management, out of which 1005 species of plants exhibited insecticidal properties, 384 with antifeedant properties, 297 with repellent properties, 27 with attractant properties and 31 with growth inhibiting properties have been identified. The earliest mention of poisonous plants occurs in the *Rigveda* and following ancient examples illustrate the use of plants as biopesticides: (i) Romans used false or white hellebore (*Veratrum album* or *V. viride*) as a rodenticides, (ii) Chinese used Derris species as insecticide, (iii) Pyrethrum was used as an insecticide in Persia and Dalmatia and (iv) tobacco was used in the Middle East. Approximately 700 species are reported to have been used for pest control during the past few decades in different parts of the world. Plant derived chemicals (e.g. pyrethrum, and nicotine) were also used for pest control in the West, but most of these biopesticides were eliminated from their use due to availability of synthetic pesticides through the growth of agro-chemical industry, after the World War II.

The combination of efficacy, speed of action, ease of use and low cost of the synthetic insecticides (organochlorines, organophosphates, carbamates and later the pyrethroids and neonicotinoids) drove many botanicals to near obscurity in most industrialized countries. Synthetic insecticides effectively relegated botanicals from an important role in agriculture to an essentially trivial position in the marketplace among crop protectants. There are more than 2300 plant species belonging to 189 plant families which are said to be rich sources for bioactive organic compounds. The srructure of more than 600 alkaloids, 300 terpenoids, thousands of phenyl propanoids, 1000 flavanoids, 500 quinones, 650 polyacetylenes and 400 unusual aminoacids was described (Metcalf and Metcalf, 1992). These secondary metabolites show various pesticidal effects like antifeedants or feeding deterrents, repellents, anti-juvenile hormone/ moulting hormone activity, antifertility or growth distruptors etc. This means that in some or the other way; these compounds inhibit normal development of insects. Botanical insecticides have many advantages over synthetic insecticides which include:

(1) Botanical pesticides in general possess low mammalian toxicity thus constitute least or no health hazards and environmental pollution.

(2) There is practically no risk of developing pest resistance to these products, when used in natural forms.

(3) These cause fewer hazards to non-target organisms and pest resistance has not been reported except synthetic pyrethroids.

(4) Mostly botanicals are compatible with insect microbials.

(5) No adverse effect on plant growth, seed viability and cooking quality of the grains.

(6) Botanical pesticides are less expensive and easily available because of their natural occurrence especially in oriental countries.

(7) They are cheaper and renewable.

8.2. CURRENT BOTANICALS IN USE

The term "botanical pesticides" refers to those compounds or materials derived from plants (excluding synthetic analogues) that demonstrate useful toxicity to pests involved in agriculture, horticulture, animal husbandry, and human health problems. The plant material is used either directly (e.g., powdered bark) or as a source of extracts for use in formulations that are often combined with other pesticides, synergists, and formulants to increase efficacy and speed of action. At present there are four major types of botanical products used for insect control (pyrethrum, rotenone, neem, and essential oils), along with three others in limited use (ryania, nicotine, and sabadilla). Additional plant extracts and oils (e.g., garlic oil, *Capsicum* oleoresin) regional use in various countries. Botanical like neem, bel, senwar, pyrethrum, tobacco, karanj, mahua and sweet flag etc. have already attained the status of potential pesticides of plant origin to be used in IPM of crop field insects as well as in storage ecosystems (Arora and Dhaliwal, 1994; Prakash and Rao, 1997).

8.2.1. Pyrethrum

Pyrethrins are highly concentrated active compounds which are extracted from the daisylike flower of *Chrysanthemum cinerariaefolium*, commercially grown in Kenya. The flowers are ground to a powder and then extracted with hexane or a similar nonpolar solvent; removal of the solvent yields an orange-colored liquid that contains the active principles (Jones, 2001). These are three esters of chrysanthemic acid and three esters of pyrethric acid. Among the six esters, those incorporating the alcohol pyrethrolone, namely pyrethrins I and II, are the most abundant and account for most of the insecticidal activity. The flower grounded into a powder, the product is called a pyrethrum. Technical grade pyrethrum, the resin used in formulating commercial insecticides, typically contains from 20 to 25 per cent pyrethrins (Casida and Quistad, 1995).

Most insects are highly susceptible to low concentrations of pyrethrins. The insecticidal action of the pyrethrins is characterized by a rapid knockdown effect, particularly in flying insects, and hyperactivity and convulsions in most insects. These symptoms are a result of the neurotoxic action of the pyrethrins, which block voltage-gated sodium channels in nerve axons. As such, the mechanism of action of pyrethrins is qualitatively similar to that of DDT and many synthetic organochlorine insecticides. Pyrethrins may be used against a broad range of pests including ants, aphids, roaches, fleas, flies, and ticks. They are available in dusts, sprays, and aerosol "bombs." Many insects can break down pyrethrins quickly, and after a brief period they may recover rather than die. To overcome this break-down ability, many pyrethrin insecticides contain a synergist such as piperonyl butoxide (PBO). A synergist enhances insecticidal action by inhibiting the enzymes in the insect's body that break down the toxin. PBO is a synthetically produced chemical substance that is not allowed by some organic certification programs. Before purchasing pyrethrins or other botanicals, check the label to see if PBO is listed among active ingredients (http://osufacts.okstate.edu). In purity, pyrethrins are moderately toxic to mammals (rat oral acute LD_{50} values range from 350 to 500 mg kg^{-1}), but technical grade pyrethrum is considerably less toxic (~1500 mg kg^{-1}).

Pyrethrins are especially labile in the presence of the UV component of sunlight, a fact that has greatly limited their use outdoors. A recent study indicated that the half-lives of pyrethrins on field-grown tomato and bell pepper fruits were 2 hours or less (Antonious, 2004). This problem created the impetus for the development of synthetic derivatives ("pyrethroids") that are more stable in sunlight. The modem pyrethroids, developed in the 1970s and 1980s, have been highly successful and represent one of the rare examples of synthetic pesticide chemistry based on a natural product model. However, the modem pyrethroids bear little structural resemblance to the natural pyrethrins, and their molecular mechanism of action differs as well. Pyrethrum is the predominant botanical in use, perhaps accounting for 80% of the global botanical insecticide market (Isman, 2005).

For many years world production of pyrethrum was led by Kenya, with lesser quantities produced in Tanzania and Ecuador. In the past five years, Botanical Resources Australia, with plantings in Tasmania, has become the second largest producer in the world (~30% of world production at present) (Dhaliwal and Arora, 2001).

8.2.2. Neem

Neem or neem oil is extracted from the seeds of the neem tree, *Azadirachta indica*, a native of India. Neem oil, obtained by cold-pressing seeds, can be effective against soft-bodied insects and mites but is also useful in the management of phytopathogens. More highly valued than neem oil are medium-polarity extracts of the seed residue after removal of the oil, as these extracts contain the complex triterpene azadirachtin. Neem seeds actually contain more than a dozen azadirachtin analogs, but the major form is azadirachtin and the remaining minor analogs likely contribute little to overall efficacy of the extract. Seed extracts include considerable quantities of other triterpenoids, notably salannin, nimbin, and derivatives thereof Azadirachtin as the most important active principle and acts as an insect feeding deterrent and growth regulator (Isman *et al.,* 1996). Neem seeds typically contain 0.2% to 0.6% azadirachtin by weight, so solvent partitions or other chemical processes are required to concentrate this active ingredient to the level of 10% to 50% seen in the technical grade material used to produce commercial products (Dhaliwal *et al.,* 2004).

Azadirachtin has two profound effects on insects. At the physiological level, azadirachtin blocks the synthesis and release of molting hormones (ecdysteroids) from the prothoracic gland, leading to incomplete ecdysis in immature insects. In adult female insects, a similar mechanism of action leads to sterility. In addition, azadirachtin is a' potent antifeedant to many insects. The discovery of neem by western science is attributed to Heinrich Schmutterer, who observed that warming desert locusts in Sudan defoliated almost all local flora except for some introduced neem trees (National Research Council, 1992). Indeed, azadirachtin was first isolated based on its exceptional antifeedant activity in the desert locust, and this substance remains the most potent locust antifeedant discovered to date. Several azadirachtin-based insecticides are sold in the United States. In California, azadirachtin-based insecticides constituted about one third of the botanicals used in agriculture in 2003 (~600 kg). Azadirachtin has systemic action in certain crop plants, greatly enhancing its efficacy and field persistence (Schmutterer, 2002). When applied as a dust to soils, neem can be taken up by the roots of some plants and translocated to other parts of the plant. Neem may remain active in the soil up to four weeks, depending on soil conditions. But when neem is applied to plant foliage, its systemic action is limited—new foliage must be sprayed periodically for adequate protection against insect pests.

Currently registered products for ornamental pest control claim activity against a variety of sucking and chewing insects. Neem is most effective against actively growing immature insects. Azadirachtin is considered nontoxic to mammals (rat oral acute LD_{50} is > 5000 mg kg^{-I}), fish (Wan *et al.,* 1996), and pollinators (Naumann and Isman 1996). The influence of azadirachtin on natural enemies is highly variable (Lowery and Isman, 1995). Like the pyrethrins, azadirachtin is rapidly degraded by sunlight. For example, on olives growing in Italy, azadirachtin has a half-life of approximately 20 h (Caboni *et al.,* 2002).

8.2.3. Plant Essential Oils

Certain plant essential oils show a broad spectrum of activity against pest insects and plant pathogenic fungi, and some oils have a long tradition of use in the protection of stored products. Recent investigations indicate that some chemical constituents of these oils interfere with the octopaminergic nervous system in insects. As this target site is not shared with mammals, most essential oil chemicals are relatively non-toxic to laboratory animals and fish in toxieological tests, and meet the criteria for "reduced risk" pesticides. Some of these oils and their constituent chemicals are widely used as flavoring agents in foods and beverages and are even exempt from pesticide registration in the United States. This special regulatory status combined with the wide availability of essential oils from the flavor and fragrance industries, has made it possible to fast-track commercialization of essential oil-based pesticides in the U.S.A. Though well received by consumers for use against home and garden pests, these "green pesticides" can also prove effective in agricultural situations, particularly for organic food production (Isman 2004). A factor favouring development of botanical insecticides based on plant essential oils is the relatively low cost of the active ingredients, a result of their extensive worldwide use as fragrances and fiavorings.

The oils are generally composed of complex mixtures of monoterpenes, biogenetically related phenols, and sesquiterpenes. Examples include 1, 8-cineole, the major constituent of oils from rosemary (*Rosmarinus officinale*) and eucalyptus (*Eucalyptus globus*); eugenol from clove oil (*Syzygium aromaticum*); thymol from garden thyme (*Thymus vulgaris*); and menthol from various species of mint (*Mentha* species) (Isman, 1999). The rapid action against some pests is indicative of a neurotoxic mode of action, and there is evidence for interference with the neuromodulator octopamine (Enan, 2001) by some oils and with GABA-gated chloride channels by others.

Some of the purified terpenoid constituents of essential oils are moderately toxic to mammals, but, with few exceptions, the oils themselves or products based on oils are mostly nontoxic to mammals, birds, and fish (Isman, 2000). However, as broad-spectrum insecticides, both pollinators and natural enemies are vulnerable to poisoning by products based on essential oils. Owing to their volatility, essential oils have limited persistence under field conditions; therefore, although natural enemies are susceptible via direct contact, predators and parasitoids reinvading a treated crop one or more days after treatment are unlikely to be poisoned by residue contact as often occurs with conventional insecticides.

8.2.4. Rotenone

Rotenone is one of several isoflavonoids produced in the roots or rhizomes of the tropical legumes *Derris, Lonchocarpus* and *Tephrosia.*

Extraction of the root with organic solvents yields resins containing as much as 45% total rotenoids; studies indicate that the major constituents are rotenone (44%) and deguelin (22%) (Cabizza *et al.,* 2004). Rotenone is commonly sold as a dust containing 1% to 5% active ingredients for home and garden use, but liquid formulations used in organic agriculture can contain as much as 8% rotenone and 15% total rotenoids.

Rotenone is a mitochondrial poison, which blocks the electron transport chain and prevents energy production (Hollingworth *et al.,* 1994). Rotenone is a broad-spectrum contact and stomach poison that is effective against leaf-feeding insects, such as aphids, certain beetles (asparagus beetle, bean leaf beetle, Colorado potato beetle, cucumber beetle, flea beetle, strawberry leaf beetle, and others) and caterpillars, as well as fleas and lice on animals. Pure rotenone is comparable to DDT and other synthetic insecticides in terms of its acute toxicity to mammals (rat oral LD_{50} is 132 mg kg^{-I}), Safety of rotenone has recently been called into question because of (a) controversial reports that acute exposure in rats produces brain lesions consistent with those observed in humans and animals with Parkinson's disease (Betarbet *et al.,* 2000), and (b) the persistence of rotenone on food crops after treatment. A study of rotenone residues on olives conducted in Italy determined that the half-life of rotenone is 4 days, and at harvest residue levels were above the tolerance limit (Cabras *et al.,* 2002). Moreover, residues were concentrated in oil obtained from the olives. As an agricultural insecticide, use of rotenone is limited to organic food production. Protective clothing and a mask should be worn while handling rotenone. It may be mixed with pyrethrins or piperonyl butoxide to improve its effectiveness.

8.2.5. Sabadilla

Sabadilla comes from the ripe seeds of the tropical lily *Schoenocaulon officinale.* In purity, the active principles, cevadine-type alkaloids are extremely toxic to mammals (rat oral LD_{50} is ~13 mg kg-I), but commercial preparations typically contain less than 1% active ingredient, providing a margin of safety. The dust formulation of Sabadilla is the least toxic of all registered botanical insecticides. However, pure extracts are very toxic if swallowed or absorbed through the skin and mucous membranes. The alkaloids in Sabadilla affect insect nerve cells, causing loss of nerve function, paralysis, and death. It breaks down rapidly in sunlight and air, leaving no harmful residues. Sabadilla is a broad-spectrum contact poison, but has some activity as a stomach poison. It is commonly used in organic fruit and vegetable production against squash bugs, harlequin bugs, thrips, caterpillars, leaf hoppers, and stink bugs. It is highly toxic to honeybees, however, and should only be used in the evening, after bees have returned to their hives. Formulations include baits, dusts or sprays.

8.2.6. Ryania

Ryania is extracted from the stems of a woody South American plant, *Ryania speciosa*. The powdered wood contains <1% ryanodine, an alkaloid that interferes with calcium release in muscle tissue. It is a slow-acting stomach poison; it causes insects to stop feeding soon after ingestion. It works well in hot weather. Ryania is moderate in acute or chronic oral toxicity in mammals. It is generally not harmful to most natural enemies, but may be toxic to certain predatory mites. Ryania has longer residual activity than most other botanicals. It is used commercially in fruit and vegetable production against caterpillars (European corn borer, corn earworm, and others), codling moth and thrips.

8.2.7. Nicotine

Nicotine, an alkaloid obtained from the foliage of tobacco plants (*Nicotiana tabacum*) and related species, has a long history as an insecticide. They cause symptoms of poisoning similar to those seen with organophosphate and carbamate insecticides (Hayes, 1982). The insecticide usually is marketed as a 40% liquid concentrate of nicotine sulfate, which is diluted in water and applied as a spray. Dusts can irritate skin and are not normally available for garden use. Nicotine is used primarily for insects with piercing-sucking mouthparts, such as aphids, whiteflies, leafhoppers, thrips, and mites. Nicotine is more effective when applied during warm weather. It was registered for use on a wide range of vegetable and fruit crops, but is no longer registered commercially (Dhaliwal and Arora, 2001).

8.3. POTENTIAL NEW BOTANICALS

8.3.1. Annonaceous Acetogenins

The Annonaceae (custard-apple family) is a large family of almost exclusively tropical trees and shrubs comprising about 130 genera and 2300 species (Cronquist, 1993). Botanical insecticides have been traditionally prepared from the seeds of tropical *Annona* species. These include the sweetsop *A. squamosa*) and soursop (*A. muricata*), important sources of fruit juices in Southeast Asia. Acetogenins – a number of long chain fatty acid derivatives extracted from tree leaves, bark and seeds have pesticidal and/or insect antifeedant properties (Ratnayake, 1992). The major acetogenin obtained from seeds of *A. squamosa* is annonin I, or squamocin, and a similar compound, asimicin, was isolated from the bark of the American pawpaw tree, *Asimina triloba*. These compounds are slow acting stomach poisons, particularly effective against chewing insects such as lepidopterans and the Colorado potato beetle (*Leptinotarsa decemlineata*). Their mode of action targets the mitochondrial electron transport with a specific action at NADH ubiquinone oxidoreductase (NADH-dehydrogenase, also known as Complex I). Furthermore, the inhibitory effects of annonaceous acetogenins have been shown to be more potent than those of classical respiratory inhibitors such as rotenone. In purity certain acetogenins are toxic to mammals (LD_{50} is <20 mg kg^{-I}), an impediment to regulatory approval, even though standardized extracts from *Annona* seeds and *Asimina* bark are much less toxic. *Annona* seed extracts may prove more useful in tropical countries where the fruits are commonly consumed or used to produce fruit juice, in which case the seeds are a waste product. For example, Leatemia and Isman (2004) demonstrated that crude ethanolic extracts or even aqueous extracts of seeds from *A. squamosa* collected at several sites in eastern Indonesia are effective against the diamondback moth (*Plutella xylostella*).

8.3.2. *Melia* Extracts

Seeds from the chinaberry tree, *Melia azedarach;* contain a number of triterpenoids, the meliacarpins that are similar but not identical to the azadirachtins, and these too have insect growth regulating bioactivities. But development of commercial insecticides has not paralleled that of the neem insecticides. The main reason for the scarce development of *M. azedarach* as a commercial insecticide in comparison to *A. indica* lies in that the fruits of the former contain meliatoxin, a triterpenoid that is toxic for mammals (Schmutterer, 2002). However, the chemical composition of *M. azedarach* varies notably between its wild and cultivated state. The fruits developed in Argentina have triterpenoids instead of meliatoxin, mainly meliartenin which is a strong insect antialimentary that could be useful for pest and disease management (Carpinella *et al.*, 2003). Whereas in other countries like South Africa using aqueous extracts of chinaberry leaves, presumably lacking meliatoxins and efficacious against the diamondback moth (Charleston, 2004) .

The extract of bark of *Melia toosendan* contains a number of triterpenoids based on toosendanin, a substance reported to be a stomach poison for chewing insects. Later studies suggest that this substance acts primarily as a feeding deterrent but can also serve as a synergist for conventional insecticides (Chen *et al.*, 1995).

8.3.3. Antifeedants

These are chemicals that inhibit feeding in insects when applied on foliage (food) without impairing their applied and gastory receptors or driving (repelling) them away from the food. They are also called phagodeterrents gustatory repellants, feeding deterrents and rejectants. With these, insects die slowly of starvation.

It affects only phytophagous insects and so does not harm beneficial parasities, predators and pollinators. As the pest is not immediately killed by antifeedant, its parasities and predators continue fed on in, and keep it under control. They produce no phtytoxicity or pollution. Only chewing type of insects are affected by antifeedants; sucking pest remain unaffected. The concept of using insect antifeedants gained strength in the 1970s and 1980s with the demonstration of the potent feeding deterrent effect of azadirachtin and neem seed extracts to a large number of pest species, particularly *Schstocerca gregaria*, which is very destructive poly-phagous insect. Pyrethrum in small doses acts as antifeedant for biting fly, Glassine spp. Alkaloid (leptine, tomatine, solanine etc) extracted from some species of solanum reduces feeding and survival time in the potato leafhopper. The leaf extract of lemon grass (*Gymbopagan citrates*), Argemone (*argemone maxicanna*), Artemissia (*Artermisia absintnium*), Sigasbekia (*Orientasis*) are strong antifeedant of caterpillar (*Crocidolamic bimotatis*) (Harborne and Chadwick, 1990).

Two main problems face the use of antifeedants in agriculture (Isman 2002). The first is interspecific variation in response-even closely related species can differ dramatically in behavioral responses to a substance limiting the range of pests affected by a particular antifeedant (Isman 1993). Some substances that deter feeding by one pest can even serve as attractants or stimulants for other pests. The second is the behavioral plasticity in insects-pests can rapidly habituate to feeding deterrents, rendering them ineffective in a matter of hours. This has been recently demonstrated not only for pure substances like azadirachtin (Bomford and Isman 1996), but also for complex mixtures (plant extracts) (Akhtar, 2003). Whereas a highly mobile (flying) insect may leave a plant upon first encountering an antifeedant, a less mobile one (larva) may remain on the plant long enough for the deterrent response to wane.

8.3.4. Attractants

Chemicals that cause insects to make oriented movements towards their source are called insect attractants. They influence both gustatory (taste) and olfactory (smell) receptor or sensilla. Iso-thiocynates from seeds of crucifera, sugar and molasses, and terpenes from barks in with pheromones are natural attractants for various pests of crucifereae and bark beetles (Birch and Haynes, 1982). In onion propylmercapton from umberlliferae and phenylacetaldehyde from flowers of *Araujia serisofera* are attracted carrot fly (*Psila rosae*) and Lepidoptera, respectively. Insect attractants can be used in three ways for the control of insect pests. In sampling or monitoring insect pest populations to assess the extent of infestation and decides the measure of control to be adopted in lusting pest to insecticide- coated traps or poison baits and in distracting insects from normal mating, aggregation, feeding or oviposition.

They do not kill the insects therefore, don't disrupt ecosystem. They can be used to mass trap the insects to be subsequently killed by insecticides. They can be used to misguide the insects to wrong host plants, wrong mating partners or wrong oviposition sites whereby their number will go down by starvation or by producing unfertilized eggs. They cannot be relied as sole control measure, used only in integrated control programs.

8.3.5. Insect repellants

Chemicals that prevent insect damage to plants or animals by rendering them unattractive or offense called repellants. They are more important in protection of man and his animals from insect attacks than in the protection plants.they can be used as dusts and sprays for personal protection against mosquitoes and biting flies. The oils extracted from Citronella contain main constituent citronellol which is used in mosquito coils to repel mosquitoes. Garlic can be intercropped to prevent pests from going near the main crop. It can be planted as a border around the field or garden also proves beneficial. It grows around fruit trees to repel borers (Kumbhar *et al.,* 2001). Other uses for repellents under investigation include perimeter treatments of buildings to exclude termites and the use of essential oils to repel cockroaches from kitchens and flies from dairy barns (Maistrello *et al.,* 2004). Another important use of plant essential oil constituents is in fumigation of beehives to manage economically important honey bee parasites, the Varroa mite *(Varroa jacobsoni)* and the tracheal mite *(A ea rap is woodi).* In North America, menthol (from peppermint) is widely used for this purpose (Delaplane, 1992), and in Europe thymol (from garden thyme) is most often used.

8.4. LIMITATIONS OF BOTANICAL PESTICIDES

(1) Most of these products are not truly insecticides since many are merely feeding deterrents and their effect is low.

(2) Botanical pesticides lack persistence and wide spectrum activity.

(3) They are rapidly degraded by UV light so their residual action is short.

(4) They are not necessarily available season long.

(5) Some botanicals do not have a knockdown effect against insects. Farmers, who used to see a quick kill of the pests with synthetic insecticides, do not consider botanicals effective and do not readily accept them.

(6) Drawback for low adoption of botanical insecticides by growers is the variation in performance of a particular product, even when prepared by the same process.

(7) The plant oil sprays may cause phyto-toxicity due to formation of cream.

8.5. FUTURE PROSPECTS

Botanicals have to face tremendous competition after invention of newer generation of synthetic insecticides with minimum risk to non target organisms, such as neonicotinoids and chitin synthesis inhibitors. Among total bio pesticides used, botanicals constitute only less than 1 per cent (California Dept. Pesticides, 2005). Advances in chemical and biotechnology are increasing the speed and ease with which man can discover and develop secondary compounds of plants as pesticides. These advances, combined with increasing need and environmental pressure are greatly increasing the interest in plant products in pest management agents (McLaren, 1986). The future pest management programs for sustainable agricultural and public health emphasizes on plant extracts called botanicals and isolated potent plant products and their derivatives in addition to safer chemical control agents and bio pesticides. In view of the resultant problems with commercial chemicals pesticides, natural pesticides are gaining importance for the pest management.

Extensive research activities are going on in the isolation of natural products and also their derivatives. This will definitely help us to escape from clutches of harmful conventional synthetic insecticides, as fortunately we were rich with variety of plants. Natural products and their derivatives, thus, will be are of effective pest management tactis in future. Looking to their increasing utility, it is expected that within 10-15 years, these compounds will capture 25 per cent share in insecticide market and they will not limited to garden areas but farmers will also accept this valuable gift of god in agricultural uses. However, a strong will and concrete efforts are required to exploit the pest management efficacy of natural products.

REFERENCES

Akhtar, Y., Rankin, C.H. and Isman, M.B. (2003) Decreased response to feeding deterrents following prolonged exposure in the larvae of a generalist herbivore, *Trichoplusia ni* (Lepidoptera: Noctuidae). *Insect Behav.* 16:811-31.

Antonious, G.F. (2004) Residues and half lives of pyrethrins on field-grown peper and tomato. *J. Env. Sci. Health* 39 : 491-503.

Arora, R. and Dhaliwal, G.S. (1994) Botanical pesticides in insect pest management. In: G.S. Dhaliwal and B.D. Kansal (eds.) *Management of Agricultural Pollution in India.* Commonwealth Publishers, New Delhi, pp. 215-245.

Betarbet, R., Sherer, T.B., MacKenzie, G., Garcia-Osuna, M., Panov, A.V. and Greenamyre, J.T. (2000) Chronic systematic pesticide exposure reproduces features of Parkinson's disease. *Nature Neurosci.* 3:1301-6.

Birch, M.C. and Hynes, K.F. (1982) *Insect Pheromones.* The Institute of Biology's Studies in Biology no. 147, E. Arnold Ltd., London.

Bomford, M.K. and Isman, M.B. (1996) Desentization of fifth instar *Spodoptera litura* (Lepidoptera: Noctuidae) to azadirachtin and neem. *Entomol. Exp. Appl.* 81:307-13.

Cabizza, M., Angioni, A., Melis, M., Cabras, M., Tuberoso, C.V. and Cabras, P. (2004) Rotenone and rotenoids in cube resins, formulations and residues on olives. *J. Agric. Food Chem.* 52: 288-93.

Cabras, P., Caboni, P., Cabras, M., Angioni, A. and Russo, M. (2002) Rotenone residues on olivies and in olive oil. *J. Agric. Food Chem.* 50: 2576-80.

California Department of Pesticide Regulation (2005) *Summary of Pesticides use Report Data, 2003.*

Carpinella, M.C., Defago, M.T., Valladares, G. and Palacios, S.M. (2003) Antefeedant and insecticide properties of a limonoid from *Melia azedarach* (Meliaceae) with potential use for pest management. *J. Agric. Food Chem.* 51: 369-74.

Casida, J.E. and Quistad, G.B. (1995) *Pyrethrum Flowers: Production, Chemistry, Toxicology and Uses.* Oxford Univ. Press, Oxford, UK.

Charleston, D.S. (2004) Integrating biological control and botanical pesticides for management of *Plutella xylostella.* Ph.D Thesis. Wageningen Univ. 176 pp.

Chen, W., Isman, M.B. and Chiu, S.F. (1995) Antifeedant and growth inhibitory effects of the limonoid toosendanin and *Melia toosendan* extracts on the variegated cutworm, *Peridroma saucia* (Lep., noctuidae). *J. Appl. Entomol.* 119:367-70.

Cronquist, A. (1993) An Integrated System of Classification of Flowering Plants. Columbia University Press, New York, NY.

Delaplane, K.S. (1992) Controlling tracheal mites (Acari: Tarsonemidae) in colonies of honey bees (Hymenoptera: Apidae) with vegetable oil and menthol. *J. Econ. Entomol.* 85: 2118-24.

Dhaliwal, G.S. and Arora, R. (2001) Role of phytochemicals in integrated pest management. In: O. Koul and G.S. Dhaliwal (eds.). *Phytochemical Biopesticides.* Harwood Academic Publishers, Amsterdam, pp. 97-117.

Dhaliwal, G.S., Arora, R. and Koul, O. (2004) Neem research in Asian Continent: Present status and future outlook. In: O. Koul and S. Wahab (eds). *Neem: Today and in the New Millenium.* Kluwer Academic Publishers, Dordrecht, pp. 65-96.

Enan, E. (2001) Insecticidal activity of essential oils: octopaminergic sites of action. *Comp. Biochem. Physiol.* 130C: 325-37.

Harborne, J.B. and Chadwick, D.J. (1990) In: March, J. (ed) *Bioactive compounds from plants.* Johny- Wiley and Sons, New York, Pp. 126-139.

Hayes, W.J. Jr. (1982) *Pesticides Studied in Man.* Baltimore: Williams and Wilkins. pp. 672.

Hollingworth, R., Ahmmadsahib, K., Gedelhak, G. and McLaughlin, J. (1994) New inhibitors of complex I of the mitochondrial electron transport chain with activity as pesticides. *Biochem. Soc. Trans.* 22: 230-33.

Isman, M.B. (1999) Pesticides based on plant essential oils. *Pestic. Outlook* 10: 68-72.

Isman, M.B. (2000) Plant essential oils for pest and disease management. *Crop Prot.* 19: 603-8.

Isman, M.B. (2002) Insect antifeedants. *Pestic Outlook* 13:152-57.

Isman, M.B. (2005) Problems and opportunities for the commercialization of botanical insecticides. In: Regnault, C.R., Philogene, B.J.R. and Vincent, C. (ed) *Biopesticides of Plant Origin.* Lavoisier, Paris. Pp 283-91.

Isman, M.B., Matsuura, H., MacKinnon, S., Durst, T., Towers, G.H.N. and Arnason, J.T. (1996) Phytochemistry of the Meliaceae. So many terepenoids, so few insecticides. In: Romeo, J.T., Saunders, J.A. and Barbosa, P. (eds) *Phytochemical Diversity and Redundancy.* pp 155-78, Plenum, New York.

Isman, M.B. (1993) Growth inhibitory and antifeedant effects of azadirachtin on six noctuids of regional economic importance. *Pestic. Sci.* 38:57-63.

Jones, G.A. (2001) Pyrethrum. *Pestic. Outlook* 12: 195-98.

Khumbhar, P.P., Kshama, M.C., Ujwala, P.B., Vidya, N.B., Bendre, R.S. and Dewang, P.M. (2001) Anti-fungal potecy of some common spice extracts. *J Pestology* 25(8):44-46.

Leatemia, J.A. and Isman, M.B. (2004) Efficacy of crude seed extracts of *Annona squamosa* against diamondback moth, *Plutella xylostella* L. in the greenhouse. *Int. J. Pest Manag.* 50:129-33

Lowery, D.T. and Isman, M.B. (1995) Toxicity of neem to natural enemies of aphids. *Phytoparasitica* 23: 297-306.

Maistrello, L., Henderson, G. and Laine, R.A. (2004) Efficacy of vetiver oil and nootkatone as soil barriers against Formoson subterranean termite (Isoptera: Rhinotermitidae) *J. econ. Ent.* 94: 1532-37.

McLaren, J.S. (1986) Biologically active substances from higher plants; status and future potential. *Pestic. Sci.* 17: 359-578.

Metcalf, R.L. and Metcalf, R.R. (1992) *Plant Kairomones in Insect Ecology and Control.* Chapman and Hall, New York, USA.

National Research Council (1992) *Neem. A Tree for Solving Global Problems.* Natl. Acad. Press, Washington, DC.

Naumann, K. and Isman, M.B. (1996) Toxicity of neem (*Azadirachta indica* A. Juss.) seed extracts to larval honeybees and estimation of dangers from field applications. *Am. Bee J.* 136:518-20.

Perry, A.S., Yamamoto, I., Ishaaya, I. and Perry, R.Y. (1998) *Insecticides in Agriculture and Environment: Retrospects and Prospects.* Springer-Verlag, Berlin.

Prakash, A. and Rao, J. (1997) *Botanical Pesticides in Agriculture.* Pp 481. CRC Lewis Publishers, Boca Raton, USA.

Purohit, S.S. and Vyas, S.P. (2004) *Medicinal Plants Cultivation, A Scientific Approach.* Agrobios, Jodhpur, India. Pp 158-170.

Ratnayake, S., Rupprecht, J.K., Potter, W.M. and McLaughlin, J.L. (1992) Evaluation of various parts of the paw paw tree, *Asimina triloba* (Annonaceae) as commercial sources of the pesticidal annonaceous acetogenins. *J. Econ. Entomol.* 85: 2353-56.

Roger, R.C., Philogene, B.J.R. and Vincent, C. (eds) (2005) *Biopesticides of Plant Origin.* Lavoisier, Paris, France.

Schmutterer, H. (ed) (2002) *The Neem Tree and other Meliaceous plants sources of unique Natural Products for Integrated Pest Management, Medicine, Industry and other Purposes.* VCH, Weinheim.

Thacker, J.R.M. (2002) *An Introduction to Arthropod Pest Control.* Cambridge University Press, Cambridge, UK.

Wan, M.T., Watts, R.G., Isman, M.B. and Strub, R. (1996) An evaluation of the acute toxicity to juvenile Pacific northwest salmon of azadirachtin, neem extract and neem-based products. *Bull. Env. Contam. Toxicol.* 56: 432-39.

Theory and Practice of Integrated Pest Management
R. Arora, B. Singh and A.K. Dhawan (Eds)
Scientific Publishers (India), Jodhpur 2012, pp. 135 - 140.

9

PESTICIDE RESISTANCE : MONITORING, MECHANISMS AND MANAGEMENT

B.K. Kang
Department of Entomology
Punjab Agricultural University, Ludhiana - 141 004, India

Pesticide Resistance is one of the principal problems facing crop production, human health and animal protection. Georghiou and Lagunes Tejeda (1991) documented the history of field observations of resistance to pesticides. They stated that the first report of resistance was published by Melander (1914), who described the resistance of orchard pests to sulfur-lime, a compound typical of inorganic chemicals used for pest management one hundred years ago. By 1989, they had counted over 500 arthropod species with strains evolving resistance in the field to toxins used against them. Within this total, 23 beneficial species were included. The resistant pests are categorized as crop pests (59%) and medical or veterinary pests (41%). By 1989, chemicals selecting of resistance included not only the modern classes of organic chemicals (cyclodienes, DDT, organophosphate, carbamate, pyrethorid) but also inorganic and elemental chemicals (e.g., arsenicals, sulfur) commonly used 1940. From the estimated 10,000 arthropod pests, 574 species are reported with resistance to insecticides. (http://www.pesticideresistance.org). Forty four per cent of the cases involve organophosphate pesticides, while organochlorine pesticides are involved in 32% of the cases of resistance (Whalon, 2008).

Resistance to pesticides is the development of an ability in a population of a pest to tolerate doses of toxicants that would prove lethal to the majority of individuals within the same species. The term behaviorstic resistance describes the development of the ability to avoid a dose that would prove lethal. Resistance is distinct from the *natural tolerance* shown by some species of pests. Here a biochemical or physiological property renders the pesticide ineffective against the majority of normal individuals (Mehrotra and Phokola, 2000).

Cross-resistance is a phenomenon whereby a pest population becomes resistant to two or more pesticides as a result of selection by one pesticide only. It must not be confused with *multiple resistance,* which is readily induced in some species with simultaneous or successive exposure to two or more pesticides.

9.1. CURRENT STATUS

Insecticide resistance cases in India were first reported in insect pests of public health importance. The first report came in 1952 when *Culex fatigans*, a transmitter of filaria was found resistant to DDT in Uttar Pradesh and Mumbai. Since then this mosquito has been reported to be resistant to both DDT and HCH in various parts of the country. The resistance in the mosquito *Anopheles stephensi* a malaria transmitter in urban areas to DDT was reported first from Erode, Tamil Nadu in 1956. A year latter *A. culicifacies*, a major vector of malaria in rural areas was reported resistant to DDT from Gujarat. The international pest diamondback moth (DBM), Plutella xylostella has been reported to be resistant to DDT in Punjab in 1968, to endrin in Hissar in 1972 and parathion and fenitrothion in Punjab in 1976. The insecticide resistance in DBM has now extended to synthetic pyrethroids like cypermethrin, fenvalerate, deltamethrin in widespread areas of the country. The pest has also developed resistance to the newer phenyl pyrazole insecticide fipronil. Resistance to these insecticides appears to be widespread and has been reported form Delhi, Panipat (Haryana), Jaunpur (Uttar Pradesh), Ranchi (Jharkhand) and Bangalore (Karnataka). Levels of resistance reported were 40-144 folds against cypermethrin, 40-200 fold against fenvalerate, 96-190 fold against deltamethrin and 80-628 fold against quinalphos (Mehrotra, 1992). Biochemical parameters indicated that the Jassid, *Empoasca kerri* and aphids *Lipaphis erysimi* and *Aphis craccivora* have most probably developed resistance to endosulfan and tolerance to malathion and dimethoate has developed in *Lipaphis erysimi* in some areas in Punjab. Apprearance of pesticide resistance in cotton boll worm, *Helicoverpa armigera* to pyrethroids in India is a serious problem (Kranthi, 2003).

Pesticide resistance in stored-product insects has become an increasingly serious problem which can be judged from the fact that the number of resistant species rose from 3 in 1960 to 17 in 1974. The pesticide resistance in stored grain pests although appeared late, has assumed serious proportions in the country. This is because of the closed environment available in the storage facilities and rapid spread of resistance through the movement of resistant strains of insects from on country to another or from one state to another through trade aggravating the problem throughout India and the world. It is fortunate that *Callosobruchus chinensis*, a serious pest of pulses, has not developed resistance to pesticides so far and is still susceptible.

A large body of literature exists in the rapidly developing field of Bt resistance mechanisms and there implications for resistance management. To date, the only two species reported to have developed resistance in the field to Bt sprays (Bt strain *kurstaki*) have been *P. xylostella* and *Trichoplusia ni*. Bt resistant strains of other species of insects have been observed in the laboratory are mosquitoes, *Heliothus virsescens* and pectinophora gossypiella. Resistance to spinosad has been documented in several insect species. Wide spread resistance to spinosad has been noted in *P. xylostella* in Hawaii and Thailand. Resistance mechanisms to imidacloprid have been observed across multiple insect species including *Nilaparvata lugens, B. tabaci*. Target site insensitively to imidacloprid has only been observed in *N. lugens*. *P. xylostella* has the distinction of being the only known pest species to evolve resistance to Bt insecticidal proteins used under field conditions and has done so in multiple world areas. As a consequence of extensive exposure to insecticides, *B. tabaci* has developed resistance to a wide range of insecticides.

9.2. BIOCHEMICAL MECHANISMS

During the 1950s, an era when biochemical knowledge developed very fast, there was a very strong belief that by finding the biochemical mechanism for resistance, it should be easy to find some substance that counteracted it, for instance, inhibitors of enzymes that detoxicate the pesticide or a new pesticide that shows higher activity toward the resistant insects. The following mechanisms have been postulated for insect resistance to insecticides.

1. Behaviour: Insects may have modified their behaviour so that they avoid the areas sprayed with the insecticide. Such behaviour may be genetically determined.
2. reduced penetration of the pesticide through the cuticle or the intestine.
3. lower transport into the target sites.
4. lowered bioactivation: Some pesticides such as the sulfur-containing organophosphates may often be bioactivated.
5. Increased storage in fat depots or other inert organs.
6. Increased excretion of active ingredients.
7. Increased detoxication or decreased bioactivation.

8. Less sensitive receptors or enzymes that are inactivated or hyperactivated by the pesticides.

Extensive research has shown that points 7 and 8 are almost always involved, but with the other factors playing a modifying or additional role.

9.3. MANAGEMENT STRATEGIES

Many scientists and administrators have addressed this problem, but there is no clear-cut simple method for preventing resistance (Georghiou, 1990). The only method to delay development of resistance is to slow down the evolutionary process, which means providing better reproduction possibilities to susceptible individuals. A rapid influx of (susceptible) immigrants will greatly enhance the chance of the resistant trait being conserved. Once a certain quantity of resistant insects escape extinction, the R-allele can quickly increase in frequency following treatments.

Refuge strategy

The strategy implies that some areas are not treated with pesticides, but are kept as a refuge for susceptible individuals. These insects can then mate without any selection pressure from pesticides. The (few) surviving, more or less resistant, individuals in the nearby treated fields will mate with the individuals from the refuge areas and the resultant offspring will be heterozygotes and susceptible to the pesticide. The refuge strategy has two critical assumptions: that inheritance of resistance is recessive, and that random mating occurs between resistant and susceptible individuals. If resistance is recessive, hybrid first-generation (F_1) offspring produced by mating between S and R-insects are killed by the pesticide. If mating is random, initially rare homozygous R-individuals are likely to mate with more abundant homozygous susceptible insects that have not been exposed to the insecticide. In most cases, where the genetics of resistance have been studied, the resistance alleles are recessive or semidominant. The primary strategy for delaying insect resistance to transgenic *B. thuringiensis* (Bt) cotton plants is to provide such refuges of cotton plants that do not produce Bt toxins.

Mixing pesticides with different modes of action and different detoxication patterns

Two pesticides where the mechanism of resistance is likely to be caused by different genes may be mixed. It is highly unlikely that individuals with resistance to one will also be resistant to the other. Using such high doses of both pesticides such that they will kill heterozygote resistance will probably be prevented when the frequency of R-alleles for both is low, and none of the insects are homozygotes for mechanisms of resistance for both pesticides. Sometimes insecticides with widely different modes of action and structure may give resistance through a common mechanism. One example is parathion-methyl resistance in houseflies, which may be caused by a certain glutathione transferase. This same enzyme, however, is also able to detoxicate lindane. Sequential or rotational use of pesticides will not prevent development of resistance, at least not if the survival of heterozygotes is better than that of normal susceptible ones. It may just make a delay because the selection pressure for each is lower.

Switching life-stage target

In many pests, for instance, in cabbage flies, the larvae, which are the real pest, will probably be much more robust and have a better repertoire of detoxifying enzymes (glutathione transferases) in order to detoxify the isothiocyanates and other natural pesticides present in the host plant (cabbage), while the adults may not eat anything during their short life span and hence do not have detoxication enzymes. The adults therefore will be more sensitive to pesticides and less likely to develop resistance. If pesticides are used against the adult stage and not against the real pest, the larvae, resistance development may be slowed down.

Increased sensitivity in resistant pests

If it were possible to develop two pesticides such that increases resistance to one of them led to increased susceptibility to the other, resistance development would at least be delayed. There is one example of this principle. The systemic fungicide diethofencarb is particularly effective against *Botrytis* spp., which are resistant against benzimidazole. Benzimidazoles like carbendazim and thiophanate bind to a site on the tubulin protein and inhibit mitosis. Resistant *Botrytis* has a tubulin that does not bind benzimidazoles, but may bind diethofencarb better.

Inhibition of detoxication enzymes

After DDT dehydrochlorinase was found as one of the causes of DDT resistance in flies, it was subsequently found that N,N-dibuty1-p-chlorobenzene sulphonamide inhibits the enzyme. This substance, called WARF-Antiresistant, was tried, and although it increased the sensitivity of resistant insects, it was of no durable practical usefulness because either the flies developed other mechanisms of resistance or the amount of DDT dehydrochlorinase increased to overcome the action.

9.4. CONCLUSIONS

Resistance has been a problem in the whole pesticide era and is seen in animals, plants, fungi, and microorganisms. Resistance is caused by evolution – a fundamental biological process. The biochemical and physiological mechanisms may be many, but resistance is often due to an insensitive target for the pesticide or to increased detoxication. It is possible to curb or delay resistance development by making reproduction better for populations or subpopulations of pests that are not under selection pressure. Cross-resistance may occur for pesticides with related modes of action, but sometimes also between nonrelated pesticides if they can be detoxified by the same enzyme.

REFERENCES

Georghiou, G.P. and Lagunes-Tejeda, A. (1991) *The Occurrence of Resistance to Pesticides in Arthropods. An index of Cases Reported Through 1989.* Food and Agriculture Organization of the United Nations, Rome.

Kranthi, K.R. (2003) Managing *Helicoverpa armigera* resistance to insecticides. In: *Proceedings of the National Symposium on Frontier Areas of Entomological Research,* Nov. 5-7, 2003, New Delhi, India, pp. 355-375.

Melander, A.L. (1914) Can insects become resistant to sprays? *J. Econ. Ent.* 7: 167-173.

Mehrotra, K.N. (1992) Pesticide resistance in insect pests – Indian scenario. In: *Pest Management and Pesticides: Indian scenario.* B.V. David (ed.), Namrutha Publications. pp. 17-27.

Mehrotra, K.N. and Phokela, A. (2000) Insecticide resistance in insect pests: current status and future strategies. In: G.S. Dhaliwal and B. Singh (eds). *Pesticides and Environment.* Commonwealth Publishers, New Delhi, pp. 39-85.

Whalon, M., Sanchez, D.M and Holling Worth, M.R. (2008) *Global Pesticide Resistance in Arthropods.* CAB International, Trowbridge, U.K.

Whalon, M., Sanchez, D.M and Duynslager, L. (2003) List of insect pests resistant to pesticides. Internet Edition, http://www.cips.msu.edu/whalonlab/rpmnews/

Theory and Practice of Integrated Pest Management
R. Arora, B. Singh and A.K. Dhawan (Eds)
Scientific Publishers (India), Jodhpur 2012, pp. 141 - 163.

10

RECENT DEVELOPMENTS IN PESTICIDE FORMULATION TECHNOLOGY

P.S Shera and Ramesh Arora*

Department of Plant Breeding & Genetics, *Department of Entomology
Punjab Agricultural University, Ludhiana - 141 004, India

The use of pesticides has been the most common method for combating pest problems in agriculture and public health. Over the last 50 years, pesticides have made an enormous contribution towards increasing the yield and quality of the world's food supply, improving public health and the provision of fibres for clothing and other essential materials (Knowles and Vander, 2008). The biological activity of a pesticide; be it chemical or biological in nature, is determined by its active ingredient (a.i. - also called the *active substance*). Pesticide products very rarely consist of pure technical material. The a.i. is usually formulated with other materials and this is the product as sold, but it may be further diluted in use. The science of formulation technology involves incorporating the active substance into a final product that is the most practical for the user, and which guarantees that a small amount of active substance is distributed evenly over a large treated area. Pesticide chemicals in their "raw" or unformulated state are not usually suitable for pest control. These concentrated chemicals (active ingredients) may not mix well with water, may be chemically unstable, and may be difficult to handle and transport. For these reasons, manufacturers add inert substances such as clays and solvents to improve application effectiveness, safety, handling, and storage. Inert ingredients do not possess pesticidal activity and are added to serve as a carrier for the active ingredient. The mixture of active and inert ingredients is called a "**Pesticide Formulation**". This formulation may consist of:

- The pesticide active ingredient that controls the target pest.
- The carrier, such as organic solvent or mineral clay.
- Surface-active ingredients, such as stickers and spreaders.
- Other ingredients, such as stabilizers, dyes, and chemicals that improve or enhance pesticidal activity.

Abbreviations are often used to describe the formulation (e.g., WP for wettable powders); how the pesticide is used (e.g., TC for termiticide concentrate); or the characteristics of the formulation (e.g., LO for a low-odor formulation). The amount of active ingredient (a.i.) and the kind of formulation are listed on the product label. For example, an 80% SP contains 80 per cent by weight of active ingredient and is a soluble powder (Fishel, 2009).

10.1. OBJECTIVES OF PESTICIDE FORMULATION

Formulation technology is now seen as an "enabling technology" which can provide safe and effective products which are convenient to use. It can also modify the toxicity of active ingredients and improve their ability to target a specific pest. Formulation is seen as an essential part of the total "delivery" system which can also include the packaging and application equipment. At a time when the discovery of new pesticide compounds is more difficult and certainly a high risk and expensive operation, formulation technology can extend the useful patent life of an active ingredient. It can also provide a competitive edge by improving product quality of existing formulations, or by introducing a new formulation of an existing active ingredient. The main objectives of formulation are to provide the user with a convenient, safe product which will not deteriorate over a period of time, and to obtain the maximum activity inherent in the active ingredient (Knowles, 2008).

- Ease of application
- Improved pesticide performance
- Stability of product
- Safety
- Compatibility
- Need based Products
- Economic, efficient and safe use

10.2. FORMULATION TYPES AND CODES

Many different types of formulations have been developed depending largely on the physico-chemical properties of the active ingredients. In recent years, the number of formulation types has increased enormously to meet the needs of operator and environmental safety or to improve the activity and persistence of the active ingredient (Knowles 2008). An international coding system was, therefore, devised by GIFAP (International Group of National Association of Manufacturers of Agrochemical Product) in 1984 (in 1996 GIFAP was renamed GCPF - Global Crop Protection Federation, based in Brussels, Belgium. This organisation is now known as CropLife International listing about 90 different types of formulation. Formulation terminology follows a 2-letter convention: (*e.g.* GR: granules) listed by CropLife International in the *Catalogue of Pesticide Formulation Types and International Coding System* (Monograph 2; 6th Edition) revised in 2008 (CropLife international 2008). The International Coding System for pesticide formulations is now well established, and is recognized by the Food and Agriculture Organization of the United Nations (FAO), the World Health Organization (WHO), and most regulatory agencies worldwide. It has been widely adopted by agrochemical companies. The types of formulations and international codes are shown in Table 10.1.

Table 10.1. Pesticides formulations with International Codes and definitions

Code	**Term**	**Definition**
AE	Aerosol dispenser	A container-held formulation which is dispersed generally by a propellant as fine droplets or particles upon the actuation of a valve.
AL	Any other liquid	A liquid not yet designated by a specific code, to be applied undiluted.
AP	Any other powder	A powder not yet designated by a specific code, to be applied undiluted.
BR	Briquette	Solid block designed for controlled release of active ing redient into water.
CB	Bait concentrate	A solid or liquid intended for dilution before use as bait.
CP	Contact powder	Rodenticidal or insecticidal formulation in powder form for direct application. Formerly known as tracking powder (TP).
CS	Capsule suspension	A stable suspension of capsules in a fluid, normally intended for dilution with water before use.
DC	Dispersible concentrate	A liquid homogeneous formulation to be applied as a solid dispersion after dilution in water. (Note: there are some formula tions which have characteristics intermediate between DC and EC).
DP	Dustable powder	A free-flowing powder suitable for dusting.

Code	Term	Definition
DS	Powder for dry seed treatment	A powder for application in the dry state directly to the seed.
DT	Tablet for direct application	Formulation in the form of tablets to be applied individually and directly in the field, and/or bodies of water, without preparation of a spraying solution or dispersion.
EC	Emulsifiable concentrate	A liquid, homogeneous formulation to be applied as an emulsion after dilution in water.
EG	Emulsifiable granule	A granular formulation, which may contain water-insoluble formulants, to be applied as an oil-in-water emulsion of the active ingredient(s) after disintegration in water.
EO	Emulsion, water in oil	A fluid, heterogeneous formulation consisting of a solution of pesticide in water dispersed as fine globules in a continuous organic liquid phase.
EP	Emulsifiable powder	A powder formulation, which may contain water-insoluble formulants, to be applied as an oil-in-water emulsion of the active ingredient(s) after dispersion in water.
ES	Emulsion for seed treatment	A stable emulsion for application to the seed either directly or after dilution.
EW	Emulsion, oil in water	A fluid, heterogeneous formulation consisting of a solution of pesticide in an organic liquid dispersed as fine globules in a continuous water phase.
FS	Flowable concentrate for seed treatment	A stable suspension for application to the seed, either directly or after dilution.
FU	Smoke generator	A combustible formulation, generally solid, which upon ignition releases the active ingredient(s) in the form of smoke.
GA	Gas	A gas packed in pressure bottle or pressure tank.
GE	Gas generating product	A formulation which generates a gas by chemical reaction
GL	Emulsifiable gel	A gelatinized formulation to be applied as an emulsion in water.
GR	Granules	A free-flowing solid formulation of a defined granule size range ready for use
GS	Grease	Very viscous formulation based on oil or fat.
GW	Water soluble gel	A gelatinized formulation to be applied as an aqueous solution.
HN	Hot fogging concentrate	A formulation suitable for application by hot fogging equipment, either directly or after dilution.
KK	Combi-pack solid/liquid	A solid and a liquid formulation, separately contained within one outer pack, intended for simultaneous application in a tank mix.

Code	Term	Definition
KL	Combi-pack liquid/liquid	Two liquid formulations, separately contained within one outer pack, intended for simultaneous application in a tank mix.
KN	Cold fogging concentrate	A formulation suitable for application by cold fogging equipment, either directly or after dilution.
LN	Long-lasting insecticidal net	A slow-or controlled-release formulation in the form of netting, providing physical and chemical barriers to insects. LN refers to both bulk netting and ready-to-use products, for example mosquito nets.
LS	Solution for seed treatment	A clear to opalescent liquid to be applied to the seed either directly or as a solution of the active ingredient after dilution in water. The liquid may contain water-insoluble formulants.
MC	Mosquito coil	A coil which burns (smoulders) without producing a flame and releases the active ingredient into the local atmosphere as a vapour or smoke.
ME	Micro-emulsion	A clear to opalescent, oil and water containing liquid, to be applied directly or after dilution in water, when it may form a diluted micro-emulsion or a conventional emulsion.
OD	Oil dispersion	A stable suspension of active ingredient(s) in a water-immiscible fluid, which may contain other dissolved active ingredient(s), intended for dilution with water before use.
OF	Oil miscible flowable concentrate	A stable suspension of active ingredient(s) in a fluid intended for dilution in an organic liquid before use.
OL	Oil miscible liquid	A liquid, homogeneous formulation to be applied as a homogeneous liquid after dilution in an organic liquid.
OP	Oil dispersible powder	A powder formulation to be applied as a suspension after dispersion in an organic liquid.
PA	Paste	Water-based, film-forming composition.
PR	Plant rodlet	A small rodlet, usually a few centimetres in length and a few millimetres in diameter, containing an active ingredient.
PS	Seed coated with pesticide	Self defining
RB	Bait (ready for use)	A formulation designed to attract and be eaten by the target pests.
SC	Suspension concentrate	A stable suspension of active ingredient(s) with water as the fluid, intended for dilution with water before use.
SD	Suspension concentrate for direct application	A stable suspension of active ingredient(s) in a fluid, which may contain other dissolved active ingredient(s), intended for direct application, to rice paddies, for example.

Code	Term	Definition
SE	Suspo-emulsion	A fluid, heterogeneous formulation consisting of a stable dispersion of active ingredient(s) in the form of solid particles and of water-non miscible fine globules in a continuous water phase.
SG	Water soluble granule	A formulation consisting of granules to be applied as a true solution of the active ingredient after dissolution in water, but which may contain insoluble inert ingredients.
SL	Soluble concentrates	A clear to opalescent liquid to be applied as a solution of the active ingredient after dilution in water. The liquid may contain water-insoluble formulants.
SO	Spreading oil	Formulation designed to form a surface layer on application to water.
SP	Water soluble powder	A powder formulation to be applied as a true solution of the active ingredient after dissolution in water, but which may contain insoluble inert ingredients.
ST	Water soluble tablet	Formulation in form of tablets to be used individually, to form a solution of the active ingredient after disintegration in water. The formulation may contain water-insoluble formulants.
SU	Ultra-low volume (ULV) suspension	A suspension ready for use through ULV equipment.
TB	Tablet	Pre-formed solids of uniform shape and dimensions, usually circular, with either flat or convex faces, the distance between faces being less than the diameter.
TC	Technical material	A material resulting from a manufacturing process comprising the active ingredient, together with associated impurities. This may contain small amounts of necessary additives.
TK	Technical concentrate	A material resulting from a manufacturing process comprising the active ingredient, together with associated impurities. This may contain small amounts of necessary additives and appropriate diluents.
UL	Ultra-low volume (ULV) liquid	A homogeneous liquid ready for use through ULV equipment.
VP	Vapour releasing product	A formulation containing one or more volatile active ingredients, the vapours of which are released into the air. Evaporation rate is normally controlled by using suitable formulations and/or dispensers.
WG	Water dispersible granules	A formulation consisting of granules to be applied after disintegration and dispersion in water.
WP	Wettable powder	A powder formulation to be applied as a suspension after dispersion in water.

Code	Term	Definition
WS	Water dispe-rsible powder for slurry seed treatment	A powder to be dispersed at high concentration in water before application as slurry to the seed.
WT	Water dispersible tablet	Formulation in the form of tablets to be used individually, to form a dispersion of the active ingredient after disintegration in water.
XX	Others	Temporary categorization of all other formulations not listed above.
ZC	A mixed form-ulation of CS and SC	A stable suspension of capsules and active ingredient(s) in fluid, normally intended for dilution with water before use.
ZE	A mixed formulation of CS and SE	A fluid, heterogeneous formulation consisting of a stable dispersion of active ingredient(s) in the form of capsules, solid particles, and fine globules in a continuous water phase, normally intended for dilution with water before use.
ZW	A mixed form-ulation of CS and EW	A fluid, heterogeneous formulation consisting of a stable-dispersion of active ingredient(s) in the form of capsules and fine globules in a continuous water phase, normally intended for dilution with water before use.

Source : *www.CropLife International.org*

10.3. CURRENT TRENDS IN PESTICIDE FORMULATION TECHNOLOGY

The early conventional formulations were simple solutions in water (SL), emulsifiable concentrates (EC) in a non-aqueous solvent or dusts (DP) and wettable powders (WP). The presence of petroleum-based solvents in EC formulations and dusty powders in DP and WP formulations can lead to safety hazards in use and a negative impact on the environment generally. These types of formulations are regarded now as 'old technology'. Increasing attention now is being paid to develop safer and more environmentally friendly pesticide formulations in more convenient packaging. The current trends (Knowles, 2008) in the development of pesticide formulations can be summarized as follows:

- To eliminate solvents wherever possible and use aqueous emulsions or microemulsions
- To replace dusty powders by suspension concentrates or water dispersible granules.
- To develop multiple active ingredient formulations where appropriate.
- To build into the formulation bioenhancing surfactant adjuvants.

- To control release rate and targeting of pesticides by encapsulation techniques or seed treatment applications.
- To develop novel formulations such as tablets or gels.
- To develop more effective spray tank adjuvants to enhance biological activity and reduce pesticide dosage.
- Specialized delivery system with no or very low operator exposure e.g. gel, tablets, powders or granules in water soluble sachets, multiple emulsions.
- Use of inert formulation ingredients which have low mammalian and aquatic toxicity, and low environment impact.

10.4. NEW GENERATION PESTICIDE FORMULATIONS

The recent developments in the formulation technology (new generation formulation) include

- Suspension Concentrate (SC)
- Non-aqueous Suspension Concentrate (OD, OF)
- Suspo-emulsions (SE)
- o/w emulsions (concentrated emulsions) (EW)
- Microemulsions (ME)
- Microcapsules (capsule suspensions) (CS) - Controlled released formulations
- Water dispersible granules (WG)
- Seed treatment formulations (DS, WS, LS, FS)
- Gel formulations (GW, GL)

10.4.1. Suspension concentrates (SC) or Wet Flowables or Flowable concentrate

Suspension concentrates are a type of formulation in which a solid active substance is finely dispersed in water. In SC-formulations, the active substance is present as a solid that is finely distributed in particles of between 1 and 4 μm in size. The particles are held in suspension in the water-based medium through the addition of so-called dispersing agents; this effect involves mutual repellency between the particles; and in combination with other substances that form net-like structures, the dispersing agents inhibiting the re-agglomeration of the suspended active substance.

Advantages

- High concentration of insoluble active ingredients

- Ease of handling and application
- Safety to the operator and environment
- Relatively low cost
- Enable water-soluble adjuvants to be built-in for enhanced biological activity.
- Water-based liquid, non flammable
- Low cost of formulation
- Can incorporate surfactant adjuvants for further biological enhancement.

Farmers generally prefer suspension concentrates to wettable powders (WP) because they are non-dusty and easy to measure and pour into the spray tank. Because they are water-based, suspension concentrate formulations offer many advantages such as ease of pouring and measuring, safety to the operator and environment, and economy. The fact that the active substance is present in crystalline form also means that its biological availability is lower than with EC, EW and SL formulations. This is why SCs are most suitable when a high degree of contact exposure of the host organism is the main aim.

Disadvantages

- Expertise required to develop a good stable formulation
- Formulation stability can be sensitive to minor changes in raw material quality
- Formulations need to be protected from freezing

One disadvantage though is that the product's high viscosity generally makes it more difficult to empty the packaging completely. The formulation physical stability problems of SC formulations are shown graphically below:

10.4.2. Emulsions (EW) - O/W — Oil in water emulsion

Emulsion formulations can be either oil-in-water (EW) or water-in-oil (EO). Oil-in-water emulsions (EW) are more useful for agrochemicals, and are becoming increasingly available as alternatives to emulsifiable concentrates because of the need to reduce or eliminate volatile organic solvents. They may also be safer to use and have reduced dermal toxicity because they are water based. However, they require careful selection of surfactant emulsifiers and stabilizers to prevent flocculation, creaming and coalescence of the emulsion droplets. Non-ionic surfactants, block copolymers and polymeric surfactants are now being used to produce stable emulsions. Emulsion formulations are most suitable for oily or semi-solid active ingredients and can sometimes be made without the addition of any solvent. However, it is more usual to add a small amount of solvent to reduce the viscosity of the active ingredient or to increase the bulk volume of the dispersed oil phase.

Advantages of EW over ECs

- Total or partial elimination of petroleum based solvents (VOC's)
- Non flammable or very low flammability
- Low dermal toxicity
- Non phytotoxic
- Safety in manufacture and use
- Low formulation cost

Disadvantages

The main formulation problem with EW formulations is to ensure long term stability by preventing settling, coalescence and creaming of the oily dispersed phase.

10.4.3. Suspo-Emulsions (SE)

Mixed combination formulations are becoming more popular because of their convenience, they ensure that the farmer applies the correct amount of each component pesticide and overcome problems of tank mix incompatibility. Suspoemulsions can, therefore, be considered to be mixtures of suspension concentrates and oil-in-water emulsions with added surfactants to prevent flocculation and thickeners to prevent separation of the dispersed phases. Surfactants used as dispersing agents for the solid phase are similar to those already mentioned for suspension concentrates.

Emulsifiers for the oily liquid phase are similar to those used for oil-in-water emulsions. As these formulations are aqueous based and generally thickened. It is a water-based product that contains both suspended solids (like an SC) and emulsion droplets (like an EC after dilution). They are designed to allow the widest possible combination of active ingredients in a single formulation, particularly where one active ingredient is a solid and the other active ingredient is a liquid. A three-way mixed formulation is possible if one of the active ingredients is soluble in water. If one active ingredient is a solid and the other is a liquid, it is necessary to produce a suspoemulsion formulation. These mixed formulations are becoming more popular because of their convenience and their broad three phases:

- Insoluble solid phase as dispersed particles
- Insoluble liquid phase as oil droplets
- Water as bulk phase

However, it is not usually possible to make a suspo-emulsion by simply mixing a suspension concentrate SC formulation with an oil-in-water EW formulation because the correct balance of dispersing and emulsifying agents may not be achieved. This may lead to preferential adsorption on one surface or the other, resulting in heteroflocculation problems. Considerable development work may be required to determine the optimum type and concentration of dispersing and emulsifying agents to prepare a stable suspoemulsion formulation.

Advantages

- Multiple solid and liquid AI's in a single formulation
- Water based —no organic solvent VOC's in external bulk phase
- Ready mixed formulation
- Easy to pour and measure
- Low flammability
- Possible reduction in skin and eye toxicity
- Avoids tank mix compatibility problems
- May be reduced cost due to low or no solvent usage

Disadvantages

- Sometimes difficult and time consuming to develop a high quality stable formulation
- Process development work required
- Pack rinsing and disposal problems similar to SC formulations

The following problems of physical instability can occur during storage, especially at elevated temperatures:

- Separation of solid particles and oil droplets
- Coalescence of oil droplets
- Heteroflocculation between particles and droplets
- Crystal growth of particles
- Phase transfer to water-in-oil

Most of these instability problems can be largely overcome by using polymeric surfactants, such as the Uniqema Atlox range, which adsorb strongly onto the particle and droplet surfaces preventing them from coming too close together.

10.4.4. Water dispersible granules (WDG)

Water dispersible granules, or dry flowables as they are sometimes known, are a relatively new type of formulation and are being developed as safer and more commercially attractive alternatives to wettable powders and suspension concentrates. They are becoming more popular because of their convenience in packaging and use, being non-dusty, free-flowing granules which should disperse quickly when added to water in the spray tank. They therefore, represent a technological improvement over wettable powders and imitate liquids in their handling characteristics. They can be packed into paper bags or cartons with the minimum of contamination and pack disposal problems (Knowles, 2005). They are like wettable powders except instead of being dust like, they are formulated as small, easily measured granules. Water-dispersible granules must be mixed with water to be applied. Once in water, the granules break apart into fine particles similar to wettable powders. The formulation requires constant agitation to keep them suspended in water. The percentage of active ingredient is high, often as much as 90 percent by weight (Fishel, 2009)

Water dispersible granules usually contain a wetting agent and a dispersing agent in the same way as a wettable powder or a suspension concentrate, but the dispersing agent is usually at a higher concentration. They may also contain a water soluble salt to act as a disintegrant in the spray tank. The remainder of the formulation is usually water soluble or water dispersible filler.

Advantages

- Being free flowing, the granules can be easily, quickly and completely transferred from the package in the spray tank,
- There is insignificant quantity of dust during handling and slurry preparation which avoids hazards to workers and their surroundings,
- Spillage does not lead to clinging of the granules to the clothings and spilled material can be readily swept,
- The packed granules can be readily and safely quantified before use,
- The granules can even be packed in standard water soluble packs, completely avoiding the problem of disposal of container,
- The dispersions obtained from such granules are often better than those obtained from Water dispersible powder,
- The transport and storage is economic and safe,
- Problems like settling, thickening etc associated with wet flowables are missing in dry flowables,
- If required, these granules can also be used like ordinary granules.

Disadvantages

- High initial capital expenditure of plant,
- Possible higher production costs than for liquids per kg of product,
- Possible low bulk density of product,
- Product quality may be sensitive to process and raw material variations,
- Re-cycling off-size material needs to be considered,
- Expensive plant to contain toxic and/or explosive dusts.

Water-dispersible granules share many of the same advantages and disadvantages of wettable powders except:

- They are more easily measured and mixed,
- Because of low dust, they cause less inhalation hazard to the applicator during handling.

10.4.5. Microemulsions (ME)

A microemulsifiable concentrate is analogous to an EC, in that it is solvent-based and contains dissolved active ingredients, but unlike an EC, an MEC forms a *microemul*sion - not an emulsion - upon dilution in water (Cush, 2006). Microemulsions are water based formulations and are thermodynamically stable (syn. soluble concentrate) over a wide temperature range due to their very fine droplet size, usually between 0.05-0.1 microns (50-100 nanometers). The very small droplet size also makes these formulations transparent, so that they are sometimes regarded as solubilised micellar solutions (Knowles, 2008). They generally consist of four components:

- Liquid active ingredient or solid dissolved in a solvent (preferably water immiscible)
- Surfactant solubilisers (blend)
- Co-surfactant
- Water

Often a co-surfactant such as a medium chain aliphatic alcohol is needed. In the case of microemulsions there is only one phase containing relatively large swollen micelles of surfactant in which the oil and the active ingredient are dissolved or solubilised. For this reason micro-emulsions appear transparent.

Advantages

- Water based transparent novel formulation type
- Good water dilution and application properties

- Packs and equipment easy to clean
- Long shelf life stability
- Low flammability due to low solvent content in a continuous water phase
- Possible bioenhancement due to better penetration or uptake of the active ingredient resulting from the high solubilising power of the surfactants.

Disadvantages

- Low active ingredient content, usually below 30%
- High concentration and cost of surfactants
- Limited number of suitable surfactant systems for active ingredients
- Penetrating formulation may present phytotoxicity and handler toxicity issues

10.4.6. Non-aqueous Suspension Concentrates (NSC)

These are concentrated dispersions of active ingredient in non-aqueous organic liquids, especially water-immiscible aliphatic hydrocarbons or vegetable oil-based liquids. As for aqueous suspension concentrates, active ingredients must satisfy the following criteria:

- High melting point
- Low solubility in the organic liquid
- Chemically stable in the organic liquid.

Advantages

- Good bioefficacy and rainfastness
- Useful for active ingredients insoluble in organic solvents or chemically unstable in water
- Low toxicity of organic liquids used as dispersing media
- Bioenhancing adjuvant system can be used as the dispersing medium

Disadvantages

- Fairly complex formulation if water-dilution required, requiring long lead times to develop a satisfactory formulation. Dispersing agent/emulsifier/active ingredient interactions may affect quality.
- Formulation can be sensitive to changes in active ingredient quality

There are many active ingredients formulated as nonaqueous dispersions. They can be divided as follows:

a. Oil-miscible flowable suspension (OF). This is a concentrated nonaqueous suspension which is designed for dilution in an organic liquid before use in, for example, low volume, ultra-low volume, controlled droplet or thermal fogging application. Since the formulation is not designed for subsequent dilution in water, no emulsifiers are needed.

b. Ultra-low volume suspension (SU). These formulations are nonaqueous suspensions for direct ultra-low volume application.

c. Oil Dispersion (OD). These formulations are non-aqueous dispersions intended for dilution into water before use, and represent the most complex of the non-aqueous suspension formulations.

10.4.7. Controlled Release Formulations

Controlled release technology has been receiving increasing attention in pesticide formulation because it has a large number of possible advantages over conventional pesticide formulations. Controlled release is a technique or method whereby active ingredients are made available to a specified target at a certain concentration and duration to produce an intended effect (Mulqueen, 2003).

Controlled-release formulations can be designed to:

- Extend the activity of the active ingredient. The rate of release of the active ingredient depends primarily on the rate of diffusion out of the capsule matrix and is a function of the formulation ingredients, the degree of capsule thickness and strength achieved during manufacture.
- Reduce evaporative losses caused by the action of wind and weather.
- Reduce pesticide levels in the environment. Most pesticide sprayed or incorporated in the soil is wasted. Controlled-release formulations use a much lower concentration of active ingredients and might reduce the damage to the environment and to non-targeted plants and animals.
- Reduce phytotoxicity with application of lower amounts of herbicide.
- Protect pesticides from environmental degradation caused by the action of sunlight, bacteria, wind and water.
- Reduce leaching in the soil.
- Reduce mammalian toxicity to those handling the product.
- Reduce solvent levels in a formulation.

- Reduce formulation odour.
- Improve formulation rainfastness.

Controlled release pesticide formulations can be divided into four main types:

- Coated pesticide granules
- Matrix systems containing physically trapped pesticides
- Polymer systems containing covalently bound pesticides
- Polymer membrane-pesticide reservoir systems, e.g. microencapsulation (CS)

The polymer membrane, or microencapsulation technique, has become popular in recent years. A well known method of microencapsulation uses the principle of interfacial polymerisation.

a. Microencapsulation (Capsule Suspension - CS). Microencapsulated formulations comprise products where the biologically active material is surrounded by a barrier or membrane of polymeric or similar material in order to give controlled or delayed release of active ingredient. This treatment often also gives benefits in operator and environmental safety and tank mix compatibility (Knowles, 2008). It is composed of small spherical particles (typically 1- 20 μm in diameter) composed of a core material of active ingredient and an outer shell wall of polymeric material are suspended in water.

b. Microcapsulation Release. The rate of release of the active ingredient can be controlled by adjusting the droplet size, the thickness of the polymer membrane and the degree of cross-linking or porosity of the polymer. The rate of release of the pesticide is, therefore, a diffusion controlled process (Perrin, 2000).

To be biologically effective, after spraying and deposition on the target species the active ingredient must be released and this may be achieved in a number of ways:

- Wall collapse / rupture on dry-down of spray deposit (especially if small capsule size).
- Mechanical rupture e.g. by crawling or feeding insect.
- Degradation of polymer wall by, for example, light, heat, pH.
- Diffusion through the capsule wall.

Release rate of active ingredient is governed by the following :

- Physical properties of the encapsulated active ingredient.
- Nature of wall material.
- Surface area of microcapsule.
- Wall thickness.

- Degree of crosslinking of capsule wall.
- Diffusion coefficient.
- Distribution coefficient.
- Permeability.
- Active ingredient concentration gradient across wall.

Advantages

- Controlled release of active ingredient can give greater persistence of effect.
- Opportunity for delayed or triggered release of active ingredient.
- Improved handler safety since active ingredient is enclosed by the microcapsule wall.
- Improved environmental effects e.g. leaching.
- Degradation of active ingredient in the environment (for example by UV light) can be reduced.
- Objectionable odours can be masked.
- Losses by vapour activity are reduced, where this is an issue.
- Phytotoxicity effects reduced.
- Can improve stability of multi-component suspensions by acting as a barrier between active ingredients where these are physically or chemically incompatible.

Disadvantages

- The process works best for liquid active ingredients or those readily soluble in suitable solvents.
- Physical or chemical incompatibility between active ingredient and monomer.
- Loss of, or reduced, bioefficacy of active ingredient due to reduced availability and uptake, or inappropriate release profile.
- Inherent physical instability of these multiphase systems on long-term storage, as for other aqueous colloidal dispersions, particularly with respect to separation and sedimentation ('claying') in containers. Any sedimentation will tend to be hard-packed, due to the essentially spherical form of the particles.

c. Examples

Syngenta, Dow AgroSciences and Monsanto appear to be in the lead with microencapsulation technology.

- Syngenta has gained a new lease of life for the insecticide, lambda-cyhalothrin, using its patented "Zeon technology" which encapsulates the active ingredient in very small capsules with thin walls. This enables quick "knockdown" of insects coupled with long-term persistence. Hallmark Zeon is safe to beneficial insects and is the only pyrethroid insecticide endorsed by the British Beekeepers' Association.
- Syngenta has also overcome toxicity problems with the insecticide tefluthrin by producing a capsule suspension as a seed treatment formulation.
- In a similar way, Monsanto has developed a safer capsule suspension of the maize herbicide acetochlor.
- Dow Agro is able to encapsulate chlorpyrifos as an emulsion and then convert this into a water dispersible granule by fluid bed spray granulation.
- Bayer CropScience have a new technology for coating solid particles with a polyurea/urethane coating.
- DuPont are looking at liquid wax coating techniques. Further innovations are expected in microencapsulation technology over the next few years that will contribute to safer pesticide use.

10.4.8. Gel Formulations

Gels are structured or thickened solvent-based emulsifiable concentrates, aqueous solutions or dispersions, and are designed to be re-dispersed into water or used undiluted as seed treatments and contact baits.

a. Gels for dilution in water — Water Soluble Packets. These are Emulsifiable gel (GL) and Water-soluble gel (GW) formulations, usually thickened emulsifiable concentrate formulations (EC's) to enable packaging in water-soluble bags, which are cast from, for example, polyvinyl alcohol. The concept was developed to:

(a) overcome safety and handling issues associated with solvents in EC's, and

(b) overcome pack disposal issues.

Thickening/gelling of the formulation is carried out by using acrylate-based polymers or silica and such products are less liable to leak through 'pinholes' in the water-soluble bags.

Gel formulations in water-soluble bags have the following advantages:

- Safety in handling compared to emulsifiable concentrates, and
- Fewer issues of pack disposal.

Main disadvantages include the following:

- Possible pack degradation,
- Leakage through pack, and
- Slow or incomplete dispersion in water.

Manufacturers package precise amounts of wettable powder or soluble powder formulations in a special type of plastic bag. Moreover, packing oil based products as gels has become an interesting method of reducing packaging waste on selected formulations. In this method, non-aqueous liquid formulations are gelled and then are suitable for packaging within polyvinyl alcohol film as water-soluble sachets or bags. Gelation of liquid formulations can be brought about by thickening agents such as polyacrylic acids, xanthan gum, silicas, clays, surfactants and combinations thereof. In addition to water of a soluble bag containing such gels, the soluble bag releases its contents within 1 min, the gel disperses homogeneously and the sachet film dissolves completely within 3 min (Malqueen, 2003).

Both types of formulations are packaged in watersoluble pouches, with product and pouch enclosed in a moisture-proof bag or carton. The applicator tears off the outside protective cover and places the water-soluble bag into the spray tank. The bag dissolves and releases the dry or gel formulation into the water. There are no risks of inhaling or contacting the undiluted pesticide as long as you do not open the packets. Once mixed with water, however, pesticides packaged in water-soluble packets are no safer than other diluted pesticides (Fishel, 2009). The benefits of water-soluble packaging include limited exposure to the concentrated pesticides, elimination of the container rinsing process, enhanced emergency spill response, and reduction of the amount of waste placed in landfills.

b. Gels as seed treatments. These formulations are designed to be applied to seed either undiluted or after dilution with a small quantity of water. An example is a combined tebuconazole / thiram fungicidal gel seed treatment ('Raxil' Gel 206) produced by Bayer CropScience.

c. Gels applied undiluted. These formulations are either applied undiluted to water or target species (mainly herbicides) or are used as contact treatment against rodents or crawling insect pests. 'Midstream' gel is used for localised aquatic weed control. The gel is slowly soluble in water and prevents too rapid a diffusion rate of active ingredient. Contact gels or baits include products where the gel physical form has a specific advantage and include use as rodenticides and control of crawling insect pests. In these cases gel formulations can be conveniently dispensed by syringe into cracks and crevices where access is difficult and activity over prolonged periods is required.

10.4.9. Seed Treatment Formulations

Traditionally, seed treatment was used to control seed- and soil-borne diseases and insects. Recently, new generation fungicides and insecticides with strong systemic properties have been commercialised that provide control of early foliar diseases and insect pests

a. Seed treatment formulation types. In the past, the majority of treatments were powders (DS) that were simply dry mixed with the seed, or powders that were slurried (WS) and then sprayed onto the seed before planting. The current trend is towards flowable suspensions (FS) and emulsions (ES) to reduce dust hazards, improve formulation stability and increase adhesion on the seed. Very occasionally solvent-based formulations (LS) are applied. Microcapsule formulations (CF) are also used. The dustiness of water slurriable powder seed treatments can be overcome by converting the formulation into a water dispersible granule (WG). For special seed applications gel formulations (GF) are sometimes used. The exact formulation requirements will depend on the products being applied, the target seed, the nature of the application and the equipment used for the application.

b. Water slurriable powders (WS). Water slurriable powders for seed treatment are similar to wettable powder formulations in that they contain wetting and dispersing agents so that the powder can be made easily into a slurry in water for application to the seed. Examples of surfactants used are lignosulphonates and aliphatic alcohol ethoxylates. However, they also contain a flocculating agent such as a polyphosphates which prevent the slurry of pesticide particles from settling too quickly during the seed treatment operation. They also contain a red dye or pigment as a safety marker for the dressed seed. Water slurriable formulations are popular in Europe, especially in France for fungicide applications.

The advantages of WS formulations are:

- Easy to produce
- Good storage stability
- Water dilutable so easy plant cleaning

The main disadvantage of WS formulations is that they can be messy when making up the slurry in water and the slurry requires constant stirring during application to the seed.

c. Liquid solution seed treatments (LS). Liquid solution formulations contain active ingredients and additives dissolved in organic solvents such as propylene glycol or glycol ethers. N-methyl pyrrolidone (NMP) was a commonly used solvent, but is now being withdrawn due to toxicological concerns. LS formulations also contain a solvent soluble red dye as a safety marker for dressed seed.

Advantages

- Easy to produce for suitable active ingredients
- Good storage stability
- Good adhesion on seeds

Disadvantages

- Toxicity problems with solvents
- Solvents may cause some germination problems
- Solvents may attack some seed treatment machinery plastic or rubber parts
- Products cannot be diluted in water
- Needs solvents to clean seed treatment machinery

Because of the general move away from VOC (volatile organic compounds) containing products, LS formulations are of limited use in a small number of countries, and their use is likely to decline. Furthermore, according to the EPA in the USA, NMP is no longer available for use in pesticide formulations due to potential human toxicity problems.

d. Flowable seed treatments (FS). Flowable concentrates for seed treatment are ready for application (RFA) products, which are usually applied by pumping the liquid suspension formulation directly onto the seed. These formulations are very similar to suspension concentrates (SC) and very often contain similar dispersing agents such as:

- Sodium lignosulphonates
- Sodium naphthalene sulphonate-formaldehyde condensates
- Aliphatic alcohol ethoxylates
- Trystyrylphenol ethoxylate phosphate esters
- EO/PO block copolymers

They also contain a red pigment as a safety marker on dressed seed.

Flowable concentrates require careful selection of gelling and thickening agents to control viscosity to prevent separation of particles, whilst at the same time having low enough viscosity to be easily pumpable directly onto the seed, even at low temperatures.

Advantages

- Water based and water dilutable
- Low potential for germination problems
- Good retention on seed

- No powder dusting problems
- Easy to clean up seed treatment machinery

Disadvantages

Require lengthy formulation development

Storage stability may be affected by temperature extremes

High loadings may cause stickiness and poor flow properties of seed

Flowable seed treatments have now become the most popular formulation type in Europe because they are concentrated formulations and safe to apply because they are water based.

e. Emulsion seed treatments (ES). These products are oil-in-water emulsion formulations (similar to EW formulations).

f. Microcapsule seed treatments (CF). A modification of flowable emulsion seed treatment formulations is to convert the emulsion droplets into capsules by microencapsulation techniques described in an earlier chapter. This can be used to reduce the operator handling exposure to active ingredients which may have potential skin irritancy problems. Syngenta have successfully produced an example of this type of product with Force ST (also known as Evict), which contains a capsule suspension of tefluthrin. These products are also available from Bayer CropScience.

g. Water dispersible granule seed treatments (WG). In order to overcome the problems of dustiness and sedimentation with water slurriable powder formulations (WS), a small number of powders have been converted into water dispersible granules (WG).

10.5. CONCLUSIONS

Increasing attention is being paid to developing safer and more environmentally friendly pesticide formulations in more convenient packaging. The early conventional formulations were simple solutions in water (SL), emulsifiable concentrates (EC) in a non-aqueous solvent or dusts (DP) and wettable powders (WP). The presence of petroleum-based solvents in EC formulations and dusty powders in DP and WP formulations can lead to safety hazards in use and a negative impact on the environment generally. These types of formulations are regarded now as 'old technology'. Hence, there is a growing demand for the formulations which are cleaner and safer for the user, have minimal impact on the environment, and can be applied at the lowest dose rate user and environment friendly. The current trends, therefore, are to replace petroleum-based solvents as much as possible with water to give water-based suspensions and emulsion formulations. This has led to the development of water-based liquid formulations such as suspension concentrates (SC), oil-in water emulsions (EW), suspoemulsions (SE) and microcapsules (CS). There has also been a move away from dusty powders towards water dispersible or soluble granules (WG/SG). Controlled release formulations are being developed to improve safe handing of some toxic active ingredients, or to extend the persistence of activity on the crop. Microencapsulation has the advantage of offering controlled or delayed release of active ingredient, often using biodegradable polymers to achieve the required effect. Water soluble packaging have added advantage for safe handling of solid formulations, particularly wettable powders, and also for gelled or thickened solvent based formulations. These developments in formulation technology and other novel formulation types like water-soluble packs, can give products a competitive advantage, add value or extend the life-cycle of active ingredients. These types of formulations are regarded as 'new technology' and, therefore, have the potential to reduce significantly the total environmental impact. While much progress has been made towards new technology formulations in the developed countries like USA, Europe and Japan and China however, pesticides in the developing countries still now are mainly available in conventional formulations.

REFERENCES

CropLife international (2008) *Catalogue of Pesticide Formulation types and International Coding System.* Technical Monograph No. 2, 6th Edition, CropLife International, Brussels, Belgium.

Cush, R. (2006) *Back to Basics: A Review of pesticide formulation types.* Syngenta Crop Protection Inc. Greensboro, N.C.

Fishel, F.M. (2009) *Pesticide Formulations.* UF/IFAS Publication PI-231, Florida Cooperative Extension Service, Institute of Food and Agricultural Sciences, University of Florida, Gainesville, FL.

Knowles, D.A. (1998) *Chemistry and Technology of Agrochemical Formulations.* Kluwer Academic Publishers, Netherland.

Knowles, D.A. (2005) *New Developments in Crop Protection Product Formulation.* AGROW Reports. T & F Informa UK. 349 pp (www.agrowreports.com)

Knowles, D.A. (2008) Recent developments of safer formulations of agrochemicals. *Environmentalist* **28**: 35-44.

Knowles, D.A. and Vander, S. (2008) Introduction. *Environmentalist* **28** : 3-4.

Mulqueen, P. (2003) Recent advances in agrochemical formulations. Advances in Colloid and Interface. *Science* **106**: 83-107.

Parmar, B.S. and Tomar, S.S. (2004) *Pesticide Formulation: Theory and Practice.* CBS Publishers & Distributors, New Delhi. 414 pp.

Perrin, B. (2000) Improving insecticides through encapsulation. *Pestic. Outlook.* **11** : 68-71.

Theory and Practice of Integrated Pest Management
R. Arora, B. Singh and A.K. Dhawan (Eds)
Scientific Publishers (India), Jodhpur 2012, pp. 164 - 194.

11

PESTICIDE APPLICATION EQUIPMENT

I.P.S. Sandhu
Department of Entomology
Punjab Agricultural University, Ludhiana - 141 004, India

11.1. INTRODUCTION

Pesticide use is an important aspect of the modern agriculture. Cultivated plants are beset by insects, fungus and virus parasites, as well as weeds, which are not favourable to their growth. Weeds can be eradicated by effective cultivation, but pests and diseases have to be kept under control with chemical spray and powder (dust) application. Many of the modern pesticides are very potent. As low as 10-12 g of active ingredient of certain pesticides are sufficient to kill the pest in an acre moreover chemicals are being used to kill weeds without damaging the crop in addition to the weeding tolls. Chemicals are either sprayed in liquid form or applied as dry powder. Now-a-days many sprayers and dusters are available in many forms for this purpose. To protect the crop farmers use a large quantity of pesticide than actually required. This causes high loss of pesticide. Nearly one-third of the pesticide is applied to the crops and requires more time for its application. This also adds to health hazards as well as ecological and environmental destruction. It is, therefore, necessary to select the most efficient equipment for securing a uniform deposit of pesticide on the substrate (leaf surface, soil, walls, ceilings and floors or enclosed space) without wastage of material, in least time, with minimum labour and maximum pest control.

Prior to 1860's, there was little knowledge regarding chemicals for the control of plant pests, and hence there was no need for application equipments. Famines caused by insect outbreaks and epiphytotics of plant diseases had come and gone but no effective control measures were known to prevent the catastrophes. In 1860's the Colorado potato beetle assumed great economic importance and threatened the sowing of potatoes which constituted one of the staple foods in the United States. Paris green diluted in lime water was found to be an effective remedy. It was applied by means of a box with wire-screen bottom or a loosely woven cloth bag, either one of which was filled with the mixture and shaken over the leaves.

Meanwhile, grape-vine mildew had spread to nearly all grape-vine growing areas of France and also inflicted severe losses in parts of Germany and Italy. In 1882, Millardet made his historic observations on the effectiveness of copper sulphate-lime combination (Bordeaux mixture). This was applied by dipping a brush in the spray fluid and then swinging it at the vines for splattering the foliage with the spray mixture from the brush.

Early in the 1940's, some very promising chemicals were developed. The plant protectionists obtained a good deal of information on the biology and ecology of the pest organisms in a better position to use these pesticides more effectively. The engineers were, however, required to make use of their technical skill and practical experience gained primarily in other fields, to fabricate appliances for applying these new chemicals most efficiently. The development of machines for the application of pesticides followed hand in hand with the development of pesticides that could give effective insect and disease control at rather low dosages.

Thus, from the discovery about 100 years back of bundle of twigs, a feather or a brush-broom for the application of pesticidal sprays and use of hands clothes and blow tubes for application of dusts and later the invention of simple syringe-type sprayer was followed by various types of hand and foot-operated high volume sprayer. By the end of the Second World War considerable progress had been made, and it was possible to apply small quantities of pesticides uniformly over large areas. The major revolution in spraying techniques came in 1951 with the development of motorized knapsack sprayers. However, it is only recently that development of plant protection equipment has made phenomenal progress in a much broader field of application, particularly in regard to low pressure and low volume spraying of field crops. It resulted in very high gain to agriculture. The makes, models, types and sizes of plant protection equipment that are available today are so numerous as to bewilder the inexperienced buyer. Manufacturers offer a wide range of selection for different conditions and uses.

Conventional knapsack sprayer which is most commonly used by the farmers require 400-600 litre/ha of the spray solution. This requirement puts considerable hardship on a farmer who has to fetch clean water sometime from a long distance. Hence, a reduction in the volume of water needed for spraying will mitigate some of the problems faced by the farmer. Low and ultra-low volume spraying techniques have reduced the spray volume requirement considerably.

Generally, in many of the pest control situations, a discontinuous but a most thorough and adequate cover of plants is desirable. Adequate cover means reasonable distribution of certain minimum level of deposit. Density of droplets per unit of the area, droplet size and the amount of chemical they carry are all important. Both the nature of the pest and the mode of action of the pesticide determine the requirements for the degree of coverage needed in effective pest control. e.g. aphids and scale insects which moves very less, requires a continuous cover of the plant if contact insecticides are used. So the equipment must be adequately powered to force the spray to all the parts of the plant. Systemic pesticides which are absorbed and distributed to various parts of the plant require only a moderate dispersion. Insecticides such as nicotine, TEPP and dichlorvos that releases toxic vapours control the pests even if they are dispersed moderately. This, however, does not mean that they may be applied carelessly.

When droplets of aqueous spray are deposited on plant surfaces, their diameter increases up to 3 times. The oil droplets spread up to 10 to 15 times. The larger the droplet size, the lower the spread factor. So to form a continuous film on one acre of foliage with 5 litre of oil, only the smallest droplets are sufficiently numerous to coalesce continuously. Since the area of leaf surface of one acre of vegetation is many acres, the coverage per unit of leaf surface will be considerably less. With the increase in the size of the droplets, their number decreases by a cubic factor. Thus, doubling the droplet size, for example, decrease the number of drops available for deposit to one eighth. A droplet of 400 μ in diameter contains sufficient liquid to make 125 droplets of 80 μ in diameter or 1000 droplets of 40 μ diameter. It follows that small droplets are valuable as their smallness improves the coverage by the spray deposit for a given application rate. Wetting agents tends to decrease droplet size and increase spread. Low volatility carriers aid in preventing evaporation of small droplets and hence ensure better coverage.

11.2. PARTICLE SIZE IN RELATION TO EFFECTIVENESS

The classification given in the Table 11.1 is based on the droplet size of the discharge. Further sub-division could occur subsequently. This would depend on such factors as velocity, volume of the air, and pressure and rate of discharge of the liquid.

Table 11.1. Classification of pesticidal particles by their size

Type of discharge	Equipment used	Particle size μ (dia.)
Coarse spray	Hydraulic sprayer	More than 400 μ
Fine spray	Hydraulic sprayer, Spray blower and Mist blower	100 to 400 μ
Mists	Mist blower and Aerosol generator	50 to 100 μ
Fogs	Aerosol generator	0.1 to 50 μ
Smokes	Smoke generator	0.001 to 0.1 μ
Vapours	Vaporizer	Less than 0.001 μ

11.3. GROUND APPLICATION OF INSECTICIDES

Sprays. The atomization devices used commercially in agriculture produce, particles ranging in size from 1 μ to 2000 μ. It is necessary to indicate the average particle size to have an appreciation of atomization. The particle size may be indicated by *mass median diameter, numerical average diameter (simple arithmetic average),* or *surface mean diameter (SMD).* The SMD is also sometime designated as *Sauter mean diameter.*

Mass median diameter (MMD), also called *mass average diameter,* is the droplet diameter that satisfies the condition that half the spray volume is of large droplets and half is of smaller droplets.

Numerical average diameter is most easily obtained and is found by adding diameter of all droplets and dividing by the number of droplets. This method tends to weigh the average in favour of the numerous small droplets. The SMD is the theoretical diameter in microns of a single droplet with the same ratio of surface to volume as the entire spray. The SMD is smaller than the MMD, being between one-half and three-quarters of it depending on droplet size distribution.

The most important factor is the optimum size for spray droplets because it affects the efficacy of the pesticidal spray. The *optimum droplet size* can be defined as the one that gives maximum pest control with minimum pesticides and minimum contamination of the ecosystem. The actual spray-droplet spectrum depends upon atomization. It is determined by the type of nozzle, operating pressure, velocity of emission of the liquid from the nozzle, orientation of the nozzle with respect to air movement, and properties of the spray fluid, such as its density, surface tension and viscosity. Thus atomization depends on the characteristics and operating conditions of the nozzle and properties of the liquid being atomized. The average droplet size from a given nozzle decreases if either the surface tension or the density of the liquid is reduced. The average droplet size produced by given nozzle handling a particular fluid varies as the inverse square root of μ the press

$$\frac{D_1}{D_2} = \frac{\sqrt{P_2}}{\sqrt{P_1}}$$

Where, D_1 is the average diameter at pressure P_1 and D_2 is the average diameter at pressure P_2.

For example, the average droplet size for a fan-type nozzle might be increased from 75 to 150 μ by reducing the pressure from 7.0 to 1.75 kg/cm^2. As already indicated above, atomization is also influenced by the size of orifice, type of whirl device, size of eddy chamber, and other features of the nozzle. Any hydraulic nozzle, however, produces a wide spectrum of droplet sizes. Increasing the average droplet size does not eliminate small droplets but only reduces their number. Recently, it has been claimed that optimum size for insecticide spray droplets is in the range of 20 μ.

Aerosols. When employing aerosols, the problem is more complex in that the droplets have to be small enough to remain air-borne but large enough to impact the insect during flight and not be carried around it by the air currents. Larger droplets having a greater terminal velocity will fall more steeply than smaller droplets and will, therefore, land closer to the point of emission. Smaller droplets will follow a more horizontal path. With aerosols emitted in to still air, the higher settling rate of droplets above 20 μ reduces air-borne concentration of insecticide and droplets below 2 μ have reduced ability of impingement. Therefore, droplet sizes outside this range (2-20 μ) are considered wasteful and undesirable. Spraying must be stopped when turbulence reaches a certain level. Turbulence is, generally, at a minimum in the early morning and again in the evening. So these are often the best time to make aerosol spraying application. Working outdoors with droplets less than 30 μ to achieve a residual deposit is practicable only at certain hours. Evenings are preferred since thermal inversion conditions are most likely to prevail. It may be noted that heat-generated aerosols of 5 to 15 μ are not applied in the field when the air at one meter above the ground is more than 3^{0}C warmer than the air at 6 meter above. Particles in this range are usually are not visible in the air or on the surface on which they settle, since the smallest particles that can be seen with unaided eye is of about 35 μ. As such the need for effective respirators to be worn by the operators and placing suitable cover over surfaces not to be contaminated cannot be over-emphasized when applying aerosols.

Dusts. The size of individual dust particle, in general, is comparatively much smaller than in case of sprays. Thus, the individual dust particles are more prone to drift. They possess less sticking power. Consequently they do not give good deposit. However, the dust cloud comprises both the individual particles and agglomerates of up to 300 individual particles. Such agglomerates are in a better position to land on the target surface though their adherence is little better than that of individual particles. Dust applied at a high velocity penetrates the air-cushion around the objects and settles on them only if they are near the blower; however, the effect of high initial velocity is lost. On account of this and gravitational force, ground application of dusts on tall trees is not very effective.

Deposition of the dust on the target surfaces is not influenced by its fineness. On the foliage it is maximum in sulphur, being in decreasing order by lead arsenate, derris and cube, pyrethrum and cryolite. The high deposit of sulphur is due to fairly large size of the particles. In case of lead arsenate, the deposit is good because the particles are heavy even though small (mostly less than 2μ) and also form agglomerates during application the least deposit in case of cryolites is because of tendency of its particles to float around. Dust particles of derris, cube and pyrethrum are light though coarse. As such their deposit is less.

Sprayer function. Sprayer is a machine to apply fluids in the form of droplets. The sprayer is used to apply herbicides to kill weeds, to apply, fungicides to minimize fungus disease, to apply insecticides to control insect pests and to apply micro nutrients in the plants. The main function of a sprayer is to break the liquid into droplets of effective size and distribute them uniformly over the plants. It regulates the amount of insecticides to avoid excessive application that might prove harmful or wasteful on crops. The sprayer should produce a steady stream of spray materials in the desired fitness of the particles so that the plants to treated may be covered uniformly. It should deliver the liquid at sufficient pressure so that it reaches all the foliages and spreads entirely over the sprayed surface. It should be light yet sufficiently strong easily workable and repairable. A sprayer that delivers droplets large enough to wet the surface readily should be used for proper application of surface residual sprays. Extremely fine droplets tent to be diverted by air currents and get wasted.

Most sprayers in use today are hydraulic types in which spray pressure is built up by the direct action of pump on liquid spray material. The pressure thus developed forces the liquid through nozzles, which break the liquid into fine size droplets and disperse them in the desired spray pattern. Also sufficient energy is imparted to the spray droplets to carry them from nozzle to the surface to be treated. The main part of hydraulic sprayer are; pump, tank containing an agitator, framework for mounting sprayer, pressure regulator and relief value, pressure gauge, strainers and screens, control values, pipes and fittings, distribution system and power source.

Volume median diameter (VMD). This parameter is usually employed to describe the droplet spectrum. It is droplet diameter that satisfies the condition that half the spray volume is of droplets smaller than a droplet whose diameter is VMD and other half of spray volume contains larger droplets. This can be obtained from a volume cumulative curve of droplets size spectrum it is the diameter corresponding to 50% cumulative. Droplets in the range of 120-300 micron VMD were found efficacious.

Number median diameter (NMD). It is an average diameter of the droplets without any references to their volume. The diameter corresponding to 50% cumulative number curve will give found efficacious.

Uniformity coefficient (UC). The ratio of VMD and NMD is called uniformly coefficient. It indicates the range of size of droplets. The more uniform the size, the nearer is the ratio to unity.

Drift. Drift is of particular concern when deposits of harmful chemicals must be avoided or minimized in the fields nearby the area where chemical is used. The parameters affecting drift are rate of fall of particles, initial height of fall, wind velocity and direction, atmosphere stability and other metrological factors. Evaporation of water or other volatile materials from droplets reduces the droplet size and thus adversely affects the deposition efficiency and drift. Employing atomizing devices that produce spray having large volume mean diameters can minimize drift. Drift is more serious with dust than with sprays because of smaller particles size.

Droplets density (DD). The number of droplets per unit area of leaf surface is called droplet density. Droplet density of 20-50 droplets of pesticide per square centimeter is most effective against cotton pests.

Leaf area index (LAI). It is the area of one side of the leaf divided by the corresponding ground area. It is considered to be a gauge to assess the crop growth.

Spread factor: The ratio of the diameter after incidence to the diameter before incidence is called spread factor for a particular object and liquid spread. It is greater than unity because after impingement the droplet spreads. Size of droplet, surface of object and formation of the spray liquid affect the spread factor.

Spray volume can be calculated by:

$$V = L.D.F.V_{.MD}.\ 100/P$$

Where,

V = spray volume (litre/ha)

L =leaf area index

D =theoretical droplet density (droplets/cm^2)

F = factor of safety

V_{VMD} =volume of droplet of VMD size (litre)

P = percentage of useful volume. It is ratio of volume deposited on the leaves to the total volume multiplied by hundred.

For a given nozzle with known discharge rate and the swath, the speed with the operator will have to walk can be worked by:

$$S = q/r\ v_0$$

Where,

S = walking speed (m/min)

q = discharge of the nozzle (litre/min)

r = row to row spacing of the crop (m)

v_0 = spray volume (litre/m^2)

Dimensions of boom depend upon the plant and the nozzle parameters. Assuming that the plant canopy is of rhombus shape and the nozzle covers both sides of leaves, boom dimensions can be calculated by:

$$L = [H_2^2 - 4R_1(R_1 - H_1/\tan ?)]^{1/2}$$

$$H_3 = \{(H_2 - L)/2\} + H_1$$

$$B = Y + e/2$$

Where,

L = boom length (cm)

$Y = \tan^{-1}[R_1/(H_1 - L_1)]$

$L_1 = (H_1 - L)/2$

q = cone angle of nozzle (degree)

$H_2 = H - H_1$ = plant canopy height (cm)

H = plant height (cm)

H_1 = height of first leaf from ground (cm)

R_1 = distance of operator from plant canopy center (cm)

H_3 = height of lower nozzle from ground (cm)

B = inclination of nozzle center with boom axis (degree)

Selection of spraying equipment. A farmer willing to buy a new sprayer is often faced with a problem of selecting a machines from a large choice of models' of varying sizes and capacities available in the market. It is important that he makes the right choice for economical and efficient use (Singh and Kandoria, 1999). A choice must be made as to which of the four: duster, high-volume sprayer, low volume sprayers or ultra-low volume (ULV) sprayers are required. The dusters are preferred in arid and semi-arid areas where water is scare. If a sprayer is the choice, select one that could be used for applying spray liquid in the high-volume as well as low-volume (Srivastava, 1990, Srivastava *et al.,* 1994, Smith and

Wilkes, 1977). High volume sprayers are selected where more than 400 litres of spray liquid per hectare is to be sprayed. If spray volume is in between 5 to 400 litres per hectare, low volumes of sprayers are selected. Ultra-low sprayers are only selected where volume of spray liquid to be applied is less than 5 litres per hectare.

Foot and rocking sprayers give satisfactory performance for small trees and bushy crops. Knapsack sprayers are good for spraying field crops. Motorized knapsack sprayers perform better for spraying field crops, bushy crops and small size fruit trees. In hilly regions and waterlogged areas, manual and power operated knapsack sprayers are preferred. The sprayer selected should be easy to operate, maintain and requires least number of operators and technical skills. The sprayer must be purchased from the reputed manufacturers using high quality nozzles. Easy availability and after-sales service should also be considered while buying the sprayer. Size and type of sprayer to be purchased will also depend on the crop, area to be covered and investment capability of the farmer. The operator must be trained in advance in safety precautions and effective application techniques.

Multi-purpose farm sprayers are available to meet the many spraying needs on diversified farms (Ojha and Michael, 1992). It has necessary pressure range to be used for weed spraying or for spraying fruit trees or livestock. These sprayers have free choice of spray materials because of type of pump and agitation provided. Wettable powders, cold water paints and other spray materials that are abrasive or difficult to keep in suspension can be used satisfactory in this type of sprayer.

Power sprayers are used for those spraying jobs, which are just too large for coverage with hand sprayers. These sprayers are popular on golf greens, large gardens or green house. Commercial growers of fruit and vegetables high pressure and volume delivery desired to get complete spray coverage of high growing fruits and shade trees in full foliage as well as the dense growth of vines and other crops. Plunger-type multiple cylinder reciprocating pump is used exclusively with large sprayers at maximum pressure of 28-69kg/cm^2 delivering 30-227 litres/min. Low-pressure, low volume sprayers have been primarily designed to meet the specific requirements for low-volume field spraying. They are very popular for use in controlling weeds and insects in field crops. Because of their relatively low cost, their use is limited to sediment-free type of spray material due to type of pump and agitation provided. These are available either tractor-mounted type or trailed type.

Calibration of sprayer. The spray volume required to cover one unit of area depends on discharge rate of nozzle, spacing of nozzles on the boom, nozzle pressure, forward travels speed and spray width. Therefore, it is essential to have a trial run with plain water on small area and record the above mentioned parameters. Based on these parameters, the quantity of spray fluid required per unit area can be worked out as:

$$\text{Spray vol. required (1/ha)} = \frac{\text{Area to be sprayed (ha)} \times \text{spray vol. used on trial plot (litre)}}{\text{Trial plot area (ha)}}$$

Nozzle discharge rate can be determined using above information by the equation:

$$Q_d = 0.001\ Q_a\ S\ d$$

Where,

Q_d = nozzle discharge rate per nozzle (litre/hr)

Q_a = application rate (litre/ha)

S = travel speed rate (km/hr)

d = nozzle spacing (cm)

11.4. TYPES OF SPRAYERS

Sprayers are available for every use, in the home, garden, orchard and field. Garden sprayers are manually operated but field and orchard sprayers are manually as well as power operated. Sprayers are of 4 types, viz., manually operated, engine operated tractor and air plane sprayers.

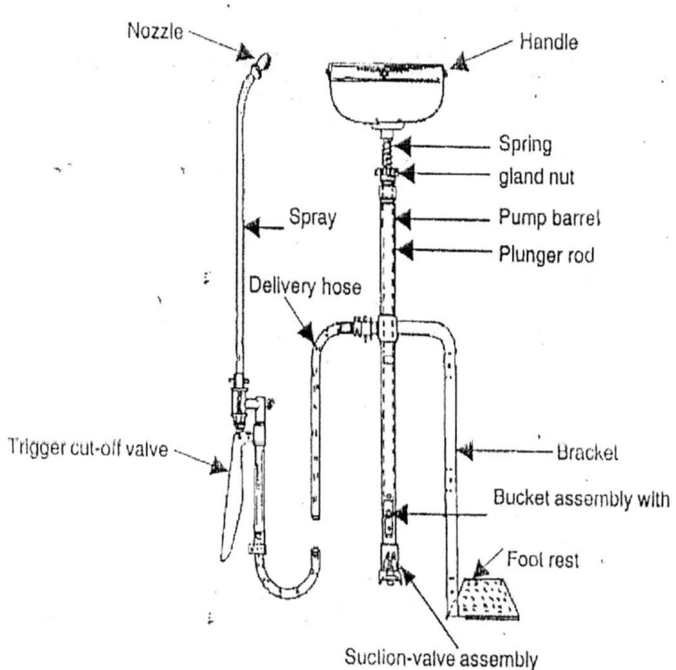

Single-barrel stirrup pump sprayer

11.4.1. Manually operated equipment

Stirrup-pump sprayer. It consists of a brass pump, a footrest (stirrup), hose, lance and nozzle. It may have a single barrel or double barrel pump. In this type of sprayer continuous pumping is necessary for getting uninterrupted discharge of spray fluid. It develops a pressure of 4-10 km/cm^2. It is suitable for small scale field spraying. It does not give uniform spray, as single barrel pump has no mechanism to retain pressure. Two persons are required for operating this sprayer; one to operate the pump and other to direct spray from the nozzle at the end of spray lance. The length of nozzle varies depending upon the requirement in the field but it is generally kept as 5 m.

Knapsack sprayer. It is generally used for spraying low crops, vegetables and trees up to 2.5 m height. It can spray about 0.5 ha area if crop with single nozzle in a day. The frame of sprayer is shaped so as to conveniently fit on the back of operator. The capacity of container is 10-14 litres. It is made of galvanized iron, brass, or stainless steel. Translucent plastic materials for visual fluid-level indication are used for spray tank. Some times it is provided with a double-barrel pump of piston type with a lever to operate the pump. In other cases, the tank is provided with a single pump and a pressure barrel having a piston pump and a mechanical agitator. Some models are equipped with externally mounted pumps and such pumps have more life because it is not immersed in fluid. A pressure of 3-5 kg/cm^2 is maintained in the pressure chamber.

Compression Knapsack sprayer. It consists of a tank for holding spray fluid with compressed air, vertical air with handle, filter hole, spray lance with nozzle and a cut-off device. The capacity of tank varies from 10-20 litres. It can develop 4-5 kg/cm^2 pressure in the tank filled with fluid to ¾ of its capacity. It is mostly used n gardens, small farms, and livestock and poultry houses. The major drawback is that pressure in the tank drops progressively as spraying proceeds.

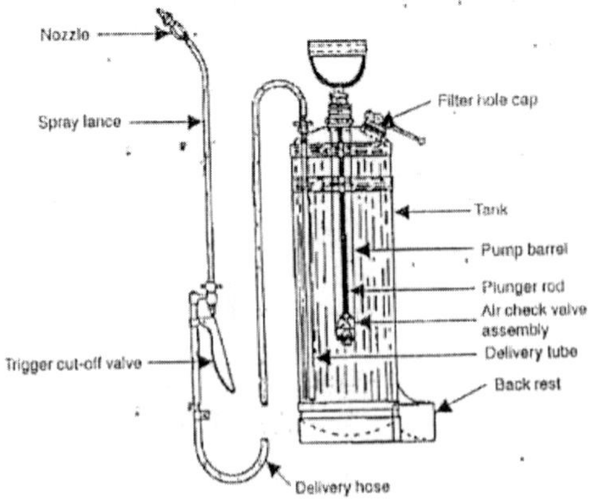

Compressor Knapsack Sprayer

Motorized Knapsack Sprayer. This is called power operated gaseous-energy knapsack sprayer. The hopper having capacity of 7-12 litres is made of high density polyethylene. Another small tank of 0.75-2.25 litres capacity is also provided for fuel. It is powered by 1-3 hp engines. The fuel consumption varies from 0.6-2.0 litre/ha. The frame is provided with shock-proof cushion that comfortably fits on the back of operator and eliminates the transfer of vibrations of the engine to the operator. It is provided with an on-off control switch and a mechanism for adjusting the discharge rate from 0.5-5.0 litre/min. The spray liquid is Blown-out by an air current generated in the tank. A part of air generated is diverted into the hopper to form an air cushion over the liquid in the tank. This ensures a uniform delivery of liquid. The liquid from tank passes through a tube to the nozzle on the spray lance by gravity. Effective swath width is 4-5 m horizontally and 3-4 m vertically. It can cover approximately 3 ha/day. The machine fitted with a rotary pump and tree-spray lance can dusting and ULV application.

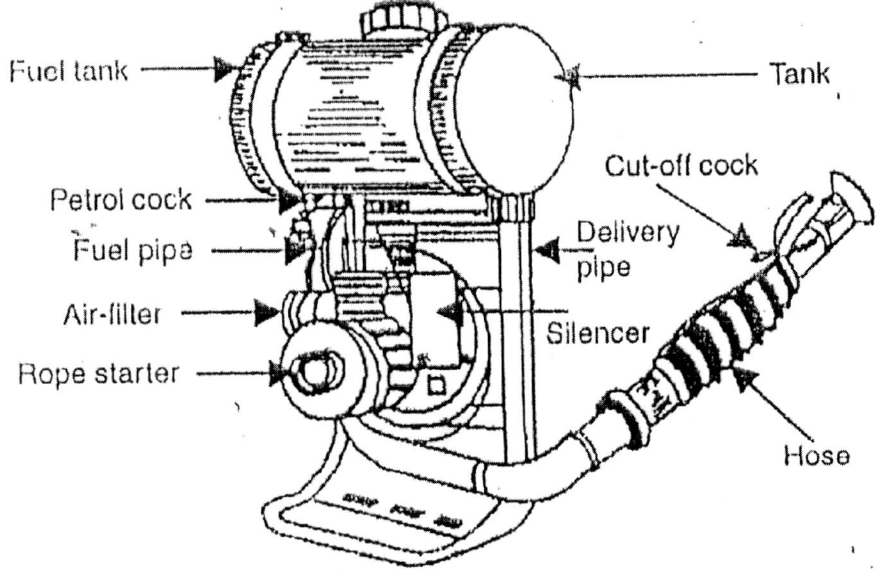

Motorized Knapsack Sprayer

Sprayer for coconut palm trees. It is a small sprayer consisting of a brass or aluminum tank of 3 litres capacity covered with air-tight brackets at the top. It has a delivery lance to which a nozzle is attached.

Motorized knapsack sprayer

The spay liquid is put in the tank and air is compressed by an air pump. The sprayer is placed on the operator's back by straps and he climbs over the coconut and palm trees or other such tall trees and applies the pesticides.

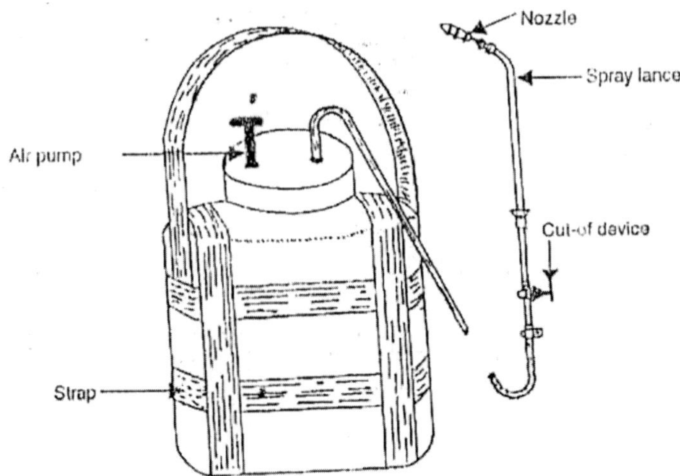

Sprayer for coconut palm trees

Foot-operated sprayer. It is also called pedal pump. It consists of a plunger assembly, stand, suction hose, delivery hose and an extension rod with nozzle. One end of suction hose is fitted with a strainer and other end with a flexible coupling.

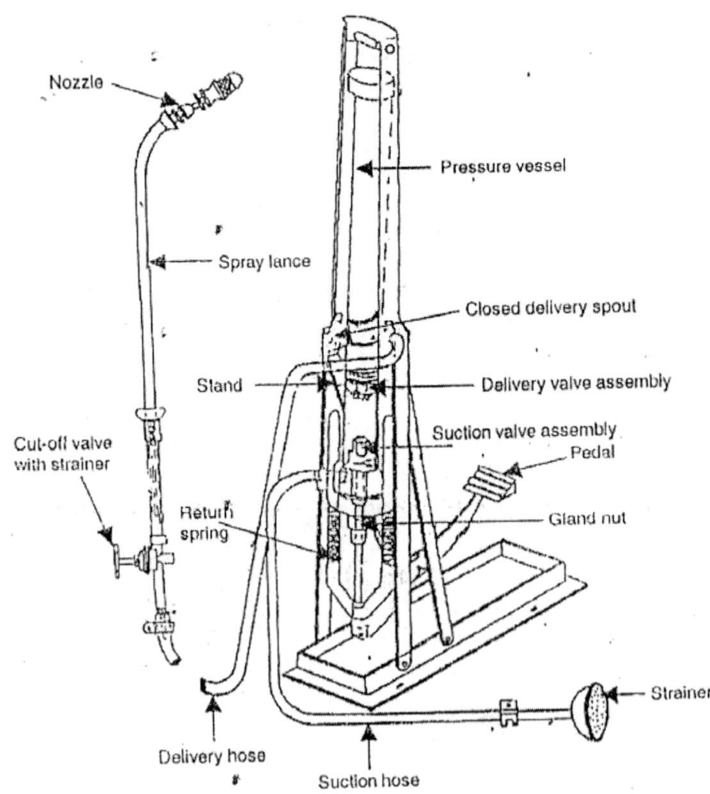

Foot-operated sprayer

Similarly, one end of delivery hose is fitted with a cut-off valve and other end with a flexible coupling. An additional container is required to hold spray fluid, as this sprayer does not have a built-in tank. Continuous pedaling is required for uniform spray. It can develop a pressure of 17-21 kg/cm^2. It is easy to operate and can be used for spraying tall crops and fruit trees up to 4m height. Sprayer can be used to spray trees up to 6 m height with additional hose.

Rocking sprayer. It consists of pump assembly, platform, operating lever, pressure chamber, and suction hose with strainer, delivery hose and an extension rod with spray nozzle. Rocking movement of handle operates the pump resulting in built up pressure in pressure chamber. This also requires an additional container to hold fluid. It can develop a pressure of 14-18 kg/cm^2 but in some cases it can develop up to 36 kg/cm^2. It can be used for spraying coconut and palm trees.

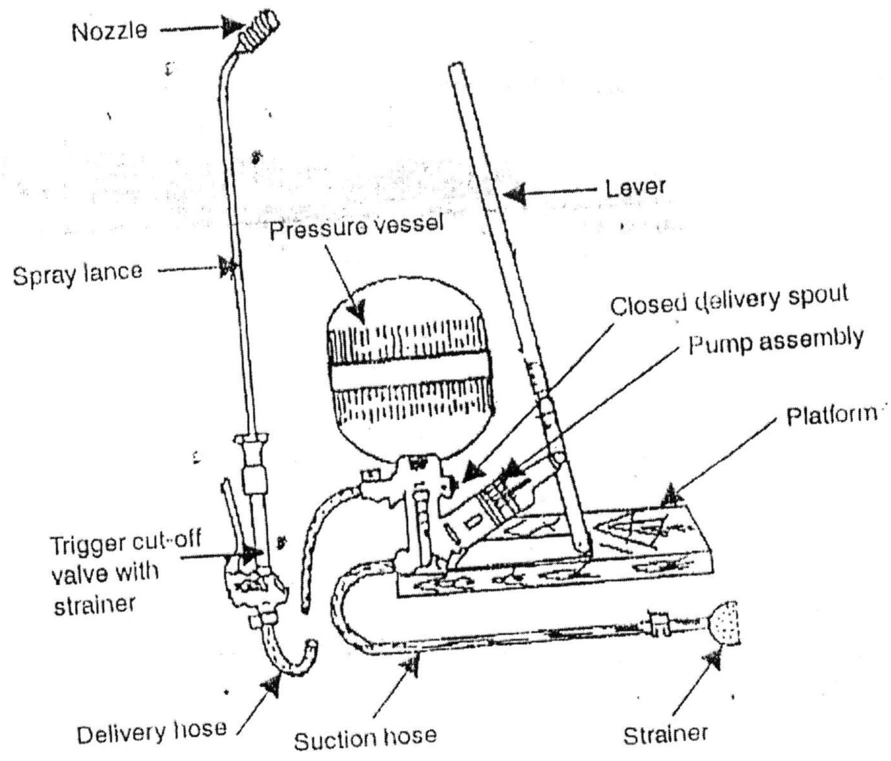

Rocking sprayer

11.4.2. Battery Operated Sprayers

Electrodyn sprayer. Recent increasing interests have been using less solvent spraying pesticides. There have been many developments in this direction and development of sprayer based on charging of droplets is one of them. Operation of an electro-dynamic sprayer is based on the droplets emerging from the delivery gun with an electric charge. Since the droplets have the same charge, droplets repel each other causing them to form a reasonable wide spray. Electro-dynamic spraying uses rotary atomization principle. The principle feature of electrodyn sprayer is the device for providing a high-tension voltage in a very space through which the liquid emerges by gravity. The source of power in this sprayer is four standard torch light batteries that last for about 60 hrs. Since there are no moving parts, the sprayer lasts for several years.

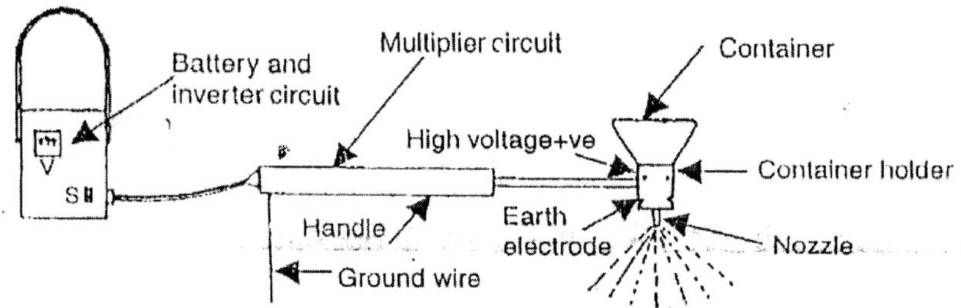

Schematic view of electrodyn sprayer

Schematic view of electrodyn sprayer

The sprayer has a backpack unit enclosing the charging circuit and battery and a long handle having multiplier unit (Ganapathy and Since 1993). The spray tank and nozzle are supported at the other end of handle. A high voltage of 20 kv is applied to the inside the nozzle, which creates

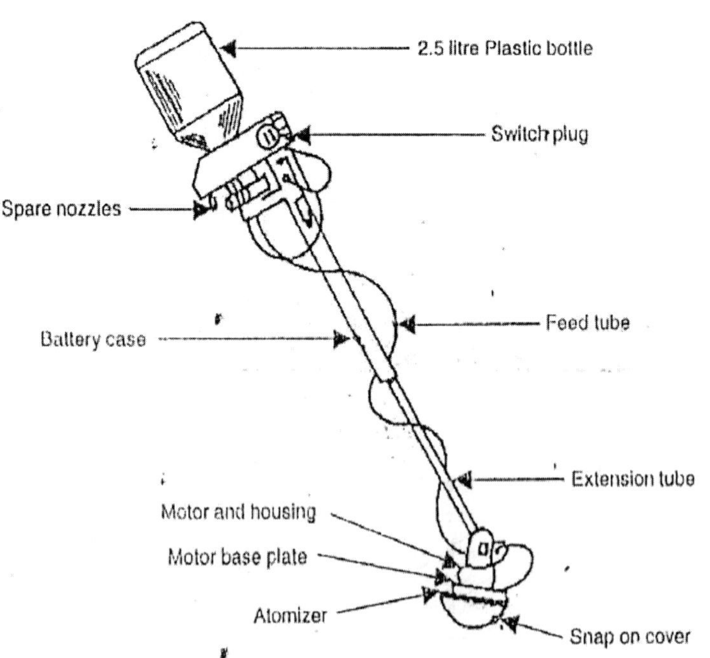

Ultra-low volume sprayer.

Ultra-low volume sprayer. Sprayer are broadly classified as high volume (HV), low volume (LV) and ultra-low volume (ULV) according to volume of spray applied per unit area. These terms describe the volume of water used as carrier. The new trend is to use minimum amount of carrier and this leads to use of low volume and ultra-low volume techniques. Various equipment and techniques are used to apply ultra-low volume of solution. The equipment use high rpm spinning disc devices.

Motorized knapsack sprayer and aircraft fitted with special spinning disc attachment are also used as ULV sprayers. ULV sprayers used blower for creating a blast of air atomizing the spray liquid and for carrying the droplets to the target. The low volume concentration (LVC) used in ULV perform form better due to higher initial deposits. LVC droplets remain in the liquid form and more readily picked up by the pest that gives higher mortality of the pests. The sprayer has a motor powered by 6 to 12 volt batteries. The spinning disc having grooves or teeth is attached to motor and rotates at 4,000-9,000 rpm. The spinning disc receives the chemical from the plastic container of approximately one litre capacity. Average droplet size varies from 35-100 micron. It is used for application of weedicide and for spraying small trees and crops. Battery-powered hydraulic, pneumatic and centrifugal sprayers are also available, which can be operated by 12-24 volt batteries (Tajuddin, 1997).

Electro-static spinning disc sprayer. Electro-static spraying technique offers good scope for reduction in drift and chemical requirement and for increasing the penetration into the crop canopy (Ganapathy and Singh, 1993). In this system, electro-static forces are used for transport and disposition of spray and some times for disintegration of spray liquid also. Electro-static charging improves deposition by more than two-fold for different crops and at different spraying conditions.

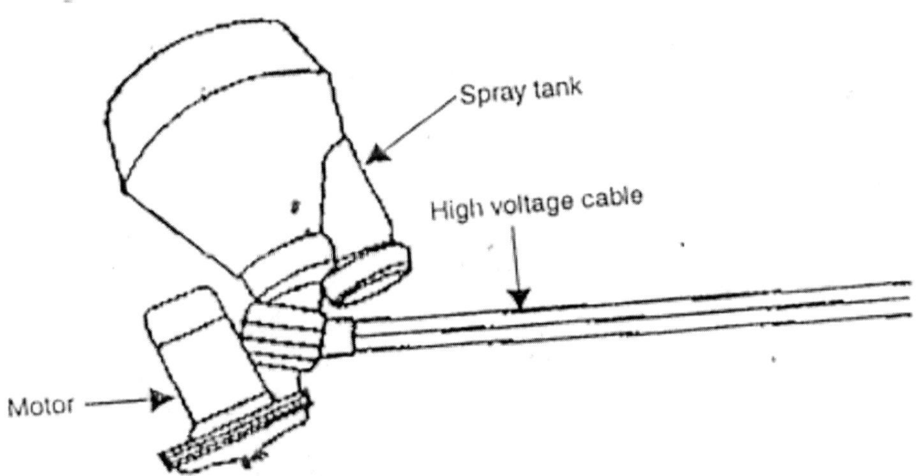

Schematic view of electrostatic spinning disc sprayer

The sprayer consists of a spray tank, high voltage cable and motor.

High voltage is applied to the liquid through a mild steel coupling placed in the flow path tank to atomizing disc. The motor housing is well insulated by a rubber casing. If the total surface charges 'q' of a drop is increased to a level where outward.

Electro-static force equals the binding surface tension, the any further increase in charge results in disruption of drop into charge fragments which can be represented by:

$$q = 8\pi \sqrt{\varepsilon_2 \sigma r^3}$$

where,

q =limiting surface charge on drop (C)

ε_2 =permittivity of air (C^2/Nm^2)

σ =surface tension of spray liquid (N/m)

r =diameter of droplet (m)

The droplet size decrease with an increase in voltage applied and increase proportionally with an increase in flow rate.

Knapsack electro-static spinning-disc sprayer. It consists of backpack frame, handle and boom (Gupta *et al.,* 1994). It contains a 5 litre plastic container for the spray solution; a 12 volt DC wet-cell rechargeable battery and a high voltage power supply unit. A tailored pocket-type cloth holder contains the battery and power supply unit. A canopy is provided at the top of the backpack frame. The backpack has belts for shoulders, chest and waist of the operator. The aluminum pipe handle and boom are connected to backpack frame through nuts and bolts. The right handle has a switch box containing one switch for sprayer motor and two switches for connecting high voltage power supply to the electrodes. The liquid lines passing through the handle has a flow control valve positioned near the operator's hand. The aluminum boom carries two electrostatic spinning discs and spacing between them can be adjusted. The height of the boom can be changed as per need. An electrical circuit reduces the input 12 to 6 volts DC required for each motor. Each electrostatic spinning-disc sprayer runs at 1.2 kv electrode potential, 50 ml/min liquid flow rate and 2,000-3,000 rpm disc speed. The machine provides a spray charge of about 1.0 mC/kg and about 165 um mean droplet size. It can cover about 0.05 ha/hr as compared to 0.1 ha/hr of the single nozzle hydraulic sprayer.

11.4.3. Power operated equipment

Sprayers for tall trees: Air blast sprayers also known as air carrier sprayers have been developed for applying spray liquid to tall trees. Air blast sprayer uses air stream to carry the droplets and utilize smaller droplets size to obtain adequate coverage with low amount of spray material per unit area. The effectiveness of sprayer depends upon its ability to disperse air in all parts of trees with spray-laden air. Deposits on leaf surfaces decrease in proportion to air velocity. Drift problems with this machine are comparable with those of aircraft spraying. Most of the large orchard-type air blast sprayers have axial flow fans with guide vanes to direct the air radially outward through a partial circumferential slot. Some sprayers have two opposite axial flow fans blowing towards each other from either side of the slot. The machine can cover one side of one row or the adjacent side of two rows. The included angle of delivery on each side is adjustable to accommodate different size of trees.

Blower sprayers. They are also called mist sprayers. They are used to apply pesticides in concentrated form. It requires 20-80% less water as a diluent and saves labour cost substantially. Further savings can be made in the quantity of chemicals used by reducing runoff from foliage when the equipment is operated properly. These sprayers are used for spraying chemicals on fruit trees, large, shade trees, vegetable and certain other crops.

Types of spray pattern for orchard spraying. Large fruit trees may requires thinning and pruning to permit proper penetration of small airborne droplets to inner and top-most branches. These machines are similar to power dusters and employ liquid chemical instead of dry powder. This sprayer also has low-pressure, low-volume pump, which forces the spray material under low pressure to the fan where it is discharged into the air stream in small spray droplets by a group of nozzle.

Tractor mounted high clearance sprayer

Tractor mounted high clearance sprayer for tall crops.

Power sprayers of tractor operated sprayer. Power sprayers are operated usually with internal combustion engines of 1-5 hp capacities or by a tractor.

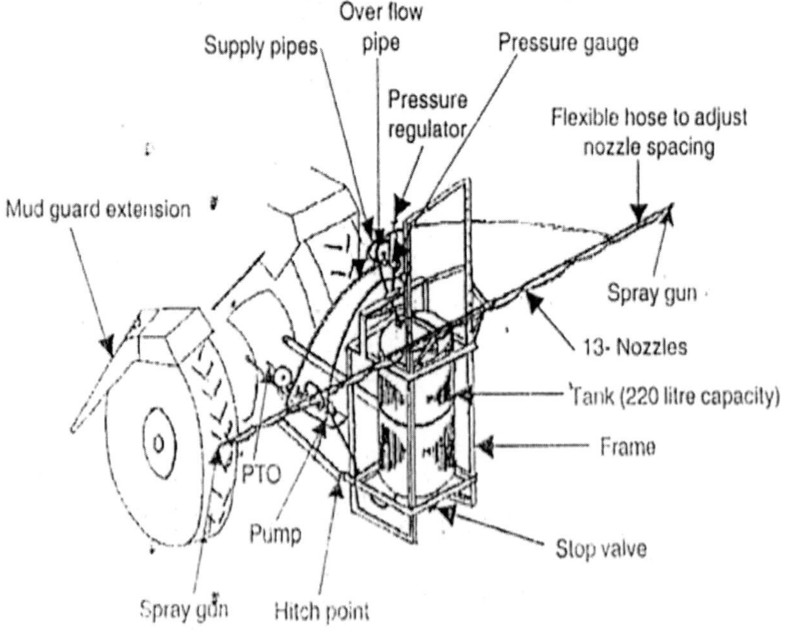

Isometric view of tractor mounted high clearance sprayer for tall crops

A small engine operated power unit ensuring a constant steady pressure operates the pressure pump. They are operated at pressure from 20-55 kg/cm^2. The pump is operated by tractor PTO. Power sprayer consists of a pump, one or more drums, control values, pressure gauge, pressure regulator, relief valve and a spray boom fitted with nozzles. The boom may be of flexible hosepipe on which nozzles are mounted to meet different row spacing. The boom is tied with a rigid beam by clamps. Inlet liquid supply to boom is provided at two points for even distribution of liquid.

Holes are provided on the frame to lower or raise the beam to suit the crop height. Normally sprayer boom has 13-15 triple action nozzles and can cover 7-8 m width. The capacity of tractor-mounted sprayer is about one hectare per hour while operating at a speed of 3.5 km/h. The tractor is operated in the same pathways and in the same direction for all sprays. Sprayers are used in row and tall crops such as cotton, maize, sugarcane etc. The tractor-operated can be used for spaying all the 6-7 sprays on the cotton crop if same path and direction is used during all the sprayings. This reduces crop damage under the tractor. The pest control is better in case the spraying is done by tractor-operated sprayer as compared to knapsack sprayer (Shukla *et al.,* 1987). There is a net increase of about 29% yield in the field sprayed by tractor-mounted sprayer as compared to that sprayed by knapsack sprayer.

Self-propelled sprayer. This sprayer has been developed as a special purpose machine to spray those fields and row crops, which are too high for conventional power sprayers and tractors. These machines usually have low-pressure, low-volume sprayer. The spray boom is mounted on the rear of the machine. The height of boom can be varied mechanically/hydraulically to permit spraying either in short or tall crops a self-propelled sprayers, developed at Punjab Agricultural University, Ludhiana, consists of a 20 hp diesel engine, gearbox, water tank, spray pump (piston type) and boom with nozzle (Anonymous, 1995-97). Engine is mounted on a chassis, which has ground clearance of 120 cm. The power from the engine is transmitted to two front logged wheels through chain from gearbox. A clutch is provided to engage or disengage the engine power to drive Wheels. The gearbox provides 4 forward and one reverse speed. The boom height can be adjusted from 31.5-168.5 cm to suit various crops. Drive to pump is given directly from engine through v belt. The spraying pressure can also be adjusted by relief valve. The machine can cover about 1.5-2.0 ha/hr. Liquid application rate can vary from 150-175 litres/ha and fuel consumption is 1.5 litre/hr.

Hydro-pneumatic sprayers. It has the same range of use as the low pressure, low-volume sprayer. The spray liquid is carried in a pressure tank and spraying pressure is developed by means of an engine-powered air compressor. The spray material therefore does not pass through the pump or contact any other moving parts. The sprayers are available in mounted types and tank size is limited to about 1000 litres because of cost and weight of pressure tank. Agitation is provided either by means of a mechanical agitator or by an air tube discharging air below the surface of liquid in the tank.

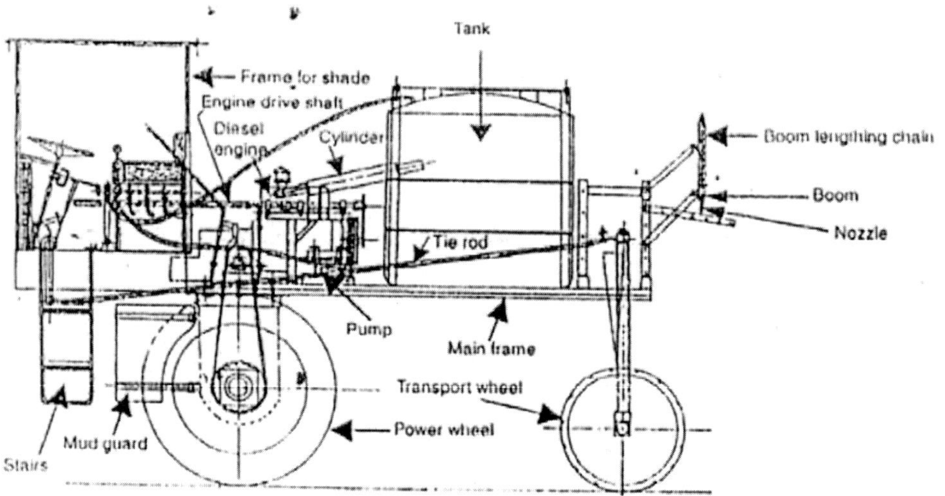

Isometric view of self-propelled high clearance sprayer

Airplane sprayer. The use of airplane to apply and distribute spray material has become one of the most popular and economic methods of applying insecticides and fungicides to agricultural crops. Crosswinds influence the spray pattern and uniform distribution of spray materials applied by airplane. Two types of nozzles are available for spraying liquids with aircraft. Each types of nozzle has a check valve to stop the flow of spray liquid when pressure is cutoff. It has been observed that, when the spray droplets are discharge from the side of the nozzle into the air stream, the shearing action of air at high velocity as it passes the nozzle tip further reduces the size of droplets.

The rotary-wing helicopter is also used for spraying orchards and areas where fixed-wing airplanes have difficulty in maneuvering. The helicopter can move slowly at a speed at which spray material may be dispensing most effectively. It is very expensive and costing much more than the fixed-wing airplane.

11.4.4. Setting sprayer for given feed rate of liquid chemical

Calculate the quantity of liquid to be delivered through spray nozzles at the given rate of consumption per hectare.

$$q = \frac{vBf}{10 \times 60}$$

For spray nozzles having variable flow of liquid

$$q = \frac{vBf}{u \times 10 \times 60}$$

Where,

q =required delivery of liquid chemical through spray nozzles (litres/min)

v = travel speed (km/hr)

f =feed rate of chemicals

u = utilization factor for nozzle delivery, taken between 0.5-0.7.

11.5. DUSTERS

Duster is a machine to apply chemicals in dust form. Dusters make use of air streams to carry pesticides in finely divided dry form on the plants. A duster essentially consists of hopper, agitator, feed control, fan or blower and delivery tube and nozzle. Shoulder-mounted and belly-carried rotary dusters are commonly used. These dusters are extensively used for dusting field crops, vegetables, bushes, small trees etc. they can cover 0.8-1.2 ha/day. The belly-type is more convenient to operate in tall crop like sugarcane as it does not get entangle in plants during dusting. The capacity of hopper is 5 kg and should not be filled more than of its capacity there are different types of dusters such as plunger type; knapsack type, rotary type and power operated dusters. Dusters can also be set for given feed rate of chemicals using the same formula as used for sprayers.

BELLY MOUNTED DUSTER

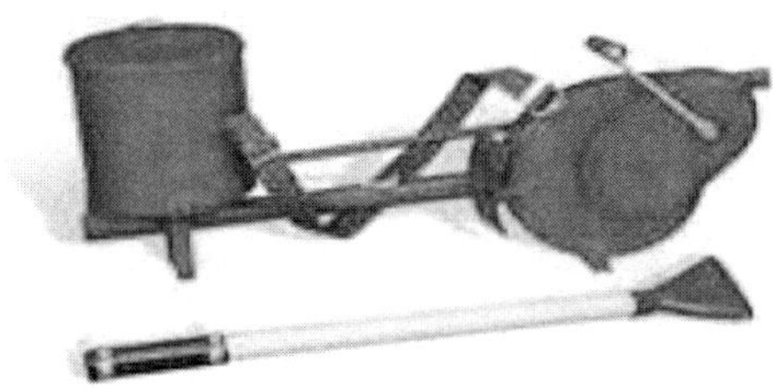

SHOULDER MOUNTED DUSTER

Field dusters or power operated duster can be mounted upon a platform bolted to the rear of tractor and operated by PTO. Operating the tractor in high gear makes possible to dust a large area. It mainly consists of a power driven fan, a hopper and a delivery spout. The fan creates strong air, which causes the dust to blow off from hopper to a considerable distance either vertically or horizontally. A movable delivery spout suitability fitted with the unit regulates the direction of dust. An agitator is provided in the bottom of hopper to keep the dust in fluffy condition to meter out uniformly through the feed system into the air stream. This type of dusters is used for large areas. Tractor-drawn trailing dusters with PTO drive and auxiliary engine drive are also available. An auxiliary engine operates most orchard dusters. They have only one large flexible metal hose, which can be turned to direct dust in any direction.

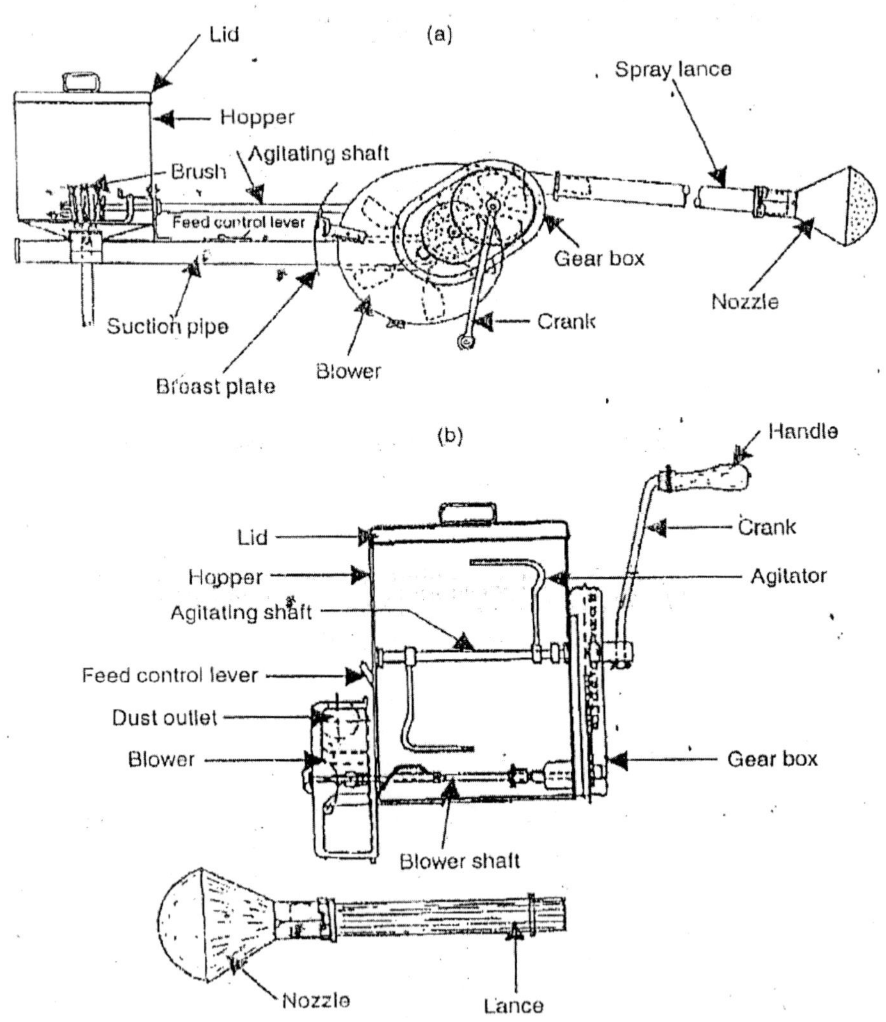

(A) Shoulder mounted rotary duster (B) Belly mounted rotary duster

Plunger type duster is a simple duster with a small piston. The piston moves current of air over the dust in the hopper. The dust is carried away through a delivery spout. This is suitable for small areas where vegetable or flowers are grown. Knapsack type duster consists of powder container mounted on the back of operator. It has a hopper through which a current of air is blown to pick up the dust. The air current is produced by lever operated leather bellows. These dusters are suitable for small areas only. Rotary duster having a hand operated rotor is mounted in front of the operator. Rotary dusters are preferred where more force of delivery is required for spraying tall crops. The dust is fed from a hopper into a current of air produced by a rotary fan and is blown out through a delivery pipe. It has a stirring device actuated by the fan crank to ensure a steady flow of dust. A valve below the hopper can regulate the rate of delivery. Controlling the speed of fan controls the delivery force.

Airplane dusters. They have been successfully used to apply the dust to both field crops and orchards. A hopper capable of holding about 225 kg of chemical is built inside the fuselage in the space occupied by front seat. A close-fitting lid hinged in front, covers the opening in the top for filling. An agitator driven by a small propeller mounted on the wing stirs the dust in hopper just from hopper. The blast of air created by plane's propeller rushes through venture nozzle at a high velocity, catching the dust and discharges it in a whirling cylindrical column that spreads and settles on the plant. The high velocity of air through nozzle creates a partial vacuum in the feed opening which aids the flow of dust.

Aerosol generators/fogging machines. These machines disperse the spray material in the form of extremely fine droplets, which remain airborne for a considerable period of time. The insect killed with this machine is dependent upon the airborne insecticide contacting the insects. This machine is employed for temporary control of adult mosquitoes, files and other similar insects in buildings and in restricted area.

11.6. PUMPS FOR SPRAYING

Mostly three types of pumps are used for spraying; viz. piston or plunger type; rotary type and centrifugal type. Most hydraulic sprayers are equipped with a positive displacement pump (reciprocating pumps-plunger and piston type) capable of developing pressures in the required range for spray jobs. It has a spring loaded by pass valve to control pressure. It is well suited for high pressure spraying. The discharge capacity of these pumps is approximately proportional to speed. The volumetric efficiency of this pump is as high 90%. Mechanical efficiency ranges from 50 to 90% depending upon size and condition of pump

$$\text{Volumetric efficiency} = \frac{\text{Actual suction capacity}}{\text{Theoretical suction capactiy}}$$

$$\text{WHP} = \frac{\text{Water horsepower (WHP)}}{\text{Shaft horsepower (SHP)}}$$

Water horsepower is the power, which is required only to deliver liquid in the pump

$$\text{WHP} = \frac{\text{QH}}{45{,}000} = \frac{\text{WHP}}{450}$$

Where,

Q =theoretical suction capacity (litre/min)

H =head (m) (1kg/cm^2=10m water)

P =pressure (kg/cm^2)

Suction capacity of power sprayer can be calculated by

$$Q = D^2\, NLn10^{-6} \qquad (7.1)$$

Where,

Q = theoretical suction capacity (litre/min)

D = diameter of plunger (mm)

N = rpm

n = number of plunger

L = stroke length (mm)

Example: Find the suction capacity of a power sprayer working at 1200 rpm if plunger diameter is 30 mm, stroke length 20 mm and number of plunger is four. Also calculate the water horsepower and shaft horsepower if the pump is operating at 28 kg/cm^2 pressure and 85% efficiency.

Solution: Using equation 7.1

$$Q = \frac{\text{x}\, 1{,}200 \times 4 \times 10^{-6}}{4} = 67.86 \text{ litre/min} = 67.86 \text{ litre/min}$$

$$\text{water Horsepower} = \frac{\text{QP}}{450} = \frac{67.86 \times 28}{450} = 4.22 \text{ hp}$$

Rotary pump is commonly used for low-pressure sprayers. They are compact, lightweight and cheap. These pumps are commonly mounted on and powered by tractor PTO shaft. Oversize units are usually specified to provide sufficient by-pass-liquid for hydraulic agitation in the tank. Most common type of rotary pumps is gear type (internal and external) and roller pumps. Gear pumps are classified as positive displacement pump and available in various sizes. Their service life is greatly reduced if used to pump abrasive type of spray material. Rollers in roller pump are commonly made of nylon, but sometimes rubber and steel are also used. Pressures above 600 kPa are generally not recommended for rotary pumps when pumping non-lubricating liquids.

Centrifugal pump works on high speed having column discharge. The discharge rate varies directly with speed. It is not a positive displacement pump. It is extensively used on sprayers. This type of pump can handle any type of spray material. They have long life and have wide range of capacity. They are directly mounted on PTO shaft of tractor. They are relatively unaffected by abrasive-type spray material. These pumps are well suited for air blast sprayers and aircraft sprayers, for which high flow rates needed at relatively low pressure. Diagram-type pumps are also used on sprayers where flow rate is limited to 19-23 1/min and pressure does not exceed 550 kPa. Since the valves and diaphragm are the only moving parts in contact with spray material, these pumps can readily handle abrasive material.

11.7. NOZZLES

Nozzle is an important mechanism, which breaks the spray liquid into the desired size of droplets for application to the surface to be sprayed. The nozzles help to control the rate, uniformity, thoroughness and safety of pesticides application. The principal parts of nozzle are body, valve, washer, disc and cap. The shut-off may be provided in the supply line or in the nozzle itself. The valve is used to impart the whirling motion to the spray liquid as it flows at high velocity through valve into eddy chamber of nozzle body. Since no single nozzle can meet various spray requirements, they are now commonly manufactured with inexpensive replaceable nozzle tips or discs, which can be selected to give, desired spray characteristics and volume for the specified job. Nozzles vary with respect to rate of discharge, the angle of spray and the type of spay pattern. Keeping these characteristics into mind, five types of nozzles commonly used in pesticides spray equipment are hollow-cone, solid-cone, triple-action, flat fan and flooding nozzles.

The hole at centre of nozzle disc results in a hollow cone spray. If there is a hole through the centre of valve, a solid cone of spray is produced. Built into some nozzles is a removable strainer with slightly smaller openings than the nozzle orifice to prevent clogging. Most of these nozzles require a minimum hydraulic pressure of 75 kPa whereas, for full development of spray pattern a hydraulic pressure of 300 kPa is required.

Hollow-cone nozzle. They are commonly used for spraying insecticides and fungicides. The pattern of spray is spray is circular with tapered edges to form a hollow cone or ring type spray pattern. This type of nozzle provides adequate coverage of pesticides from all angels. The liquid is fed into a whirl chamber through tangential entry or through a fixed spiral passage to give a rotating motion. The liquid comes out in the form of a hollow conical sheet, which then breaks into small drops.

Solid-cone nozzle. These nozzles produce a solid-cone spray pattern and are used for spot spraying of herbicides in lawns. They cover entire circular area evenly. The construction is similar to hollow cone nozzle with the addition of an internal jet, which strikes the rotating liquid just within the orifice of discharge. The breaking of spray droplet is mainly due to impact action.

Flat-fan nozzle. The flat-fan nozzle produces a spray pattern in the form of a flat sheet and is used for spraying herbicides. The spray pattern has been designed in such a way that 30-50% overlap can be achieved for even distribution of spay liquids. The nozzle tip is provided with an outlet that is elliptical. This nozzle is mostly used for low pressure spraying.

Flooding nozzle. It makes a wide-angle flat spray pattern. It works at lower pressure than other nozzles. The spray pattern is fairly uniform across its width. It is used for spraying herbicides and fertilizer solutions. The flooding flat nozzles result in less drift than the other type of nozzles.

Triple-action nozzle. These nozzles are very popular because they can produce a jet, a solid-cone and a hollow-cone, three types of spray pattern by adjustment. They are used for high-volume as well as low-volume spray applications and are mostly used with tractor-drawn sprayers. The triple action nozzle consists of a swirl plate, swirl chamber, a nozzle disc and a cap. Nozzles are usually made of brass, stainless steel or plastic. Pressure has a considerable effect on nozzle performance. For a given size and design of nozzle and depth of swirl chamber, the higher the pressure, the higher the throughput, the wider the spray angle and the finger the spray droplets.

Selection of nozzles. Most of the nozzle manufacturers give discharge of various nozzles manufactured by them at different pressures. This information can be used for selection of correct nozzle for spray job to be done. Before selecting the right nozzle, following points should be considered:

- Type of spray job i.e. spraying of weedicides, insecticides etc.
- Total amount of spray solution to be applied
- Row spacing
- Number of nozzles to be used per row and nozzle spacing
- Type of spray pattern desired i.e. fan or cone type
- Speed of travel, and
- Approximate pressure to be sued in spraying

It has been observed that the use of single hollow cone nozzle directly over the crop row gives better insect control and results in higher yields. The fan and bloomless-type spray nozzles do not give as good results as hollow nozzles. The higher yields are obtained from the rows nearest to the nozzle and where more complete coverage is achieved with smaller droplets.

Flat fan nozzle

Flood jet nozzle

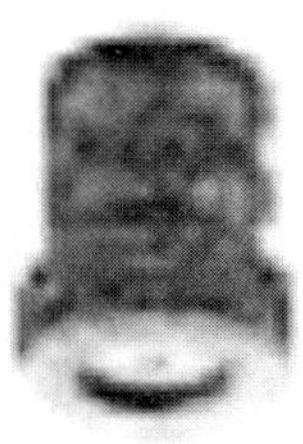

Cone type nozzle

Triple action nozzle

11.8. AGITATORS FOR CHEMICALS

Positive agitation of spray material in the tank is essential for proper mixing of insecticides and to permit using full range of spray materials including wettable powders, emulsions, fungicides, cold water paints or any other spray material. A propeller or paddle type mechanical agitator in sprayers is equipped with reciprocating type pumps. Hydraulic agitation is used in sprayers provided with pumps mounted on PTO shaft of tractor. Hydraulic agitation usually is not as thorough as mechanical agitation. Mechanical agitation is obtained by means of either flat blades or propellers on a shaft running length-wise in the near the bottom. The speed of agitator is kept in the range of 100-200 rpm. The peripherals speed of paddles and shaft input at any peripheral speed can be obtained for round-bottom tanks with flat, I-shaped paddles sweeping close to the bottom from the relations (Kepner *et al.,* 1987)

$$Sm = 5.39 \times \frac{A^{0.422}}{R^{0.531}} \times Fe^{0.293}$$

$$Sp = 3.26 \times 10^{11}\ R^{0.582}\ S^{3.41}\ L$$

Where,

Sm = minimum peripheral speed of paddles (m/min)

Sp = shaft input at any peripheral speed S (kW)

R = total combined width of all paddles divided by tank length

A = depth of liquid above agitator shaft centerline (mm)

L = Length of tank (mm)

S = peripheral speed of shaft (m/min)

Fe = factor indicating relative difficulty of agitating a given oil-water emulsion

= 0.83 for 60% oil in oil-water emulsion

= 1.00 for 40-50% oil in oil-water emulsion

= 0.89 for 10% oil in oil-water emulsion

Paddle tip speeds in excess of about 150 m/min may cause serious foaming of some of the mixtures. The minimum tip should be 1.22 Sm for mechanical agitation of emulsion in flat bottom tanks with rounded corners. This increase in tip speed doubles the power requirement of input shaft.

11.9. SEED TREATMENT EQUIPMENT

One of the constraints in adopting improved and efficient implement for cotton cultivation is the difficulty being faced in mechanical sowing, which otherwise makes easy the operations like inter-culture, weeding and plant protection. Cottonseeds bear considerable quantity of fuzz around them, which chokes the seed metering and dropping mechanism of sowing machines. Viable and healthy seeds can not be separated from the lot of fuzzy seeds with the result that there is a poor crop stand.

The fuzz around the seeds can be removed by acid delinting that involves treating the cottonseeds with concentrated sulfuric acid in specific quantities and time and subsequently washing the mix in abundant clean water. In this process, the light and impotent seeds start floating and easily separated. Heavy and healthy seeds, which sink, are later dried and sown.

Small quantity of cottonseeds can be acid-delinted in a plastic bucket. The seeds to be delinted are put into bucket and required amount of sulfuric acid is added and mixed till all the fuzz is burnt. The seeds then are transferred to a sieve tray washed thoroughly in water. The capacity of this system is very low. The hand operated small size delinter with an enameled drum can be used by small farmers (Bahl *et al.,* 1982). Agitators provided on a vertical shaft and rotated manually mix the seeds and acid in the drum properly. Then the treated seeds are taken out from the drum and washed thoroughly with clean water and dried. The hand operated commercial grade cotton delinter is a available for large-scale acid delinting of cottonseeds. It consists of a mixing drum, which is eccentrically mounted on a horizontal shaft. Agitators are fixed to the shaft inside the drum for thorough mixing drum. A suitable gate is provided in the drum for filling and taking out seeds. Fuzzy cottonseeds are put into the drum and specific quantity of concentrated sulfuric acid is added to it. The machine is rotated manually till the fuzz of cottonseeds burnt. The seed is taken out from drum, washed with clean water, rinsed with limewater and dried under the sun.

Acid delinting of cottonseeds is a precise process and requires precautions. If the quantity of acid is more or mixing time is more, the whole of seed may get damaged because of overreaction of acid. The outer coat of seed may get ruptured and result in reduced germination. On the contrary, if less acid is added or mixing time is short, delinting may be incomplete and some fuzz may still remain on the seed. Every care should be taken to ensure required quantity of sulfuric acid to be added to seed and treated till delinting is complete. The person working with this operation should be careful in handling sulfuric acid. The residual water after cleaning the treated seed should be disposed off to suitable place with necessary care.

11.10. SAFETY MEASURES

All chemicals used for dusting and spraying plants are very strong poisons and should be handled with utmost care. The following safety precautions be taken while handling and spraying the insecticides and pesticides. Do not allow youngsters, pregnant women and nursing mothers to take part in seed dressing, dusting and spraying of plants; make sure that all workers are familiar with safety rules governing the handling of chemicals and are supplied with respirators, goggles and overalls; keep a supply of clean water, soap, towels and first-aid-kit with necessary antidotes on the working site; do not use leaky sprayers., during spraying avoid smoking and do not take food; at the end of the working day wash the face and hands; wash the clothes after use ; store the left over chemicals in safe place under lock and key and destroy the container properly. For dry dressing of seeds, choose an open place at a distance from the living quarters, cattle sheds; storehouse and water storage pools and pastures grounds.

Transport the chemicals in tight and well closed containers. Supply all workers with overalls and protective devices while handling and spraying chemicals. Should there be case of poisoning, send the victim to immediately to the nearest hospital. If the cause of poisoning is arsenic, fluorine, mercuric bichloride or phenol, give the victim some warm milk with egg. If the cause of poisoning is copper compounds, do not give milk.

REFERENCES

Anonymous (1995-97) Biennial Report. Department of Farm Power & Machinery, Punjab Agricultural University, Ludhiana.

Bahl, V.P., Sharma, D.N. and Jain, M.L. (1982) Acid delinting of cottonseed made easy. *Agril. Engg. Today.* **6**(3): 17-22.

Garg, I.K. and Singh, Surendra (2002) *Farm Equipments for Punjab Agriculture.* Department of Farm Power and Machinery, Punjab Agricultural University.

Ganapathy, S. and Singh, G. (1993) Development of an electro-static sprayer. *Agricultural Mechanization in Asia, Africa and Latin America* **24**(1):17-22

Gupta, C.P., Alamban, R.B. and Dante, E.T. (1994) Development of knapsack electrostatic- Spinning disc sprayer for herbicide application in rice. *Agricultural Mechanization in Asia, Africa and Latin America* **25**(4):31-34.

Kepner, R.A., Bainer, R. and Barger, E.L. (1987) Principles of Farm Machinery. Third Indian Edition. CBS Publisher & distributors, Delhi.

Ojha, T.P. and Michael, A.M. (1992) *Principles of Agricultural Engineering.* Vol. 1, Jain Brothers, New Delhi.

Shukla, L.N., Sandhar, N.S., Singh, Surendra and Singh, Joginder (1987) Development and Evaluation of wide-swath, tractor mounted sprayer for cotton crops. *Agricultural Mechanization in Asia, Africa and Latin America* **18**(2):33-36.

Singh, Surendra and Kandoria, J.L. (1999) *Selection, use and Maintenance of Pesticide Application Equipment.* Indian Council of Agricultural Research, New Delhi.

Srivastava, A.C. (1990) *Elements of Farm Machinery.* Oxford & IBH Publishing Co. Pvt. Ltd. New Delhi.

Smith, H.P. and Wilkes, L.H. (1997) *Farm Machinery and Equipment.* Tata Mcgraw-Hill Publishing Co. Ltd., New Delhi

Srivastava, A.K., Goering, C.E. and Rohrbach, R.P. (1994) *Engineering Principles of Agricultural Machines.* ASAE, textbook No. 6, ASAE, 2950 Niles Road, St. joseph. .Michigan, USA.

Tajuddin, A. (1997) Comparative performance evaluation of battery-operated hydraulic, pneumatic and centrifugal sprayers. *J. Inst. Engineers (India), Agril. Engg.* 78 September : 14-16.

Theory and Practice of Integrated Pest Management
R. Arora, B. Singh and A.K. Dhawan (Eds)
Scientific Publishers (India), Jodhpur 2012, pp. 195 - 212.

12

ROLE OF ENTOMOPATHOGENIC NEMATODES IN INSECT PEST MANAGEMENT

Naveen Aggarwal
Department of Entomology
Punjab Agricultural University, Ludhiana - 141 004, India

12.1. INTRODUCTION

Indian agriculture is the backbone of economy, owing to its climatic conditions and its particular environment, suffers severe losses of agricultural produce due to attack by various insect pests. Agriculture in India has gone through immense changes in the second half of the twentieth century. During the last decade, chemointensive integrated pest management modules are being widely advocated for management of the insect pests. Use of chemical fertilizers and pesticides has played a positive role in increasing agricultural productivity and in making India self-sufficient in food grain production. About 3 per cent of the total pesticides used in the world are utilized in India. The use pattern of pesticides reflects that cotton crop alone consumes 44.5% pesticides followed by rice, which accounts for 22.8% of pesticide consumption. In recent years the use of synthetic insecticides in crop protection programmes around the world has resulted in disturbances of the environment, pest resurgences, pest resistance to pesticides and lethal effect to non-target organisms in the agro-ecosystems in addition to direct toxicity to users. Growing public concern over the consequences of the use of chemical pesticides, particularly ground water contamination, residues on food, resistance development, have fueled an intense search of some safer alternatives. Ultimately IPM got the momentum and of which biological control of crop pests is considered as its important component and found to be most viable alternative, is economical has long term control, without risk to human beings and other non target organisms. In this method, microbes or entomopathogens used as microbial control measure are the microscopic or submicroscopic entities with their toxins cause diseases in insects and their by kill damage or render the pests harmless to crops. The annual growth increase in insecticides use is 1% and that of microbial insecticides is 10-25%. When used as part of IPM programs microbial pesticides (bacteria, fungi, nematodes, protozoans and viruses) are viable, safer and ecofriendly alternatives.

Nematodes associated with insects, referred to as entomophillic, entomogenous or entomopathogenic are known to parasitize, cause disease and kill the insects. The entopathogenic nematodes (EPNs) are potential agents as they serve as as vectors of bacteria, achive quick kill of target insect pests. The first nematode (*S. carpocapsae*) used successfully in the control of an insect pest was reported 30 years ago from Australia; Commonwealth Scientific and Industrial Research Organization (CSIRO) was the first in the world to use EPNs commercially against black vine weevil in ornamentals and against currant borer moth in black currants (Georgis, 2002). There are several species of EPNs used around the world against a variety of pests in niche markets (e.g. fungus gnats in nurseries, hydroponics, and mushroom; weevils on ornamentals, strawberries, cranberries, citrus and bananas; scarabs of turf, ornamentals and blue berries; cutworms, webworms, billbugs and mole crickets; termites in wooden articles and trees; peach borer moth in apples and carpenter worm in shade trees in China, and fig trees in USA) (Poinar, 1990). Recently out of 13% bio-insecticide sale in industrialised countries, EPN sale was only second to Bacillus thuringiensis at 80% (Lisansky and Coombs, 1994). Considerable progress has also been made during the last 20 years on the subject dealing with taxonomy, biology, genetics, ecology, host range, production, application technologies, laboratory and field trials and commercialisation of EPNs and their symbiotic bacteria. Eighty three described EPN species have been identified (64 species of Steinernema and 8 species of Heterorhabditis and 1 species of Neosteinernema) from various insects or from the soil worldwide (Grewal *et al.,* 2001).

Entomopathogenic nematodes (EPN) in the genera *Steinernema* Travassos and *Heterorhabditis* Poinar (Nematoda: Steinernematidae and Heterorhabditidae) are obligate and lethal parasites of insects (Grewal *et al.,* 2005). These nematodes have a unique symbiotic relationship with enteric bacteria. Bacteria of the genera *Xenorhabdus* and *Photorhabdus* (Enterobacteriaceae), with the exception of *Photorhabdus asymbiotica* associate mutualistically with nematodes of the families Steinernematidae and Heterorhabditidae, respectively. The bacteria produce toxins for killing insect, protect cadavers from secondary infections and provide congenial conditions for nematode survival and reproduction. Nematodes in turn protect bacteria by harbouring them in their gut and act as a vector for transporting bacteria in to insect host. EPN together with their symbiotic bacteria show great promise as biological alternatives to chemicals for the control of economically important insect pests (Grewal and Georgis, 1999). Species of *Steinernema* and *Heterorhabditis* have shown to be effective in biological control of several important insect pests. Their impressive list of positive attributes including high virulence (ability to kill host within 48 h), availability of scalable mass production techniques, broad host range, host search ability, safety to non-target organisms, and exemption from registration in many countries has generated a great deal of scientific and commercial interest in these insect-killing nematodes. Species and isolates of EPN exhibit considerable differences in host range, virulence, tolerance to environmental stresses, foraging behaviour, longevity and amenability to mass production and formulation (Grewal *et al.*, 1994; Somasekhar *et al.*, 2002).

12.2. BIOGEOGRAPHIC DISTRIBUTION

The species of *Steinernema* and *Heterorhabditis* have worldwide distribution. Steinernematids have been reported from all the continents except Antarctica. Among the species of *Steinernema,* some species like S. *carpocapsae* and *S. feltiae* have a global distribution while other species show restricted geographic distribution (Hominick *et al.,* 1996). However, with increasing number of surveys in many countries, the geographical distribution range of many known species is expanding besides discovery many new species. The distribution of heterorhabditids is relatively narrow compared to the steinernematids because fewer number of species. Among *Heterorhabditis* species, *H. bacteriophora* has widest geographical distribution and is found in North and South America, Europe, Australia and Asia. *H. indica* which was originally described from India also has wide distribution including, several countries in tropics and subtropics (Hominick *et al.,* 1996). However, *H. zealandica* and *H. marelatus* have a limited distribution and are found only in New Zealand (Akhurst, 1987) and USA (Liu and Berry, 1996), respectively.

12.3. BIOLOGY

The parasitic cycle of *Steinernema* nematodes is intiated in by the third-stage infective juvenile (IJs) or dauer juvenile, which is resistant to environmental conditions. These non-feeding juveniles locate their insect hosts in soil by detecting chemical cues such as excretory products and carbon dioxide gradient and invade suitable host insects through natural body openings (i.e, anus, mouth and spiracles). Once inside the host, nematodes invade the haemocoel and release the symbiotic bacteria that are held in nematode's intestine. The nematodes, together with their symbiotic bacteria form a unique biocontrol system. Toxins produced by the developing nematodes (Burman, 1982) and bacteria (Akhurst and Boemare, 1990) cause speticemia, killing the host. The IJs feed on the rapidly multiplying bacteria and disintegrated the host tissues. They undergo successive molts to develop into adult males and females. Steinemematids are amphimitic in all generations. Large females and smaller males mate and produce progeny. About 2-3 generations of the nematodes are completed within the host cadaver.

When food reserves are depleated, the nematode reproduction ceases and the offspring develop into resistant IJs that disperse from dead host, and are able to survive in the environment and seek out new hosts. From a medium-sized insect cadaver 100,000 to 500,000 IJs are produced and they do not feed and can survive in the soil for several months. The cycle of entry of infective juveniles in to the insect host to emergence of new batch of infective juveniles usually requires 10 to 14 days.

The IJs are capable of tolerating stresses fatal to other developmental stages, and therefore, can be effectively formulated and preserved for several months and shipped for application. The symbiotic association of EPN with specific bacteria facilitates rapid mass production of the nematodes (bacteria serve as food) and successful pathogenicity. Nematodes acts as vectors to transport the bacteria into a host within which they can proliferate, and the bacteria create conditions necessary for nematode survival and reproduction within the insect cadaver. The life cycle of *Heterorhabditis* is also similar to *Steinernema,* but in this case the infective juveniles develop into hermaphrodites in the first generation while second generation juveniles develop in to adult males and females inside the insect host. *Heterorhabditis* carries a bacterial symbiont *Photorhabdus* instead of *Xenorhabdus.* Further, *Heterorhabditis* can enter the insect haemocoel directly through the insect's integument (Wang and Gaugler, 1998).

12.4. HOST RANGE AND ECOLOGICAL CONSIDERATIONS

The entomopathogenic activity of steinernematid and heterorhabditid species has been documented against a broad range of insect pests in a variety of habitats (Kaya and Gaugler, 1993). They have been used inundatively in a number of high value cropping systems (Koppenhofer *et al.,* 2000). The nematode-bacterium complex kills insects so rapidly that the nematodes do not form the intimate, highly adapted, host-parasite relationship characteristic of other insect nematode associations like the mermithids. This rapid mortality permits the nematodes to exploit a range of hosts that spans nearly all insect orders, a spectrum of activity well beyond that of any other microbial control agent. In laboratory tests, *S. carpocapsae* alone infected more than 250 species of insects from over

75 families in 11 orders (Poinar, 1983). The nematodes attack a far wider spectrum of insects in the laboratory where host contact is assured, environmental conditions are optimal, and no ecological or behavioural barriers to infection exist (Gaugler, 1981). However, their efficacy under field conditions is influenced by wide range of biotic (foraging strategies of nematodes, behavioral responses of insects, etc) and abiotic factors (DV radiation, temperature, relative humidity, etc) (Gaugler *et al.,* 1988). Foliage feeding lepidopteran larvae are seldom impacted in the fields though infected in the laboratory reason being quick inactivation by dessication, radiation and temperature. The ecological and behavioural barriers restrict their natural host range. Temporal and spatial factors, such as synchronous life cycles and differing habitat preferences, also appear to determine host range. They should be normally applied in the late evening hours to avoid exposure to intense sunlight and UV radiation with suitable adjuvants like glycerol, liquid paraffin, etc. Entomopathogenic nematodes require adequate soil moisture for survival and movement, but too much moisture may cause oxygen deprivation and restrict movement (Kaya, 1990; Koppenhofer *et al.,* 1995). Optimum moisture levels will vary by nematode species and soil type. A light irrigation before and after application of nematodes provides optimum soil moisture for nematode activity in case of soil applications. Isolation and identification of native nematode strains with high level of tolerance to environmental stresses and development of stress tolerant strains through genetic improvement are necessary to increase the efficacy of these nematodes under field conditions. Some nematode species may search for hosts at or near the soil surface (e.g., *S. carpocapsae* and *S. scapterisci*), whereas others are adapted to search deeper in the soil profile (e.g., *H. bacteriophora* and *S. glaseri*). The former group has been referred to as "ambusher", which remains nearly sedentary while waiting for the mobile surface dwelling hosts (Campbell and Gaugler, 1993). The latter group has been referred to as "cruiser" which is highly mobile, responds strongly to long-range host chemical cues, and is therefore best adapted to find sedentary hosts (Grewal *et al.,* 1994).

12.5. MASS PRODUCTION TECHNOLOGY

EPN can be mass produced by *in vivo* or *in vitro* methods. Insect host serve as small biological reactor for mass production of EPN in the *in vivo* production method. The larvae of wax moth *Galleria mellonella* are commonly used for *in vivo* nematode production in laboratory. The basic *in vivo* production technique was first described by Dutky *et al.* (1964). Later some improvements were suggested by other workers (Woodring and Kaya, 1988; Lindgren *et al.,* 1993). The yield of infective juveniles varies from 0.5 x 105 to 4 x 105 IJs/larva. Several lepidopterous insect larvae have also been used for mass production of *Steinernema* and *Heterorhabditis* species. The yield of infective juveniles varies with nematode species/isolate, inoculum level, insect host species, environmental conditions and nutrient status of host. The *in vivo* technique is very simple, requires minimum equipment and investment. However, this method is not suitable for large scale' commercial production as it lacks economy of scale and is sensitive to biological variations.

EPN are cultured *in vitro* on a variety of solid or liquid substrates such as dog biscuit-agar, potato mash, chicken offal, synthetic liquid media, etc. The *in vitro* culture method was first described by Glaser as early as in 1931. However, the first successful large scale *in vitro* production method using solid media was developed by Bedding (Bedding, 1981). This method is now widely used. In this method, polyurethane sponge pieces are impregnated with artificial media along with symbiotic bacterium. *In vitro* culture using solid media generally yields 6 x 10^5 to I 0 x 10^5 infective juveniles/g media. Friedman (1990) developed a liquid fermentation technique for large scale mass production of EPN. Several steinemematids *(S. carpocapsae, S. riobravae, S. scapteresci,* S. *feltiae* and *S. glaseri*) and heterorhabditids (*H. bacteriophora, H. megidis* and *H. indica*) were efficiently produced using monoxenic liquid fermentation technique and it is commonly used in large-scale commercial mass production of EPN in industries with fermentors of 80000-300000 litre capacity. *In vitro* culture using liquid media generally yields 50 x 10^{12} infective juveniles/g. This method is very economical for large scale production and gives consistent out put and quality. Nematode species and isolates exhibit considerable differences in suitability for *in vitro* production. Therefore this technique needs to be fine-tuned to suit the nematode species for getting optimum yield.

12.6. FORMULATIONS

Formulations of Entomopathogenic nematodes are developed basically to extend their shelflife and deliver live nematodes to the customers. Active nematodes are immobilized to prevent depletion of their lipid and glycogen reserves. Immobilization has been accomplishied by maintenance in aqueous suspension at low temperature (5-15°C). Steinerematids can be stored at 4-10°C for 6-12 months without much loss of activity whereas heterorhabditids donot store well and 2-4 months of storage at 4-10°C is considered good. A major constraint of commercialization of these nematodes is their poor storage capacity. They can be stored under refrigeration in oxygenated water for up to 5 years (Friedman, 1990). But this methodology is not commcially feasible. Nematode metabolism is temperature driven and warm temperature (20-30°C) increase metabolic activities, reducing nematode viability. Simple formulations were developed using carriers like polyurethane sponge, peat, vermiculite, etc which have sufficient interstial space to facilitate gas exchange required for nematode survival. In these formulations carrier substances are impregnated with nematodes. Nematodes from the sponge formulations are extracted by squeezing them in to water before applications. Although these formulations are simple and easy to produce, they are labor intensive and needs constant refrigeration thus useful for only for small scale applications. Later on, newer formulations with improved shelf life and scalability were developed. These formulations are aimed at extending the shelf life by conserving storage reserves of infective juveniles by restricting their movement (physical immobilization) or oxygen consumption by inducing a state of partial anhydrobiosis (physiological immobilization). The formulations based on principle of physical immobilization include activated charcoal, alginate gels, flowable gels, etc. (Yukawa and Pitt, 1985; Kaya and Nielsen, 1985; Georgis and Manweiler, 1994). Nematodes are extracted by dissolving these gels in water containing sodium citrate before application. The alginate gel formulations of *S. carpocapsae* possessed the shelflife of 3-4 months even at room temperature. However, preparation of these formulations and disposal of consumables used is a cumbersome process and may not be suitable for large scale use. A significant advance in formulation of EPN was made with the development of novel water dispersible granule (WDG) formulations based on the principle of physiological immobilization. In these formulations IJs are encapsulated in 10-20 mm diameter granules made up of a complex mixture of silica, clays, cellulose, lignin and starch. The granular matrix allows supply of oxygen to nematodes in the center of the granule in a partially desiccated state (Silver *et al.,* 1995). The WDG formulations have a shelf life of several months at room temperature. Nematodes can be extracted by dissolving these granules in water before application. The shelf life of a formulation varies depending on nematode species, storage temperature and type of formulation.

12.7. APPLICATION TECHNOLOGY

Entomopathogenic nematodes should be applied at the first sign that a pest population is initializing to cause damage. Reapplying nematodes depends on the success of the first nematodes released. Their survivorship and success are based on environmental condition, moisture and soil type and percentage of living nematodes actually released during the first application. Nematodes should be reapplied on seven day intervals if damage continues. In order to ensure maximum effectiveness, it is crucial to apply them at the optimum environmental conditions needed for their better survival. Therefore, it is best to irrigate the target site, both before and after application because they need moist conditions to prevent desiccation and aid with movement to find hosts. Also, the best results are obtained when the relative humidity is high, ambient temperature is neither extremely hot or cold, soil temperature is between 10 to 35°C, soil is moist and direct sunlight is minimal. All of these factors help prevent the nematodes from drying out and increase their survival and virulent.

EPN can be applied using a wide range of application equipment including ground and aerial spray equipment like pressurized sprayers, mist blowers, and electrostatic sprayers (Georgis, 1990) and irrigation systems like drip and sprinkler irrigation systems. The efficacy of application. influenced by the nematode species, type of application equipment and the manner in which the nematodes are applied. A general recommendation for entomopathogenic nematodes has been common nozzle type sprayers with openings larger than 50 μ and operating pressures less than 2000 kPa (290 psi) (Georgis, 1990). However, some nematode species can tolerate operating pressures up to 2000 kPa while other species like. *H. megidis* can not withstand operating pressure above 1380 kPa. EPN can be used as soil application, foliar sprays in the field (Grewal and Georgis, 1999); they can be applied to trap crops (Purcell *et al.,* 1992) or propagation material like cuttings and root stalks (Bedding and Miller, 1981; Bari, 1992) or they can be used in insect traps (Parkman and Frank, 1993).

Carbanallis and Raulstan (1996) found that the efficacy of *S. riobrave* was higher on *Helicoverpa zea* in corn when applied at 200,000 IJ/mt^2 through furrow irrigation or post irrigation than when nematode were sprayed on to soil before irrigation. Thus *S. riobrave* was effective for suppressing corn earworm under field conditions of high temperature with irrigation. For foliar application of EPN, antidesiccants have been used successfully to retard evaporation of the nematode suspension on foliage and thus to reduce dessication of nematodes. Excellent control of Japenese beetle, *Popillia japonica* and the northern masked chaffer, *Cyclocephala borealis* was obtained with 1.25-5.0 billion IJs/ha when plots were irrigated before application and again within 24 hrs after treatment (Downing, 1994). Baurer *et al.* (1997) studied the effect of a number of adjuvant on entomopathogenic nematode persistence and efficacy against *Plutella xylostella.* In laboratory TX7719, Rodspray oil and Nufilm P provided the best antidessicant activity and in greenhouse, TX 7719+ Blankophor BBH increased nematode persistence on watercress leaves and efficacy against *P. xylostella* significantly. Addition of phagostimulants enhanced the effectiveness of *S. carpocapsae* against the 4th instar larvae of *Spodoptera litura* feeding on sunflower head in laboratory and field tests.

12.8. SUCCESSFUL STORIES OF EPNS AGAINST AGRICULTURAL PESTS

Several weevils attack root and tuber crops, among them those of the genus Cylas and Euscepes species are the most important on sweet potato and cassava in Brazil and other South American countries. Mannion and

Jansson (1992) assessed the virulence of ten entomopathogenic nematodes to *Cylas formicarius*. Most nematodes were more virulent to larvae than to pupae. Adults were less susceptible to nematodes than other stages, and adult males were more susceptible than females. Under field conditions *S. carpocapsae* and *H. bacteriophora* were shown to reduce weevil densities upto 83% and 81% on plants treated with the two species, respectively (Jansson *et al.,* 1990).

Hypothenemus hampei is a major pest of coffee seeds in Brazil and other South American countries (Waterhouse, 1998). Infestations of *H. hampei* occur in coffee seeds while they are enclosed in berries on the trees and in berries that fall to the ground. Spraying of nematodes on fallen berries might remove the need to collect them, leaving them to produce mulch. Dispersal of infected adults may also spread nematodes into the pest population. According to Waterhouse (1998) there appears to be only one record of nematodes attacking *H. hampei* in the field in India (Varaprasad *et al.,* 1994). Allard and Moore (1989) showed that a *Heterorhabditis* sp. could cause high mortality of both adult and larvae and that infective juvenile were produced from adults and large larvae. Castillo and Marban-Mendoza (1996) reported differences in the infectivity of eight nematode strains to *H. hampei* larvae and found that three strains of Heterorhabditis sp. and one of *S. carpocapsae* caused high mortality of the larvae.

Several species of cutworms, Agrotis spp., *Spodoptera frugiperda, S. exigua* and *S. litorallis* cause serious problems to agricultural, vegetable and forage crops, worldwide. Cutworms are highly susceptible to a number of entomopathogenic nematode species and strains (Morris and Converse, 1991). Control of *Agrotis segetum*, with *S. feltiae* (=*N. bibionis*) in lettuce was equivalent to endosulfan (Lossbroek and Theunissen, 1985) under field conditions. *A. ipsilon* has been effectively managed with *S. carpocapsae* on golf course greens. Larvae and pupae of armyworms are very susceptible to entomopathogenic nematodes (Kaya and Grieve, 1982), and can be effectively managed by nematodes. Richter and Fuxa (1990) reported 33-43% infection of *S. frugiperda* by *S. carpocapsae* in field corn. They also found that spraying of nematodes onto corn ears caused up to 71% infection of *S. frugiperda* and they concluded that *S. carpocapsae, S. riobrave*, and *H. megidis* have potential for controlling *S. frugiperda*. The corn rootworms, *Diabrotica* spp., are important pests of corn. In North America *D. virgifera virgifera* and *D. barberi* are the two dominant species that cause significant economic losses to maize. Nematode applications for rootworm suppression were ineffective in the early experiments (Munson and Helms, 1970), but more recently in field studies, *S. carpocapsae* significantly reduced maize root damage (Ellsbury *et al.,* 1996), reduced rootworm larval population (Jackson, 1996), and rootworm adult emergence (Ellsbury *et al.,* 1996). In some cases, nematode performance was equal to, or better than, insecticides (Wright *et al.,* 1993). The limiting factors of efficacy are the need for timing of application to coincide with the phenology of susceptible stages of *Diabrotica* spp. (Jackson and Brooks, 1995) and the adverse effects of desiccation on survival of the nematodes. Recent work (Nishimatsu and Jackson, 1998) showed that the combined use of insecticides (tefluthrin) with entomopathogenic nematode may offer an integrated approach to increase nematode of rootworm.

The worst pest of China's one million hectares of apples is a moth (*Carposina niponensis*) rather like codling moth except that it over-winters as a grub in the soil rather than in crevices in the bark and it was found that EPNs could be used to control this pest better and more cheaply than conventional insecticides. *Teinernema carpocapsae* IJs sprayed beneath apple tree canopies just after the first summer rains (a signal for the caterpillars to migrate from their over winter sites deep in the soil to near the surface to pupate) gave well over 95% kill and the few moths emerging were insufficient to produce economically damaging levels of infestation during that season. That the Chinese replaced most of their older trees (that favour infestation with the moth) with new ones has so far stopped the EPNs being a major success story. However, the new trees will grow and the moth will once again require controlling. One of the first insects to be controlled commercially using EPNs was the black currant borer moth, the caterpillars of which bore through the stems of blackcurrants often halving the yield and also halving the vitamin-C content of the remaining berries (Bedding and Miller, 1981). Over a million cuttings were stacked in walls and sprayed all over with a concentrated suspension of *S. feltiae* infective juveniles and nearly 100% of the larvae had been killed by nematodes. Several plantations that had serious infestations of currant borer moth were treated by blast spraying the currant bushes with a suspension of *S. feltiae* infective juveniles and this reduced borer moth populations by over 70% (Miller and Bedding, 1982).

12.9. SAFETY AND REGULATORY PROCEDURES

EPN had no adverse long term impact on non target insects and other invertebrates (Bathon, 1996). Although EPN could infect wide range of insect in laboratory studies with high dosages, their impact on field populations of non target invertebrates was found to be negligible (Bathon, 1996). Investigations on safety of EPN and their symbiotic bacteria to vertebrates and humans also showed no negative effects despite widespread use during the past 20 years. Some isolates of *Photorhabdus luminiscens* were isolated from human wounds (Farmer *et al.,* 1989). However, detailed studies based on DNA analyses clearly showed that these clinical isolates are different from isolates having symbiotic association with nematodes and later they were grouped under a new species *P. asymbiotica* (Fischer-Le Saux *et al.,* 1999). Inundative applications of EPN showed no adverse effects on plants. In some cases, plant vigor increased following soil application of EPN. EPN were found to suppress populations of plant parasitic nematodes that cause serious damage to plants (Smitley *et al.,* 1992; Jagdale *et al.,* 2002). Further, they were found to suppress plant parasitic nematodes selectively i.e. without adversely effecting the populations of free-living microbivorus nematodes that are beneficial to plant growth and hence this phenomenon is referred to as beneficial non-target effect (Somasekhar *et al.,* 2002b). Environmental Protection Agency of USA has exempted even the genetically modified EPN also from registration. However, the introduction of exotic strains of EPN is regulated in many countries to avoid biological pollution.

12.10. STATUS OF EPNs IN INDIA

Research on EPNs in India was started in 1960s with an exotic strain (DD136) of *S. carpocapsae.* Several important insect pests of rice, sugarcane and apple were found susceptible to this nematode in laboratory (Rao and Manjunath, 1966). However, attempts to expand their use under field conditions were not successful due to problems in adaptability of this exotic strain to local conditions and lack of mass production and application techniques. Rahman *et al.* (2000) reviewed the research done on EPNs in India. EPN research in India got slowed down during 1970s-80s due to greater emphasis given for use of chemical pesticides and fertilizers to support high yielding varieties released during green revolution period.

Organized research on EPNs with special emphasis on isolation and identification of indigenous strains was initiated in late 1980s at Sugarcane Breeding Institute (SBI), Coimbatore and Tamilnadu Agricultural University, Coimbatore. Several indigenous isolates of EPN were collected from tropical and subtropical sugarcane growing regions of India in surveys carried out by SBI, Coimbatore. A new isolate of *Heterorhabditis* was recovered from sugarcane top borer, *Scirpophaga excerptalis* (Pyralidae: Lepidoptera) in the vicinity of Coimbatore, India in one of these surveys. This isolate was later described as a new species *Heterorhabditis indica* Poinar, Karunakar and David, 1992 (Poinar *et al.,* 1992). Several insect pests of sugarcane including borers (*Scirpophaga excerptalis, Chilo infuscatellus and Chilo sacchariphagus indicus*) and white grubs (*Hototrichia serrata*) were found susceptible to EPNs (Karunakar *et al.,* 1999 and 2000; Sankarnarayanan *et al.,* 2003 and 2006). Research on biological control of white grubs in sugarcane using indigenous EPNs strains was carried out at SBI, Coimbatore under an international collaborative research project involving leading research laboratories from Ireland and Germany. During this period detailed studies on molecular characterization of isolates of *Heterorhabditis indica* isolates from India and other countries (Stack *et al.,* 2000), development of large scale mass production technique using fermentors for Indian isolate of *H. indica* (Ehlers *et al.,* 2000) and field evaluation of EPN for biological control of white grub *H. serrata* in sugarcane were carried out. A new species *Steinernema thermophilum* was described by Ganguly and Singh from Indian Agricultural Research Institute, New Delhi, India (Ganguly and Singh, 2000). Occurrence of several isolates of known species of EPN were also reported from India including *S. carpocapsae* and *S. bicornutum* (Hussaini *et al.,* 2001), *S. riobrave* (Ganguly *et al., 2002), H. bacteriophora* (Sivakumar *et al.,* 1989). Recently, *S. asiaticum* and a rhabditid nematode *Oscheius* sp were reported to be infecting rice yellow stem borer (*Scirpophaga incertulas*) and leaf folder *(Cnaphalocrosis medinalis)* (Katti *et al.,* 2006; Prasad *et al.,* 2006; Padmakumari *et al.,* 2007).

Progress made in mass production and formulation technology coupled with increasing demand to reduce pesticide usage in recent years gave boost to EPN research in India. Currently several other laboratories in India are working on EPN including Project Directorate of Biological Control (PDBC, Bangalore), Gujarat Agricultural University (GAU), Anand and Tamil Nadu Agricultural, University (TNAU), Coimbatore, Directorate of Rice Research, Hyderabad. Commercial formulations of EPN (Soil commander and Green commander) were released in market in India by Ecomax company during 1980s but they were not successful. The failure of these formulations may be due to the poor adaptability of exotic strains used in these formulations or production or formulation problems and lack of awareness about proper use of nematodes in field at that time. Recently, Multiplex company from Bangalore started producing commercial formulations of *Steinernema* and *Heterorhabditis.* Information on field efficacy of these formulations is yet to be available.

12.11. CONCLUSIONS

The great advantages of using entomopathogenic nematodes as biopesticides for insect pest control are in operator and end-user safety, absence of withholding periods, the advantage of minimising the treated area by monitoring insect populations, minimal harm to natural enemies and lack of environmental pollution. The establishment of entomopathogenic nematodes in India would bring several benefits such as firmly establishing insect nematodes will promote the sustained use of agriculture and develop a better understanding of biodiversity. Due to the exceptional successes made with other biological control agents, India is poised for developing innovative ideas to implement the use of EPNs. This may be accomplished through holding workshops on entomopathogenic nematodes; promote cooperative international projects involving scicntists could providc training for Indian scientists and students in university programs. National and international agreements should express the mutual interest of both parts, in terms of exchanging experiences, material and information and respecting the national and international legislation related to conservation of biodiversity and exchange of biological control agents.

Future research efforts should be focused on discovery of native strains of EPN with high virulence and environmental stress tolerance, genetic improvement of native strains using conventional or genetic engineering methods, improvement of formulations with public-private partnership model involving research laboratories and industry. Further, creating awareness about these beneficial worms among farmers and extension agencies and promoting them as an ecofriendly component of integrated pest management (IPM) programs is vital for successful use of entomopathogenic nematodes.

REFERENCES

Akhurst, R.J. (1987) Use of starch gel electrophoresis in the taxonomy of the genus Heterorhabditis (Nematoda: Heterorhabditidae). *Nematologica* **33**:1-9.

Akhurst, R.J., Boemare, N.E. (1990) Biology and taxonomy of Xenorhabdus. In: Gaugler, R. and Kaya H.K. (eds) *Entomopathogenic Nematodes in Biological Control*. CRC Press, Boca Raton, FL. Pp 75-90.

Allard, G.B. and Moore, D. (1989) Heterorhabditis sp. nematodes as control agents for the coffee berry borer, *Hypothenemus hampei* (Scolytidae). *J. Invertebr. Pathol.* **54**: 45-48.

Bari, M.A. (1992) Disinfestation of artichoke stumps with entomopathogenic nematode against the artichoke plume moth. In: *Proc 19th International Congress of Entomology*. Abstracts. Beijing, China. pp. 319.

Bathon, H. (1996) Impact of entomopathogenic nematodes on non-target hosts. *Biocont. Sci. Tech.* **6**: 421- 34.

Baurer, M.E., Kaya, H.K., Gaugler, R. and Tabashnik, B. (1997) Effects of adjuvants on entomopathogenic nematode persistence and efficacy against *Plutella xylostella. Biocont. Sci. Tech.* **7**:13-525.

Bedding, R.A. (1981) Low cost in vitro mass production of *Neoaplectana* and *Heterorhabditis* species (Nematoda) for field control of insect pests. *Nematologica* **27**:109-14.

Bedding, R.A. and Miller, L.A. (1981) Use of a nematode *Heterorlzabditis heliothidis*, to control black vine weevil, *Otiorhynchus sulcatus*, in potted plants. *Ann. Appl. Biol.* **99**: 211-16.

Bedding, R.A. and Miller, L.A. (1981a) Disinfesting blackcurrant cuttings of *Synanthedon tipuliformis*, using the insect parasitic nematode, *Neoaplectana Bibionis. Env. Ent.* **10**: 449-53.

Burnnan, M. (1982) *Neoplectana carpocapse*: toxin production by axenic insect parasitic nematodes. *Nematologica* **28**:62-70.

Campbell, J.F. and Gaugler, R. (1993) Nictation behaviour and its ecological implications in the host search strategies of entomopathogenic nematodes (Heterorhabditidae and Steinernernatidae). *Behaviour* **126** :3-14.

Carbanillas, H.E. and Raulstan, J.R. (1996) Evaluation of *Steinerema riobravis, S. carpocapsae* and irrigation timing for the control of corn earworm, *Helicoverpa zea. J. Nemt.* **28**:75-82.

Castillo, A. and Marban, N.M. (1996) Laboratory evaluation of Steinernematid and Heterorhabditid nematodes for biological control of the coffee berry borer, *Hypothenemus hampei* Ferr. *Ematropica* **26**:101-09.

Dowing, A.S. (1994) Effect of irrigation and spray volume on efficacy of entomopathogenic nematodes against white grubs. *J. econ. Ent.* **87**: 643-646.

Dutky, S.R., Thompston, J.V. and Cantwell, G.E. (1964) A technique for the mass production of the DD-136 nematode. *J. Insect Pathol.* **6**:417-422.

Ehlers, R.U. and Hokkanen, H.M.T. (1996) Insect Biocontrol with non-endemic entomopathogenic nematodes (*Steinemema* and *Heterorhabditis* spp.): Conclusions and recommendations of a combined OECD and COST workshop on scientific and regulatory policy issues. *Biocont. Sci. Tech.* **6**: 295-302.

Ellsbury, M.M., Jackson, J.J., Woodson, W.D., Beck, D.L. and Stange, K.A. (1996) Efficacy, application distribution, and concentration by stem flow of *Steinernema carpocapsae* (Rhabditida: Steinernematidae) suspensions applied with a lateral move irrigation system for corn rootworm (Coleoptera: Chrysomelidae) control in maize. *J. econ. Ent.* **89**:71-81.

Farmer, III J.J., J.H. Jorgensen, P.A.D. Grimont, R.J. Akhurst, G.O. Poinar Jr., E. Ageron, G.V. Pierce, J.A. Smith, G.P. Carter, K.L. Wilson and F.W. Hickman-Brenner (1989) *Xenorhabdus luminescens* DNA hybridization group 5 from human clinical specimens. *J. Clin. Microbiol.* **27**: 1594-1600.

Fischer-Le Saux, M., Viallard, V., Brunei, B., Normand, P. and Boemare, N.E. (1999) Polyphasic classification of the genus *Photorhabdus* and proposal of new taxa: *P. luminescens* subsp *luminescens* subsp. nov., *P. luminescens* subsp. *akhurstii* subsp. nov., *P. luminescens* subsp. *Laurnondii* subsp.nov., *P. temperata* sp. nov., *P. temperata* subsp. *temperate* subsp. novo and *P. asymbiotica* sp. novo *Int. J. Syst. Bacteriol.* 1645-56.

Friedman, M.J. (1990) Commercial production and development. In: Gaugler, R, and Kaya, H.K. (eds) *Entomopathogenic Nematodes in Biological Control.* Boca Raton, Fl, USA, Pp 153-72.

Ganguly, S. and Singh, L.K. (2000) *Steinernema thermophilum* (Rhabditida: Steinernematidae) from India. *Int. J. Nemat.* **10**: 183-9l.

Ganguly, S., Singh, M., Lal, M., Singh, L.K., Vyas, R.V. and Patel, D.J. (2002) New record of an entomopathogenic nematode, *Steinernema riobrave* Cabanillas, Poinar and Raulston, 1994 from Gujarat, India. *Indian J. Nemat.* **32**: 223.

Gaugler, R. (1981) Biological control potential of neoaplectanid nematodes. *J Nemat.* **13**: 241-49.

Gaugler, R. (1988) Ecological considerations in the biological control of soil inhibitory insects with entomopathogenic nematodes. *Agric. Ecosys. Environ.* **24**:351.

Georgis, R. (1990) Formulation and application technology. In: Gaugler, R. and Kaya, H.K. (eds) *Entomopathogenic Nematodes in Biological Control.* Boca Raton, FL, USA, CRC Press. pp 173-194.

Georgis, R. (2002) The Biosys experience: an insider's perspective. In: Gaugler, R. (ed) *Entomopathogenic Nematology.* CABI Publishing, Wallingford, UK. pp 357.

Georgis, R. and Manweiler, S.A. (1994) Entomopathogenic nematodes: a developing biological control technology. *Agric. Zool. Rev.* **6**: 63-64.

Grewal, P.S. and Georgis, R. (1999) Entomopathogenic nematodes. In: Hall, F.R. and Menn J.J. (ed), *Methods in Biotechnology,* Vol. 5; *Biopesticides: Use and Delivery*. Humana Press, Totowa, NJ. Pp 271-99.

Grewal, P.S., Gaugler, R. and Wang, Y. (1994) Enhanced cold tolerance of the entomopathogenic nematode *Steinernema feltiae* through genetic selection. *Ann. Appl. Biol.* **129** : 335.

Grewal, P.S., Nardo, E.D. and Aguillera, M.M. (2001) Entomopathogenic Nematodes: Potential for exploration and use in South America. *Neotropical Ent.* **30**:191-205.

Grewal, P.S., Ehlers, R.U. and Shapiro-Ilan, D.J. (ed.) (2005) *Nematodes as Biocontrol Agents.* CAB Publishing, Wallingford, UK.

Hominick, W.M, Reid, A.P., Bohan, D.A. and Briscoe, B.R. (1996) Entomopathogenic nematodes: biodiversity, geographical distribution, and convention on biological diversity. *Biocont. Sci. Tech.* **6**: 317-31.

Hussaini, S.S., Ansari, M.A., Ahmad, W. and Subbotin, S.A. (2001) Identification of some Indian populations of *Steinernema* species (Nematoda) by RFLP analysis of ITS region of rDNA. *Int. J. Nemat.* **11**, 73-76.

Jackson, J.J and Brooks, M.A. (1995) Parasitism of western corn rootworm larvae and pupae by *Steinernema carpocapsae. J. Nemat.* **27**:15-20.

Jagdale, G.B., Somasekhar, N., Grewal, P.S. and Klein, M.G. (2002) Suppression of plant parasitic nematodes by application of live and dead entomopathogenic nematodes on boxwood (*Buxus* spp). *Biol. Contr.* **24**:42-49.

Janson, R.K., Lecrone, S.H., Gaugler, R. and Smart, G.C. (1990) Potential of entomopathogenic nematodes as biological control agents of sweet potato weevil (Coleoptera: Curculionidae). *J. econ. Ent.* **83**:1818-26.

Karunakar, G., Easwaramoorthy, S. and David, H. (1999) Susceptibility of nine lepidopteron insects to *Steinernema glaseri, S. feltiae* and *Heterorhabditis indicus* infection. *Int. J. Nemat.* **9**:68-71.

Katti, G.R., Prasad, J.S., Padmakumari, A.P. and Sankar, M. (2006) Effect of storage period on survival and infectivity of indigenous entomopathogenic nematodes of insect pests of rice. *Nematolog. Mediter.* **34**:37-41.

Kaya, H.K. (1990) Soil Ecology. In: Gaugler R and Kaya H.K. (ed) *Entomopathogenic Nematodes in Biological Control.* CRC Press, Boca Raton, FL, USA.

Kaya, H.K. and Nelsen, C.E. (1985) Encapsulation of steinernematid and heterorhabditid nematodes with calcium alginate: a new approach for insect control and other applications. *Env. Ent.* **14**: 572-74.

Kaya, H.K. and Gaugler, R. (1993) Entomopathogenic nematodes. *A. Rev. Ent.* **38**: 181-206.

Kaya, H.K. and Grieve, B.J. (1982) The nematode *Neoplectana carpocapsae* and the beet armyworm *Spodoptera exigua*: Infectivity of pre pupae and pupae in soil and of adults during emergence from soil. *J. Invertebr. Pathol.* **39**: 192-97.

Koppenhofer, A.M., Kaya, H.K. and Taormino, S.P. (1995) Infectivity of entomopathogenic nematodes (Rhabditida: Steinemematidae) at different soil depths and moistures. *J. Invertebr. Pathol.* **65**: 193-99.

Koppenhofer, A.M., Grewal, P.S. and Kaya, H.K. (2000) Synergism of imidacloprid and entomopathogenic nematodes against white grubs: The mechanism. *Entomol. exp. Appl.* **94**: 283.

Lindgren, J.E., Valero, K.A. and MacKey, B.E. (1993) Simple in vivo production and storage methods for *Steinemema carpocapsae* infective juveniles. *J. Nemat.* **25**:193-97.

Lisansky, S.G. and Coombs, J. (1994) Developments in the market for biopesticides. *Brighton Crop Protection Conference - Pests and Diseases.* pp 1049-1054.

Liu, J. and Berry, R.E. (1996) *Heterorhabditis marelatus* n.sp. (Rhabditida: Heterorhabditidae) from Oregon. *J. Invertebr. Pathol.* **67**:48-54.

Lossbroek, T.G. and Theunissen, J. (1985) The entomogenous nematode Neoplectana bibionis as a biological control agent of *Agrotis segetum* in lettuce. *Exp. Appl. Nematol.* **39**: 261-64.

Mannion, C.M. and Jansson, R.K. (1992) Comparison of ten entomopathogenic nematodes for control of sweetpotato weevil (Coleoptera: Apionidae). *J. econ. Ent.* **85**:1642-50.

Miller, L.A. and Bedding, R.A. (1982) Field testing of the insect parasitic nematode, *Neoaplectana bibionis* (Nematoda: Steinernematidae) against currant borer moth, *Synanthedon tipuliformis* (Lep.: Sesiidae) in blackcurrants. *Entomophaga* **27**: 109-14.

Morris, O.N. and Converse, V. (1991) Effectiveness of steinernematid and heterorhabditid nematodes against noctuid, pyralid, geometrid species in soil. *Can. Entomol.* **123**:55-61.

Munson, J.D. and Helms, T.J. (1970) Field evaluation of a nematode (DD-136) for control of corn rootworm larvae. (Diabrotica). *Entomol. Soc. Amer. N. Cent. Br. Proc.* 25th : 97-99.

Nishimatsu, T. and Jackson, J.J. (1998) Interaction of insecticides, entomopathogenic nematodes, and larvae of the western corn rootworm (Coleoptera: Chrysomelidae). *J. econ. Ent.* **91**: 410-18.

Padmakumari, A.P., Prasad, J.S., Katti, G.R. and Sankar, M. (2007) Efficacy of Rhabditis (Oscheius) sp. against rice yellow stem borer, *Scirpophaga incertulas* (Walker). *Indian J. Pl. Prot.* **35**:255-58.

Parkman, J.P. and Frank, J.H. (1993) Use of a sound trap to inoculate *Steinernema scapterisci* (Rhabditida : Steinernematidae) into pest mole cricket populations (Orthoptera: Gryllotalpidae). *Fla Entomol.* **76**: 75- 82.

Poinar, G.O. (1983) *The Natural History of Nematodes.* Prentice Hall, New Jersey. pp: 323.

Poinar, G.O. (1990) Taxonomy and biology of Steinernematidae and Heterorhabditidae. In: Gaugler, R. and Kaya, H.K. (eds.) *Entomopathogenic Nematodes in Biological Control.* CRC Press: Boca Raton, Florida. Pp 23-61.

Poinar, Jr. G.O., Karunakar, G.K. and David, H. (1992) *Heterorhabditis indicus* n. sp. (Rhabditida, Nematoda) from India: separation of *Heterorhabditis* spp. by infective juveniles. *Fundam. Appl. Nematol.* **15**:467-72.

Purcell, M., Johnson, M.W., LeBeck, L.M. and Hara, A.H. (1992) Biological control of *Helicoverpazea* with *Steinernema carpocapsae (Rhabditida:* Steinernematidae) in corn used as trap crop. *Env. Ent.* **21**:1441-47.

Rahrnan, P.F., Sharma, S.B. and Wightman, J.F. (2000) A review of insect-parasitic nematodes research in India: 1927-1997. *Int. J. Pest Manag.* **1**:19-28.

Rao, V.P. and Manjunath, T.M. (1966) DD-136 nematode that can kill many insect pests. *Indian Farmg* **16**:43-44.

Richter, A.R. and Fuxa, J.R. (1990) Effect of *Steinernema feltiae* on *Spodoptera frugiperda* and *Heliothis zea* (Lepidoptera: Noctuidae) in corn. *J. econ. Ent.* **83**: 1286-91.

Sankaranarayanan, C., Easwaramoorthy, S. and Somasekhar, N. (2003) Infectivity of entomopathogenic Nematodes *Heterorhabditis* and *Steinernema* spp. to Sugarcane Shoot Borer, *Chilo infuscatellus* Snellen at Two Temperatures. *Sugar Tech.* **5**:167-71.

Sankaranarayanan, C., Somasekhar, N. and Singaravelu, E. (2006) Biocontrol potential of entomopathogenic nematodes *Heterorhabditis* and *Steinernema* against pupae and adults of white grub *Holotrichia serrata F. Sugar Tech.* **8**: 268-71.

Silver, S.C., Dunlop, D.B. and Grove, D.J. (1995) Granular formulation of biological entities with improved storage stability. *Int. Patent WO* 95/05077.

Sivakumar, C.Y., Jayaraj, S. and Subramanian, S. (1989) Observations on Indian population of the entomopathogenic nematode *Heterorhabditis bacteriophora. J. Biol. Con.* **2**: 112-13.

Smitley, D.R., Warner, F.W. and Bird, G.W. (1992) Influence of irrigation and *Heterorhabdtis bacteriophora* on plant parasitic nematodes in turf. *J. Nemat.* **24**:637-41.

Somasekhar, N., Grewal, P.S. and Klein, M.G. (2002) Genetic diversity in stress tolerance and fitness among natural populations of *Steinernema carpocapsae. Biol. Cont.* **23**: 303-10.

Stack, C.M., Easwaramoorthy, S., Mehta, U.K., Downes, M.J., Griffin, C.T. and Bumell, A.M. (2000) Molecular characterization of *Heterorhabditis indica* isolates from India, Kenya, Indonesia and Cuba. *Nemat.* **2**: 477-87.

Varaprasad, K.S., Balasubramanian, S., Diwakar, B.J. and Rao, C.V.R. (1994) First report of an entomogenous nematode, *Paragrolaimus* sp. from coffee-berry borer, *Hypothenemus hampei* (Ferrari) from Karnataka, India. *Pl. Prot. Bull.* **46**:2-3.

Wang, Y. and Gaugjer, R. (1998) Host and penetration site location by entomopathogenic nematodes against Japanese beetle larvae. *J. Invertebr. Pathol.* **72**: 313-18.

Waterhouse, D.F. (1998) *Biological Control of Insect Pests: Southeast Asian Prospects.* Canberra, Australian Center for International Agricultural Research, ACIAR Monograph Series, No. 51:548

Woodring, J.L. and Kaya, H.K. (1998) *Steinernematid* and *Heterorhabditid Nematodes: A Handbook of Biology and Techniques.* Southern Cooperative Series Bulletin 331, Arkansas Agricultural experiment Station, Fayettesville, AR. USA.

Wright, R.J., Witkowski, J.F., Echtenkamp, G. and Georgis, R. (1993) Efficacy and persistence of *Steinernema carpocapsae* (Rhabditida: Steinernematidae) applied through a center-pivot irrigation system against larval corn rootworms (Coleoptera: Chrysomelidae). *J. econ. Ent.* **86**:1348-54.

Yukawa, T. and Pitt, J. (1985) Nematode storage and transport. *Int. Patent W085/03412.*

Theory and Practice of Integrated Pest Management
R. Arora, B. Singh and A.K. Dhawan (Eds)
Scientific Publishers (India), Jodhpur 2011, pp. 213 - 229.

13

IMPLICATIONS OF BIOTYPE DEVELOPMENT IN HOST PLANT RESISTANCE

D.R. Sharma and R.S. Gill*

Department of Horticulture, *Department of Entomology
Punjab Agricultural University, Ludhiana - 141 004, India

13.1. INTRODUCTION

Thousands of years ago, humans have been choosing the plants that are tolerant to pests but the arthropods have been evolving mechanisms to combat defences in plants. In this process, during the 18th and early 19th centuries, several insect resistant cultivars were developed and cultivated but breeding of arthropod resistant plants became more scientific in the late 19th century with the discovery of Gregor Mendels basic tenets of hereditary and plant hybridization. However, ever since the publication of first comprehensive review by Snelling (1941), the host plant resistance (HPR) has become an important component of IPM and is defined differently by different scientists. Painter (1951) defined it as “the relative amount of heritable qualities possessed by the plant which influence the ultimate degree of damage done by the insect in the field”, but Panda and Khush (1995) further described it as “any degree of host reaction less than full immunity”. The breeding of resistant cultivars is a continuous process as resistance genes in arthropod-resistant cultivars may be overcome by the development of resistance-breaking arthropod biotypes that possess an inherent genetic capability to overcome the properties of plant resistance genes. Biotypes have been defined as “populations within an arthropod species that differ in their ability to utilize a particular trait in a particular plant genotype”. Biotypes are routinely detected by exposing a set of plant cultivars, each possessing a different arthropod resistance gene or gene combination that reacts differentially to a given arthropod biotype. With the exception of the brown planthopper, *Nilaparvata lugens* (Stal), existing arthropod biotypes occur in association with individual plant resistance genes inherited as dominant traits, and exhibit virulence genes expressed after selection, recombination or mutation to nullify plant resistance genes.

The gene-for-gene relationship between arthropod virulence genes and plant resistance genes is similar to that described by Flor (1971) for plant pathogen resistance genes and pathogen virulence genes. The virulence or avirulence of an arthropod biotype to a plant resistance gene depends on how resistance genes in the host plant and virulence genes in the arthropod interact. When gene products of the avirulent arthropod are recognized by the defense system of the resistant plant, the arthropod is unable to infest a resistant plant. Conversely, when a normally resistant plant does not recognize arthropods gene products, the virulent arthropod biotype overcomes the plant resistance gene or genes. Puterka and Burton (1990) reasoned that biotypes develop from mutations or a pre-existing variability for virulence, from sexual recombination or plant resistance gene selection pressure that may change arthropod virulence gene frequency. The intensity and duration of virulence gene expression in turn depend on the category of resistance exhibited by the plant resistance gene, the initial virulence gene frequency, and the interaction between the cultivar, the pest, and the environment. The use of the term biotype has some problems because of the lack of knowledge about the genetic structure of arthropod biotypes. A major tenant of Flors (1971) concept assumes the single gene relationships shown above between the pest and the host. However, arthropod biotypes may be populations of individuals exhibiting a particular set of virulence genes or arthropod populations responding similarly to a set of plant differentials with more than one gene (Wilhoit, 1992). Mitchell-Olds and Bergelson (2000) suggested that because of the current developments in plant genomics, the use of a gene-for-gene concept may be oversimplified, and that a "gene-for-genome" concept will allow simultaneous evaluation of several resistance genes involved in potentially overcoming a pest virulence gene.

13.2. CATEGORIES OF BIOTYPES

Biotype may be classified into one or more of the following five categories: *(a)* nongenetic polyphenisms, *(b)* polymorphic or polygenic variation within populations, *(c)* geographic races, *(d)* host races, and *(e)* species (Diehl and Bush, 1984).

Nongenetic polyphenisms. Polyphenisms also called phenocopies or ecomorphs, occur when the same genotype produces different phenotypes under different environmental conditions.

Polymorphic or polygenic. Genetic polymorphisms consist of discontinuous phenotypes among individuals within a freely interbreeding population that are the result of allelic variation of a frequency higher than can be maintained by recurrent mutation. Allelic differences at a single locus may appear continuous and overlapping due to environmental variation.

Geographic races. The status of geographically isolated biotypes, called geographic races (semi-species or subspecies), is difficult to establish because the differences observed may or may not be sufficient to ensure reproductive isolation if the races become sympatric. The related term "ecotype" (ecological race) is occasionally used when populations within a plant or animal species in the same habitat in different localities are similar in morphology, ecology or behavior, presumably because of convergent selection.

Host races. Host races are a more controversial subject and their definition warrants careful consideration. It is defined as a population of a species that is partially reproductively isolated from other conspecific populations as a direct consequence of adaptation to a specific host. The basis of isolation may involve genetically based differences in host preference (when mating occurs on the host) or many other factors such as allochronic barriers that arise as a direct result of phenological differences among hosts.

Species. Species are generally regarded as natural populations that are reproductively isolated from one another and that follow distinct and independent evolutionary paths. Species that morphologically resemble each other so closely that they can be recognized only after careful study of biochemical, cytological, or behavioral traits are called sibling species. The existence of sufficient genetic isolation between biotypes to warrant species designation can be established with certainty only in situations where the different populations occur sympatrically without substantial interbreeding or introgression

13.3. DISTRIBUTION OF BIOTYPES

Eighteen different species of Diptera, Homoptera, Acari and one species of Coleoptera have developed virulent biotypes to plant resistance genes (Table 13.1). Ten of the eighteen are aphid species, in which parthenogenic reproduction contributes greatly to the successful development of resistance-breaking biotypes. Three out of five species are sexually dimorphic Diptera with very high reproductive potentials whereas green leafhopper, *Nephotettix virescens* (Distant), green rice leafhopper, *Nephotettix cincticeps* Uhler and *Nilaparvata lugens* (Stal) occur continuously on large monocultures of rice.

13.4. DEVELOPMENT OF BIOTYPES

Grapes. *Vitus* spp. resistant to the grape phylloxera, *Daktulosphaira vitifoliae* (Fitch) one of the most classical example, has an extensive history of phylloxera biotype development. After the re-establishment of the wine grape industry in France in the late 1880s, studies on Vitus-phylloxera interaction became intensely serious. The extensive literature on various aspects of Vitus resistance to phylloxera and phylloxera suggest the possible existence of phylloxera biotypes, termed races at the time. Riley (1872) was the first to hypothesize that phylloxera biotypes would develop, after observing high levels of resistance in the *Vitus vinifera* x *Vitus labrusca* hybrid Concord grape to phylloxera in the U.S. (Missouri). Although biotypes did not develop at that time but thereafter reports of biotype variation began to pour from France, Italy, Ukraine and Germany. All of them reported two presumed phylloxera biotypes based on biological, geographic or morphological parameters. However, none of their claims could be substantiated.

Table 13.1. Arthropod biotypes associated with plant resistance

Crop	Arthropod	Number of Biotypes	Crop	Arthropod	Number of Biotypes
Malus spp.	*Eriosoma lanigerum*	2	*Triticum aestivum*	*Aceria toschiella*	5
	Dysaphis devecta	3		*Macrosiphum avenae*	3
	Rhagoletis pomonella	2		*Mayetiola destructor*	16
	Acyrthosiphon kondoi	2		*Diuraphis noxia*	2
Medicago sativa	*Acyrthosiphon pisum*	4			
	Therioaphis maculata	6		*Schizaphis graminum*	8
	Nephotettix cincticeps	2	Vegetables	*Brevicoryne brassicae*	2-4
Oryza sativa	*Nephotettix virescens*	3		*Bemisia tabaci*	2
	Nilaparvata lugens	4		*Trialeurodes vaporariorum*	2
	Orseolia oryzae	4	*Vitus* spp.	*Daktulosphaira vitifoliae*	2
Rubus idaeus	*Amphorophora idaei*	5	*Zea mays*	*Rhopalosiphum maidis*	5
Rubus nigrum	*Dasineura tetentsi*	?			

Source: Smith (2005)

In the late 1970s, phylloxera infestations intensified on Concord and Niagara grapes in the north-eastern U.S. and based on biological studies two phylloxera biotypes Clinton biotype unable to feed on Concord and six other grape species, and the Concord biotype, unable to feed on Clinton and nine other grape species were demonstrated (Table 13.2). Results of isozyme analyses supported the existence of two biotypes. The rootstock AXR 1, a *Vitus vinifera* x *Vitus rupestris* hybrid, was initially resistant to phylloxera, has been overcome by virulent phylloxera populations in Europe, South Africa and in the U.S. AXR 1 was the rootstock in the majority of the Napa and Sonoma county wine producing areas of California and phylloxera from these areas grow much better on AXR 1 than phylloxera from other areas. Phylloxera populations avirulent to AXR 1 were named biotype A and the population virulent to AXR 1 was named biotype B (Table 13.2). The rootstock *SC,* a *Vitus berlandieri* x *Vitus riparia* hybrid, possesses both antibiosis and antixenosis to biotypes A and B.

In France, a phylloxera biotype originating on the *Vitus berlandieri* x *Vitus vinifera* hybrid 41B. The California A and B phylloxera biotypes are unaffected by the resistance in 41B and grow equally well on it.

Table 13.2. Genes from *Vitus* germplasm for resistance to *Daktulosphaira vitifoliae*

Genes	Biotype		
	1	*2*	*3*
A_1	R	S	R
A_2	S	R	R
A_3 A_4	S	R	S
A_5	R	S	S
A_6	R	S	S
A_7	R	S	S
A_1+ A_2	R	R	R
A_1+ A_3	R	R	R
A_1+ A_4	R	S	R
A_{10}	R	R	R

R=Resistant, S= Susceptible; Granett *et al.* (1985, 2001)

Raspberry. At least 12 major genes in raspberry, *Rubus phoenicolasius* express resistance to five different biotypes of raspberry aphid, *Amphorophora idaei* Borner in the United Kingdom. There are four *A. idaei* biotypes and specific *A. idaei* resistance genes. The A_1 gene and several uncharacterized minor genes from red raspberry are present in approximately one-third of all U.K. cultivars. These genes confer resistance to biotype 1, but not biotype 2 or X. The AIO gene from North

American black raspberry is also present in approximately one-third of all cultivars and is effective against all three biotypes. The prolonged use of A_1-based cultivars led to a shift in *A. idaei* virulence to A_1 from 3 per cent in the 1960s to 75 per cent in the 1990s. *A. idaei* clones virulent to AIO have also been developed. Ribosomal DNA probes of biotypes 1, 2 and X have identified large genetic variation both within and between the three biotypes, suggesting that virulence to A_{10} may lead to the selection of further *A. idaei* biotypes.

Apple. Two biotypes of the wooly aphid, *Eriosoma lanigerum* (Hausmann) exist on the foliage of different clones of apple, *Malus domestica* in South Australia. The Clare (native) biotype is avirulent to resistant apple cultivars such as Northern Spy, but the Blackwood biotype is virulent to several cultivars normally resistant to the Clare biotype. *E. lanigerum* biotypes have also been shown to exist in response to different North American apple cultivars.

Resistance in English apple cultivars to the rosy leaf curling aphid, *Dysaphis devecta* (Walker), follows a gene-for-gene relationship. Four resistance genes govern resistance to three aphid biotypes. Biotypes 1 and 2 are avirulent on Coxs Orange Pippin, which is protected by the *Sd1* gene. Resistance to biotype 1 only exists in Northern Spy, via the *Sd2* gene, and a single gene *(Sd3)* in *Malus robusta* and *Malus* x *zumi* selections controls resistance to biotype 3. Several cultivars also possess precursor genes necessary for biotype-specific genes for resistance to be effective.

Pea. Three pea aphid, *Acyrthosiphon pisum* (Harris) biotypes exist on peas in the field and greenhouse. It was determined that morphological differences exist between two of the three biotypes in North America, and that a fourth biotype exists in a pink color morph from France. Differences among the four biotypes are also evident when they are reared on an artificial diet.

Alfalfa. The first biotype of the spotted alfalfa aphid, *Therioaphis maculata* (Buckton), was identified in Arizona. Since then, six additional biotypes (A, B, E, F, H, N) have been elucidated (Table 13.3) on the bases of differential responses to clones of Moapa alfalfa, *Medicago sativa.* Biotypes C, D, and G are resistant to organophosphorous insecticides. Biotype development in *T. maculata* also follows a gene-for-gene progression, although genetic analysis of alfalfa is difficult due to its heterozygosity and tetraploidy.

The blue alfalfa aphid, *Acyrthosiphon kondoi* Shinji, became a pest of alfalfa in the U.S. and the first alfalfa cultivar with *A. kondoi* resistance, CUF 101 was released in 1977. Germplasm previously resistant to *A. kondoi* was later observed to be susceptible in New Mexico and Oklahoma. Comparisons of the response of CUF 101 to *A. kondoi* showed that the frequency of resistance in CUF 101 declined from 50 to 19 per cent. The virulent *A. kondoi* biotype was named as BAOK90.

Table 13.3. Resistance of Moapa, *Medicago sativa* clones to biotypes of *Therioaphis maculata*

Medicago clone	*Aphid biotype*					
	A	B	E	F	H	N[a]
C-903	S	R	S	S	S	R
C-904	R	R	S	I	I	R
C-905	R	R	S	S	S	R
C-906	R	R	R	S	S	R
C-907	R	R	R	R	S	R
C-908	S	R	S	S	S	R
C-909	R	R	R	I	I	R
C-910	R	R	R	I	S	R
C-911	S	R	S	S	S	S

R -resistant; I - Intermediate; S - susceptible; biotypes C, D and G are resistant to organophosphorous insecticides; [c]Nebraska population.
Source: Nielson *et al.* (1980)

Maize. Biotype development in the corn leaf aphid, *Rhopalosiphum maidis* (Fitch), first detected in its population was denoted as KS-2, capable of surviving on cultivars of sorghum, Sorghum bicolor, previously resistant to the parent aphid population KS-l. KS-3, a biotype capable of infesting wheat, *Triticum aestivum*, and KS-4, a biotype capable of reproducing on barley, *Hordeum vulgare,* at low temperature (13°C) was also detected. KS-5, detected in *R. maidis* field populations, displays high temperature (29°C) tolerance as well as the ability to feed on the wheat species *Triticum timopheevi* that is resistant to biotypes 1-4.

Rice. Four biotypes of *N. lugens* exist due to their abilities to destroy different rice cultivars with specific genes or gene combinations for hopper resistance (Table 13.4). Studies revealed that a Philippine *N. lugens* population unable to destroy the cultivar Mudgo is denoted as biotype 1; a population damaging Mudgo but avirulent to the cultivar ASD7 is biotype 2 and a laboratory population created in the Philippines that destroys ASD7 is biotype 3. Biotypes 1 and 2 were originally termed Philippine biotypes, but are now widely distributed in Southeast Asia. *N. lugens* biotype 4 occurs in Bangladesh, Sri Lanka and Southern India, based on virulence to the resistance genes in Mudgo and ASD7. *N. lugens* biotypes vary considerably in both physiological and behavioral aspects of feeding.

Biotypes 1, 2, and 3 have been differentiated based on the weight of honeydew collection, amino acids excretion and morphological characters.

The dominant gene *Bph-1* in Mudgo confers resistance to Philippine biotypes 1 and 3 (Table 13.4).

Table 13.4. Relationships between genes from different *Oryza sativa* cultivars to damage by biotypes of *Nilaparvata lugens*

Cultivar	Gene	Biotype			
		1	2	3	4
Taichung Native 1	None	S	S	S	S
Mudgo	*Bph-1*	R	S	R	S
ASD7	*Bph-2*	R	R	S	S
Rathu Heenati	*Bph-3*	R	R	R	R
Babawee *Bph-4*	R	R	R	R	
ARC 10550	*Bph-5*	S	S	S	R
Swamalata	*Bph-6*	S	S	S	R
Tl2	*Bph-7*	S	S	S	R
Chin Saba	*Bph-8*	R	R	R	--
Balamawee	*Bph-9*	R	R	R	--

R - resistant; MR - moderately resistant, S - susceptible, (--) - reaction not known
Source: Khush and Brar (1991)

ASD7 containing a single recessive gene *bph-2,* is resistant to biotypes 1 and 2, but susceptible to biotype 3. *Bph-3* a single dominant gene in the cultivar Rathu Heenati and *bph-4,* a single recessive gene in the cultivar Babawee, provide resistance to all known biotypes. ARC 10550, a cultivar susceptible to biotypes 1, 2 and 3, is resistant to biotype 4, and contains the single recessive gene *bph-5,* that segregates independently of *Bph-1, bph-2, Bph-3* or *bph-4.* A single dominant gene in Swarnalata, also resistant to biotype 4, is independent of *bph-5* and has been termed *Bph-6.* A recessive gene in the cultivar T12, also resistant to biotype 4, is non-allelic with *bph-5,* and has been termed *bph-7.* A single recessive gene in Chin Saba, controls resistance to biotype 1, 2 and 3 and is non-allelic to *bph-2* or *bph-4.* This gene has been designated *bph-8.* Similarly, a single dominant gene in the cultivar Balamawee that controls resistance to biotypes 1, 2 and 3, is non-allelic to *Bph-1* and *Bph-3.* The Balamawee gene has been designated *Bph-9.* susceptible to biotypes 1, 2 and 3, is resistant to biotype 4, and contains the single recessive gene *bph-5,* that segregates independently of *Bph-1, bph-2, Bph-3* or *bph-4.* A single dominant gene in Swarnalata, also resistant to biotype 4, is independent of *bph-5* and has been termed *Bph-6.* A recessive gene in the cultivar T12, also resistant to biotype 4, is non-allelic with *bph-5,* and has been termed *bph-7.* A single recessive gene in Chin Saba, controls resistance to biotype 1, 2 and 3 and is non-allelic to *bph-2* or *bph-4.* This gene has been designated *bph-8.* Similarly, a single dominant gene in the cultivar Balamawee that controls resistance to biotypes 1, 2 and 3, is non-allelic to *Bph-1* and *Bph-3.* The Balamawee gene has been designated *Bph-9.*

Shifts in the frequency of different *N. lugens* biotypes have been observed. *N. lugens* populations in China and Japan shifted from biotype 1 to biotype 2 between 1987 and 1992 whereas Malaysian populations are a mixture of biotypes 2 and 3. The biotype composition of populations from Japan, Malaysia and Thailand were further compared and it was determined that populations in tropical areas of Malaysia, Thailand and South Vietnam are predominantly biotype 4 (virulent to *bph-2)* and that populations in Japan and North Vietnam are biotype 2 (avirulent to *bph-2).*

After intensive cultivation of IR 26, which contained the *Bph-1* gene, in the Philippines and Indonesia in the 1970s, biotype 2 became predominant. IR 36 cultivated on about 10 million hectares in the Philippines, Indonesia, the Solomon Islands, and Vietnam, from the late 1970s to the early 1980s has displayed greater durability to *N. lugens* infestations than many other rice cultivars. However, that resistance has now been overcome and IR36 is now fully susceptible to biotype 2. IR64 contains *Bph-1* and several minor genes maintained a moderate level of resistance to biotype 2 in the field for more than 10 years and in greenhouse experiments involving forced selection for virulence. The stability of both IR 36 and IR64 resistance is thought to be due to the effects of *Bph-1* (IR36) or *bph-2* (IR64) with several minor genes.

Biotypes 1, 2 and 3 can be effectively converted from themselves to each of the other two biotypes, depending on the rice cultivar on which they are reared. These findings are supported by results that indicate *N. lugens* virulence to different rice cultivars is variable within and among populations and is controlled by a system of polygenes. This is not surprising, given the multigenic nature of rice resistance to *N. lugens,* as well as the involvement of both the dominant and recessive modes of inheritance of resistance in rice. Results of several studies have shown that there is little correlation between virulence to *Bph-1* and *bph-2* in different *N. lugens* populations, and that shifts from biotype 1 to biotype 2 and 3 occur independently. Scientists have selected populations virulent to cultivars carrying either *bph-8* or *Bph-9.*

Nephotettix virescens, a vector of rice tungro virus, has shown the ability to overcome genes for hopper resistance in several improved rice cultivars in both laboratory experiments and after field infestations for several years.

Orseolia oryzae is a major pest of rice in Asia. As the planting of *Orseolia*-resistant cultivars has increased, so has the development of *Orseolia* biotypes. Two *O. oryzae* biotype groups exist i.e. one in China and one in India (Table 5).

Six biotypes from India have been identified and four biotypes exist in China. Two gall midge biotypes exist in Sri Lanka and others may exist in Cambodia, Indonesia, Laos, Thailand and Vietnam. The original biotype determinations were based on midge development and survival on a differential set of cultivars with various midge resistance genes. At least seven distinct rice *Orseolia* resistance genes have been identified but none are resistant to all midge biotypes (Table 13.5). The *Gm6* gene in Duokang #1 is resistant to all Chinese biotypes. An unnamed gene in IR36 *(N. lugens* resistant) is resistant to four of six Indian biotypes. All *Gm* genes are inherited as dominant traits with the exception of *gm3.*

Table 13.5. Reaction of *Oryza sativa* cultivars with various genes for resistance to biotypes of *Orseolia oryzae* in India and China

Resistance gene	Cultivar origin	Orseolia oryzae country of origin									
		India						China			
Gm1	Samridhi	R	S	R	S	R	R	MR	S	R	S
Gm2	Phalguna	R	R	S	S	R	S	R	R	S	S
Gm3	RP20681853	R	-	S	S	R	-	-	-	-	-
Gm4	Abhaya	R	R	S	R	S	-	-	-	-	-
Gm5	ARC5984	R	R	S	S	R	-	R	S	-	R
Gm6(t)	Duckang #1	S	S	S	S	-	-	R	R	R	R
Gm7	ARC10659	R	R	S	R	-	S	S	S	S	-
Gm?	IR36	R	R	R	R	S	S	S	S	S	S

R = resistant; MR = moderately resistant, S = susceptible, - reaction not known
Source: Katiyar *et al.* (2001)

Vegetables. Though not as pronounced as biotypes occurring on some other crops, biotypes exist on various resistant cultivars of different vegetables. The best known example is that of a biotype of the cowpea weevil, *Callosobruchus maculatus* (Fabricius), occurring on a landrace of cowpea, *Vigna unguiculata* with weevil resistance in Africa. First evidence of an African *C. maculatus* biotype was identified in 1983, when approximately 15 per cent of a weevil population from Niger could overcome the resistance in TVu2027, a weevil-resistant cowpea landrace. Several additional studies have determined that *C. maculatus* populations from Nigeria and Yemen are virulent to the resistance in TVu2027 but that Brazilian populations are avirulent.

Three biotypes of the cowpea aphid, *Aphis craccivora* (Koch), also occur in Africa. Biotypes A and B occur on different sources of aphid resistance in Niger and biotype K occurs on resistant germplasm in Upper Volta.

Dutch and Italian biotypes of the western flower thrips, *Frankliniella occidentalis* (Pergande) on accessions of cucumber, *Cucurbita sativus*, normally resistant to *F. occidentalis* were detected. Biotypes of the cabbage aphid, *Brevicoryne brassicae* (Linnaeus), exist among different cultivars of rape, *Brassica napus* in New Zealand and among clones of Brussels sprouts, *Brassica oleracea* var. *gemmifera* in England.

Wheat. Development of improved wheat cultivars in the U. S. has been closely paralleled by the development of morphologically indistinguishable biotypes of *M. destructor* and the first biotype was referred as biological strains. Sixteen current *M. destructor* biotypes are differentiated by how various combinations of fly virulence genes interact with wheat for *M. destructor* resistance (Table 13.6). These interactions are the most thoroughly studied for gene-to gene relationships between an arthropod and its host plant. The Great Plains (GP) fly biotype, initially isolated in western Kansas, now occurs in Texas, and carries the homozygous recessive condition for virulence only in association with the wheat cultivar Turkey. The GP biotype is avirulent to cultivars with the H3, H5 and H6 genes or the H7H8 gene combination. Biotype A also found primarily in Texas is similar to the GP biotype, but is virulent to H7H8. Biotype B now occurs in the north-eastern U.S. and possesses an additional gene for virulence to H3. Biotype C, found occasionally in New York and Texas, has the opposite reaction of biotype B to H3 and H6. Biotype D, found in the Midwestern, mid-Atlantic, and north-eastern U.S., is similar to biotype B, but is also virulent to *H6.* Biotype E, detected in Georgia, now occurs in Georgia, Florida and Texas and is avirulent to *H7H8.* Biotypes F and G occur in the south and south-eastern U. S. and are avirulent to *H7H8.* Biotype F is avirulent to *H3, H5* and *H7H8,* while biotype G is avirulent to *H7H8,* but virulent to *H3* and *H6.* Biotypes J and L developed in field populations in Indiana in response to *H5* resistance in the cultivar Arthur 71. Biotype J is avirulent to *H1H2* and *H7H8.* Biotype L now occurs in the eastern two-thirds of the United States wheat producing areas, and is avirulent to *H9, H10, H12-19,* and *H21-27.* Biotypes M and O occur in the south-eastern U. S and have similar patterns of virulence to *H3,* H5, *H6, H11* and *H19.* Based on results of United States national surveys, biotypes I and N do not presently occur at detectable levels. A Spanish *M. destructor* biotype recently characterized is avirulent to *H3, H6, H9, H11, H12, H13, H18, H21,* and *H30* from the wild wheat *Aegilops triuncialis.*

Table 13.6. Cereal genes for resistance to *M. destructor* and virulence genes in *M. destructor* biotypes

Resistance gene(s), source, cultivar, PI or CI number	*Biotype*												
	GP	A	B	C	D	E	F	G	J	L	M	*O*	*SP*
None, *Triticum aestivm,* Turkey	S	S	S	S	S	S	S	S	S	S	S	S	-
H1H2, T. aestivum, Dawson, C13342	R	S	S	S	S	S	S	S	R	S	S	S	-
H3, T. aestivum, Monon, CI 12061	R	R	S	R	S	S	R	S	S	S	S	S	$
H4, T. aestivum, Java, CI10051	?	?	?	?	?	?	?	?	?	?	?	?	?
H5, T. aestivum, Magnum, PI 562068	R	R	MR	R	R	S	R	S	S	S	S	S	-
H6, T. turgidum, Caldwell, PI 94587	R	R	MR	S	S	MR	MR	S	S	S	S	S	R
H7H8, T. aestivum, Seneca & CI 12529	R	S	S	S	S	R	S	R	S	S	R	R	-
H9, T. turgidum, CI 17714	R	R	R	S	R	R	R	MR	-	R	MR	MR	R
H10, T.turgidum, CI 17714	R	R	MR	R	R	S	MR	S	-	R	MR	MR	R
H11, T.turgidum, PI 94587	R	R	R	R	R	R	S	R	S	S	S	S	R
H12, T. aestivum, Lusco	R	R	MR	R	R	S	MR	R	-	MR	MR	MR	R
H14, T. aestivum, CI 1764	R	-	R	R	R	R	R	R	S	R	MR	MR	R
H15, T.turgidum, CI 1764	R	-	R	R	R	R	R	R	S	S	R	R	-
H16, T.turgidum, PI94587	R	-	R	R	R	R	-	R	-	R	R	R	-
H17, T.turgidum, PI 428435	R	-	-	R	R	R	R	R	R	R	R	R	-
H18, T.turgidum, CI6887	R	R	MR	R	R	R	R	MR	-	R	MR	MR	R
H19, T. turgidum, PI 422297	R	R	MR	R	R	MR	MR	MR	-	R	S	S	-
H20, T.turgidum, Jori	-	-	-	R	-	-	-	-	?	R	-	-	R
H21, Secale cereale, Chaupon	-	-	-	-	R	-	-	-	-	R	-	-	R
H22, T. tausehii, TA1644	-	-	-	-	R	-	-	-	-	R	-	-	-
H23, T.tausehii, TA1642	-	-	-	-	R	-	-	-	-	R	-	-	-
H24, , T.tausehii, TA2452	-	-	-	-	R	-	-	-	-	R	-	-	-
H25, S.cereale, Chaupon	-	-	-	-	-	-	-	-	-	R	-	-	-
H26, , T.tausehii, TA2473	-	-	-	-	R	-	-	-	-	R	-	-	-
H29, T. turgidum, PI 422297	-	-	-	-	-	-	-	-	-	R	-	-	-
H30, Ae, triumcialis, TR-3531	-	-	-	-	-	-	-	-	-	-	-		

R-resistant, MR-moderately resistant, S- Susceptible
Source: Smith (2005)

Biotype surveys have documented major shifts in the occurrence and degree of *M. destructor* virulence since the mid-1980s. The once resistant genes *H3, H5, H6* and *H7H8* are ineffective, as virulence exists to each in biotype L, which is now predominant in the eastern United States. Virulence to *H3* and *H6* has also been detected in *M. destructor* populations in Idaho and Washington. Numerous fly populations in the south-eastern and Midwestern United States have also begun to exhibit virulence to several un-deployed resistance genes, including *H9, H10, H11, H12, H13, H14, H15, H18* and *H19*. Although potentially ineffective across wide geographic areas some of these genes may be useful against avirulent biotypes in selected geographic areas. The *H16* and *H17* genes from durum wheat, *Triticum turgidum* remain resistant to all *M. destructor* biotypes currently in the United States.

Schizaphis graminum Rondani infest barley, oat, *Avena sativa*, rye, *Secale cereale*, sorghum, wheat and numerous grasses and have been regarded as a cereal pest in the U.S. since the late 1800s. Although there were no specific biotype designations but differences were observed in Mississippi and Oklahoma *S. graminum* populations as early as 1946. The first *S. graminum* biotype B was named due to its ability to overcome the resistance in Dickinson 28A wheat to biotype A (Table 13.7). Biotype B tolerates higher temperatures than biotype A but cannot survive on Piper sudan grass, *Sorghum vulgare* var. *sudanese*. In the late 1960s, *S. graminum* populations began to severely damage sorghum in the southwestern U.S. These populations were named biotype C, and forced the development of biotype C-resistant sorghum and wheat cultivars. Amigo wheat, possessing the *Gb2* gene from rye, is resistant to biotypes B and C. Another biotype D is resistant to the insecticide disulfoton. gene from rye, is resistant to biotypes B and C. Another biotype D is resistant to the insecticide disulfoton.

Table 13.7. Wheat genes governing resistance to biotypes of *Schizaphis graminum* in North America and Asia

Source	Origin	Gene	*Reaction to biotype*									
			B	*C*	*E*	*F*	*G*	*H*	*I*	*J*	*CH*[b]	*PK*[c]
Dickinson	*Triticum turgidum*	*Gb1*	R	S	S	R	S	S	-	R	S	S
Amigo	*Secale cereale*	*Gb2*	R	R	S	S	S	S	-	R	R	R
Largo	*Aegilops tauschii*	*Gb3*	S	R	R	S	S	R	R	R	R	S
CI17959	*Aegilops tauschii*	*Gb4*	S	R	R	S	S	S	R	R	S	-
CI17882	*Aegilops speltoides*	*Gb5*	S	R	R	S	S	S	-	R	MR	-
GRS1201	*Secale cereale*	*Gb6*	R	R	R	S	R	S	R	R	MR	-

Biotype B-J, North America; [b]China; [c]Pakistan; R-resistant, MR-moderately resistant, S- Susceptible

Source: Smith (2005)

In the late 1970s, biotype E capable of feeding on Amigo wheat, became prevalent in the high plains of Texas. Some biotype C resistant wheat cultivars have retained antixenosis resistance to biotype E, but most barley and sorghum cultivars are susceptible. Biotype E resistance also exists in sorghum plant introductions with the bloomless character in Largo wheat containing *Gb3* and CI17959 wheat containing *Gb4.* Biotype E resistance also exists in CI17882, which contains *Gb5,* presumably donated by the *Ae. speltoides* parent of CI17882. Biotype E exhibits normal feeding behavior and fecundity on wheat cultivars resistant to biotype B but has increased digestive enzyme activity, enabling it to reproduce at approximately twice the rate of biotype C.

Schizaphis graminum biotypes exhibit significant morphometric differences, the most prominent of which is the length of the first flagellar segment. Biotype E individuals also require a longer scotophase to induce the production of males than do biotype C. Few isozyme differences exist between biotypes B, C or E.

Biotype F, isolated from turfgrass in Ohio, lacks a dorsal stripe and is avirulent to *gb1.* Biotype F can feed on and kill Canada bluegrass, but biotype E cannot. Biotypes G (from Oklahoma) and H (from Texas) have the appearance of biotype F, but differ in their reactions to barley and wheat. Biotype G survives on all known sources of resistant wheat, but is avirulent to *Gb6* in GRS 1201, a *T. aestivum/S. cereale* hybrid. Biotype H has the same effect on wheat cultivars as biotype E, but is avirulent to *Gb3.* Biotypes I and K are virulent to *S. graminum*- resistant sorghums but avirulent to *Gb3, Gb4,* and *Gb6.* Biotype J, a unique population from Idaho, is avirulent to all *Gb* genes but feeds effectively on Post barley, which is normally resistant to all *S. graminum* biotypes. *Schizaphis graminum* biotypes have also been identified in China and Pakistan. The Chinese biotype (CRN-l) is unique in contrast to U. S. biotypes E and I, in that it is virulent to *Gb2* but avirulent to *Gb4* (Table 13.7). The Pakistan biotype (PK-l) differs from U. S. biotypes E and I in that it is avirulent to *Gb3.*

The Russian wheat aphid, *Diuraphis noxia* (Mordvilko), is a serious worldwide pest of wheat and barley. The injection of *D. noxia* salivary enzymes into plant tissues results in plant leaves rolling longitudinally around the main leaf vein, forming tubular refuge for feeding aphids from predators, parasites, and insecticides. Eleven *Dn (Diuraphis noxia)* genes from *Aegilops tauschii*, rye or wheat control the resistance to *D. noxia* (Table 8). *Dn1, Dn2, Dn4, Dn5, Dn6* and *Dnx* are from wheat. A recessive gene *dn3* is present in SQ24, a line derived from crosses between *Ae. tauschii* and durum wheat. *Dn7* is a rye gene derived from a whole arm rye chromosome translocation to wheat. *Dn8* and *Dn9* are co-expressed with *Dn5.* The chromosome location of *Dny,* in the aphid resistant cultivar Stanton is not known. The first *D. noxia*-resistant U. S. wheat cultivar, Halt, containing *Dn4,* was released in Colorado 1994. Differences in were first noted in *D. noxia* populations on aphid resistant wheat lines in Texas. Comparisons of populations from different countries using plant differential cultivar responses indicated virulence to *Dn4* in aphids from Syria and Russia (Table 13.8). Virulence to *Dn1, Dn2, Dn4,* and *Dn5* has been extensively documented in Hungary, although the aphid is presently not a pest of cereals there. Populations occurring in areas of Colorado, Nebraska and Texas were identified in 2003 as virulent to all *Dn* genes except *Dn7*. Populations from Chile, the Czech Republic and Ethiopia are also virulent to *Dn4* but avirulent to *Dn6*. The Czech population is also virulent to *Dnx* and the Ethiopian population is virulent to *Dny*.

Table 13.8. Reaction of *Diuraphis noxia* biotypes to wheat genes controlling *D. noxia* resistance

Source	*Origin*	*Gene*	*D. noxia* source						
			Czr	Chi	Eth	Hun	USA (CO)	USA (KS)	ZA
PI137739	*Triticum aestivum*	*Dn1*	-	-	-	S	S	R	R
PI262660	*Triticum aestivum*	*Dn2*	R	-	R	S	S	R	R
SQ24	*Aegilops tauschii*	*dn3*	-	-	-	-	S	R	
PI372129	*Triticum aestivum*	*Dn4*	S	S	S	S	S	R	R
PI294994	*Triticum aestivum*	*Dn5*	R	R	S	S	R	R	
PI243781	*Triticum aestivum*	*Dn6*	R	R	R	-	S	R	
Turkey 77	*Secale cereale*	*Dn7*	-	-	-	-	R	R	R

R - resistant, S-susceptible

Source: Basky *et al.* (2002); Nkongolo *et al.* (1989); Holey *et al.* (2004); Smith *et al.* (1991)

The Chilean, Ethiopian and Hungarian populations are unaffected by the antibiosis resistance expressed by *Dn4* plants. Wheat cultivars and germplasm containing *Dn1, Dn2* and *Dn5* maintain resistance to the *D. noxia* in South Africa. Research is in progress to develop advanced breeding lines containing genes from new sources of resistance as well as from *Dn7*.

Genetic plant resistance to the mite has proven effective in controlling both wheat curl mite, *Aceria tosichella* Keifer and the virus. Resistance was first demonstrated in *Triticum x Thinopyrum* (wheat-grass) hybrids, where the *Thinopyrum* gene, designated *Cmc2,* is inherited as a dominant trait. Mite resistance from the *Ae. tauschii* gene *Cmc1* is also inherited as a dominant trait. Several wheats with translocations from rye also effectively reduce *A. tosichella* populations and the incidence of wheat streak mosaic virus. However, *A. tosichella* adapts quickly to new hosts. It was observed that an *A. tosichella* biotype that became virulent in only 7 years to TAM107, a popular mite-resistant cultivar grown in western Kansas. *Aceria tosichella* biotypes from the U.S. states of Kansas, Nebraska, Montana, Alberta Canada, South Dakota, and Texas display differential virulence expression to a differential set of resistance genes in different grasses (Table 13.9). Results indicated that *Cmc1* and *Cmc2* are resistant to *A. tosichella* biotypes from Kansas, Montana, South Dakota, Texas, and Alberta, Canada, but *A. tosichella* from Nebraska are virulent to *Cmc1* and *Cmc2.* The rye-based resistance gene in TAM107 and PI475772 *(Cmc3)* is effective against biotypes from Montana, Nebraska, and Canada, but biotypes from South Dakota, Texas, and Kansas, where *A. tosichella* has adapted to TAM107, are virulent to *Cmc3.* The *Cmc4* gene from *Ae. tauschii* in TA2397 and a *Triticum timop-heevi* gene in wheat PI222655 are resistant to all *A. tosichella* biotypes. Wheat PI222655 contains a gene for resistance to *A. tosichella* from Canada and Nebraska, as well as a gene for resistance to *D. noxia.*

Table 13.9. Reaction of *Aceria tosichella* biotypes to cereal genes controlling *A. tosichella* resistance

Gene Source (Germplasm name)	Gene	*A. tosichilla* colony origin					
		KS	MT	NE	SD	TX	CN
Aegilops tauschii (ACPGR 16635)	*Cmc1*	R	R	S	R	R	R
Thinopyrum ponticum (PI525452)	*Cmc2*	R	R	S	R	R	R
Secale cereale (TAM107)	*Cmc3*	S	R	R	S	S	R
Aegilops tauschii (WGRC40)	*Cmc4*	R	S	R	R	R	R
Triticum aestivum (PI222655)	-----	S	S	R	S	S	R
Triticum aestivum (Lamed)	-----	S	S	S	S	S	S

R - resistant, S-susceptible
Source: Harvey *et al.* (1999) and Malik *et al.* (2003)

REFERENCES

Basky, Z. (2002) Biotypic variation in Russian wheat aphid *Diuraphis noxia* Kurdjumov (Homoptera: Aphididae) between Hungary and South Africa. *Cereal Res. Commun.* **30**: 133-139.

Diehl, S.R. and Bush, G.L. (1984) An evolutionary and applied perspective of insect biotypes. *A. Rew. Ent.* **29:**471-504.

Flor, H.H. (1971) The current status of the gene-For-gene concept. *Ann. Rev. Phytopathol.* **9**:275-296.

Granett, J., Timper, P. and Lider, L.A. (1985) Grape phylloxera *(Daktulosphaira vitifoliae)* (Homoptera: Phylloxeridae) biotypes in California. *J. econ. Ent.* **78**:81463-1467.

Granett, J., Walker, M.A., Kocsis, L. and Omer, A.D. (2001). Biology and management of grape phylloxera. *Ann. Rev. Ent.* **46**:387-412.

Haley, S.D., Peairs, F.B., Walker, C.B., Rudolph, J.B. and Randolph, T.L. (2004) Occurrence of a new Russian wheat aphid biotype in Colorado. *Crop Sci.* **44**: 1589-1592.

Harvey, T.L., Seifers, D.L., Martin, T.I., Brown-Guedira, G.L. and Gill, B.S. (1999) Survival of wheat curl mites on different sources of resistance in wheat. *Crop Sci.* **39**: 1887-1889.

Katiyar, S.K., Tan, Y., Huang, B., Chandel, G., Xu, Y., Zhang, Y., Xie, Z. and Bennett, J. (2001) Molecular mapping of gene *Gm-6(t)* which confers resistance against four biotypes of Asian rice gall midge in China. *Theor. Appl. Genet.* **103**:953-961.

Khush, G.S. and Brar, D.S. (1991) Genetics of resistance to insects in crop plants. *Adv. Agron.* 45: 223-274.

Malik, R., Smith, C.M., Harvey, T.L. and Brown-Guedira, G.L. (2003) Genetic mapping of wheat curl mite resistance genes *Cmc3* and *Cmc4* in common wheat. *Crop Sci.* 43: 644-650.

Mitchell-Olds, T. and Bergelson, J. (2000) Biotic interactions Genomics and coevolution. Curr. Opin. *Plant Biol.* **3**: 273-277.

Nielson, M.W. and Lehman, W.F. (1980) Breeding approaches in alfalfa. In: F.G. Maxwell and P.R. Jennings (Eds.) *Breeding Plants Resistant to Insects.* John Wiley, New York. pp. 279-311.

Nkongolo, K.K., Quick, J.S., Meyers, W.L. and Peairs, F.B. (1989) Russian wheat aphid resistance of wheat, rye, and triticale in greenhouse tests. *Cer. Res. Comm.* **17:** 227-232.

Painter, R.H. (1951) *Insect Resistance in Crop Plants.* University of Kansas Press. Lawrence. pp. 520.

Panda, N. and Khush, G.S. (1995) *Host Plant Resistance to Insects.* CAB International, Wallingfort (U.K.) pp. 444.

Puterka, G.J. and Burton, R.L. (1990) Aphid genetics in relation to host plant resistance. In: D.C. Peters, J.A. Webster and C.S. Chlouber (Eds.), *Proc. Aphid- Plant Interactions: Populations to Molecules. Okla. Agric. Exp. Stn. MP*-132: 59-69.

Smith, C.M. 2005. *Plant Resistance to Arthropods — Molecular and Conventional Approaches.* Springer Dordrecht, The Netherlands. pp: 413

Smith, C.M., Belay, T., Stauffer, C., Stary, P., Kubeckova, I. and Starkey, S. (2004) Identification of Russian wheat aphid (Homoptera: Aphididae) biotypes virulent to the *Dn4* resistance gene. *J. econ. Ent.* **97**:112-117.

Snelling, R.O. (1941) Resistance of plants to insect attack. *Bot. Rev.* **7:** 543-586.

Wilhoit, L.R. (1992) Evolution of virulence to plant resistance: Influence of variety mixtures. In: R.S. Fritz and E.L. Simrns (Eds.). *Plant Resistance to Herbivores and Pathogens: Ecology, Evolution and Genetics.* University of Chicago Press, Chicago, pp. 91-119.

Theory and Practice of Integrated Pest Management
R. Arora, B. Singh and A.K. Dhawan (Eds)
Scientific Publishers (India), Jodhpur 2011, pp. 230 - 247.

14

ROLE OF INSECT SURVEILLANCE IN INTEGRATED PEST MANAGEMENT

D.S. Brar and Vijay Kumar
Department of Entomology
Punjab Agricultural University, Ludhiana - 141 004, India

Insect pests associated with agricultural crops cause varied level of damage to them at different growth stages and places. Since middle of the 20th century pesticides played a crucial role in controlling specific pests on crops displacing indigenous biological and cultural control (Anonymous, 2010a). During mid-sixties, the green revolution technology relied on the use of dwarf and semi dwarf high yielding varieties of crops with increased use of agro chemicals and irrigation has considerably increased the food production. These practices caused several ecological, environmental and socio economic problems. If we talk of crop insect pests, these practices has also favoured their build-up, with the result that intensity and severity of many insect pests has increased, many minor pests have assumed the status of major ones and several new pest problems have appeared on some crops in certain regions. Inspite of a variety of control measures applied against pests, crop losses have consistently shown an increased trend (Dhaliwal and Koul, 2010) in post-green revolution era (Dhaliwal *et al.,* 2004) than in the pre-green revolution era (Pradhan, 1964). The crop losses to major agricultural crops of India, based on production and minimum support prince (MSP) during 2007-08 (Anonymous, 2010b), have been estimated by Dhaliwal *et al.* (2010) as 17.5 per cent worth 8,63,884 crores of rupees whereas overall losses were only 7.2 per cent in early 1960's.

During the VIIIth five year plan (1992-97) the Government of India adopted the concept of Integrated Pest Management (IPM) as the cardinal principal of crop protection technology. This laid emphasis on eco-friendly strategies to contain pests at sub-economic levels since environment pollution is the primary concept of the country today. To keep track of the activities of insect pests, their numbers and extent of damage during different growth stages of crops in different regions the regular assessments are needed. Therefore, to achieve the principles of IPM and for its effective implementation, the knowledge about surveillance and its role in IPM must be known to all concerned.

14.1. SURVEILLANCE

It involves the continuous type of activity for measuring pest densities in the field, with the intention of using the information to determine if any control measures are needed. Continuous observations and assessments are required for timely application of control measures (Appukuttan, 1993). It differs from other terms like survey and monitoring. Survey is planned activity which measures and records factors observed either at a single period or over an extended period of time. In monitoring the site visits may not be enough to discover all pest species. It is long term record with the help of some monitoring devices i.e pheromone/attractant/lure traps of appropriate types which record the seasonal cycles and population densities of insect pest (Eavy, 2004). The presence of pest in a monitoring trap gives us a good indication of sizes of pest population. The Monitoring trap also acts as a useful tool to track and monitor the pest activity. According to Mathur (1986) surveillance is a tool to obtain any or all information required for application to any control measure if necessary for implementation of an Integrated Pest Management (IPM) programme. From surveillance and research, information of value to prediction or economic threshold level is obtained.

14.1.1. Pest Surveillance — a prerequisite for IPM

Crop pest surveillance and monitoring implies a complete vigilance on biotic factors affecting crop growth and development in other words, it aims at analyzing possible impacts of various living and non-living components on crop health in a particular crop ecosystem. The crop intelligence provides essential and relevant information and data which not only helps in developing and devising models fore warming pest/ disease development and outbreaks in advance but also keeps abrest with the trend, status and resurgence of pests and diseases. The information so generated over the years on pest and disease incidence and intensity on a certain crop surves very useful purpose for developing IPM strategy and approach for specific pests and crops in a given region (Diwakar, 1999). Field information on the status of pest, occurrence of biotypes of pest species or change of virulence, stages of prevalent at a given time, reaction of crop varieties to the prevalent pest, extent of incidence and or damage status of an insect species etc. are obtained through carefully designed surveillance programmes (Mathur, 1986).

14.1.2. Approaches to surveillance

The pest and disease surveillance in India has been taken up with the following two basic approaches:

1. The first approach is based on particular pest problem. In this approach considerable attention is given to the various parameters responsible for population explosion outbreak. The parameters are identified and evaluated after detailed studies on the lifecycle, biology, ecological requirements, biotic and abiotic factors. Generally such studies are conducted by elaborate sampling at permanently fixed sites with a attempt to obtain representative information for a crop area. A number of correlations between different affecting factors and pest activity are developed. These specialized studies are particularly useful in cases where crop pest environment is fairly stable and association lasts for considerable length of time. These efforts help to develop long term forecasting system.
2. The second approach is related to looking at the crop as a whole at periodical intervals and to for suggest action in terms of control measures as and when necessary. It is short term forecasting and is a prime requisite for the success of IPM

14.1.3. Type of surveillance

Rapid Roving Surveillance. It is made to monitor pest and incidence qualitatively in short time over large area for determining the need, locality and proper times for pest control decisions. By and large the sample pest monitoring on predetermined routes with a definite periodicity and intervals of 8, 10 or 15 days has been found quite satisfactory (Diwakar, 1999).

Fixed Plot surveillance. It is made to generate quantitative field data for developing fore warn system after simulating with weather conditions. The fixed plot observations supported by light traps, yellow traps, sticky traps, spore traps and pheromone traps also serve useful purpose for forewarning and prognosis (Diwakar, 1999).

14.1.4. Objectives of surveillance

The IPM programmes aim at minimizing the use of hazardous chemical pesticides and the surveillance and monitoring is corner stone for success of IPM. The pest surveillance is like a mobile diagnostic unit in the health service schemes. It is also comparable to the signal core of army and intelligence wings such as CBI. It is not essential for the effectiveness of any of pest control methods. Without the basic and essential information required for adoption of the pest control techniques, it will not be possible to adopt IPM approach. Such an information is possible only through regular and intensive surveillance programme by highly qualified and trained technical staff (Diwakar, 1999).

The many fold objectives of surveillance are as under:

- To monitor and foreworn the likely build-up of pest/disease in order to facilitate planning and adoption of timely suitable control measures based on tentative or established economic threshold (ETH) levels and to avoid prophylactics operations as for as possible.
- To study the trend and sequences of pest developments in a given crop ecosystem to identify the resurgence and potential problems irrespective of their current economic status. This enables the determination of research priorities and studies to find out the solutions before the problems assume serious proportions.
- To generate field data on ecology and epidemiology of major pests for some years which could be correlated with various crop and weather parameters for developing the fore warn system.
- To generate vital information about pests such as their occurrence in traces, their bio types, change in virulence, reaction of crop varieties to pests in a given region, estimation of crop losses etc.
- To provide information and data on status of possible and potential role of bio control agents under field conditions which helps in determining the type of measures to be taken in the over all Integrated Pest Management.
- To detect the invasion of exotic pests, to determine the rate of spreads of exotic pests that have already got established and to provide information for later management phases.

For accurate population estimation of pest or natural enemies, we should depend on reliable or standardized sampling method, sampling equipment (Table 14.1), sample size and number of samples.

Table 14.1. Sampling method and equipment needed for important groups of insect pests

S No	Sampling Method	Fauna	Equipment
1.	Sweet net method	Leaf hoppers, Thrips	Sweep net
2.	Square metre method	Stem borers, Armyworm	Square metre frame
3.	Plant Count method	Gram pod borer	
4.	Leaf count method	Whiteflies, Thrips	
5.	Linear metre method	Gram pod borer, Blister bettle	
6.	Water pan method	Aphids	Water pan
7.	Light trap method	Moths, Gall midge (adult)	Light trap
8.	Pheromone trap method	Moths	Pheromone trap
9.	Sticky trap method	Aphids, Whiteflies	Sticky traps
10.	Sieves method	Stored grain pests	Sieves
11.	Boll count method	Boll worm complex of cotton	

14.2. ECONOMIC THRESHOLD LEVELS

For timely adoption of need based pest management measures, Economic Threshold (ETH) Levels for different insect pests on different crops differ from place to place depending upon the prevailing biotic and abiotic conditions as given below by Dhaliwal and Arora (2001) for major insect pests of cotton at different locations in India (Table 14.2).

Table 14.2. ETH levels of major insect pests of cotton at different locations in India

Insect pest	ETH level	Location (States)
Jassid	1 nymph/leaf	Ludhiana (Punjab)
	II grade injury symptoms	Faridkot (Punjab)
	2 nymphs/leaf	Coimbatore (TN)
	1-2 nymphs/leaf	Coimbatore (TN)
	3.5-5.1 nymphs/leaf	Surat (Gujarat)
	50-100 nymphs or adult/50 terminals	Coimbatore (TN)
	10 nymphs/leaf	Surat (Gujarat)
Whitefly	6-8 adults/leaf	Ludhiana (Punjab)
	20 nymphs/leaf	Ludhiana (Punjab)
	Stickiness due to honeydew excreted on leaves in 50% of plants	Ludhiana (Punjab)

	8-10 adults or 20 nymphs/leaf	Coimbatore (TN)
	5-10 adults/leaf	Bapatla (Andhra Pradesh)
Thrips	10 thrips/leaf	Coimbatore (TN)
	1-2 nymphs or adults/leaf	Coimbatore (Tamil Nadu)
Aphids	15-20% affected plants	Coimbatore (TN)
Mites	10 mites/cm^2	Coimbatore (TN)
Pink bollworm	8 male moths/trap/night	Hisar (Haryana)
	4-6 moths/trap/night	Ludhiana (Punjab)
	4% bollworm incidence in bolls on plants	Ludhiana & Bathinda (Punjab)
	5% bollworm incidence	Coimbatore (TN)
	10% bollworm incidence	Ludhiana (Punjab)
	Spray initiation at 75 days on LH 900 variety	Ludhiana (Punjab)
Spotted bollworms	1 larva/10 plants	Surat (Gujarat)
	10% incidence in squares	Coimbatore (TN)
	10% incidence in shoots or reproductive parts	Coimbatore (TN) & Bathinda (Punjab)
	5 larval units/ 10 plants	Surat (Gujarat)
	10 larval units/10 plants	Anand (Gujarat)
American bollworm	25 adults/trap/night or 1 larva/10 plants or 20 bolls	Baptala (AP) & Coimbatore (TN)
	1 larval/plant	Coimbatore (TN)
	10% incidence in reproductive parts	Bathinda (Punjab)
	2 larval units/10 plants	Surat (Gujarat)
	7 larval units/plants	Surat (Gujarat)
Bollworm complex	5% bollworm damage	Akola (Maharashtra)
	10% bollworm incidence	Bhubaneswar (Orissa)
	5% bollworm incidence in freshly shed fruiting bodies	Coimbatore (TN)
	Spray initiation at squaring in 25% of plants in *arboreum* cotton	Bathinda (Punjab); Ludhiana (Punjab)
Stem weevil	10% plant infestation	Coimbatore (TN)
Tobacco cutworm	4 egg masses/50 m row length	Coimbatore (TN)

Source: Modified after Simwat (1994)

14.3. STATUS OF PEST SURVEY/SURVEILLANCE

According to Mathur (1976) there was practically no systematic crop pest surveillance in India till late sixties except for locust monitoring in Scheduled Desert Area. The Locust Control and Research, a division of Directorate of Plant Protection, Quarantine and Storage (DPPQS), is being implemented through organisation known as Locust Warning Oraginsed (LWO) established in 1939 to monitor, forewarn and control Desert Locust (an international pest), It has 10 circle offices located at Bikaner, Jaisalmer, Barmer, Palanpur, Bhuj, Jalore, Phalodi, Nagpur, Suratgarh and Churu with its Field Headquarter at Jodhpur and Central Headquarter at Faridabad. The LWO has a satisfactory locust control potential being maintained by its own wireless network for exchange for information on locust not only in India but also in Pakistan (at Karachi) every year during June-November. Further more, it has pesticides, plant protection equipments, a fleet of 143 vehicles, 64 wireless sets, 52 GPSs and trained staff (Anonymous, 2010c). A very effective tool has become available to the locust scientists is Radar which can follow up the activities of desert locust swarms as well as single locust both during day and night (Sinha and Yadava, 1985). However, due to high costs and various difficulties of operation of Radar in interior of the desert it did not become a popular tool for survey and detection of locust. Another method attempted is the aerial photography in which the high resolution cameras are mounted in air craft to take pictures of locust habitat below, which when processed indicate the information on area and locusts present in the area. The latest tool is satellite. The above given methods are costly and time consuming for detecting locust but the satellite can be used to locate areas likely to contain locusts themselves or potentially inapproachable habitats. From satellite imageries it is also possible to determine vegetation cover characteristics and to locate locust and estimate rainfall situation. As locust generally concentrates in green areas, therefore, mapping green areas is as good as mapping potential locust habitats. The satellite thus precises pinpointing of actual breeding/development habitats of locust.

Before mid-fifties information on crop pests and diseases used to be collected but in an *ad-hoc* manner and no systematic records were maintained for meaningful interpretation. The first attempt in wheat disease surveillance with a special emphasis on the rust diseases was initiated in *rabi* 1966-67 by IARI in collaboration with Ford Foundation, Agricultural Universities, Research Institutes and State Plant Pathologists. The wide spread occurrence of Rice Tungro Virus and green leaf hopper epidemics on paddy during *Kharif* 1968 and 1969 in Bihar, UP, West Bengal causing heavy crop losses, led to the establishment of pest surveillance programme. The organised government-driven surveillance was commenced during late sixties (*Kharif,* 1970) on rice in the wake of

Rice Tungro Virus outbreaks on rice under the agies of DPPQS in collaboration with Indian Council of Agriculture Research, State Departments of Agriculture of UP, Bihar, West Bengal and Orissa and the Ford Foundation. The Government of India initiated following two steps simultaneously to monitor pest and disease situation on crops for suggesting timely control measures:

During 1969-70, a Pilot Project for Pests and Diseases Surveillance in 2 blocks with each of 4 innovative Intensive Agricultural District Programme (IADP) i.e. Raipur (MP), Sambalpur, (Orissa), Thanjavur (Tamil Nadu) and West Godavari (AP) on important pests and diseases of selective crops was started by Government of India with the help of concurred State Governments, the Ford Foundation and the Indian Meteorological Department, Subsequently, 4 more districts i.e Aligarh (UP), Nandya (Karnataka), Burdwan (West Bengal) and Palghat (Kerala) joined IADP.

The other action taken was to set up five separate survey teams of expects to survey vulnerable areas of paddy by following prescribed routes in Bihar, UP, West Bengal, MP and Orissa. Later on Punjab was also included. These roving surveys were commenced by the DPPQS in collaboration with Ford Foundation, Indian Council Agricultural Research and Concerned States. From 1973-74, the surveys have been taken up by the DPPQS with the help of States

14.3.1. National Pest and Disease Surveillance Scheme

The Government of India established a National Pest and Disease Surveillance Scheme during the 4th Five Years Plan and 12 Central Surveillance Stations in 9 States were opened in 1973-74. This scheme was district based and was also taking help of IADP districts for obtaining additional information on from other areas. The following is the list of Central Surveillance Stations in 9 different states:

S.No	State	Centrol Surveillance Stations (s)
1.	Uttar Pradesh	Luckhnow and Deoria.
2.	Andhra Pradesh	Nizamabad and Eluru
3.	Madhya Pradesh	Billaspur and Indore
4.	Punjab	Jallandhar
5.	Orissa	Cuttak
6.	Tamil	Trichirapulli
7.	Karnataka	Shimoga
8.	West Bengal	Malda
9.	Bihar	Samstipur (transferred from Pusa)

During VIIIth Plan, Integrated Pest Management scheme was launched by the Union Ministry of Agriculture, Government of India by merging 19 Central Surveillance Stations, 11 Central Biological Control Stations and 13 Central Plant Protection Stations together in their respective states to form the Central Pest Management Centre (CIPMC) and at present there are 26 such Stations in the country. However, the States of UP, MP, AP and J&K have two while others have only one such Station (Anonymous 2010a).

14.3.2. Surveillance service in the states

Certain State Governments organised Pest and Disease Surveillance in other districts of their states which were not covered under National Pest and Disease Surveillance Scheme from their own recourses. The State Pest and Disease Situation Report was compiled and sent tot the headquarters of DPPQS and the States organised control operations or advised the farmers accordingly when the situation warranted so. Some states like Assam, AP, Bihar, Kerala, MP, Orrisa, UP and Tami Nadu have developed their own Pest Surveillance Programmes. Other states also deploy their plant protection staff for pest surveillance during peak crop seasons on an *ad-hoc* basis. The Agricultural Universities also take up surveillance on important crops in limited areas for research work or conduct special surveillance at their own or in collaboration State Departments of Agriculture under some special situations. The Punjab Agricultural University, Ludhiana also conducts such surveillance through team of scientists of University; and its Regional Stations, district level Farm Advisory Service Scheme (FASS) and *Krishi Vigyan Kendras* (*KVK,s*) sometimes in collaboration with State Department of Agriculture Punjab and allied agencies under some special situations (Anonymous 1991, 97, 98, 99, 2001 and 2005).

The Department of Agricultural Entomology of Dr. Punjabrao Deshmukh Krishi Vidya Peeth, Akola (Maharashtra) has a separate Pest Surveillance Unit (one of the Centers of "Network Project on Insect Biosystematics") funded by ICAR, New Delhi. It makes the surveillance of crop pests and defenders. It is well equipped with the requisite man power as one Chief Plant Protection Officer and two Senior Research Assistants and provided with a vehicle to have swift mobility. This unit is also provided with supporting staff i.e. one Section Assistant and one driver. The unit takes immediate note of the outbreak, if any, and acts upon immediately to inform the farming society for the remedial measures to be adopted. It also releases the regular information bulletins to enlighten the cultivators from time to time (Anonymous, 2010d).

14.3.3. Surveillance in different crops

Surveillance for pests is mandatory for application of control measures against biotic fauna in the given crop ecosystem. Pesticides, in cotton need to be applied based on economic thresholds, which can be achieved through surveillance. Adoption of surveillance at micro level to apply right pest management strategies will hopefully reduce the cost of plant protection and will avoid contamination of ecosystem through indiscriminate application of costly toxic chemicals. Pest monitoring helps in forecasting the outbreak of a pest. In the absence of such system, the cost of plant protection fails to sustain the profit margin, which in fact, has declined in near times. The methodology and procedure of surveillance of some important field crops are given below:

14.3.4. Surveillance of cotton insect pests

Site selection. Site should be selected in different directions and also in central area (under cultivation) of a village. The village must have reasonable area under cotton at least not less than 50 per cent of the crop area. The field should be examined at least twice a week to collect information pertaining to the incidence of insect-pests, natural enemy complex and plant protection measures followed. An area of one acre should be marked at each site for recording the population and incidence of all categories of insect pests such as sucking insect pests, foliage feeders, bollworms and natural enemies of insect pests. For bollworm complex data should be recorded twice a week once the pests crosses economic threshold level for adoption of pest management strategies, as it is one of the major limiting factors in successful raising of cotton crop.

Surveillance Procedure. Surveillance procedure for major insect pests and natural enemies of cotton is given below:

Jassid. Population of nymphs should be made from 3 fully opened leaves from the upper canopy of 12 plants selected at random. The selected plant should be observed for damage symptoms i.e. yellowing curling of leaves along margins.

Whitefly. Counts of whitefly adults should be made from upper canopy leaves, which are taken into account to record the population of jassid nymphs. The counts should be made before 10 a.m. However, nymphal and pupal counts should be made from lower surface of middle canopy, if possible from an area of 2.5 x 2.5 cm^2 per leaf. Three leaves are selected from each plant. The damage symptoms i.e. appearance of honeydew on leaves should be observed on 50 plants.

Yellow pan sticky traps. The 5-8 traps should be put to use for monitoring adult whitefly population.

Aphid. The nymphs adults of aphid should be counted from 3 top and middle canopy leaves. The damage symptoms i.e. appearance of honeydew on leaves should be observed on 50 plants.

Thrips. The nymphs and adults of thrips are recorded from 3 fully formed leaves from the upper canopy.

Cotton leaf roller and green semi looper. Counts of larval and pupal population from selected plants should be made on whole plant basis.

Bollworms. The damage due to bollworms should be recorded on freshly shed fruiting bodies i.e. squares, buds, flowers and young bolls (upto 3-week age) under 12 plants selected at random from one location. For recording the damage among freshly shed fruiting bodies, the whole area should be divided into 4 quarters and from each quarter 25 freshly shed fruiting bodies should be counted. The healthy and bollworm damaged fruiting bodies should be counted separately to work out the percent shedding due to bollworms. The total number of growing points of plants damaged by spotted bollworms should also be counted from 50 plants selected at random during the vegetative phase.

Monitoring of bollworms with Sex Pheromones. Pheromone is a chemical or a mixture of chemicals that is secreted by an organism to the environment and causes one or more specific reactions in the receiving organism of same species. Sex pheromones of bollworms proved extremely useful for:

(a) Survey work
(b) Pest destruction in traps
(c) Mass Trapping
(d) Mating disruption

The monitoring of bollworms is done with the initiation of flowering stage of crop. Observations on moth traps are recorded on every alternate day. The monitoring strategy helps in making decision for effective management of bollworms.

Pink bollworm. The Sticka/Delta traps with at least 10 microlitre of gossyplure should be used @ one trap per hectare and placed 15 cm above the crop canopy. The lures should be replaced after 15 days.

Spotted/Spiny bollworms. The Sleeve/Moth traps should be used @ two traps per hectare and placed 15 cm above the crop canopy. The lures should be replaced after 15 days.

American bollworm: The Sleeve/Moth traps with at least 10 microlitre of helilure should be used @ two traps per hectare and placed 15 cm above the crop canopy. The lures are replaced after 15 days.

Tobacco caterpillar. The Sleeve/Moth catch trap for tobacco caterpillar should be used @ two traps per hectare and placed 15 cm above the crop canopy. The lures should be replaced after 15 days.

Natural enemies. Population of nymphs and adults of *Chrysoperla* spp., predatory bugs and spiders should be recorded from 25 plants selected at random. Similarly counts of grubs and adults of coccinellids should be made on whole plant basis from 25 plants selected at random. Twenty five or more eggs (one sample/acre) should be collected from field to record observations on egg parasitoids of spotted bollworms (SBWs) and American bollworm (ABW). For larval parasitization 25 or more larvae (one sample/acre) of spotted bollworms (SBWs), pink bollworm (PBW) and American bollworm (ABW) should be collected from the field to observe their parasitoids and per cent parasitism. The field collected eggs and larvae should be kept in laboratory at temperature range of 25-30°C and RH 70-80%.

14.3.5 Surveillance of rice insect pests

Stem borers and leaf folder. For this purpose 100 plants per acre should be selected at random from a minimum of 5 spots. The data on dead hearts caused by the stem borers or leaves damaged by the leaf folder should be recorded from these plants.

Dead heart - a plant whose central shoot is dead/dry and can be easily pulled out.

Leaf damage - a leaf with more than one-third damaged area should be referred as a damaged leaf.

Plant hoppers, White backed plant hopper and brown plant hopper. One month after transplanting, 100 plants should be observed at random, from a minimum of 5 spots per acre to record the presence of the planthoppers. A few plants in the field should be slightly tilted and tapped 2-3 times at the base. The hoppers floating on water should be counted.

Natural enemies. For this purpose 100 plants should be selected at random from one acre field. The natural enemies (adults and nymphs) present on these planted should be counted on selected plants (on whole plant basis).

14.3.6. Surveillance of sugarcane insect pests

Early shoot borer. The population of early shoot borer should be assessed by per cent incidence based on dead hearts recorded in post germination phase at 30 days interval up to 120 days. For this purpose, 500 randomly selected canes should be observed from one acre field (100 each from 4 quadrates and center) and per cent incidence should be calculated.

Top borer. The population of top borer should be assessed by per cent damage incidence recorded during the 3rd and 4th broods (July,

August and September). For this purpose, 500 canes selected at random should be observed from one acre field (100 each from 4 quadrates and center) and per cent incidence should be calculated.

Stalk borer. The population of the stalk borer should be assessed by percent pest incidence. For this purpose 500 randomly selected canes should be observed for pest incidence from one acre field (100 each from 4 quadrates and center) from one acre field and per cent incidence damage should be calculated.

Pyrilla. The population of nymphs, adults and egg masses should be recorded from a unit of 10 canes (20 leaves) from five places (two each from 4 quadrates and from center).

Whitefly. The population of nymphs and puparia should be recorded from a unit of 10 canes (20 leaves) from proximal, middle and distal region and average population per cm square should be calculated.

Thrips. The population of nymph and adults should be recorded from a unit of 10 canes (20 leave tips) from five places (two each from 4 quadrates and center).

14.3.7. Surveillance of oilseed insect pests

Rapeseed-Mustard

Mustard Aphid. The number of aphids per 10 cm central terminal shoot on 10 plants selected at random should be counted. Monitoring of alate aphids should be done with yellow sticky traps @ 5 traps per acre.

Cabbage caterpillar. The Number of larvae per 10 plants selected at random should be counted. To record larval parasitization about 50 larvae from 10 locations should be collected and reared on natural food to record parasitization.

Mustard Sawfly. The number of larvae per 10 plants selected at random should be counted.

Painted bug. The number of bugs (adults + nymphs) per 10 plants selected at random should be counted.

Coccinellid beetles. The number of adults and grubs per 10 plants selected at random should be counted (on whole plant basis).

Chrysoperla carnea. The number of larvae per 10 plants selected at randomed should be counted (on whole plant basis).

Sunflower

Head borer. The number of larvae per 10 plants selected at random should be counted.

Jassid. The number of nymphs per 6 randomly selected leaves (Two leaves each from top, middle and lower canopy of plants) should be counted.

Whitefly. The number of adults per 6 randomly selected leaves (Two leaves each from top, middle and lower canopy of plants) should be counted.

Coccinellids. The number of adults and grubs per 10 randomly selected plants should be counted (on whole plant basis).

Spiders. The number per 10 randomly selected plants should be counted (on whole plant basis).

Chrysoperla. The larvae per 10 randomly selected plants should be counted (on whole plant basis).

Reduviids, Mantids. Count the number per 10 randomly selected plants should be counted.

Groundnut

Aphids. Monitoring of winged aphids should be done by using yellow sticky traps @ 5 traps per acre. The number of apterous aphids should be counted per plant on 10 randomly selected plants.

Jassids and Thrips. For jassids, five sweeps of the sweep nets should be taken in 5m length during the vegetative, pod filling and maturity stages of the crop.

Thrips. The number of thrips from 5 compound leaves and 10 flowers selected at random should be counted.

Spodoptera litura, Spilosoma obliqua **and** ***Helicoverpa armigera.*** Twenty compound leaves should be collected from 5 places in a plot randomly and foliage damage and larval population should be noted.

Leaf miner. Twenty leaves should be collected from 5 places in a plot and the foliage damage should be recorded.

Natural Enemies. Five sweeps of the sweep net should be taken while taking the observations on thrips and jassids and the number of natural enemies trapped should be recorded. The number of lady bird beetles, Syrphid flies from 10 main shoots (when the aphids are congregated) should be recorded.

For natural enemies of leaf miner, 20 infested compound leaves collected random from five places from the plot after 15, 45, 60 and 75 days of incidence should be kept separately in plastic jars covered with muslin cloth and the emergence of parasitoids should be recorded. For *Spodoptera litura* and *H. armigera* about 50 larvae of these insect pests should be collected depending upon the availability and reared on the natural food to record parasitization, if any. The population of predators such as spiders etc. should be recorded from 5 plants selected at random.

14.3.8. Surveillance of Insect Pests of maize

Borers. The observations on total number of plants, leaf injury and dead hearts present in infested plants caused by maize borer *C. partellus* are recorded.

Leaf injury. The total number of plants after germination and plants exhibiting leaf injury showing damage grade upto 6.0 should be recorded. The plant damage caused by *C. partellus* is rated according to 1-9 scale by observing individual plant. Percentage of leaf injury should be calculated on 100 plant basis.

Dead hearts. The plants showing dead hearts (grade 7-9) caused by maize stem borer should be counted and per cent dead heart formation was worked out on 100 plant basis.

The observations on percentage dead hearts and leaf infestation should be recorded from the central rows at 15 and 35 days old crop. The severity of damage should be recorded by leaf injury rating scale (1-9) given below.

Score	Description
1	Apparently healthy plant;
2	Plant showing slightest damage on leaf or a few holes on 1-2 leaves;
3	Plant showing more pin holes or shot holes on 3-4 leaves;
4	Plant showing injury (pin holes, shot holes, slits) in about one-third of total number of leaves and mid-rib tunneling on 1-2 leaves, if any;
5	Plant showing about 50 per cent leaf damage (pin holes, shot holes, slits streaks) and mid-rib damage if any;
6	Plant showing varied type of leaf injury in about two third of the total number of leaves;
7	Plant with every type of leaf injury and almost all the leaves damaged;
8	The entire plant showing maximum leaf injury and likely to form dead heart (such plants usually show stunted growth);
9	Dead-heart.

Shoot fly. The observations on total number of plants, leaf injury and dead hearts present in infested plants caused by maize shoot fly should be recorded.

Leaf injury. The total number of plants after germination and plants exhibiting leaf injury (crumpling and crinkling) should be recorded by observing individual plant. Percentage of leaf injury should be calculated on 100 plant basis.

Dead hearts. The plants showing dead hearts caused by shoot fly should be counted and per cent dead heart formation should be worked out on 100 plant basis.

The observation on percentage dead hearts and leaf infestation should be recorded from central rows at 7 and 15 days old crop. The severity of damage should be recorded by dead hearts incidence based injury scale given bellow.

Aphid. The number of aphid per five plants should be counted and mean should be worked out.

Jassid. The number of jassid per three sweeps should be counted and mean is worked out.

Pyrilla. The number of pyrilla per plant fro ten plants should be counted and mean is worked out.

Thrips. The number of thrips per five plants should be counted and mean should be worked out.

Surface grasshopper. The numbers of grasshoppers per plants should be counted and mean should be worked out. For recording incidence of grasshoppers the total number of plants after germination and plants exhibiting leaf injury should be recorded. Percentage of leaf injury should be calculated.

Grey weevil. The number of grey weevils per five plants should be counted and mean should be worked out.

Silk cutter. The number of silk cutters per plant should be counted and mean should be worked out.

Corn earworm. The number of cobs infested with corn earworm should be recorded and percentage should be worked out.

Spiders. The number of spiders per plant should be counted and mean should be worked out.

Coccinellids. The number of coccinellids per plant should be counted and mean should be worked out.

14.4. CONCLUSION

From the above discussion, it is quite evident that the success of Integrated Pest Management approach depends on well organized, planned and efficient insect pest surveillance at micro and macro level. Surveillance programme acts as a catalyst for energizing all major components of Integrated Pest Management Programme. The practical trainings at block and village level should also be organised to make the farmers practically aware about the identification of insect pests of crops, their damage, proper methods of surveillance and requirement of need based Pest Management as per recommendations. The extension functionaries should also be trained though regular refresher courses about the new emerging pest problems and recent pest management strategies to be followed. The surveillance is already being taken both at Central and State levels but urgently needs to be strengthened in the light of changing scenario of pest complexes and weather conditions. As in Dr. Punjabrao Deshmukh Krishi Vidya Peeth, Akola (Maharastra) there should be the provision of a separate "Pest Surveillance units" in all the State Agricultural Universities with adequate trained staff infrastructure, supporting staff, vehicles etc. to conduct regular surveillance for timely collection of information to take immediate and proper remedial measures for the farmers of the respective states.

REFERENCES

Anonymous (1991) Weekly surveillance reports about the incidence of *Helicoverpa armigera* and other important insect pests on cotton, gram, late wheat, tomato, berseem, arhar and sunflower submitted to the Head, Department of Entomology, PAU, Ludhiana.

Anonymous (1997) Weekly surveillance reports about the incidence of *Helicoverpa armigera*, other insect pests, and natural enemies on cotton submitted to the Head, Department of Entomology, PAU, Ludhiana.

Anonymous (1998) Weekly surveillance reports about the incidence of *Helicoverpa armigera,* other insect pests and natural enemies submitted to the Head, Department of Entomology, PAU, Ludhiana.

Anonymous (1999) Weekly surveillance reports about the incidence of *Helicoverpa armigera*, other insect pests and natural enemies submitted to the Head, Department of Entomology, PAU, Ludhiana.

Anonymous (2001) Present Status of American bollworm on cotton and Emergency Steps for its Proper Management (September 4-5, 2001) (Submitted to the Director, Agricultural, Punjab).

Anonymous (2005) Evaluation of Bt Cotton hybrids recommended for cultivation in Punjab during 2005 *kharif* season. Regional Research Station (Punjab Agricultural University), Bathinda. pp 18 (submitted to Government of Punjab).

A Anonymous (2010a) Central Integrated Management Centre, Jammu and Kashmir. http: //cidmcjk.nic.in / profile. asp.

nonymous (2010b) Economic Survey 2009-2010 Kalyani Publishers, New Delhi. Ministry of Finance and Company Affairs, Government of India, New Delhi.

Anonymous (2010c) Locust control and Research, Directorate of Plant Protection Quarantine and Storage. https:/ www. ippc. int / file_ uphaded 11146658.

Anonymous (2010d) Dr. Punjabrao Deshmukh Krishi Vidya Peeth, Akola. pdkv.ac.in / Entomology Dept. pup cached.

Appukuttan, K. (1993) Pest surveillance. Lecture presented on specialized Workshop-cumseminar on Pest Survaillance held at Central Plant Protection, Training Institute (CPPTI), Hyderabad from July 28 to 30, 1993.

Dhaliwal G.S., Jindal, V. and Dhawan, A.K. (2010) Insect pest problems and crop losses: changing trends. *Indian J. Ecol,* **37** (1) : 1-7.

Dhaliwal, G.S. and Arora, R. (2001) *Integrated Pest Managemen t: Concepts and Approaches*. Kalyani Publishers, New Delhi.

Dhaliwal, G.S. and Koul, O. (2010) *Quest for Pest Management : From Green Revolution to Gene Revolution*. Kalyani Publishers, New Delhi.

Dhaliwal, G.S., Arora, R. and Dhawan, A.K. (2004) crop losses due to insect pests in Indian Agriculture: An update. *Indian J. Ecol.* **31** (1) : 1-7.

Diwakar, M.C. (1999) crop pest surveillance and monitoring in India - An over view. *Pl. Prot. Bull.* **51**(1-2) : 14-16.

Eavy, A.L. (2004). Pest insect population survey and surveillance for a predictive data base used in IPM for tropical vegetable production http : //www. nmc net.edu / community.efm/compass/are/and http://www.crees.org.

Mathur, K.C. (1986) Pest surveillance methods for IPM. *Pl. Prot. Bull.* **38**(1-4) : 65-67.

Mathur. K.C. (1976) Pest Forecasting Surveillance for Crop Pests. In *Proc. Summer Insitute Methods and Techniques in Pest fore Casting.* July 1-2, 1976. Punjab Agricultural University, Ludhiana.

Sinha, P.P. and Yadava, V.K. (1985) Sensing desert Locust breeding areas through satellite imageries. *Pl. Prot. Bull.* **37** (1) 27-29.

Theory and Practice of Integrated Pest Management
R. Arora, B. Singh and A.K. Dhawan (Eds)
Scientific Publishers (India), Jodhpur 2011, pp. 248 - 261.

15

ROLE OF INSECT-PEST FORECASTING IN INTEGRATED PEST MANAGEMENT

M.S. Mahal, Beant Singh*, P.S. Sarao* and Sandeep Singh
Department of Entomology, *Department of Plant Breeding and Genetics
Punjab Agricultural University, Ludhiana - 141 004, India

15.1. INTRODUCTION

Insect-pest forecasting is the foundation of a sound integrated control. Pest Forecast helps us in guiding the timing of pest incidence to eliminate blanket applications, reduce pesticide amounts, and achieve quality results. Insect-pest forecasting is the foundation of integrated control in advance, the important prerequisite of a bumper harvest as a result of effective control. However, such prediction is affected not only by the characteristic of insect pests but also the restriction of the meteorological factors. For a pest manager, being able to predict pest species abundance and distribution (pest pressure), and its timing and level, is crucial to both strategic planning and tactical decision-making (Maelzer & Zalucki, 2000). Phenological models, based on insect physiological timescales, have been relatively successful at predicting the timing of population peaks (Mohandass & Zalucki, 2004) and are useful for timing sampling and control measures. Forecasting pest pressure is more problematic because many factors influence abundance (Yonow *et al.,* 2004). Timely forecasts would be useful for determining insecticide budgets, or making strategic decisions on which crops to plant. Long-term forecasts of abundance would be especially useful for major pest species such as *Helicoverpa* spp.

In precision farming systems, insect-pest forecasting provides opportunity to the farmers for preparedness and takes timely action to apply bio-agents and pesticides, which ultimately cut down the cost of production and serve as an important tool in agriculture. Pest forecasting models are essential tools and pre-requisite of IPM as the insect population levels are deciding factors in pest management. In India, one of the earliest works on pest forecasting was started by Dr. S. Pradhan when he gave the concept of biotic theory of locust population explosion. According to him, temperature was one of the most important factors, which determines the duration of various stages of insect life cycle and the number of generations. Remote sensing has also been used for forecasting of locusts (*S. gregaria*) in India. The locusts oviposit in the sandy deserts where moisture is present in the soil. The satellite detection of moist soil sand as breeding site and the amount of green vegetation as food in the desert give the possible amount of locust population. The migration of locusts is forecasted using the wind direction.

15.2. PRE REQUISITE FOR THE DEVELOPMENT OF FORECASTING MODELS

The following basic information and methods are required for the development of a dynamic pest forecasting and prediction system.

(i) Quantitative seasonal studies. Sampling of population, spatial and geographical distribution of pest abundance, seasonal counts in relation to climate, phenology of the crop and topography need to be provided for pest modelling.

(ii) Life history and pest biology. The duration of the life cycle, number of generations in a year, growth index, life span, survival rate, food, habitat, over-wintering, alternate hosts and intrinsic growth rate in the field and simulated laboratory conditions should be available with the working group. The effect of abiotic factors on pest development and spread needs to be worked out accurately.

(iii) Ecological studies of the pest. Life-table studies of pest are important for better understanding of pest population build-up, natural mortality factors, critical stage, and intrinsic growth rate. Life table of a pest can be used to find the favourable period for mating and emergence. Besides these, the spatial pattern can also provide clues to find the suitable sampling schemes to be used for collecting pest population samples. The migration and immigration of pests can also be used for forecasting of pests.

(iv) Crop phenology. Different crop cultivars, fertilizer dosages, irrigation and plant spacing influence the phenology of the crop. Therefore, the population of the pest on various phonological stages needs to be studied.

(v) Natural enemies. The population of natural enemies present on various time intervals in the crop needs to be worked out precisely for different locations, since they play an important role in suppression of pest population.

(vi) Agro-ecosystems. The changing cropping pattern and crop diversification involving different crops with a wide range of varieties of different maturity groups serve as a suitable niche for supporting the build-up of *Helicoverpa*. Therefore, the information on cropping pattern and crop sequences needs to be included in the pest modelling.

15.3. MODELLING OF PEST POPULATIONS

Based on the data available, various modelling options are available to model the pest population with other factors such as weather, natural enemies present, and crop biology. The relationship between the different components is shown in Fig. 15.1.

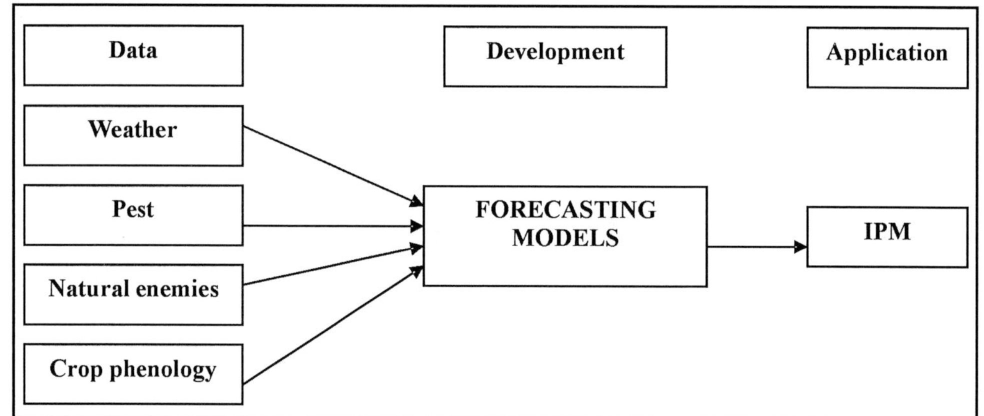

Fig. 15.1. Relationship of components of pest forecasting

Some of the successful modelling techniques used in pest forecasting include statistical, mathematical and simulation models. Statistical techniques such as simple linear regression techniques, multiple regression, non-linear models, group data handling and time-series analysis are use din pest forecasting. Mathematical models such as simple growth models are widely used in pest forecasting.

Modeling plays an important role in improving pest management in a number of ways viz., widening the range of control measures, pest forecasting, and improving crop protection decision rules (Ram Niwas, 2007). Modeling techniques fall into four broad categories: Statistical, Mechanistic, Optimization and Bioclimatic, depending upon their functions. Different models are discussed as under.

Statistical Models. These include regression and multivariate models, experimental designs, and standard theoretical statistical distributions, which are based on empirical observations and do not describe the processes. They include predictive models such as regression models which have been used in UK.

Mechanistic Models. These are constructed to represent knowledge concerning the biological and ecological processes that underline pest population dynamics, damage, and/or control. Analytical models, simulation models, inferential models, modeling with spreadsheets and rule-based models fall into this category.

(a) Analytical Models. These are mathematical models, based on differential equation. They provide general information about the system by describing the characteristics in a handful of relationships and parameters. Analytical models have been used in developing host-parasitoid relationship for sustainable biocontrol strategies and for determining population changes.

(b) Simulation Models. Simulation is the process of designing a model of a real system and conducting experiments within the model for the purposes of understanding the behaviour and operation of the system. Simulations models are used in pest management mainly in relation to the population dynamics of crop, pests and their natural enemies.

(c) Inferential Models. They have been proposed for modeling development of an insect population that passes through a series of discrete developmental stages. The relationship between an insect, its hosts and predators is often dependent upon the relative phonologies of each.

(d) Optimization Models. These models are constructed in such a way that the optimal strategy of pest management can be determined either by finding a mathematical solution or by some numerical optimization procedures. Simulation and optimization are complementary tools in pest management.

Bio-climatic Models. These are climate-driven models used for various purposes of pest management, such as the CLIMAX and BIOCLIM models. CLIMAX is a climate-driven dynamic simulation model that enables the prediction of an animal or plant's potential relative abundance and distribution as determined by the climate. CLIMAX can be used for assessing the risk associated with the introduction of pest in new areas and appraising the likely changes in pest status associated with climatic change.

BIOCLIM is a bioclimatic model used for climatic analysis and mapping the potential distribution of any biological entity in relation to climate.

15.4. PEST FOREWARNING MODELS

Some of the pests forewarning models have been developed in different parts of the world including India for different pests (Trivedi *et al.,* 2000, 2002, Das *et al.,* 2001).

15.4.1. *Helicoverpa armigera* Forecasting Models

Forecasting *Helicoverpa* populations in Australia. Long-term forecasts of pest pressure are central to the effective management of many agricultural insect pests. In the eastern cropping regions of Australia, serious infestations of *Helicoverpa punctigera* and *H. armigera* are experienced annually. Regression analyses of a long series of light-trap catches of adult moths were used to describe the seasonal dynamics of both species. The size of the spring generation in eastern cropping zones could be related to rainfall in putative source areas in inland Australia. Subsequent generations could be related to the abundance of various crops in agricultural areas, rainfall and the magnitude of the spring population peak. As rainfall figured prominently as a predictor variable, and can itself be predicted using the Southern Oscillation Index (SOI), trap catches were also related to this variable. CLIMEX, a software package developed to estimate the effect of climate on *Helicoverpa* sp. is generally used to predict the suitability of a site for a species based on long-term average climate and the species estimated responses to temperature and moisture requirements. The geographic distribution of each species was modeled in relation to climate and CLIMEX was used to predict temporal variation in abundance at given putative source sites in inland Australia using historical meteorological data. These predictions were then correlated with subsequent pest abundance data in a major cropping region. The regression-based and bioclimatic-based approaches to predicting pest abundance are compared and their utility in predicting and interpreting pest dynamics undertaken (Zalucki and Furlong, 2005).

(ii) A computer-based simulation model MOTHZV has been developed for predicting the population dynamics of *Helicoverpa* species. The model uses early season number of eggs, larvae or adults to forecast the timing and size of later potentially damaging population. Pheromone trap has provided the means for measuring early season numbers of *Helicoverpa* adults. This trap data along with climatic variability and crop phenology are used as input to the MOTHZV model to predict the timing of future *Helicoverpa* generations.

(iii) In India, the monthly pheromone trap moth data of 1983-90 along with weather data of corresponding period collected from PAU, Ludhiana have been analyzed at NCIPM by Das *et al.* (2000) and it was found that there were two major peaks of moth catches in a year. The main peak was during March-April and the second peak was during

October every year. During March-April peak, chickpea was the main crop and during October, cotton was grown almost over entire area. The peak population of *Helicoverpa* moth during March-April and October can be predicted in advance by multiple regression models using different weather parameters along with previous season's population of *Helicoverpa.*

The population density during March-April (P_{M-A}) has been regressed with different weather parameters as well as pest population density of previous five months separately and cumulatively as well as pest population. P_{M-A} depends on total moth catch per trap during previous October to February (P_{O-F}), mean monthly R.H. recorded in the afternoon during previous February (RHE_F) and mean monthly minimum temperature of previous February ($Tmin_F$) as follows:

$$P_{M-A} = —1032.65 + 2.06\ P_{O-F} —34.26\ RHE_F + 516.74\ Tmin_F\ (R^2=0.75)$$

Similarly, the population density during October (P_o) can be predicted in advance with multiple regression using weather parameters as well as pest population of preceding months as follows:

$$P_o = —.39 + 0.12\ P_{M-A} + 1.28\ P_J —0.51\ R_{J-J}\ (R^2=0.87)$$

P_{M-A} = Pest population density of March and April

P_J = Pest population density of June

R_{J-J} = Total amount of rainfall during June and July

NCIPM has also developed weather-based insect-pest forecasting systems against *Helicoverpa* on rice, cotton, pigeonpea, sugarcane, groundnut and mustard ecosystems.

***Helicoverpa armigera* forecasting on cotton in Punjab.** Singh *et al.* (2003) reported that information on likely timing of *Helicoverpa armigera* incidence can be useful for growers planning to cultivate cotton in the Indian Punjab. A simple population dynamics model was developed to predict the incidence of *Helicoverpa* on cotton using Dymex (Software for modelling insects). Dymex makes use of different growth constants such as development threshold (Temperature, humidity and photoperiod etc), stress factors (heat and cold stress, rainfall etc) and weather data to predict the probable adult population of *H. armigera.* The prevailing weather conditions during Rabi 2002-2003 were found to be similar to Rabi 1993-94. Therefore, Dymex was run for 1993-94 and 2002-03 Rabi season. The observed peaks of *Helicoverpa* incidence coincided with peaks simulated using Dymex in 1993-94. The expected population peaks for *Helicoverpa* incidence were obtained during mid February to mid March in both the years. The validity of the model was checked by comparing the expected and observed moth population during Rabi 2002-03 season. The results show that expected number of adult counts was statistically non-significant ($P = <0.01$) with the observed ones. Assuming that the weather conditions during 2003 might follow similar trends as in 1994 for rest of year, the predictions have been made for Khraif 2003 season using weather file of 1994. The model predicts that season will experience peaks of *Helicoverpa* population during July and September-October.

15.4.2. Potato aphid (*Myzus persicae*) Forecasting Model

Various non-linear models have been developed for predicting *Myzus persicae* based on various weather parameters (Max., Min. temperatures, RH etc.). Validation of this model revealed that the values of both the forecasted and actual population of *Myzus persicae* were below critical level values (20 aphids/100 compound leaves of potato) and that no intervention of pest management was required during the season.

15.4.3. Potato Tuber Moth (PTM) Forecasting Model

The population of PTM is affected due to heavy rain and large sunny days whereas cloudy and dark days favored the multiplication of the pest. The spatial distribution and relationship with weather also serve as a basic tool for the prediction of PTM in South India.

15.4.4. SOPRA

It is a phenology-model based forecasting tool for insect pests has been developed with the objective to optimize the timing of monitoring, management and control measures in Swiss apple orchards. Applying the time-varying distributed delay approach, air and soil temperature driven phenology-models were developed for three major insect pests of apple: The rosy apple aphid (*Dysaphis plantaginea*), the apple sawfly (*Hoplocampa testudinea*) and the smaller fruit tortrix (*Grapholitha lobarzewskii*). SOPRA predicts the age structure of rosy apple aphid, apple sawfly and smaller fruit tortrix populations, computes temperature sums for other pests and announces crucial events for management activities. Provided air and soil temperature data are available, SOPRA can be applied as a decision support system for insect pest management on a local and a regional scale (Graf *et al.,* 2002).

15.4.5. FLYPAST

The development of an information management system (FLYPAST) for the suction traps of the Rothamsted Insect Survey is described. FLYPAST runs on IBM-compatible personal computers. Prototype decision support systems that use suction trap data have been incorporated into the system for forecasting outbreaks of the black bean aphid (*Aphis fabae*) on field beans and of potato virus Y levels in crops intended for use as home-saved seed (Knight *et al.,* 1992).

15.4.6. BioSIM

BioSIM is a software tool for use in forecasting events in the seasonal biology of insect pests. Forecasts are made by simulation models provided by the system and are based on regional air temperature and precipitation interpolated from nearby weather stations, adjusted for elevation and location differentials with regional gradients (Anawati and Regniere, 2009). BioSIM is of interest to any organisation responsible for monitoring or managing insect pest populations in outdoor situations (forestry, agriculture, horticulture). Currently, models are linked to BioSIM for spruce budworm, spruce budmoth, gypsy moth, eastern hemlock looper, jack pine budworm and yellow-headed sawfly. BioSIM also provides general-purpose simple or multi-stage degree-day summation models. The system can make forecasts or predictions for individual (point) locations, but can also generate maps of phenological or other events when provided with a digital elevation model of the area under consideration.

BioSIM is an integrated environment that provides seasonal simulation models with geographically specific temperature input, either historical or forecast. BioSIM controls the execution of simulations and can extract information from model outputs for presentation or further analysis. BioSIM assembles weather data for simulations at each of a large set of locations, from 3 geographically-referenced databases. The Normals database contains long-term mean and extreme monthly minimum and maximum temperatures and precipitation. Five-day forecasts are contained in a separate database. Current observed daily weather data are contained in a Real Time database. BioSIM selects the "best" sources of weather data for each point in the location list, adjusts the data for differences in elevation, latitude and longitude, and restores stochastic variation to long-term averages (normals). The weather time series sent to the simulation model is composed of real-time data whenever available (or requested), 5-day forecasts for short-term prediction (when available), and normals for prediction over the longer term or to fill gaps in observed weather.

BioSIM. It can be used to plan the timely deployment of sampling/surveying crews and materials (e.g. pheromone traps) without the need for extensive phenology monitoring. Also, it helps time the application of pest control substances for optimal results. Thus, the use of BioSIM helps optimize the use of pest management resources in a cost-effective manner.

15.4.7. Forecasting of thrips with NAPPFAST

A weather-based mapping tool, NAPPFAST, was used to predict potential establishment of *Scirtothrips dorsalis* in North America. The analysis was based on a degree-day model and cold temperature survival of *S. dorsalis*. The results demonstrated that *S. dorsalis* could potentially produce upto 18 generations and was likely to survive in the southern and western coastal plains of the United States. They concluded that *S. dorsalis* is likely to be a serious economic pest in the southern United States.

15.4.8. Forecasting of corn earworm

Sandstron and Changnon (2009) reported that the Insect Migration Risk Forecast, or IMRF, is a proactive forecasting product issued daily Monday through Friday from mid-May to the end of September (except holidays) that outlooks regions at risk for insect migration in the Midwest and Ontario, Canada related to weather patterns in the central United States. The forecast is specifically geared toward corn earworm (CEW), but is applicable to all other migratory insects as well.

15.4.9. Forecasting mustard aphid in Punjab

Dhaliwal *et al.* (2005) conducted study in Ludhiana from 1988-89 to 1997-98 to evaluate the use of agrometerological indices for forecasting mustard aphid on Raya, *Brassica juncea.* A formula for measuring the relationship between weekly aphid population and humid-thermal ration was provided. The year 1992-93 and 1996-97 were characterized with high infestation, with aphid populations ranging from 700 to 1300 aphids/plants. The growing degree days accumulated from 1 December for all the years. These were 287, 329 and 839 for first, peak and last aphid incidences, respectively. A pest weather diagram constructed during the high infestation years showed that low minimum temperature, high morning relative humidity, and reduced solar radiation or sunshine hours favoured aphid population. Dhaliwat *et al.* (2007) predicted the forecasting of mustard aphid, *Lipaphis erysimi* Kaltenbach and reported that the incidence, growth and multiplication of this aphid are largely influences by meterological parameters like temperature, relative humidity, rainfall, wind speed and cloudiness. The historical data on mustard aphid and mean temperature were analyzed from 1988-89 to 2003-2004. Mean temperature during January for both the low and high aphid attack years was compared. Mean temperature during January remained below 13°C during high aphid attack years while it remained above 13°C during low aphid attack years. The years with low aphid attack were comparatively warmer than the years with high aphid attack. Growing degree-days (GDD) were accumulated from 1 December for the low and high aphid attack years. GDD model indicated that GDD accumulation and aphid attack were negatively correlated.

15.4.10. Use of remote sensing in pest forecasting

Kullari *et al.* (2007) reviewed surveillance of Arthropod Vector-Borne Infectious Diseases Using Remote Sensing Techniques. Epidemiologists are adopting new remote sensing techniques to study a variety of vector-borne diseases. Associations between satellite-derived environmental variables such as temperature, humidity, and land cover type and vector density are used to identify and characterize vector habitats. The convergence of factors such as the availability of multi-temporal satellite data and georeferenced epidemiological data, collaboration between remote sensing scientists and biologists, and the availability of sophisticated, statistical geographic information system and image processing algorithms in a desktop environment creates a fertile research environment. The use of remote sensing techniques to map vector-borne diseases has evolved significantly over the past 25 years. They reviewed the status of remote sensing studies of arthropod vector-borne diseases due to mosquitoes, ticks, blackflies, tsetse flies, and sandflies, which are responsible for the majority of vector-borne diseases in the world.

15.4.11. Citrus forecasting model

Du *et al.* (2008) developed an airborne multispectral technique for identifying tree health problems in a citrus grove that can be combined with variable rate technology (VRT) for required pesticide application and environmental modeling for assessment of pollution prevention, in an effort to promote precision farming in citrus production. The PRZM-3 model was used to estimate environmental impacts that contribute to non-point source pollution with and without the use of multispectral remote sensing and VRT. Their research findings using site-specific environmental assessment clearly indicated that combination of remote sensing and VRT may result in benefit to the environment by reducing the non-point source pollution by 92.15%. Overall, this study demonstrates the potential of precision farming for citrus production in the nexus of industrial ecology and agricultural sustainability.

15.4.12. Monitoring of mites in orchards using remote sensing

Luedeling *et al.* (2009) reported that remote sensing techniques can decrease pest monitoring costs in orchards. To evaluate the feasibility of detecting spider mite damage in orchards, they measured visible and near infrared reflectance of 1153 leaves and 392 canopies in 11 peach orchards in California. Pairs of significant wavelengths, identified by Partial Least Squares regression, were combined into normalized difference indices. These and 9 previously published indices were evaluated for correlation with mite damage. Eight spectral regions for leaves and two regions for canopies (at blue and red wavelengths) were significantly correlated with mite damage. These findings were tested by calculating normalized difference indices from the Red and Blue bands of six multispectral aerial images. Index values were linearly correlated with mite damage (R^2 = 0.47), allowing identification of mite hotspots in orchards. They further observed that better standardization of aerial imagery and accounting for perturbing environmental factors will be necessary for making this technique applicable for early mite detection.

15.4.13. Use of GPS and GIS in remote sensing for pest monitoring

Lan *et al.* (2009) reported that remote sensing is being used with Global Positioning Systems (GPS), Geographic Information Systems (GIS), and variable rate technology to ultimately help farmers maximize the economic and environmental benefits of crop pest management through precision agriculture. Airborne remote sensing is flexible and versatile because fields can be flown at variable altitude depending on the spatial resolution required. Although the use of airborne hyperspectral remote sensing in agricultural research and applications has been steadily increasing in the last decade, the airborne multispectral technique is still a good source of crop, soil, or ground cover information. The MS-4100 is a multispectral camera that produces and aligns images from different bands with a built-in prism. Data can be analyzed from the composite image or individual band images. The camera system evaluated herein uses a camera control system to physically compensate for roll, pitch, and yaw and maintain the camera at vertical nadir orientation. They have described the use of automated airborne multi-spectral imaging system and image processing using sample imagery to demonstrate the capability and potential of the system for crop pest management.

15.4.14. Monitoring of wheat aphid with Cropscan radiometer

Yang *et al.* (2009) studied the feasibility of using remote sensing techniques to detect stress in wheat caused by aphid infestation in the Great Plains of USA. Early detection of aphid infestation is a critical part of IPM for wheat production. The purpose of this greenhouse study was to characterize and differentiate stress in wheat due to infestation by greenbugs and Russian Wheat aphids using a hand-held Cropscan radiometer. Reflectance data and derived vegetation indices from the 16 bands of the radiometer were analyzed using SAS PROC MIXED statistical analysis procedure. Their results showed that it is possible to detect the stress caused by the two aphid species and to discriminate between the two aphid-induced stresses in wheat using remote sensing. Ratio-based vegetation indices (based on 800/450 nm and 950/450 nm) were found useful in differentiating the two stresses in wheat. However, more canopy-level-studies are needed to identify bands and indices that might have potential to differentiate the two stresses on wheat under field conditions.

15.5. CONCLUSIONS

- In the present scenario of pest resistance to pesticides, resurgence and high cost of pesticides, it become imperative to develop forecasting models.
- Development early forecasting system helps in providing caution to farmers regarding pest occurrence, peak activity and migration.
- Forecasting is helpful in developing sustainable pest management system, which may be economically viable and socially acceptable.
- Validation of forecasting models is the main drawback which need to be immediate attention and models should be used only after proper validation in farmers field.
- Well established forecasting models in different parts of the world should adopted and fine tuned to the local conditions of that area.

REFERENCES

Anawati, J.N. and Regniere, J. (2009) BioSIM: Pest management planning decision support. Canadian Forest Service, Natural Resources Canada.

Das, D.K., Trivedi, T.P., Singh, J. and Dandapani, A. (2000). Weather based prediction model of *Helicoverpa armigera* for Integrated Pest Management in cotton-chickpea based agro-ecosystem of Punjab. *Paper presented in International Conference on Managing Natural Resources for Sustainable Agricultural Production in the 21st Century*. Voluntary Papers Natural Resources, Vol. 2, February 14-18, 2000, New Delhi, pp. 621-22.

Das, D.K., Trivedi, T.P. and Srivastava, C.P. (2001) Simple rule to predict attack of *Helicoverpa armigera* on crops growing in Andhra Pradesh. *Indian J. Agr. Sci.*, **71**(6): 421-23.

Dhaliwal, L.K., Hundal, S.S. and Kular, J.S. (2005) Use of agrometerological indices for forecasting of mustard aphid (*Lipaphis erysimi* Kalt.). *J. Agromet* **7 (2):** 304-306.

Dhaliwal, L.K., Hundal, S.S., Kular, J.S., Chahal, S.K. and Aneja, A. (2007) Temperature based indices for mustard aphid (*Lipaphis erysimi* Kalt.) forecasting under Punjab conditions. *Envir. Ecol.* **25 S (2):** 340-343.

Du, O., Bin, N., Yang, C.C. and Srilakshmi, K.R. (2008) Combination of multispectral remote sensing, variable rate technology and environmental modeling for Citrus pest management. *J. Environ. Mgmt.* 86 (1)L14-26.

Graf, H.U., Höpli, H., Höhn, P.H. and Blaise (2002) SOPRA: a forecasting tool for insect pests in apple orchards. *Acta Hort* 584: VI International Symposium on Computer Modelling in Fruit Research and Orchard Management.

Kalluri, S., Gilruth, P., Rogers, D. and Szczur, M. (2007) Surveillance of Arthropod Vector-Borne Infectious Diseases Using Remote Sensing Techniques: A Review. *PLoS Patholg.* 3(10): e116.

Knight, J.D., Tatchell, G.M., Norton, G.A. and Harrington, R. (1992) FLYPAST: an information management system for the Rothamsted Aphid Database to aid pest control research and advice. *Crop Prot.* **11**(5): 419-26.

Kuzmanov, D. and Dimitrov, N. (2009) Forecasting the necessity of grain fumigation during storage. *Czech J. Food Sci.*, **27**: 210.

Luedeling, E., Hale, A., Zhang, M., Bentley, W.J. and Dharmasri, C.L. (2009) Remote Sensing of Spider Mite Damage in California Peach Orchards. *International Journal of Applied Earth Observation and Geoinformation.* 11 (4): 244-255.

Maelzer, D.A. and Zalucki, M.P. (2000) Long range forecasts of the *Helicoverpa* in using the SOI and the SST. *Bull. Ent. Res.* **90:** 133-146.

Mohandass, S. and Zalucki, M.P. (2004) DBM Development: Are we measuring the right temperatures? *The Management of Diamondback Moth and other Crucifer Pests: Proceedings of the Fourth International Workshop* (eds. N.M. Endersby and P.M. Ridlands), pp.117J122. 26J29 November, Melbourne 2001, The Regional Institute, Gosford.

Nicholas, C., Coops, Richard, H., Waring, Michael, A., Wulder and Joanne, C. White (2009) Prediction and assessment of bark beetle-induced mortality of lodgepole pine using estimates of stand vigor derived from remotely sensed data. *Remote Sensing Environ.* 113 (5): 1058-1066.

Nicholas, R. Goodwin Nicholas, C., Coops, Michael, A., Wulder, Steve Gillanders, Todd A. Schroeder and Trisalyn Nelson (2008) Estimation of insect infestation dynamics using a temporal sequence of Landsat data. *Remote Sensing of Environment.* 112 (9): 3680-3689.

Nietschke, B.S., Borchert, D.M., Magarey, R.D. and Matthew, A.C. (2008) Climatological potential for *Scirtothrips dorsalis* (Thysanoptera: Thripidae) establishment in the united states. *Florida Ent.* 91 (1): 79-86.

Ram Niwas (2007) Modelling for predicting insect populations. In: J.P. Bhanot, R.K. Saini and G.S. Yadav (eds). *Application of Ecological Methods for Management of Crop Pests.* Centre for Advanced Studies, Department of Entomology, CCS HAU, Hisar.

Singh, B., Singh, B., Sethi, A. and Dilawari, V.K. (2003) Prediction and validation of *Helicoverpa* incidence on cotton in Punjab. National Seminar on Agrometeorology in the new millennium: Prospective and Changes, October 29-31, 2003, PAU, Ludhiana.

Steiner, U., Bürling, K. and Oerke, E.C. (2008) Sensor Use in Plant Protection. *Gesunde Pflanzen* 60 (4) 131-141.

Trivedi, T.P. 2009. Role of information and dissemination of integrated pest management. In: *IPM Strategies to Combat Emerging Pests in the Current Scenario of Climate Change* (Eds. Ramamurthi, V.V., Gupta, G.P. and Puri, S.N.). Proceedings of the National Symposium on IPM Strategies to Combat Emerging Pests in the Current Scenario of Climate Change held at College of Horticulture and Forestry, Central Agricultural University, Pasighat 791002 Arunachal Pradesh, January 28-30, 2009 (invited lectures), pp.48-55.

Trivedi, T.P., Paul Khurana, S.M., Puri, S.N., Bhar, L.M., Mehta, S.C., Jain, R.C., Singh, G., Choudhary, S.M. and Mohasin M., Dandapani A. and Chona Naveen (2002). Development of forewarning system of aphid *Myzus persicae* (Sulzer) on potato (*Solanum tuberosum*) in India. *Indian J. Agr. Sci.* **72** (6): 315-19.

Yang, Z., Rao, M.N., Elliott, N.C., Kindler, S.D. and Popham, T.W. (2009) Differentiating Stress Induced by Greenbugs and Russian Wheat Aphids in Wheat Using Remote Sensing *Computers and Electronics Agric.* 67 (1-2): 64-70.

Yonow, T., Sutherst, R.W., Dominiak, B., Maywald, G.F., Zalucki, M.P. and Maelzer, D.A. (2004) Modelling the population dynamics of the Queensland fruit fly, *Bactrocera (Dacus) tryoni*: a cohort-based approach incorporating the effects of weather. *Ecol. Modelling*, **173**: 9-30.

Zalucki, M.P. and Furlong, M.J. (2005) Forecasting *Helicoverpa* populations in Australia: A comparison of regression based models and a bioclimatic based modelling approach. *Insect Sci.* **12**: 45-56.

Theory and Practice of Integrated Pest Management
R. Arora, B. Singh and A.K. Dhawan (Eds)
Scientific Publishers (India), Jodhpur 2011, pp. 262 - 273.

16

INFORMATION TECHNOLOGY IN INTEGRATED PEST MANAGEMENT

A.K. Dhawan, G.K. Taggar* and R.S. Gill

Department of Entomology, *Department of Plant Breeding & Genetics
Punjab Agricultural University, Ludhiana - 141 004, India

16.1. INTRODUCTION

Information Technology (IT), as defined by the Information Technology Association of America (ITAA), is "the study, design, development, implementation, support or management of computer-based information systems, particularly software applications and computer hardware." IT deals with the use of electronic computers and computer software to convert, store, protect, process, transmit, and securely retrieve information.

Information technology is set to become important in agriculture in the coming years. It is because information forms the basic input for successful analysis of the current situation, to evaluate the possible options and select the action to be taken. In the last few decades, the application of computers and communication technologies has led to the development of a plethora of information of all types in the area of plant protection science. These information bases are available on CD-ROM, on-line form in various hosts and via mail over internet, multimedia data bases, Decision-Support Systems (DSS)/Expert Systems (ES), Simulation Models (SM), etc. to cite a few of the latest developments. Further, cyber space is fast emerging as an area of information technology, particularly in the field of crop pest management (Srivastava and Swain, 1998).

The application of information technology in pest management has been one of the most exciting developments in the field over the last decade. From the 1960s, the entomologists began using computers for spray scheduling (Watt, 1961), population modeling (Watt, 1962; Ruesink, 1976), exploring control strategies (Shoemaker, 1973; Conway *et al.*, 1975; Birley, 1977) and managing pest monitoring data (Welch & Croft, 1979; Heong, 1988). More recently, computers are being utilized in pest management delivery systems (Welch, 1984; Naegele *et al.*, 1985) and decision support systems (DSS) (Rykiel *et al.*, 1984) designed to enhance decision making.

Information technology has improved the efficiency of data collection and analysis, pest identification, pest biology, biological monitoring, modeling damage as well as weather forecasting, pinpointing different control options including pesticide/control agent selection, pesticide field applications, computer simulations for predator-prey interactions, determination of economic effectiveness, decision making, IPM scheduling, advisory services, etc. Additionally, IT has enhanced our ability in research, training, education, and information dissemination and management (Stinner, 1999).

The use of computers and computer software in certain fields relevant to pest management has quite a long history. Mainframe computers were used for data collection, simulation, and modeling in ecology in the early 1960s (Pielou, 1969; Watt, 1963). Later, expert systems became another pest management field where researchers relied on the power of computers for. However, these early applications of computers were mainly limited to research and academic arenas, and were utilized by few highly trained professionals. There are few successful cases in which computers or computer software helped in solving real world pest management problems (Jones, 1989).

In the area of plant protection, computers are useful in:

- Storing, retrieving and analyzing data such as survey, surveillance and maintenance of pest incidences.
- Modeling and simulation of pest incidence/forewarning.
- Developing Decision Support Systems/Expert Systems to pest problems leading to decision making (Dandapani *et al.*, 2005).

Knowledge and information is the key to correct pest management decisions. Integrated Pest Management (IPM), a system that emphasizes appropriate decision making, is information intensive and depends heavily on accurate and timely information for field implementation by practitioners. Additionally, it is critical to strengthen the communication links between researchers and extension professionals and their clientele to expedite multi-way exchange of information and technology transfer. In addition, researchers and extension specialists need the most up-to-date information to design new projects and set future research goals and directions. There is already a large volume of useful IPM information available on the Internet, however, the information is scattered all across the globe. These resources range from topics such as pest identification, biology, control tactics, IPM definitions and basic concepts, to modeling and systems analysis. As awareness of the Internet increases worldwide, more people are participating not only as users of the information but also as creators of new information; as a consequence, the number of both IPM Internet servers and clients is increasing rapidly, perhaps slightly lagging but generally accompanying the exponential growth of the internet itself (Bajwa *et al.,* 2003).

16.2. WEB-BASED INTERACTIVE SYSTEMS

The ability of web-based applications to access and query databases is a key to interactive applications on the Internet. Web databases may contain any type of data including extended text fields and binary objects such as graphics, sound, and video. These databases are generally used to manage dynamic information/data. Web databases can be used for information/data collection, distribution, or data exchange. Some of the uses of web databases include records of pest phenology, occurrence and distribution, pest control recommendations or pesticide information.

At present, most online IPM information is static (based on text and graphics only), such as electronic versions of informational brochures, pesticide recommendations, fact sheets, extension guides, and bibliographic references. The number of online dynamic and interactive IPM resources, however, also is growing. Such systems facilitate a user access to the decision making process. The dynamic online IPM tools include databases, degree-day and sampling calculators, and phenology/population dynamics models. These interactive tools require end-users to respond to and share information online. Generally, these tools work efficiently if they are integrated with other online resources. Several IPM lists in the form of digests, electronic journals, and discussion groups are now available.

The plant protection recommendations of various states in India have been compiled and put in a database to make it available as a reference tool. This is also available on the internet site of NCIPM so that it is available to all. This database will also be helpful to compare the key pests of each crop in different states and the management options recommended by the individual state (Dandapani *et al.,* 2005). The software COT-IPM Pro (Sharma *et al.,* 2001) has been developed to enter, store and view the agronomic, pest and weather details for cotton. COT-IPM Pro was developed for the use of any institute or a group of agencies in a network, engaged in implementing IPM in an area. A single farmer may use this software for keeping the farm records. The software can generate reports for individual or group of cotton pests. The reports can be viewed in graphical mode and the pests which cross the ETL level are marked so that appropriate decisions can be made for their management. The software can act as a repository for storing the pest and weather data from multi-locations and can provide long felt need for data to develop pest-forecasting models. The software can also be used for pest monitoring purposes.

16.3. INFORMATION TECHNOLOGY-AIDED IPM DECISION MAKING

Decision Support Systems (DSS) represent the first computer-based systems for organizing information to deal with unstructured problem-solving. A Decision Support System is an interactive computer-based system designed to help decision-makers utilize data, narrative information, and models to solve unstructured problems. The key feature that distinguishes a DSS from other systems described so far is the degree to which the information-processing task is specified beforehand by the user. In the other systems, information processing is specified in advance, while in DSS it is not. The DSS is designed to define, retrieve, process, and organize information in a way that is meaningful for the user's particular problem (Rykiel *et al.,* 1984).

Software systems have been developed that mimic the deductive and inductive reasoning of a human expert. These knowledge-based systems are called "Expert Systems" (ES) and rank among the most significant technologies to emerge in the twentieth century. Expert systems capture the knowledge of a domain expert by encoding the reasoning process used by that expert for solving problems. An expert system can explain the logical sequence of its reasoning process. As a result, a decision-maker can review the reasoning steps of a human expert rather than simply rely on outputs consisting of numerical values or recommendations.

Because inexactitude is a necessary fact of life, expert systems have inherent weaknesses. This circumstance has led to the concept of an "Expert Support System", which is a computer system that represents a blend of attributes from DSSs and ESs. The need for this type of system is evident when we consider that much of the information base in a particular problem domain may be wrong, imprecise, or incomplete.

16.4. APPLICATIONS OF DECISION SUPPORT SYSTEMS (DSS) IN INSECT PEST MANAGEMENT

An example of the application of DSS concepts to an entomological problem domain is the Southern Pine Beetle Decision Support System (SPBDSS) (Rykiel *et al.,* 1984). This system, which is a FORTRAN-based program housed in a mini-computer, was designed to aid forest managers and pest management specialists in decision-making about IPM of the southern pine beetle, *Dendroctonus frontalis*. This system provides access to much of the modeling technology and abstracted technical information derived from accelerated research and development programs on IPM of *D. frontalis*.

In SPBDSS, the user clients include private landowners, forest industry personnel, and USDA Forest Service employees. Four general problem categories were elucidated by surveying forest industry personnel. These categories are:

- Prevention of SPB infestations.
- Suppression of existing infestations.
- Utilization of beetle-killed timber.
- Prediction of potential outbreaks by evaluation of stand hazard and risk.

The ultimate goal of SPBDSS is to improve decision-making, and thereby enhance the ability to manage natural resources wisely.

A Decision Support System for the management of cotton pests in Punjab has been initiated by Punjab Agricultural University, Ludhiana, with an objective to transfer Cotton-IPM technology in the cotton belt of Punjab and to further strengthen training and information dissemination systems. This DSS provides the users all necessary information including cotton pest identification/disease diagnosis, pest/pathogen life histories (cycles), sampling and decision-making criteria, sampling threshold calculators, pest/disease development models linked to weather networks, biorational pest control methods and currently available pesticides with their safety issues and environmental impacts. The end users of this DSS are the farmers, who put their queries on the website (*http://www.paucottonipmdss.com*) through an interactive system and get the appropriate answers for the same.

This DSS also aims at strengthening the link between researcher and extension personnel. The DSS also focuses on transfer of valuable information available on cotton pest management and areawide adoption of the well proven cotton IPM technology of the state amongst the farming community. Thus, this DSS acts as an educative tool for the farmers to improve cotton production with precise use of inputs, besides helping experts to identify gaps in the information and the ways to resolve those (Dhawan *et al.,* 2008).

A Spatial Decision Support System (SDSS), designated was developed for controlling Medfly or the Mediterranean Fruitfly, *Ceratitis capitata* (Diptera: Tephritidae) in Israel (Cohen *et al.,* 2008). The main objective of the present study was to develop an SDSS for Medfly control in citrus in Israel, which would assist the coordinators to reach optimal control decisions aimed at minimizing Medfly damage while also minimizing adverse environmental effects. The criteria found to be most relevant for control decision-making were the number of flies and the presence of a 'blue eye' in the nearest trap, the host-species susceptibility, the relative development of the Medfly based on accumulative day-degree model, the history of trapping, and the Medfly population in the nearby traps. It was shown that the MedCila provides recommendations that are generally accepted by the coordinators; it reduces the number of unnecessary spray actions in the absence of a Medfly threat in space and in time; and it reduces the number of plots for which the coordinator needs to make a decision for.

CottonLOGIC DSS has been developed to communicate research findings effectively to the cotton industry and to support best management adoption in both conventional and transgenic cotton crops in Australia. EntomoLOGIC, part of the CottonLOGIC software suite, is one example. To operate EntomoLOGIC, users select sample areas in their cotton fields and collects information on the types of beneficial and 'pest' insects present, their stage of development and quantity. Based on that data, CottonLOGIC predicts what is likely to happen to pest numbers given the existing situation and indicates when pest control may be necessary. The software is then used to predict future pest numbers, using weather data, and indicates when pest numbers are over defined economic thresholds. Cotton pest managers can then use this information to make their own decisions on when and how to control pests (Bange *et al.,* 2004). Field evaluations of CottonLOGIC (formally EntomoLOGIC) have demonstrated that compared to commercial insect management, an average of 1.3 sprays can be saved every season. Many features are now available in CottonLOGIC including the *Helicoverpa* population prediction model (*Helicoverpa* Diapause Induction and Emergence Tool). Information about when diapause is induced is useful for identifying 'high risk' fields i.e. those fields likely to have diapausing pupae and which are essential targets for pupae busting. Knowing the timing of spring emergence from diapause identifies the latest opportunity for pupae busting operations and provides an alert of potential moth activity. CottonLOGIC also includes mite population and yield loss model, and NutriLOGIC, a tool for nitrogen fertiliser management. The program enables managers and consultants to store a wide range of agronomic data including crop development, pest pressures and farming operations. Numerous reports and graphs can be easily generated from this data.

SIRATAC is a computer-based dial-up crop management system that has been developed to assist cotton growers make good tactical decisions in the use of insecticides (Hearn, 1987). The SIRATAC Pest Management Program consists of several simulation models (HELIOTHIS, FRUIT, FEEDING models) and a decision model (DECISION). The decision model supports two decisions: (i) whether or not to spray a pesticide and (ii) if so, which pesticide to spray. In order to determine if an infestation needs to be controlled, the DSS first determines the appropriate action thresholds for each pest and then notes which pest requires control. In order to select an appropriate pesticide, the effect on *Heliothis* population of each group of pesticides is simulated, starting with the softest, until one is found that will reduce the population to 25% of the threshold. The SIRATAC user collects and enters two sets of data regularly: (i) Number of insects, counted every third day and (ii) Numbers of fruits (squares and bolls), counted weekly. Using the SIRATAC DSS, the cost of sprays was reduced by 37% by a maximum of 72% (Hearn *et al.,* 1981; Ives *et al.,* 1984).

Another decision support system, Stored Grain Advisor Pro (SGA Pro) was developed to provide insect pest management information for wheat stored at commercial elevators. The program uses a model to predict future risk based on current insect density, grain temperature and moisture. SGA Pro correctly predicted for 71-80% of bins whether the grain was safe or at high risk of dense infestation and grain damage. This resulted in more efficient insect pest management because fumigating bins only when insect densities exceeded economic thresholds and treating only the bins that required fumigation minimized the risk of economic losses from insects, reduced the cost of pest management, and reduced the use of grain fumigant (Flinn *et al.,* 2007).

16.5. APPLICATIONS OF EXPERT SYSTEMS (EXS) IN INSECT PEST MANAGEMENT

A web-based expert system called GyMEs (Gypsy Moth Expert System) to estimate the risk a forest stand faces from gypsy moth. GyMEs is an enhanced version of the GypsES expert system component (Thomas *et al.,* 1998). The user of the system is prompted to answer a number of questions related to the stand itself, such as the age of the stand, its defoliation and drainage history etc. Then, the system processes the user's responses and determines the risk assessment. The most significant characteristic of GyMEs is that it is web-based, that is, it runs over the Internet. A forest manager can "surf" into the GyMEs web site and interact with the system to assess a particular forest stand. In this stsyem, the defoliation prediction value is determined using field data indicating gypsy moth egg mass counts. Based on egg mass counts, stands are categorized according to the level of defoliation expected in the coming year.

An expert system was constructed to help diagnose brown planthopper (BPH) (*Nilaparvata lugens*) in Zhejiang Province, China and determine insecticide recommendations. The main aim is to develop a tool that can be used by country and village level extension staff to provide

BPH control recommendations, tailored to the particular rice crop in question. The expert system provided advice on insecticide treatment that was at least as good, if not better than provided by the rules used by two groups of extension workers. It is a user-friendly manual and training tool, it can improve the recomme-ndations that agricultural advisors make, and provide additional information on BPH and its control. Secondly, as a means of increasing communication between research and extension staff, this expert system can help in the identification of research, surveillance and monitoring needs (Holt *et al.,* 1990).

The Potato Insect Expert System (PIES) is a knowledge-base system for Colorado potato beetle (CPB) management in commercial potato fields on the eastern shore of Virginia. The system uses insect life-stage, potato growth stage, percentage of defoliation, and other factors to determine if an insecticide application is necessary to prevent tuber yield loss from Colorado potato beetle feeding. The recommendations from PIES were compared with conventional spray thresholds based on the number of Colorado potato beetles per stem. Tuber yields were not significantly different from the two methods, although on average, PIES recommended 203 fewer insecticide applications than the conventional method. In addition, visual defoliation ratings for PIES required less time than conventional sampling for Colorado potato beetle on potato stem (Vencill *et al.,* 1995).

A web-based fuzzy Expert System SOYPEST (Soybean Pest Expert System) has been developed with an objective to provide IPM decision support to the farmers through the internet. This has been used for the crops grown in different regions of India. The system involves the application of the fuzzy logic in uncertainty management during pest identification as well as for estimating pest activity level using a new adjustable operator. Web-base SOYPEST Expert System has been developed to identify the active pests and to calculate the activity of active pests followed by their diagnosis (Saini *et al.,* 2002).

A rule-based Expert system to diagnose pests of honeybees (*Apis mellifera* L.) and to suggest the appropriate treatments has been developed. The system can be used as a diagnostic tool for beekeepers and as for educational and extension purposes in bee pathology. It provides a diagnosis based on the description of the external appearance or behavior of the affected colony. Corresponding pictures accompany the most important symptoms and certain measures to be taken are proposed. The expert system was evaluated following the conventional expert system evaluation methodologies. The system was implementing using EXSYS or Microsoft Windows environment (Mahaman *et al.,* 2002).

16.6. APPLICATIONS OF EXPERT SUPPORT SYSTEMS (ESS) IN INSECT PEST MANAGEMENT

The integration of the concepts of EXS and DSS has given rise to "Expert Support Systems" (EXSS). An EXSS represents a combination of the features of expert consultative systems and those of the traditional DSS. This type of system is also referred to as an "Executive Decision Support System" (EDSS) or Integrated Expert Systems. Like DSS, EXSS enables the user to make the actual decision regarding a particular problem or question. Moreover, like EXS, EXSS interprets the significance of information presented. EXSS are particularly suited for problems associated with delivery of information on IPM (Coulson and Saunders, 1987). In Texas, the Farm Level Expert (FLEX) concept was developed for implementation in two crops, cotton (COTFLEX) and rice (RICEFLEX).

COTFLEX is a prototype integrated expert system shell for cotton production that incorporates production-level, marketing-level, and policy-level decisions. The general design of COTFLEX has five modules in its knowledge base, which addresses most of the major decisions relating to cotton production (Stone *et al.,* 1986; Frisbie *et al.,* 1987). These five knowledge bases are:

- Agronomic Practices Advisor (APA)
- Pest Management Advisor (PMA)
- Crop Marketing Advisor (CMA)
- Farm Management Advisor (FMA)
- Farm Policy Advisor (FPA)

These knowledge bases are integrated in the FLEX system through an inference engine, the user interface and front end. The PMA makes recommendations about pest control for insects, weeds and diseases. Among the recommendations it makes are: when to apply pesticide, when to scout for pests, at what time no action is the best action, etc. PMA also has access to pest models such as boll weevil model, cotton leafhopper model and a root rot model. Also associated with the system is an economics analyzer that converts yield into dollars for comparison with pest control costs. The users of this system are cotton growers, financial institutions, extension personnel, research administrators and strategic planners.

16.7. CONCLUSION

Historically, the application and use of decision support systems has been limited by several factors including a) the lack of sufficient computer power for data analysis and interpretation, b) the cost of collecting site-specific weather data and disease observations, c) cost and speed of the communications, and d) the ease of DSS use. Technological advances have greatly reduced many of these problems but the cost of collecting disease observations may continue to be a significant limitation for some DSS applications. Reduction in the cost coupled with increased awareness about user friendly software would enhance the use of computer as a modern tool in the field of agriculture and benefit the farmers and research workers world over. Suitable evaluation, area-wide validation and performance monitoring of the expert system knowledge base is necessary to ensure their expertise in solving real pest problems. The availability of new information technology will definitely provide the needed competitive edge for world agriculture.

REFERENCES

Bajwa, W.I., Coop, L. and Kogan, M. 2003. Integrated pest management (IPM) and internet-based information delivery systems. *Neotrop. Ent.* **32**: 373-383.

Bange, M.P., Deutscher, S.A., Larsen, D., Linsley, D. and Whiteside, S. 2004. Handheld decision support system facilitates improved insect pest management in Australian cotton systems. *Comput. Electron. Agric.* **43**: 131-147.

Birley, M.H. 1977. A transfer function pest management model. In: Norton, G.A. and Holling, C.S. (eds). *Proc. Conf. Pest Management* CP77-6, International Institute for Applied System Analysis, Luxemburg, Austria, pp. 201-215.

Cohen, Y., Cohen, A., Hetzroni, A., Alchanatis, V., Broday, D., Gazit, Y. and Timar, D. 2008. Spatial decision support system for Medfly control in citrus. *Comput. Electron. Agric.* **62**: 107-117.

Conway, G.R., Norton, G.A., King, A.B.S., Small, N.J. 1975. A systemic approach to the control of sugarcane froghopper. In: Galton, G.R. (ed), *Study of Agricultural Systems.* Applied Science Publishers, London, UK.

Coulson, R.N. and Saunders, M.C. 1987. Computer-assisted decision-making as applied to Entomology. *A. Rev. Ent.* **32**: 145-37.

Dandapani, A., Singh, Niranjan and Sharma, O.P. 2005. Computer-based decision support systems for IPM. In: Singh, A., Sharma, O.P. and Garg, D.K. (eds). *Integrated Pest Management: Principles and Applications, Vol. 1: Principles.* CBS Publishers & Distributors, New Delhi, India.

Dhawan, A.K., Singh, B. and Bal, H.K. 2008. Development of decision support system for cotton pests in Punjab. In: Joia, B.S., Sharma, D.R., Dilawari, V.K. and Pathania, P.C. (eds). *Proceedings of 2nd Congress on Insect Science: Contributed Papers,* February 21-22, 2008, INSAIS, PAU, Ludhiana, India, pp. 69-70.

Flinn, P.W., Hagstrum, D.W., Reed, C.R. and Phillips, T.W. 2007. Stored Grain Advisor Pro: Decision support system for insect management in commercial grain elevators. *J. Stored Products Res.* **43**: 375-383.

Frisbie, R.E., Stone, N.D., Richradson, J.W. and Sansone, C. 1987. COTFLEX: Organizing, developing, and implementing a cotton expert system for total crop management. In: *Proceedings Beltwide Cotton Production Research*

Conference, January 4-8, 1987, Dallas, Texas, National Cotton Council of America, Tenn., pp. 193-194.

Hearn, A.B., Ives, P.M., Room, P.M., Thomson, N.J. and Wilson, L.T. 1981. Computer-based cotton pest management in Australia. *Field Crops Res.* **4**: 321-332.

Heong, K.L. 1988. Computerized surveillance systems in pest management. In: *Movement of Pests and Control Strategies*, PLANTI, Serdang, Malaysia, pp. 375-386.

Holt, J., Cheng, J.A. and Norton, G.A. 1990. A systems analysis approach to brown planthopper control on rice in Zhejiang Province, China. III. An Expert System for Making Recommendations. *J. App. Ecol.* **27**: 113-122.

Ives, P.M., Wilson, L.T., Cull, L.T., Palmer, W.A., Haywood, C., Thomson, N.T., Hearn, A.B. and Wilson, A.G.L. 1984. Field use of SIRATAC: An Australian computer-based system for cotton. *Prot. Ecol.* **6**: 1-21.

Jones, P. 1989. Agricultural applications of expert system concepts. *Agricultural Systems* **31**: 3-18.

Mahaman, B.D., Harizanis, P., Filis, I., Antonopoulou, E., Yialouris, C.P. and Sideridis, A.B. 2002. A diagnostic expert system for honeybee pests. *Comput. Electron. Agric.* **36**: 17-31.

Naegele, J.A., Coulson, R.N., Stone, N.D. and Frisbie, R.E. 1985. The use of expert systems to integrate and deliver IPM technology. In: Frisbie, R.E. and Adkisson, P.I. (eds). *Integrated Pest Management in Major Agricultural Systems.* Texas Agricultural Experimental Station, College Station, Texas, pp. 692-711.

Pielou EC 1969. An Introduction to Mathematical Ecology. Wiley-Interscience, New York.

Ruesink, W.G. 1976. Status of the system approach to pest management. *Annu. Rev. Ent.* **21**: 27-44.

Rykiel, E.J., Saunders, M.C., Wagner, T.L., Loh, D.K., Turnbow, R.H., Hu, L.C., Pulley, P.E. and Coulson, R.N. 1984. Computer-aided decision making and information accessing in pest management systems, with emphasis on the southern pine beetle (Coleoptera: Scotylidae). *J. econ. Ent.* **77**: 1073-1082.

Saini, H.S., Kamal, Raj and Sharma, A.N. 2002. Web-based fuzzy expert system for integrated pest management in soybean. *Int. J. Info. Tech.* **8**: 54-74.

Sharma, O.P., Dandapani, A. and Nagarajan, S. 2001. COT-IPM Pro: User's Guide. NCIPM, New Delhi, India.

Shoemaker, C.A. 1973. Optimization of agricultural pest management. *Mathemat Biosci.* **16**: 143-175.

Srivastava, A.N. and Swain, S.C. 1998. Role of computers in integrated pest management. In: Prasad, D. and Gautam, R.D. (eds). *Potential IPM Tactics*, Westvill Publishing House, New Delhi, India, pp. 449.

Stinner, R.E. 1999. Information management: Past, present, and future. In: Kennedy, G.C. and Sutton, T.B. (eds). *Emerging Technologies for Integrated*

Pest Management-Concepts, Research, and Implementation, Conference Proceedings, March 8-10, 1999, Raleigh, NC, USA, pp. 474-481.

Stone, N.D., Frisbie, R.E., Richardson, J.W. and Coulson, R.N. 1986. Integrated expert system applications for agriculture. *Proceedings of the International Conference on Use of Computers in Agricultural Extension Programs,* February 5-6, 1986, Lake Buena Vista, Florida, USA.

Thomas, S.J., Fosbroke, S.L.C. and Cumming, A.B. 1998. GypsES: Decision support and project management-User's Guide, Version 1.0. USDA Forest Service. Northeastern Forest Experiment Station, USA.

Vencill, A.M., Zehnder, G.W. and Heatwole, C.D. 1995. Potato insect expert system: Computerized approach to insecticide management for Colorado potato beetle (Cleoptera: Chrysomelidae). *J. econ. Ent.* **88**: 944-954.

Watt, K.E.F. 1961. Use of computer to evaluate alternative insecticidal programs. *Science* **133**: 706-707.

Watt, K.E.F. 1962. Use of mathematics in population ecology. *A. Rev. Ent.* **7**: 243-260.

Welch, S.M. 1984. Developments in computer-based IPM extension delivery systems. *A. Rev. Ent.* **29**: 359-381.

Welch, S.M. and Croft, B.A. 1979. *The Design of Biological Monitoring Systems for Pest Management.* Center for Agricultural Publications, Wageningen.

Theory and Practice of Integrated Pest Management
R. Arora, B. Singh and A.K. Dhawan (Eds)
Scientific Publishers (India), Jodhpur 2011, pp. 274 - 297.

17

MANAGEMENT OF INSECT PESTS OF COTTON: RETROSPECT AND PROSPECT

A.K. Dhawan, Vijay Kumar and P.S. Shera
Department of Entomology
Punjab Agricultural University, Ludhiana - 141 004, India

17.1. INTRODUCTION

Cotton (*Gossypium* sp.) is most important commercial crops of India. About 60 million people of the country are involved directly or indirectly in cotton production, processing, textiles and related activities. Among the cotton growing countries of the world, India has a place of prominence with respect to area under this crop. The production of cotton increased from 3.0 million bales from 5.88 million hectares (m ha) in 1950-1951 to 14.23 million bales from 9.12 m ha in 1996-1997. Then started the downward slide in cotton production up to 2003, as the area under cotton decreased from 9.34 m ha in 1998-1999 to 7.60 m ha in 2002-2003 and the production was only 13.6 million bales (Singhal, 2003). During recent years, the production of cotton has shown significant increase from 13.6 million bales in 2002-2003 to 17.9 and 24.3 million bales during 2003-2004 and 2004-2005, respectively (AICCIP Report). During 2009 the total area under cotton was 10.1 m ha with a production of 29.5 million bales. In 2010 adverse climatic condition coinciding the time of sowing throughout India (high temperature and lack of irrigation water in North, long rainy spell in Centre and even delayed rains in South) may affect the total area and productivity. The average cotton productivity in India with 502 kg lint/ha is far less than other cotton growing countries, viz. Brazil (1298 kg/ha), China (1245 kg/ha), USA (918 kg/ha), Uzbekistan (770 kg/ha) and Pakistan (686 kg/ha) (AICCIP 2006-07).

Besides other causes, major bottleneck in cotton cultivation is biotic stresses due to attack of insect pests, diseases and competition from weeds which plays a significant role in achieving optimum yield potential. As many as 1326 species of insects have been listed on the cotton crop worldwide (Hargreaves, 1948) which also includes those of little or no economic importance, sporadic, casual or accidental visitors so far as cotton is concerned. In India, cotton ecosystem harbours about 162 insect species, of which 9 are of utmost importance inflicting significant losses in yield. The monetary value of yield losses due to insect pests has been estimated to be Rs. 2, 87,600/- million annually (Dhaliwal *et al.,* 2004). The insect pest complex of cotton crop (Table 1) broadly categorized into three categories, viz. sucking pests, foliage feeders and bollworms cause dam-age to various plant parts at different growth stages through out the cropping season in the Punjab. The extent of losses caused by sucking pests, bollworms and both sucking pests as well as bollworms together have been recorded up to 12, 44 and 52 per cent, respectively on *hirsutum* cotton and 47 per cent due to bollworms alone on *arboreum* cotton (Dhawan *et al.,* 1988). In the absence of effective genetic resistance against these pests especially bollworms, farmers solely relied on insecticides for their effective management. Cotton accounts for around 50 per cent of pesticide consumption in the country, despite being grown on area of 5 percent of total cropped area. The consumption of insecticides (technical grade material) per hectare in Punjab is 923 g which is the highest in India (Agnihotri, 2000).

The total consumption of pesticides in Punjab has increased from 3200 MT technical grade in 1980-81 to 7200 Mt in 2001-02, cotton being the maximum shareholder. However, there has been reduction in pesticide load ever since transgenic cotton has been commercialized in Punjab. The total consumption of pesticides was 4500 MT during 2005-06 in Punjab (Anonymous 2006). However, there has been reduction in pesticide load ever since transgenic cotton has been commercialized in Punjab.

The maximum increase in losses occurred in cotton from 18-50% from early 1960 to 2000 (Dhaliwal *et al.,* 2007). Besides increasing cost of production and environmental problems, the excessive and indiscriminate use of insecticides for the control of these pests has resulted in development of insecticidal resistance particularly in *Helicoverpa amigera* [*Helicoverpa armigera* (Hubner)] (Sidhu *et al.,* 1994; Kapoor *et al.,* 2002; Gill and Dhawan, 2006) decline in natural enemies' population and resurgence of the pests like whitefly [*Bemisia tabaci* (Gennadiaus)] (Dhawan and Simwat, 1997a; Butter and Kular, 1999; Virk *et al.,* 2004) and jassid [*Amrasca biguttula biguttula* (Ishida)] (Dhawan and Simwat, 1997a). Keeping in view these problems, the need of the hour is give up strategy of over reliance on chemicals in this era of World Trade Organization (WTO) by switching over traditional method of chemical control to ecological and economical sound integrated pest management (IPM) strategies involving minimum use of pesticides, thereby, conserving natural biodiversity. The introduction of genetically engineered Bt cotton hybrids, viz. Bollgard (Cry 1 Ac) and Bollgard-II (Cry 1 *Ac* + Cry 2 *Ab*) conferring in- built resistance to bollworms and both bollworms as well as tobacco caterpillar, respectively in Punjab has given a new impetus to IPM programme with significant reduction in number of sprays against these pests. The integration of Bt cotton as a component of IPM module augmented with suitable cultural, mechanical control strategies, sowing of *refugia* around the main field and need-based application of pesticides will not only reduce pesticide load and make cotton cultivation economical but will also get opportunity to access global market. With a change in pest scenario losses incurred due to insect pests remains as major yield a limiting factor that calls for a better, innovative and sustainable pest management. Considerable effort has been directed toward such alternatives, and new technology has been implemented. However, the major trend has been toward the use of modern insecticide chemistries to replace traditional pesticides with less hazardous chemicals or nontoxic biologically based products; but these serve to be therapeutic (Wright, 1996 and Benbrook, 1996).

17.2. CHANGING INSECT PEST SCENARIO

The pest scenario in cotton ecosystem is changing rapidly. American and spotted bollworms attained secondary pest status, and tobacco caterpillar, pink bollworm, mirids and mealy bugs are emerging as major pests. Sap sucking pests like aphids, jassids, thrips and whiteflies are major pests and economically important (Arora *et al.,* 2010). Adoption of *Bt* cotton has not only changed the cultivation profile, but also the pest scenario (Table 17.1). While there is a decline in the pest status of boll - + worms; the sap feeders, viz. aphids, jassids, mirids and mealy bugs are emerging as serious pests (Vennila, 2008). There are many factors which have been found responsible for a change in scenario of cotton insect-pests.

- Cultivation pattern of diploid and tetraploid cotton.
- Adoption of Bt-cotton.
- Changing ecological scenario.
- Adoption of large number of cultivars with different habits.
- Use of unrecommended insecticides.
- Practice of mixing of synthetic pyrethroids with conventional insecticides earlier and mixing of new chemistries now
- Faulty spray technology.

Table 17.1. Changing status of cotton insect pests in cotton-wheat cropping belt

YEAR	MAJOR PEST	MINOR PEST	NEW PEST	PEST OUTBREAK
Up to 1970	Jassid, PBW	Whitefly, Aphid, Thrips	SBW	ABW in some pockets (1978)
1971-1980	Jassid, PBW, SBW	Whitefly, Aphid, Thrips	ABW	ABW (1983)
1981-1985	Jassid, PBW, SBW	Whitefly, Aphid, Thrips	ABW (1990)	
1986-1990	Jassid, PBW, SBW	Whitefly, Aphid, Thrips	Tobacco, Caterpillar	ABW (1995)
1991-1995	Jassid, PBW, SBW, ABW	Whitefly, Aphid, Thrips	Leaf Miner, CLCuV	Whitefly (1995)
1996-2000	Jassid, Whitefly, SBW, ABW, PBW	Aphid, Thrips	-	ABW (1997, 1998)
2001-2004	Jassid, Whitefly, SBW, ABW	Aphid, Thrips	-	ABW (2001)
2004-2006	Whitefly, Jassid, Tobacco Caterpillar	Aphid, Thrips	Mealybug	Tobacco Caterpillar in some pockets
2006-2010	Whitefly, Jassid, Tobacco caterpillar	Aphid, Thrips, Mealybug	Lint Stainers (Dusky & Red Cotton bug)	Tobacco Caterpillar (2005) and Mealybug in some pockets (2006, 2007), whitefly in 2010 (in few pockets)

Source: Modified after Dhawan *et al.*, 2008 (Dhawan *et al.*, 2004).

Gossypium arboreum, which was resistant to jassid (*A. biguttula*) and having lower pink bollworm [*Pectinophora gossypiella* (Saunders)] incidence due to its short duration, was predominant type in early fifties in the zone. A large area came under *G. hirsutum* by mid fifties which helped in multiplication of these two pests (Singh and Khush, 1993). These two insect species, *A. biguttula* and *P. gossypiella* which were the only serious pests of cotton upto mid 1970s due to cultivation of susceptible, medium and long duration varieties. However, the insect species, which is accountable for devastating the cotton sustainability in the North since 1990s is *H. armigera*. Before 1977, it was pest of certain *rabi* crops like gram, *barseem* and tomatoes and was first reported on cotton in 1977 in localized area of North Zone resulting in its outbreak in some pockets in 1978. In 1983, it appeared in a serious form and attained status of major pest in early 1990s. The changing agroclimatic conditions since early 1990s, especially duration and magnitude of rainfall, large scale cultivation of susceptible varieties that do not form part of a recommendation, excessive and indiscriminate use of insecticides especially synthetic pyrethroids, development of resistance against most of the conventional insecticides led to increase in incidence of this pest resulting in severe outbreaks in the years 1990, 1995, 1997, 1998 and 2001 rendering huge economic losses to cotton growers in the Punjab State (Dhawan *et al.,* 2004).

Incidence of bollworm complex has declined due to the introduction of genetically engineered Bollgard cotton, weather conditions, IPM as well as IRM strategies and introduction of insecticides with new chemistries, *i.e* spinosad and indoxacarb. Though *Bt* cotton (Bollgard with cry1*Ac* gene) provides effective management of bollworms requiring less number of sprays it is not effective against sucking pests and tobacco caterpillar which continue to possess a threat to cultivation of Bt cotton. Moreover, the intensity of *S. litura* is likely to further increase under the potential climate change, as it has been found to consume more than 30 percent cotton leaves at elevated CO2 levels (Kranthi *et al.,* 2009).

Besides, *A. biguttula* and *B. tabaci,* other sucking pests like thrips [*Thrips tabaci* (Lindemann)] hitherto occurring during May-June and aphids [*Aphis gossypii* (Glover)] that occur at the end of the crop season are also gaining importance. Recently during 2006, a new sucking pest, mealy bug [*Phenacoccus solenopsis* (Tinsley)] appears in few pockets of Bathinda, Ferozepur and Muktsar districts. The serious damage of this pest was observed in cotton fields near roads, bunds and channels having abundance of weeds flora. The maximum damage was observed in Sangat block of Bathinda district, but in 2007, it spread to other cotton growing areas of Punjab and emerged in a serious proportion causing huge loss to the cotton crop in Punjab (Dhawan *et al.,* 2007c). Since 2003, tobacco caterpillar [*Spodoptera litura* (Fabricius)] a polyphagous pest has also started causing serious damage to cotton crop. Mirid bug (*Campylomma livida* Reuter) (Heteroptera : Miridae) nymphs and adults of both species were observed to feed on squares and small developing bolls. Though the damage potential of mirids is confined to central part of the country it has now been noticed in Sirsa also (NCIPM Newsletter, Jan-June 2010). Pink bollworm a stenophagous, insect has not developed high resistance levels to insecticides as a result of a temporal refuge that larvae find within the developing bolls. Kranthi *et al.* (2005) reported expression of Cry1Ac to be the highest in leaves followed by squares, bolls and flowers. A low level of expression of the delta endotoxin in plant parts such as the boll rind, squares and ovary was reported in the Indian Bt cotton hybrids for the first time that was reported to influence bollworm incidence on Bt cotton particularly the pink bollworm.

To combat the problem of *S. litura*, a second generation of genetically modified cotton, i.e. Bollgard II (Cry 1 *Ac* + Cry 2 *Ab*) containing an additional gene conferring in-built resistance to both bollworms as well as tobacco caterpillar has been approved by GEAC in Punjab. But major concern for the sustainability of Bt cotton is evolution of resistance in American bollworm to Bt toxin. Though no report of its resistance has been reported yet from the field, but continuous efforts are required from research point of view w.r.t monitoring and managing resistance through development and implementation of proper insect resistance management strategies for cotton in Punjab. The concept of the multi gene cotton is to enhance the expression of toxins and delay resistance development and also to increase the range of pests targeted by Cry toxins.

17.3. MANAGEMENT OPTIONS

For sustaining the production and productivity of cotton crop, an imperative necessity is to adopt management practices according to their modification in relative pest status. The detailed research work on their effectual management strategies based on resistant varieties, cultural and mechanical control, utilization of bio-agents, botanical pesticides, microbial control, monitoring and mass trapping through pheromone traps, ETH based application of insecticides, spray technology etc has been carried out and rectified from time to time for the benefit of the farmers. In transgenic cotton, the need to spray against bollworms especially *Helicoverpa* spp. is dramatically reduced thereby allowing other pests, normally controlled inadvertently by sprays targeting *Helicoverpa* spp., to emerge as pests. This includes aphids, jassids, whiteflies, mealybug, mirids and other foliage feeders and lint stainers. Use of insecticides against these surviving and emerging pests may disrupt natural enemies, creating outbreaks of mites, whitefly, thrips or aphids. Hence, in both conventional (nontransgenic) and transgenic cotton there remains a strong incentive to use IPM to help reduce reliance on insecticides. Furthermore, an IPM approach helps to avoid creating situations that increase the risk of losses due to these pests.

17.3.1. Varietal Resistance

The cultivation of resistant cultivars is a key component in pest management programmes as it helps in suppressing the pest population with least disturbance to ecosystem. The arduous efforts have been made by cotton scientists to identify the sources as well as bases of resistance to various cotton insect pests (Table 2). In the absence of any effective source of resistance to bollworms in the existing cultivated germplasm and related wild species of cotton, the commercial exploitation of transgenic cotton, *i.e.* Bollgard having Cry1*Ac* gene from *Bacillus thuringenesis* Berliner expressing delta endotoxin conferring in-built resistance to bollworms as a component of IPM has revived cotton cultivation in the zone. Besides this, second generation genetically modified cotton, *i.e.* Bollgard II (Cry1*Ac* + Cry2*Ab*) containing an additional gene conferring in-built resistance to both, bollworms as well as tobacco caterpillar, is an additional boon for farmers, thus contributing to the scaling down of pesticide use against these pests. The resistance source identification has become more pertinent in present context as large number of hybrids from different companies involving many events have been released. During 2009, 53Bt cotton (BG-I and BG-II) hybrids approved for North India by GEAC were tested under the Technology Mission on Cotton for their association with emerging and key pests, and a significant difference among the cultivars insisted for their susceptibility to whitefly infestations.

Role of approved, released, single and dual gene transgenics. Several genes coding for insecticidal toxins have been isolated and are being used to develop insect resistant transgenic crops. Currently, four GM crops (cotton, maize, potato and tomato) incorporating nine Cry (crystal) toxin genes (*cry1Ab, cry1Ac, cry1F, cry34Ab, cry35Av cry3A, cry3B, cry2Ab, cry9C*) and *vip-3A* gene isolated from *Bacillus thuringiensis* and one protease inhibitor gene are under commercial cultivation in 22 countries. Gene stacking offers an alternate option for delaying resistance especially, if two or more genes are available with high toxicity levels having independent modes of action and not having cross resistance. Protein is expressed at high dose for both the pink and spotted bollworms and moderate dose for the cotton bollworm. Stacked genes are expected to increase the spectrum of toxicity to wide range of lepidopterous insects including *Spodoptera litura* which is known to survive Cry1Ac. It has been summarized through several experiments that Cry1Ac and Cry2Ab proteins complement each other are resistance management programme for all three cotton bollworm species.

17.3.2. Cultural, Mechanical and Physical Control

With the awareness of environmental problems, exploitation of different cultural and mechanical practices, viz. planting time, plant geometry, tillage, crop sanitation, fertilizer as well as water management, intercropping, crop rotation, manipulation of carryover sources, barrier/trap crops etc. have been advocated as a vital approach to curb pest population, as these are simple and easy to follow. The extensive studies on these physical, cultural and mechanical control practices have resulted in recommendation of best suited ones as a component of IPM module for cotton pests (Table 17.2).

17.3.3. Behavioral / Biorational Methods

The utilization of chemicals that affect insect behaviour, growth or reproduction, for suppression of insect populations is often referred to as biorational control (Dhaliwal, 2005) which include insect growth regulator, chemicals based on endocrine system and semio-chemicals (pheromones, allelochemicals). Use of pheromone is still found indispensable for monitoring of bollworm infestation Chamberlain *et al.* (1992) showed successful mating disruption with *Earias* pheromone [(using the common components of the sex pheromones of *E. insulana* (Boisduval) and *E. vitella* (F.)]. But the high costs of production of the unstable diene pheromone made the technology unviable. In India, traps baited with lures have been shown to be important decision making tools in determining effective control action (Dhawan and Sidhu, 1984). Mating suppression within fields was demonstrated but the mobility of the species resulted in no significant reduction in oviposition in the area (Sohi *et al.*, 1998, Chamberlain *et al.*, 2000, Sundaramurthy, 2003).

Table 17.2. Impact of different cultural and mechanical control strategies on cotton insect pests in the Punjab and in the North Zone

Physical/Cultural/Mechanical practices	**Target Pest (s)**	**Impact**
(1)	(2)	(3)
Deep ploughing before sowing and after harvest	American bollworm; Tobacco caterpillar	Pupal mortality
Early sowing of crop (Whole of April)	American bollworm	Escape attack
Optimum plant spacing	Entire pest complex	Closer spacing increases incidence; hinders spraying operations
Growing non-Bt as *refuge* around Bt cotton	American bollworm	Delay in resistance to Bt cotton
Excessive use of nitrogenous fertilizers	Jassid, whitefly, bollworms	Conducive for population build up
Non judicious use of irrigation water	Entire pest complex	Conducive for population build up
Cultivation of undescript material	Entire pest complex	More damage
Cultivation of adjoining crops in and around cotton crop		
• *Okra, moong* and *arhar*	Jassid, whitefly, spotted & American bollworm	Most preferred; increase in population build up and incidence on cotton crop
• *Okra, moong, castor, dhaincha*	Tobacco caterpillar	
• *Okra, moong* and *Guara*	Mealy bug	
Growing Barrier crops in and around cotton crop		

(1)	(2)	(3)
• Bajra, maize, *jowar*	Mealy bug	Least preferred; reduces population build up & restricts its migration to cotton crop.
Uprooting and destruction of weeds during in and off season		
• *kanghi buti* and *peeli buti*	Whitefly, CLCuV	Reduces carry over / inoculum & population build
• *Itsit*	Tobacco caterpillar	
• *Congress grass, puthkanda, itsit, gutputna, bhakhra, kanghi buti, peeli buti,* and *tandla*	Mealy bug	
Do not throw uprooted infested plant in cotton fields/water channels	Mealy bug	Check further spread
Collection & destruction of egg masses and early instar larvae feeding gregariously along with leaves	Tobacco caterpillar	Reduces population build up & incidence
Removal of infested terminal shoots	Spotted bollworm	Reduces incidence
Detopping *arboreum* cotton plants attaining height beyond 1.5 m	Pink, spotted bollworm	Reduces incidence; increase seed cotton yield
Irrigation beyond end September	Bollworms	Increase incidence; no increase in seed cotton yield
Early termination of crop	Bollworms	Reduces carry-over
Grazing of cattle, sheep, goat after last pick to feed on plant debris & unopened bolls	Pink bollworm	Reduces carry over
Before stacking, dislodge and burn the burs and unopened bolls	Pink bollworm	Reduces carry over
Stacking of cotton sticks in open in the villages	Pink Bollworm	Reduces carry over
After last pick, beating of infested sticks against ground	Mealy bug	Collection and destroying dislodged mealy bugs by burying reduces carry over
• Stacking the infested sticks separately and using them on priority for fuel purpose up to end of February.	Mealy bug	Reduces carry over
• Do not allow the movement of farm animals in infested fields	Mealy bug	Check further spread
• Restrict the movement of workers in infested fields	Mealy bug	Check further spread
• Prevent the movement of sticks from the infested areas to the new areas	Mealy bug	Check further spread from infested areas to new areas

***Source*:** Package of Practices, PAU, Ludhiana

17.3.4. Microbial Control

Various disease causing organisms perform an important role in dynamics and natural regulation of insect population. Pathogens may exert controlling effect by means of their invasive proportions, toxin enzymes and substances (Arora *et al.,* 2000, Lewis, 2006). Among different microbial agents, the utilization of nuclear polyhedrosis virus (NPV) and bacteria *Bacillus thuringiensis* Berliner (Bt) formulations have been tested extensively for the control of cotton pests. But after the introdu-ction of Bt-cotton and as a part of insect resistant management (IRM) strategy avoidance of Bt based biopesticides that contributes to selection of a broad spectrum resistance for bollworm to several Bt gene of interest is advocated. Use of NPV has been advocated in IRM strategies at specified windows. Dhawan and Simwat (1998) observed that biolep, dipel and bioasp formulations of Bt @ 1.5 kg or l/ha were as effective as insecticide quinalphos against young larvae of *H. armigera* however, against grown up larvae, they were significantly inferior. Dhawan *et al.* (2001) recorded that NPV sprays applied alone resulted in significantly less seed cotton yield and higher larval population and extent of damage as compared to NPV in combination with insecticides or NPV alternated with insecticides which was on par with insecticidal sprays. The entomo-pathogen *Fusarium pallidoroseum* and *Verticillium lecanii* were found compatible with many insecticides used for the management of mealy bug and sucking pests (Monga *et al.,* 2009, Rishi *et al.,* 2010). Due to the inconsistent performance, commercial products have not been popular with the cotton farmer. The technology is, however, invaluable and needs to be fine-tuned to the point at which mass production costs are low and efficacy is rapid and reliable before being popularized. Efficacy of several insect growth regulators, viz. diflybenzuron, teflubenzuron, lufenuron, flufenouron, RH-2485, buprofezin have been tested against different insect pests under laboratory and field conditions in Punjab (Dhawan and Simwat, 1998; Butter and Kular, 1998; Singh *et al.,* 2005; Aulakh and Butter, 2006). Among the most recent novel insecticides with selective properties, novaluron, a chitin synthesis inhibitor has been found very effective and is recommended for the control of *S. litura* in Punjab (PAU, 2008) which can be fit ideally in IPM strategies as a potential component thereby reducing dependence on conventional insecticides.

17.3.5. Biological Control

A rich fauna of naturally occurring bio agents have been reported on insect pests of cotton crop. India is very rich in biodiversity of parasitoids and predators. From cotton alone 106 species of spiders were recorded (Singh, 2003). To conserve natural enemies, an approach of no spray during first three months after sowing to conserve natural enemies and if needed, use of relatively safe insecticide, i.e. endosulfan and avoiding use of broad spectrum organophosphates such as monocrotophos, acephate especially as early season sprays have been advocated as a part of insecticide resistance management strategy in Punjab (PAU, 2008). The work on mass rearing of useful parasitoids and predators under controlled conditions and releasing them on large scale in the field for the control insect pests especially bollworms have also been exploited in cotton agro-ecosystem. Brar *et al.* (2003, 2004) observed that efficacy of *T. chilonis* alone to be less effective under Punjab conditions, but can be integrated with insecticides. According to Singh *et al.* (2004), field releases of bio agents, i.e. *T. chilonis* and *Chrysoperla carnea* failed to provide desired kill of bollworm complex. Hence, the strenuous efforts have been made by cotton entomologist to evaluate different bio-control based IPM module for the control of bollworm complex of cotton.

Eight releases of *Trichogramma chilonis* at 150,000/ha weekly during July-August to October + one release of *Chrysoperla carnea* at 10,000/ha + need based application of insecticides for the control of sucking pest (3 sprays) and for bollworms (3 sprays) during September-October was found effective and on par with spray schedule (3 sprays for sucking pests and 6 sprays for bollworms) in reducing bollworm damage and increasing the yield (Brar *et al.,* 2001). Eleven to thirteen releases of multiple insecticide tolerant/Bathinda strain of *T. chilonis* @ 1,50,000/ha at weekly interval during July to October along with 6-9 insecticidal sprays proved significantly better than spray schedule (6 to 9 sprays) for the control of bollworms and increasing egg parasitism and seed cotton yield (Brar *et al.,* 2007). Sucking pests may be checked by releasing *Chrysoperla carnea* at 1 lakh/ha at fortnightly intervals.

To conserve native natural enemies an approach of no spraying during first three months and if required, use of relatively safe biopesticides like neem oil and prevention of use of broad spectrum organophosphate as early season sprays have been advocated under IRM project in north India (Kranthi, 2005). These diverse groups of natural enemies have greater potential to offer natural control of emerging sucking insects. Apart from increasing diversity, report of maximum abundance of coccinellids, chrysopids and syrphids on Bt over non-Bt cultivars and IPM over insecticide-treated plots is available (Kulkarni *et al.,* 2004). Though the cotton mealy bug is exotic, it has a number of predators existing in cotton fauna, *Aenasius bambawalei* Hayat, a parasitoid with parasitization efficiency 57.2% under field condition and 60.6% under laboratory conditions in cotton (Rishi Kumar *et al.,* 2009) and *Brumus suturalis*, a predator has effectively managed the mealybugs.

17.3.6. Botanical Pesticides

In view of the environmental hazards of broad-spectrum synthetic insecticides, studies on the exploitation of plant origin insecticides like neem and its formulations have also been carried out for the control of cotton insect pests as they are comparatively safe to beneficial arthropods without any environmental impact. The spray of Neemark, Indne and RD 9 was found significantly at par with dimethoate for the control of *A. biguttula* (Dhawan and Simwat, 1992). Similarly spray of these formulations was also par with dimethoate at 7 days after spray at higher doses for the control of *B. tabaci* (Dhawan and Simwat, 1992). Mann *et al.* (2001) reported two azadirachtin enriched neem based insecticides, NeemAzal (1%) and RakshakGold (1%) to be effective for *B. tabaci* whitefly control during the flowering phase. In addition, the alternate application of neem products and conventional insecticides can reduce 50 per cent of synthetic insecticide load for the its control. Against pink and spotted bollworm complex, Neem rich, Navneem, and RD-9 Replin were found equally effective as quinalphos (Dhawan *et al.,* 1992; Dhawan and Simwat, 1992; Dhawan and Simwat, 1993). However, number of commercial neem formulations, i.e. Neemrich, Repelin, Neemark, Rakshak Nimbecidine, Neemgold, Neemazal, Achook, Neemax, Neem powder were comparatively less effective than conventional insecticides and did not give desired level of kill against *H. armigera* (Simwat and Dhawan, 1992; Dhawan and Simwat, 1998; Anonymous, 2002). Application of neem oil for the management of initial infestation of *Spodoptera* and whitefly has been advocated in IRM strategies (Kranthi and Russell, 2009) Neem oil @ 1.0 litre/acre during the early part of the season has been advocated for sucking pests under IRM strategies for Bt cotton and validated at farmers field under the Technology Mission during 2008 and 2009.

17.3.7. Chemical Control

The use of insecticides is an indispensable tool of pest management. Pesticide poisoning, insecticide resistance, resurgence of pests, effect on non target organisms and pesticide residues are some of the factors to be kept in mind while selecting and using the pesticide. To use insecticides judiciously the economic threshold values established for individual insect should be followed (Table 17.3), however, very strong surveillance/ monitoring is of utmost necessity to explore these ETH values. Different group of insecticides, viz. nitroguadine, organophosphates, synthetic pyrethroids, carbamates, organochlorines, naturalyte, oxadizine, miscellaneous and insect growth regulator have been recommended based on the performance against different insect pests of cotton in Punjab (Table 17.4 & 17.5).

Different group of insecticides *viz.,* nitroguadine, organophosphate, synthetic pyrethroids, carbamates, organochlorines, naturalyte, oxadizine, and insect growth regulators have been recommended against different insect pest of cotton in light of their safety to natural enemies and efficacy against insecticides resistant bollworms. Seed treatment of cotton with imidacloprid provides some respite from jassids up to 30 days. The insecticides have been recommended as per the window based management strategies where biopesticides have been provided an important place.

Table 17.3. Economic threshold levels (ETLs) of major insect pests of cotton cropping system

Insect pests	Economic thresholds
Jassid	Appearance of 2nd grade injury — curling & yellowing at margins in some of the fully formed leaves in the upper canopy on 50% of plants.
Whitefly	Population reaches 6 adults/leaf in the upper canopy before 10 AM or Appearance of honeydew on 50% of plants.
Aphid	Appearance of honeydew on 50 % of plants.
Bollworms	(For Conventional Cotton).
❖ *G. hirsutum*	5% damage among freshly shed fruiting bodies (squares, buds, flowers and young bolls) during effective boll formation period.
❖ *G. arboretum*	Initiate spray when 25% plants start producing squares. Repeat subsequent spray at 10 days interval.

Table 17.4. Insecticides recommended for sucking pests on cotton in Punjab

Sr. no.	**Insecticides**	**Dose/ha**
JASSID		
A. Nitroguadine group		
1.	Imidacloprid 200 SL	100 ml
2.	Imidacloprid 555	100 ml
3.	Imidacloprid 17.8 SL	100 ml
4.	Thiomethoxam 25 WG	100 g
WHITEFLY		
B. Organophosphate group		
1.	Triazophos 40 EC	1.5 litre
2.	Ethion 50 EC	2.0 litre
MEALY BUG		
C. Carbamate group		
1.	Carbaryl 50 WP	2.5 kg
2.	Thiodicarb 75 WP	625 g
D. Organophosphate group		
3.	Profenophos 50 EC	1.25 litre
4.	Quinalphos 25 EC	2.0 litre
5.	Acephate 75 SP	2.0 kg
6.	Chlorpyriphos 20 EC	5.0 litre

Table 17.5. Insecticides recommended for bollworms and tobacco caterpillar on cotton in Punjab

Sr. no.	Insecticides	Dose/ha
PINK AND SPOTTED BOLLWORMS		
A. Synthetic pyrethroids group		
1.	Alphamethrin 10 EC	250 ml
2.	β-cyfluthrin 0.25 SC	750 ml
3.	Cypermethrin 10 EC	500 ml
4.	Cypermethrin 25 EC	200 ml
5.	Deltamethrin 2.8 EC	400 ml
6.	Fenpropathrin 10 EC	750 ml
PINK, SPOTTED AND YOUNG LARVAE OF AMERICAN BOLLWORM		
B. Carbamate group		
1.	Carbaryl 50 WP	2.5 kg
2.	Thiodicarb 75 WP	625 g
C. Organochlorinate group		
3.	Endosulfan 35 EC	2.5 litre
C.	**Organophosphates**	
4.	Profenophos 50 EC	1.25 litre
5.	Monocrotophos 36 SL	1.25 litre
6.	Quinalphos 25 EC	2.0 litre
7.	Triazophos 40 EC	1.5 litre
8.	Ethion 50 EC	2.0 litre
D. Miscellaneous group		
9.	Flubendiamide 480 SC	100 ml
GROWN UP LARVAE OF AMERICAN BOLLWORM		
E. Organophosphate group		
1.	Acephate 75 SP	2.0 kg
2.	Chlorpyriphos 20 EC	5.0 litre
F. Naturalyte		
3.	Spinosad 48 SC	150 ml
G. Oxadiazine		2.0 litre
4.	Indoxacarb 15 SC	500 ml
5.	Indoxacarb 15 EC	500 ml

TOBACCO CATERPILLAR		
H. Carbamate group		
1.	Thiodicarb 75 WP	625 g
I. Organochlorinated group		
2.	Endosulphan 35 EC	2.5 litre
J. Organophosphate group		
3.	Acephate 75 SP	2.0 kg
4.	Chlorpyriphos 20 EC	5.0 litre
5.	Quinalphos 25 EC	2.0 litre
K. Insect growth regulator		
6.	Novaluron 10 EC	375 ml

17.3.8. Integrated Pest Management

Keeping in view the ill effects of pesticide use in cotton, validation and promotion of IPM technology through farmer's participatory approach became a necessity to sustain productivity of cotton with least disturbance to the ecosystem (Arora *et al.,* 2010). Since 1975, continuous efforts have been made in validation and dissemination of IPM technology at farmer's field. The Indian council of Agricultural Research, New Delhi sponsored the village level operational Research Project (ORP) to evaluate and demonstrate the efficacy, practicability and economics of IPM on cotton in 1975, initially adopting 2 or 3 villages for a period of 3 years. However, total of 15 villages (Faridkot, Bathinda and Ferozpur districts) were covered during the period of 15 years (Dhaliwal and Arora, 1996). The adoption of IPM technology resulted in 73.7 and 12.4 per cent reduction in number of insecticidal sprays for sucking pests and bollworms respectively (Sidhu *et al.,* 1984) with 35.5 per cent reduction in bollworm incidence, 23.1 per cent higher yield and 31.1 per cent higher net income in ORP villages as compared to non ORP villages (Sidhu *et al.* 1990). The adoption of IPM strategy in cotton in 11 villages based on cultivation of early maturing rapid fruiting and tolerant varieties/hybrids, adoption of cultural as well as physical control and judicious application of insecticides using efficient spray technology resulted in 21.35-41.13 per cent reduction in insecticidal load and increased the seed cotton yield by 23.0 - 49.47 per cent during 1995-2000 (Anonymous, 2002). During 2002-2004, dissemination of cotton IPM technology in 6 villages over an area of 796 hectares involving 166 participatory farmers in Ferozepur and Muktsar districts resulted in 55.7 per cent reduction in number of sprays, 48.5 per cent higher yield and 83.8 per cent higher returns by participatory farmers over non participatory farmers (Sohi *et al.,* 2006). According to Singh *et al.* (2008) the adoption of insect pests resistance management strategies on Bt cotton in 112 and 120 villages of Ferozepur,

Bathinda and Mansa districts covering an area of 2591 and 5720 hectares during 2005 and 2006, respectively resulted in additional profit of Rs. 4725 and 4267 per hectare to the farmers of IPRM villages as against non IPRM villages.

17.3.9. Insecticide Resistance Management

Based on the status and performance of various insecticides against different insect pests and their safety to natural enemies, four window based IRM strategies have also been developed for dissemination among the cotton growing farmers of the Punjab to reduce/delay insecticide resistance and increase functional life of the insecticides (PAU 2008). For the implementation of the these strategies under the IRM project, villages in the cotton growing districts of Punjab, viz. Bathinda, Muktsar, Mansa, Ferozepur and Faridkot have been adopted in which farmers are given proper trainings about proper identification of insect pests as well as natural enemies, monitoring pest population, economic threshold levels, selection of appropriated insecticides, when to spray, how to spray and other recommended production and protection technologies etc. The dissemination for IRM strategies in IRM adopted villages in Ferozepur districts during 2002-2004 resulted in 39.3 and 41.3 per cent reduction in number and cost of sprays, respectively, besides increase in 23.9 per cent seed cotton yield as compared to non-IRM villages (Singh *et al.* 2005). Similarly, adoption of 84 villages during 2003 and 2004 resulted in lower bollworm incidence, higher population of natural enemies, 25.7 to 58.7 per cent reduction in number of sprays with net profit of 10400 to 13114 per hectare during 2003 and 2004, respectively in IRM villages (Chandi and Dhawan 2005). Singh *et al.* (2005) reported significant decease in the resistance level in *Helicoverpa armigera* to insecticides with execution of IRM strategies in adopted villages of Mansa district during 2003 and 2004. Moreover, higher population of predators has been observed through exploitation of insecticides on window based IRM strategies in IRM villages as compared to non IRM villages. (Kumar *et al.* 2005). Singh and Dhawan (2006) recorded fewer numbers of sprays (9.61 to 9.56) in villages of Punjab that were subjected to four window insecticide resistance management (IRM) programmes for the management of insect pests of cotton compared with 12.86 to 16.20 in non-IRM villages. Dhawan *et al.* (2006) also reported 24.4 and 42.0 per cent reduction in number of sprays, 25.6 and 25.8 per cent in cost of sprays and 19.2 and 15.5 per cent increase in seed cotton yield during 2002 and 2003 after following IRM strategies in Ferozepur district. Similarly, reduction in insecticidal application upto 46.9 and 34.5 per cent in IRM villages over non IRM during 2004 and 2005 crop season, respectively in Punjab (Dhawan *et al.* 2007b). The overall changing scenario of pest management strategies has been compiled in Table 14.

17.4. FUTURE MANAGEMENT STRATEGIES

It is emphasized that pest problems should be tackled with IPM approach based on sound ecological principles involving integration of all the available control strategies with minimum use of insecticides so that pest population can be managed below damaging level for achieving sustained production in the cotton ecosystem. In addition, insecticide and insect resistance management strategies as a part of IPM module will help in delaying the resistance in insects to insecticides and transgenic cotton, respectively.

- Resistant varieties provide an innate control that involves no expense, no ecological problems and are generally compatible with other insect control methods. Therefore, studies on screening, identification and utilization of sources of resistance to major insect pests including potential pest's needs to be strengthened so as to develop pest resistant/tolerant varieties. Further, evaluation of transgenic cotton (BG-I and BG-II) especially against sucking pests needs intensive study both under field and epizootic conditions.
- Changing crop scenario, ecological conditions, elimination of natural enemies have resulted in uplift of new and secondary pest problems. Therefore, studies on monitoring, bio-ecological parameters, assessing damage potential and formulation of effective management strategies for tackling problems of these potential pest challenges in the cotton ecosystem is of utmost importance.
- The severity of the cotton insect pests in primarily governed by various weather parameters. Therefore, thorough and long-term studies on the role of various abiotic factors are of prime importance in order to develop suitable forecasting models as they have profound impact on the survival, development and population build of insect pests in cotton ecosystem.
- Potential concern among the researchers now is sustainability of Bt technology keeping in view widespread and prolonged exposure of Bt toxin to target species. Though resistance to Cry *1Ac* has not been reported in field populations in any part of the world but capability of American bollworm to develop resistance has already been worked out by some workers under laboratory conditions. Therefore, proper monitoring for forewarning of pest adaptation is of utmost necessity so that strategies for the management of resistance can be formulated, implemented, reviewed and modified accordingly in order to conserve this effective technology for longer period of time.
- Several species of parasitoids and other predatory fauna in cotton ecosystem play a major role in regulation pest population.

During the last 2-3 years, with the adoption of Bt cotton in Punjab, there has been significant increase in its area. Therefore, studies on impact of transgenic cotton on arthropod biodiversity in cotton ecosystem needs immediate attention.

- Genetically modified cotton having genetic resistance provides innate control of bollworms and tobacco caterpillar but sucking pests are still posing a great threat. Therefore, special attention is required for the validation and promotion of IPM technologies in transgenic cotton keeping in view the upliftment of new pests, i.e. mealy bugs and secondary pest problems like thrips etc. Further, the study on impact analysis of IPM approach on socio-economic status of farmers' of utmost importance.
- Indiscriminate use of insecticides in Punjab has been a major constraint in successful utilization of biocontrol agents. However, with the adoption of Bt cotton, there has been drastic reduction in number of sprays. Therefore, studies on the development of bio-intensive IPM module in transgenic cotton need to be encouraged.
- IPM strategy based on proper monitoring, observing ETH level and use of pesticides as a last resort will help in conserving natural biodiversity in cotton ecosystem. The need of the hour is to explore the possibility of utilizing eco-friendly and effectual insect growth regulators (IGR's) with novel mode of action which can fit idyllically in IPM programme.
- The problems of resurgence and high level of resistance to conventional insecticides needs immediate attention and caution in the usage of older chemistries of insecticides. There is need to identify chemicals with new chemistry to combat the pest problems in cotton as it can be one of the most important strategy for the management and delaying resistance in insect to insecticides.
- Studies on risks associated with the application of insecticides through monitoring of resistance and resurgence in major insect pests of cotton needs to be strengthened with new molecules also.
- Farmer's perception of problem, assessing potential crop loss, judgement regarding efficacy of control tactics available, constraints with respect to resources available affect the decision making, a major component of IPM approach to a great extent. Therefore, there is need to identify factors affecting decision making on pest control strategies and to develop information technology based decision support system (DSS) for the management of insect pests of cotton.

REFERENCES

Agnihotri, N.P. (2000) Pesticide consumption in India update. *Pestic. Res. J.* **12**: 1250-1255.

AICCIP (2007) *All India Coordinated Cotton Improvement Project.* Annual Report 2006-07 Volume - 1. pp 212. Central Institute for Cotton Research, Regional Station, Coimbatore.

Anonymous (2000) *QRT Report (1995-2000).* All India Co-ordinated Cotton Improvement Project. Punjab Agricultural University, Ludhiana.

Anonymous (2002). *Report on Cotton Research in Punjab.* Punjab Agricultural University, Ludhiana.

Anonymous (2006) *Agriculture at a glance 2005-06.* Department of Agriculture, Punjab.

Arora, R., Battu, G.S. and Ramakrishan, N. (2000). Microbial pesticides: Current status and future outlook. In: G.S. Dhaliwal and B. Singh (eds) *Pesticides and Environment.* Commonwealth Publishers, New Delhi, pp. 344-395.

Arora, R., Jindal, V., Rathore, P., Kumar, R., Singh V. and Bajaj, L. (2010) Integrated Pest Management of Cotton in Punjab, India. In: E.B. Radeliffe, W.D. Hutchisow and R.E., Cancelado (eds). *Radcliffe's IPM World Text book.* url.hhtp://ipmworld.umn.edu, University of Minnesota, St. Paul, MN.

Aulakh, S.S. and Butter, N.S. (2006) Toxicity of insect growth regulator, Lufenuron to *Helicoverpa armigera* (Hubner) on Cotton. *J. Insect Sci.* **19**(2): 178-80.

Benbrook, C.M. (1996) *Pest Management at the Crossroads.* Yonkers, NY: Consumer's Union.

Brar, K.S., Sekhon, B.S., Singh, J., Shenhmar, M. and Bakhetia, D.R.C. (2001) Evaluation of Biocontrol-based IPM modules for the control of boll worm complex on Cotton. *J. Insect Sci.* **14** (1&2): 19-22.

Brar, K. S., Singh, J., Shenhmar, M., Singh, J. and Singh, J. (2003) Management of cotton bollworms with egg parasitoids, *Trichogramma chilonis* Ishii in Punjab. In : Ignacimuthu, S. and Jayaraj, S. (eds) *Biological Control of Insect Pests.* Phoenix Publishing House Pvt. Ltd. New Delhi. pp 20-24.

Brar, K.S., Virk, J.S., Jalali, S.K., Sohi, A.S. and Shenhmar, M. (2007) Evaluation of multiple insecticide tolerant strain of *Trichogramma chilonis* Ishii for the management of cotton bollworms. *J. Insect Sci.* **20** (2): 208-211

Brar, K.S., Virk, J.S., Shenhmar, M., Singh, J. and Sohi, A.S. (2004) Integration of *Trichogramma chilonis* and insecticides for the management of bollworms. Fourth International Congress of Entomological Sciences. University of Agriculture, Faislabad, Pakistan. September 22-23, 2004. pp 24-25.

Butter, N.S. and Kular, J.S. (1998) Field evaluation of chitin inhibitor, Teflubezuron against bollworm complex on cotton. *J. Insect Sci.* **11** (2): 152-53

Butter, N.S. and Kular, J.S. (1999) Resurgence of whitefly in cotton and its management *Indian J. Ent.* **61** : 85-90.

Butter, N.S. and Vir, B.K. (1989) Morphological basis of resistance in cotton to whitefly *Bemisia tabaci*. *Phytoparasitica* **17** (4): 251-61.

Chakravarthy, A.K. and Sidhu, A.S. (1980) Resistance to insect pests damage in four varieties of cotton in Ludhiana. *Insect Sci. Applic.* **7** (5): 647-52

Chamberlain, D.J., Brown, N.J., Jones, O.T. and Casagrande, E. (2000). Field evaluation of a slow release pheromone formulation to control the American bollworm, *Helicoverpa armigera* (Lepidoptera : Noctuidae) in Pakistan. *Bull. Ent. Res.* **90**: 183-190.

Chamberlain, D.J., Critchley, B.R., Campion, D.G., Attique, M.R., Rafique, M. and Arif, M.I. (1992). Use of multi-component pheromone formulations for the control of cotton bollworms (Lepidoptera: Noctuidae and Gelechiidae) in Pakistan. *Bull. Ent. Res.* **82**: 449-458.

Chandi, R.S. and Dhawan, A.K. (2005) Impact of insecticide resistance management strategies on pest management and productivity of cotton in Punjab (Abstract). In : *Proc. 1st Congress on Insect Science.* December 15-17, 2005. Punjab Agricultural University, Ludhiana. pp 163-64.

Dhaliwal, G.S. and Arora, R. (1996) *Principles of Insect Pest Management.* Kalyani Publishera, New Delhi.

Dhaliwal, G.S. Arora, R. and Dhawan, A.K. (2004) Crop losses due to insect pests in Indian Agriculture: An update. *Indian J. Ecol.* **31** (1): 100-109.

Dhaliwal, G.S. (2005) Prospects of biorational approaches to pest management. In: B.S. Chhillar, K.K. Dahiya and J.S. Rana (eds.) *Adcances in Biorational Approaches to Pest Management* CCS Haryana Agricultural University, Hisar, India, pp. 18-28.

Dhawan, A.K. and Simwat, G.S. (1992) Management of sucking pets of cotton with natural plant products. In: *Proc. First National Symposium on Allelopathy in Agroecosytems (Agriculture & Forestry)*, February 12-14, 1992, Hisar. pp 154-55.

Dhawan, A.K. and Simwat, G.S. (1997) Insecticide induce resurgence of sucking pests in upland cotton in Punjab, India. In: *Third International Congress of Entomological Sciences*, March 18-20, 1997 (Abstract). National Agricultural Centre, Islamabad.

Dhawan, A.K. and Simwat, G.S. (1998) Evaluation of different biopesticides against cotton bollworms, *Helicoverpa armigera* (Hubner). In: Dhaliwal, G.S., Randhawa, N.S., Arora, R. and Dhawan, A.K. (eds) *Ecological Agriculture and Sustainable Development*, IES & CRID, Chandigarh. pp. 274-280.

Dhawan, A.K., Butter, N.S. and Narula, A.M. (2007a) *The Cotton Whitefly Bemisia tabaci (Gennadius).* Punjab Agricultural University, Ludhiana.

Dhawan, A.K., Sidhu, A.S. and Simwat, G.S. (1988) Assessment of avoidable loss in cotton due to sucking pests and bollworms. *Indian J. Agric. Sci.* **58** (4): 290-92.

Dhawan, A.K, Sidhu, A.S., Simwat, G.S. and Madan, V.K. (1986) Pink bollworm damage in cotton at different levels of loculi infestation. *Indian J. Agric. Sci.* **56** (10): 738-42.

Dhawan, A.K., Simwat, G.S. and Sidhu, A.S. (1991) Field reaction of some varieties of Asiatic cotton (*Gossypium arboreum* L.) to sucking and bollworm pests. *J. Res. Punjab Agric. Univ.* **28** (1): 57-62.

Dhawan, A.K., Singh, K., Arora, P.K. and Kumar, T. (2006) Insecticide resistance management (IRM) strategies: Their impact on arthropod fauna and economics in cotton agroecosystem. *Indian J. Ecol.* **33:** 158-62.

Dhawan, A. K., Singh, K., Saini, S., Mohindru, B., Kaur, A., Singh, G. and Singh, S. (2007c) Incidence and damage potential of mealy bug *Phenacoccus solenopsis* Tinsley on cotton in Punjab. *Indian J. Ecol.* **34** (2): 166-72.

Dhawan, A.K., Singh, S., Kumar, S. and Kumar, T. (2007b) Impact of resistance management strategy on insecticides use pattern during different growth stages of irrigated cotton in Punjab. *Pestic. Res. J.* **19:** 67-72.

Dhawan, A.K, Sithanantham, Baumgartner J. and Rao Ranga, G.V. (2001) Evaluation of nuclear polyhedrosis virus against *Helicoverpa armigera* (Hubner) on upland cotton, *Gossypium hirsutum* Linnaeus. (Abstract). *Conference on Biopesticides: Emerging trends* February 7-9, 2001, Chandigarh. pp 119.

Dhawan, A.K., Sohi, A.S., Sharma, M., Kumar, T. and Brar, K.S. (2004) *Insecticide Resistance Management in Cotton.* pp 56. Department of Entomology, Punjab Agricultural University, Ludhiana, India.

Dhawan, A.K. and Saini, S. (2009) First record of Phenacoccus solenopsis Tinsley (Homoptera: Pseudococcidae) on cotton in Punjab. *J. Insect Sci.* **22** (3): 309-310.

Dhawan, A.K. and Sidhu, A.S. (1984) Assessment of capture thresholds of Pink bollworm moths for the timing insecticidal applications on *Gossypium hirsutum* Linn. *Indian Journal of Agricultural Sciences,* **54:** 426.

Dhillon, H.S., Mahal, M.S. and Singh, R. (1998) Antixenosis response in *Hirsutum* cotton against cotton jassid, *Amrasca biguttula* (Ishida). *J. Insect Sci.* **11** (2): 162-65.

Dhillon, H.S., Mahal, M.S. and Singh, R. (1999) Field resistance to jassid, *Amrasca biguttula* (Ishida) in cultivars/lines of *hirsutum* cotton. *J. Insect Sci.* **12** (2): 114-17.

Gill, H.K. and Dhawan, A.K. (2006) Global status of insecticide resistance in *Helicoverpa armigera* on cotton. *J. Cotton Res. Dev.* **20** (2): 226-31.

Hargreaves, H. (1948) *List of Recorded Cotton Insects of the World.* Commonwealth Institute of Entomology, London. 50 pp.

Jindal, V. (2004) Studies on the mechanism of resistance to whitefly, *Bemisia tabaci* (Gennadius), in cotton, *Gossypium hirsutum* Linnaeus. Ph.D. Thesis, Punjab Agricultural University, Ludhiana.

Jindal, V., Arora, R. and Kumar, R. (2007) Screening of cotton genotypes for resistance to sucking pests. *Ann. Pl. Prot. Sci.* **15**(1): 26-29.

Kapoor, S.K., Singh, J., Russel, D., Singh, B. and Kalra, R.L. (2002) Susceptibility change of *Helicoverpa armigera* (Hubner) to different insecticides in Punjab. *Pestic Res. J.* **14** : 177-80.

Karanthi, K.R., Kranthi, S., Gopalakrishnan, N., Asokan, R. and Maye, C.D. (2009) Bt resistant-Its management and prospects in the current context of climate change. In: V.V. Ramamurthy, G.P. Gupta and S.N. Puri (eds) *Proc. Natn. Symp. IPM Strategies to Combat Emerging Pests in the Current Scenario of Climate Change.* January 28-30, 2009, Pasighat, Arunachal Pradesh, pp. 237-261.

Kranthi, K.R. and Russell, D.A. (2009) *Changing Trends in Cotton Pest Management. In* Integrated Pest Management: Innovation-Development Process, R. Peshin and A.K. Dhawan (eds.) DOI 10.1007/978-1-4020-8992-3 17.

Kranthi, K.R., Dhawad, C.S., Naidu, S.R., Mate, K., Behere, G.T., Wadaskar, R.M., Kranthi, S. (2005) Inheritance of resistance in Indian *Helicoverpa armigera* (Hubner) to Cry1Ac toxin of *Bacillus thuringiensis. Crop Prot.* **25**: 119-124.

Kular, J.S. and Butter, N.S. (1997) Screening of cotton germplasm. against the whitefly, *Bemisia tabaci* Genn. *J. Ent. Res.* **19** (4): 341-344.

Kulkarni, K.A., Kambrekar, D.N., Gundannavar, K.P., Devaraj, K. and Udikeri, S.S. (2004) *In International Symposium on 'Strategies for Sustainable Cotton Production - A Global Vision' 3,* Crop Protection, University of Agricultural Sciences, Dharwad, pp. 149-151.

Lewis, L.C. (2006) *Ecological considerations for the use of entomopathogens in Integrated Pest Management.* In: O. Kanl and G.W. Cuperius (eds). *Ecologically based Integrated Pest Management.* CAB. International, Wallingford, U.K, pp. 249-268.

Mann, G.S., Dhaliwal, G.S. and Dhawan, A.K. (2001) Effect of alternate application of neem products and insecticides on population of *Bemisia tabaci* Gennadius and its impact on bollworm damage in upland cotton. *Ann. Pl. Prot. Sci.* **9**(1): 22-25.

Monga, D., K.C. Kumhar and Rishi Kumar 2009. Integrated management of mealy bug, *Phenacoccus solenopsis* Tinsley on cotton. *International conference on emerging trends in production, processing and utilization of natural fibres* in 16-19 April, 2009 at Mumbai, India. Pp 132-137.

PAU (2008) *Package of Practices for Crops of Punjab (Kharif 2008).* Punjab Agricultural University, Ludhiana.

Sidhu, A.S. and Dhawan, A.K. (1980) Incidence of some insect pests on the different varieties of cotton. *J. Res. Punjab Agric. Univ.* **17**: 152-56.

Sidhu, A.S., Chahal, B.S., Sandhu, M.S., Arora, R., Singh, G., Singh, C. and Garg, B.R. (1984) Experience of integrated pests control of cotton under operational research project. *Paper presented at Bienniel Workshop of AICCIP,* February 8-11, 1984, Coimbatore.

Sidhu, A.S., Sandhu, M.S., Arora, R., Brar, D.S., Dhaliwal, C.S. and Bal, R.S. (1990) Operational research project on the integrated control of cotton pests in the Punjab. *Paper presented at ICAR-Transfer of Technology Projects Workshop.* May 27-28, 1990, Ludhiana.

Sidhu, A.S., Singh Joginder, Arora, R., Brar, D.S., Dhaliwal, C.S., Bal, R.S. and Singh, J. (1994) Adoption of plant protection measures vis-à-vis effectiveness of recommended insecticides against *Helicoverpa armigera* (Hubner) during 1990-a year of cotton failure in Punjab. *J. Insect Sci.* **7**: 10-13.

Simwat, G.S. and Dhawan, A.K. (1992) Use of plant insecticides against *Helicoverpa armigera* (Hubner) infesting cotton and chickpea. In: *Proc. First National Symposium on Allelopathy in Agroecosytems (Agriculture & Forestry)*, February, 12-14, 1992, Hisar. pp 158-59.

Simwat, G.S. and Gill, B.S. (1992) Population build up of cotton jassid and whitefly on *Gossypium arboreum* varieties and its correlation with weather factors. *J. Res. Punjab Agric. Univ.* **29** (1):70-76.

Singh, B. (1970) Studies on resistance in cotton to Jassid *Amrasca devastans* (Distant). Ph D Thesis, Punjab Agricultural University, Ludhiana.

Singh, D., Singh, K. and Singh, G. (2005) Survival of cotton whitefly, *Bemisia tabaci* (Gennadius) following application of buprofezin *J. Insect Sci.* **18** (1): 140-43.

Singh, J. and Sandhu, B.S. (1996) Field evaluation of *Gossypium arboreum* genotypes for resistance to bollworms. *J. Insect Sci.* **9**(2): 162-63.

Singh, J. and Sidhu, A.S. (1984) Comparative study of resistance in two American cotton varieties against pink bollworm. *J. Res. Punjab Agric. Univ.* **21**(2): 211-19.

Singh, J., Sohi, A.S. and Russel, D. (2004) Magnitude of *Helicoverpa armigera* (Hubner) problem on cotton in Asia. In : *Proc. Seminar on Recent Advances in Spray Technology in Irrigated cotton in Asia*, Ludhiana Punjab. April 2, 2004. pp 34-44.

Singh, K. and Dhawan, A.K. (2006) Impact of window insecticide resistance management strategies for cotton pest. *Ann. Pl. Prot. Sci.* **14** (2): 310-14.

Singh, K., Dhawan, A.K., Kumar, T., Arora, P.K. and Singh, R. (2008) RM on Bt cotton: its adoption and impact on production and economics (Abstract) pp 102-03. In : *Proc 2nd Congress on Insect Science.* February 21-22, 2008. Punjab Agricultural University, Ludhiana.

Singh, K., Dhawan, A.K., Singh, R. and Arora, P.K. (2005) Insecticides resistance management on cotton: Impact on pests and seed cotton yield (Abstract) pp 158-59. In : *Proc 1st Congress on Insect Science.* December 15-17, 2005. Punjab Agricultural University, Ludhiana.

Singh, R., Singh, D., Garg, H.R. and Rathore, P.K. (2002) Search for sources of resistance to cotton jassid *Amrasca biguttula* in upland cotton. *J. Cotton Res. Dev.* **16** (1): 108.

Singh, T.H. and Khush, G.S. (1993) *Resume of Cotton Research in the Punjab. Silver Jublee 1967-1992.* All India Coordinated Cotton Improvement Project (ICAR), Punjab Agricultural University, Ludhiana.

Singh, T.H., Singh, G., Sharma, K.P. and Gupta, S.P. (1972) Resistance of cotton (*Gossypium hirsutum* L.) to cotton jassid, *Amrasca devastans* (Distant) (Homoptera : Jassidae). *Indian J. Agric. Sci.* **42** (5): 421-425.

Singhal, V. (2003) *Indian Agriculture.* Indian Economic Data Research Centre, New Delhi, India 310-334.

Sohi, A.S., Brar, K.S., Sharma, S. and Bhullar, H.S. (2006) Dissemination of integrated pest management technology in irrigated cotton with farmers's participation. *J. Insect Sci.* **19** (2): 166-70.

Sohi, A.S., Singh, J., Brar, D.S. and Russell, D.A. (1998) Mating disruption of pink bollworm, *Pectinophora gossypiella* (Saunders) with sex pheromone as a component of IPM in the cotton agro-ecosystem. *Proceedings of the Workshop, Non-Pesticidal Management of Cotton and Pigeonpea Pests,* April 10, 1998, Hyderabad, India. Centre for World Solidarity and National Institute of Agricultural Extension Management, India, pp. 46-51.

Sukhija H.S., Singh J. and Butter N.S. (1983) Flowering pattern and incidence of pink bollworm (*Pectinophora gossypiella* Saunders) in different varieties of *hirsutum* cotton under sprayed conditions. *Indian J. Ecol.* **10**(1): 106-12.

Sukhija, H.S., Singh, J., Butter, N.S. and Singh, T.H. (1980) The field screening of promising cultivars of *hirsutum* cottons against bollworms. *Indian J. Ecol.* **7** (2): 314-16.

Sundaramurthy, V.T. (2003) An integrated management system for managing the pink bollworm. *Pectinophora gossypiella* (Saunders), (Lepidoptera: Gelechiidae) and its effects on the productivity of cotton in the polycrop agro-system. *Proceedings of the World Cotton Conference-3,* March 9-13, 2003, Cape Town, South Africa, pp. 988.

Vennila, S. (2008) Pest management for cotton ecosystem or ecosystem management for cotton production? *Curr. Sci.* **94** (11): 1351-1352.

Virk, J.S., Brar, K.S. and Sohi, A.S. (2004) Impact of insecticides on the resurgence of jassid and whitefly in cotton. *Indian J. Ent.* 66 (4): 319-22.

Wright, L. (1996) *New Yorker,* Jan. 15th, 42-48, 50-53, 55.

Theory and Practice of Integrated Pest Management
R. Arora, B. Singh and A.K. Dhawan (Eds)
Scientific Publishers (India), Jodhpur 2011, pp. 298 - 320.

18

INTEGRATED MANAGEMENT OF INSECT-PESTS IN RICE

P.S. Sarao and M.S. Mahal[1]

Department of Plant Breeding & Genetics, [1]Department of Entomology
Punjab Agricultural University, Ludhiana - 141 004, India

18.1. INTRODUCTION

Rice is the staple food for around half the world's people and about three-quarter of a billion of the world's poor depends on rice. Low land rice provides more than 75 per cent of the world's annual supply, but this supply is also vulnerable due to threats from rice pests. The frequent pest outbreaks also caused further loss in yield. Global rice production has been estimated at 650 m tonnes from an area of 156 m ha (Anonymous, 2008). In Punjab rice was cultivated over an area of 27.35 lakh hectares with total production of 110 lakh tones (Anonymous, 2010). It being a crop of warm humid environment and this is also conducive for development and proliferation of insect-pests. Among more than 100 species of insects known to attack this crop, about 20 are of economic importance (Pathak and Dhaliwal, 1981). The cumulative losses due to insect-pest damage in rice were estimated at 25 per cent (Dhaliwal *et al.*, 2004). The insect pests attacking the rice crop can be categorized according to their feeding habits like stem borers (yellow, white and pink stem borers, striped borer, dark headed stem borer), leaf feeders (leaf folder, rice hispa, rice caseworm, armyworm, rice horned caterpillar, rice skipper, whorl maggot and grasshopper), sap suckers (Brown planthopper, whitebacked planthopper, green leafhoppers, rice bug and rice thrips), root feeders (rice root weevil and gall midge). Among these biotic constraints, yellow stem borer, *Scirpophaga incertulas* (Walker); white stem borer, *S. innotata* (Walker); pink stem borer, *Sesamia inferens* (Walker); brown planthopper, *Nilaparvata lugens* (Stal); whitebacked planthopper, *Sogatella furcifera* (Horvath); and leaf folder, *Cnaphalocrocis medinalis* (Guenee) cause extensive damage to the crop. Protection of the rice crop from these pests has been an integral part of crop management practices. What is IPM and why the need arise for the implementation of IPM in rice is discussed below:

Geier and Clark (1961) proposed the idea of managing insect pest population by the concept as, "protective management of noxious species" or pest management, further Geier (1966) proposed the term integrated pest control (IPC) but it also does not clarify the concept. A number of definitions were proposed for the terms but the definition by FAO (1967) is widely accepted. It define Integrated pest management (IPM) as an pest management system that in context of associated environment and population dynamics of the pest species, utilizes all suitable techniques and methods in as compatible a manner as possible and maintains pest population at levels below those causing economic injury.

18.2. COMPONENTS OF IPM

18.2.1. Varietal resistance

Painter (1951) described plant resistance as the relative amount of heritable qualities that influence the ultimate degree of damage done by the insect. In practical agriculture, resistance represents the ability of a certain variety to produce a larger crop of good quality than do ordinary varieties at the same level of insect population. He grouped mechanisms of resistance into three main categories, viz., nonpreference, antibiosis and tolerance. Kogan and Ortman (1978) proposed the term antixenosis to describe the host plant properties responsible for nonpreference. Antixenosis refers to undesirability i.e. avoidance by the insect whereas antibiosis refers to unsuitability i.e., adverse effects on the insect after feeding on the host plant, while, tolerance refers to the ability of the host plant to withstand an insect population sufficient to damage severely the susceptible plants. It is generally attributed to plant vigour.

Insect resistant varieties have played a major role as core of IPM in rice. Several concerted efforts were made to evaluate rice germplasm against major insect pests and to identify donors (Kalode *et al.,* 1983, Siddiq *et al.*1998, Reddy and Bentur, 1999). Of the 690 rice varieties developed and released for commercial cultivation in India since 1965, 20 per cent were specifically bred fro insect resistance (Prasada Rao *et al.* 2001). The genetics of BPH resistance indicated that a single dominant gene govern resistance in Krishnaveni, ARC 6031 & MTU 1J 206-7-4-1; a single recessive gene in ARC 10550, Chaitanya and ARC 11285; two duplicate dominant genes in Vijetha; two dominant complimentary genes in Ptb 33 (Padmavati *et al.,* 2006). Santhanalakshmi and Kadirvel, (2007) reporetd that five introgressed line derived from the wild species, *viz. Oryza australiensis, O. minuta* and *O. rufipogon* exhibited strong reaction to BPH. Till date twenty one major genes for BPH resistance have been identified by using standard evaluation methods, out of these eighteen genes have been localized on specific regions of six rice chromosomes using molecular genetics analysis and genomics tools (Jena and Kim, 2010). Some of popular varieties resistant to insect-pests are listed in Table 18.1. Genetic engineering is another tool to enhance insect resistance and the most important advantage of transgenic technology is the capacity to mobilize useful gene from non-rice source top rice with least disruption to rice genome (Ranjekar *et al.,* 2003). Genes encoding several insecticidal products are now being incorporated into rice to confer insect resistance (Table 18.2). In india, Nayak *et al.,* 1998 reported the development of first transgenic IR 62 rice with *Bt* gene. Since then, several groups transformed rice varieties IR 62, Karnal local with *Bt* gene such as *Cry 1 Ab, Cry 1 Ac* to get resistance against yellow stem borer (Khana and Raina, 2002). Rice variety Chaitanya with snowdrop (*Galanthus nivalis*) lectin gene, *Gna*, has conferred resistance against sucking pests like BPH, WBPH and GLH (Nagadhara *et al.,* 2003).

Table 18.1. Details of some of the rice varieties resistant/ moderately resistant to insect-pest

Pest / Variety	State / CVRC	Source of resistance
Whitebacked Planthopper		
PR 108, PAU 201	Punjab	Vijaya/Ptb 21; PR 103/ PAU 1126
HKR 120	Haryana	Ptb 33
IET 8116	Karnataka	Andrewsali
Brown Planthopper		
Jyothi (Ptb 39), Bhadra (M 11-57-51)	Kerala	Ptb 10, Ptb 20
Manasarovar	CVRC	Lab Muey Nahng
Neela, Udaya	Orissa	Ptb 21; Ptb18, Ptb 21
IET 7575	Karnataka	Manoharsali
Vajram, Chaitanya	Andhra Pradesh	ARC 6650, ARC 5984
Gall midge		
Kakatiya, Surekha, Phalguna, Abhaya	Andhra Pradesh	W 1263, Siam 29, Siam 29, Ptb 10
Jyothi (Ptb 39), Panchami, Pavithra	Kerala	Ptb 10, W 1263, Siam 29
Vikram	Karnataka	Siam 29

(1)	(2)	(3)
Shakti, Lalat	Orissa	Ptb 18, Ptb 21; Siam 29, Obs 677
R 636-405	Chhattisgarh	Madhuri
Stem borer		
Ratna, Vikas, Sasyasree	CVRC	TKM 6
Green leafhopper		
Vikramarya	Andhra Pradesh	Ptb 2
Nidhi	CVRC	Ptb 2
Multiple rseistance		
Suraksha, Shaktiman, Triguna	CVRC	MR 1523, Ptb 21, Siam 29

Table 18.2. Important genes used for conferring resistance in rice

Variety	Methodology	Gene	Pest
Indica (IR 64)	Biolistic	*Cry1Ac* under maize ubiquitin 1 promoter	Stem borer
Indica, aromatic (Tarom Molaii)	Biolistic	*Cry1Ab* (Under PEP C promoter)	Stem borer
Indica	Biolistic/ Protoplast	*Cry1Ab* under constitutive, tissue specific promoter with/ without antibiotic selector gene	Stem borer
Indica (M7, Basmati 370)	Biolistic	*Cry2A*	Stem borer, leaf folder
Indica (MS restorer line of Minhui 63)	Biolistic	*Cry1Ab & cry1Ac* fusion gene under rice actin 1 promoter	Stem borer + leaf folder
Indica (M7 Basmati 370)	Biolistic	*Cry1Ac+Cry2A+Gna*	Stem borer+ Leaf folder + BPH
Indica (Chaitanya)	Agrobacterium	*Gna*	Sap sucking pests like BPH, WBPH, GLH
Japonic (Taipei 309)	PEG mediated protoplast transformation	*CpTi* (cowpea trypsin inhibitor) under actin 1 promoter	Stem borers
Japonica (Nagdongbyeo)	PEG mediated protoplast transformation	*SKTI* (Soybean Kunitz trypsin inhibitor) under CaMv35S	BPH
Japonica varieties	Biolistic	*Pin2* (Potato proteinase inhibitor 2) under wound inducible promoter	Pink stem borer

18.2.2. Cultural methods

Changes in agronomic practices directly affect the dynamics of pest population. These practices are simple and easy to follow but have a greater impact when practiced on a community basis on larger area. Some of cultural practices in reducing pest incidence are given below:

Timely sowing. Restrict to timely sowing of nursery of non-basmati (second fortnight of May) and timely transplanting schedule (second fortnight of June). This will reduce the population build-up of the stem borers particularly on basmati crop, which has the transplanting time within first fortnight of July to second fortnight of July (Bas. 370 & Bas. 386).

Effect of shade. Avoid sowing non-basmati and basmati under shade as this may increase the incidence of leaf folder. These places serve as hot spot for further spread of the pest.

Lopping/ Detopping. Egg masses of yellow and white stem borers are laid on the tip of the leaf blades. Lopping of tall varieties of basmati (Basmati 386 and Basmati 370) at half of its height 45 days after transplanting destroys good number of egg masses and young larvae of stem borer without any loss in yield. To reduce the attack of rice hispa, gall midge and case worm, clip-off and destroy the leaf tips before transplanting, if attack starts in the nursery itself.

Alleys. Alley or paths of 30 cm width after every 2-3 m of planting form a significant component of IPM. At panicle initiation stage, alleys provide proper aeration to the crop, which help in reducing hopper population. This also facilitates effective insecticide spraying by allowing free movement of spraying person.

Fertilizer application. Excessive nitrogen application is not only uneconomical but also increases the vulnerability of the rice crop to attack by many insect-pests such as stem borer, leaf folder, gall midge, rice ear-cutting caterpillar. Hence, only recommended dosage of nitrogenous fertilizers should be applied that too in splits.

Water management. Water management is one of the most important cultural practices, which can be used for checking the incidence of many insect-pests in the field. Drain out the excess water from the rice field for 3-4 days control the brown planthopper, whitebacked planthopper and whorl maggot. On the other hand, flooding of rice fields reduce soil borne insects-pests such as, termites, rice ear-cutting caterpillar and thrips in the field.

Harvesting. Yellow and white stemborers over winter in rice stubbles, while larvae of pink stemborer are found in the internodes/stubbles of basmati crop at the time of harvesting. Harvesting of crop should be done as close to the soil as possible to destroy stem borer larvae present in the internodes/stubbles. This will also expose the insects to the birds.

Destruction of stubbles. Yellow and White stem borer larvae hibernate in the stubble of the rice crop and serve as source of primary infestation for succeeding rice crop. Thoroughly flooding of the field with water and ploughing it with discs or rotavators will kill hibernating larvae of stemborers present in stubbles. Ploughing of the fields after harvesting also exposes the larvae and pupae of rice ear-cutting caterpillars hidden in the soil to biotic and abiotic factors.

Crop rotation. Prerequisite for this practice is that the crop for rotation should be a non-host of the insect-pest. It is one of the most effective methods for breaking life cycle of a pest. In areas where pink stem borer and rice ear-cutting caterpillar are problems, paddy-wheat and paddy-winter maize crop rotation should be avoided. Sowing of sorghum and maize crop around the paddy fields should be avoided as these crops also act a host for pink stem borer and rice ear-cutting caterpillar.

Dislodging. Use of rope to dislodge leaf feeders e.g. larvae of leaf folder and case worm in the field helps to reduce the population of the insect-pests.

Ecological engineering. Ecological engineering has emerged as a new direction for pest management (Gurr *et al.,* 2004). It is defined as, "the design of human society with its natural environment for the benefit of both." It is the management strategy to enhance ecosystem services such as pest invasion resistance, pest regulation, and pollination. Habitat manipulation is a form of ecological engineering (Mitsch and Jorgensen, 2005). Changes in habitat structure can directly influence ecosystem services by enhancing biodiversity. Floral biodiversity greatly affects the insect guilds, particularly predators and parasitoids by providing them both food and refuge resources. Populating bunds with beneficial flowers (*Crassocephalum crepidioides, Tridax procumbens, Vemonia cinerea, Ageratum conyzoides, Emilia sonchifolia*) to serve as sources of nectar and shelter for predators and parasitoids. Through ecological engineering, ecosystem will be enhanced through increased habitat biodiversity, providing shelter, food resources and corridors for natural enemies.

18.2.3. Biological control

Rice fields have a teemingly rich community of beneficial insects, spiders and pathogens that attack insect pests of rice. These natural biological control agents are friends of the farmers and are responsible in controlling the population build-up of insect pests. Otherwise, the pests would multiply at a faster rate that would result to pest outbreaks. In the management of rice insect pests whether integrated (IPM) or ecological

(EPM) pest management, it is of paramount importance to have correct identification of natural enemies (NE) and understanding of NE behaviour in order to optimize the strategies for the control of insect pests.

Conservation of natural enemies. which is a pre-mediated action for protecting and maintaining natural enemies like spiders (*Pardosa, Oxyopes, Tetragnatha, Argiope, Clubiona, Thomisus, Neoscona* spp.), drynids, water bugs, mirid bugs, damsel flies, dragon flies, meadow grasshoppers, staphylinid beetles, carabids, coccinellids, *Apanteles, Tetrastichus, Telenomus, Trichogramma, Bracon, Xanthopimpla, Temelucha, Stenobracon, Brachymeria, Charops, Macrocentrus, Platygaster* by:

- Reducing direct mortality of natural enemies by
- delaying/ avoiding application of insecticides when natural enemies are abundant
- Spot/ strip application of selective insecticides
- Growing trap crops
- Providing additional niches for the natural enemies by making available their alternate hosts during time of low pest densities
- Providing food sources such as pollen, nectar and honeydew and spraying nutrient mixtures to attract natural enemies

Augmentation of natural enemies. Action taken to increase the population of natural enemies by artificial means. The population can be supplemented by mass multiplication and releases in the field.

- Inoculative releases: made as infrequently as once in a year to re-establish a species of natural enemy which was periodically killed out in an area due to unfavourable conditions during part of the year.
- Inundative releases: involve mass culture and release of natural enemies to suppress the pest population directly as in the case of conventional insecticides. These are most economical against pests that have only one or at the most a few discrete generations per year. *Trichogramma* cards should be stapled at 40 spots in one acre for 5 to 6 times at weekly interval, starting 30 days after transplanting for control of stem borer and leaf folder (Anonymous, 2010).

Pest defender ratio. 2:1 ratio may be useful to avoid application of pesticides. Some examples of natural control agents for rice insect-pest, *viz. S. incertulas* (Table 18.3), *N. lugens* & *S. furcifera* (Table 18.4) and *C. medinalis* (Table 18.5) are given below:

Table 18.3. Parasitoid complex of *S. incertulas*

Stage	Parasitoids
Egg	*Telenomus* sp., *T. rowani* (Gahan); *T. dignoides* (Nixon), *T. dignus* (Gahan); *T. beneficiens* (Zehntner); *Trichogramma* sp., *T. japonicum* Ashmead, *T. chilonis* Ishii, *T. exiguum* Pinto & Planter, *T. minutum* Riley, *T. evanescens* Westwood; *Tetrastichus schoenobii* Ferriere; *Trichogrammatoidea nana* (Zehntner)
Larva	*Stenobracon nicevillei* (Bingham); *Bracon chinensis* Szepligeti; *Cotesia flavipes* Cameron; *Tropobracon schoenobii* (Viereck); *Rhaconotus oryzae* Wilkinson; *Goniozus indicus* Muesback; *Isotima javensis* Rohwer; *Allorhogas pyralophagus* Marsh
Pupa	*Xanthopimpla stemmator* (Thunberg), *X. flavolineata* Cameron; *Tetrastichus ayyari* Rohwer; *Brachmeria* sp.; *Charops bicolor*, *Glytomorpha* sp.

Source: Ooi and Shepard (1994)

Table 18.4. Natural enemies of nymph-adult stages of *N. lugens* & *S. furcifera*

Parasitoid/pathogen	Hosts
(1)	(2)
Hymenoptera: Dryinidae	
Echthrodelphax fairchildii Perkins	*N. lugens, S. furcifera*
Haplogonatopus apicalis Perkins	*N. lugens, S. furcifera*
Gonatopus yasumatsui Olmi	*N. lugens*
Pseudogonatopus sarawaki Moczar	*N. lugens, Sogatella* sp.
Pseudogonatopus nudus Perkins	*N. lugens, S. furcifera*
Pseudogonatopus fulgori (Nakagawa)	*S. furcifera*
Pseudogonatopus flavifemur Esaki & Hashimoto	*N. lugens, S. furcifera*
Pseudogonatopus hospes Perkins	*N. lugens, S. furcifera*
Strepsiptera: Elenchidae	
Elenchus japonicus Esaki & Hashimoto	*N. lugens, S. furcifera*
Acarina: Erythraeidae	
Charletonia sp.	*N. lugens, S. furcifera*
Nematoda: Mermithidae	
Agamermis sp.	*N. lugens*
Amphimermis sp.	*N. lugens*
Hexamermis sp.	*N. lugens*
Moniliales: Moniliaceae	

(1)	(2)
Beauveria bassiana (Balsamo) Vuillemin	*N. lugens*
Metarhizium anisopliae (Metchnikoff) Sorokin	*N. lugens, S. furcifera*
Moniliales: Stilbaceae	
Hirsutella strigosa Petch	*N. lugens, S. furcifera*
Entomophthorales: Entomophthoraceae	
Erynia delphacis Hori	*N. lugens*
*Conidiobolus*sp.	*N. lugens*
Conidiobolus coronatus (Constantin) Batko	*N. lugens*

Source: Dupo and Barrion, (2009)

Table 18.5. Natural enemies of *C. medinalis*

Stage/natural enemy	Parasitoid/ predator
Egg	*Copidosomopsis nacoleiae (Eady); Trichogramma* sp.*; T. chilonis Ishii; T. chilotraeae Nagarkatti & Nagaraja; T. evanescens Westwood; T. japonicum Ashmead; T. nubilale Ertle & Davis; Trichogrammatoidea* sp.*; T. nana Zehntner*
Larva & pupa	*Scoliophthalmus micans Lamb; Dohrniphora cornuta (Bigot); Megaselia* sp.*; M. scalaris (Loew); Argyrophylax fransseni (Baranov); A. nigrotibialis (Baranov); Ceromosia* sp.*; Chaetexorista javana Brauer & Bergenstamm; Nemorilla floralis (Fallen); N. floralis maculosa (Meigen); Zygobothria cilliata (Wulp); Goniozus* spp.*; Apanteles* spp.*; A. angaleti Muesebeck; A. angustibasis Gahan; A. cypris (Nixon); A. flavipes (Cameron); A. opacus (Ashmead); A. ruficrus (Haliday); A. syleptae Ferriere; A. schoenobii Wilkinson; Bracon* sp.*; B. gelechiae Ashmead; B. ricinicola Ashmead; Cardiochiles* sp.*; C. philippinensis Ashmead; Cedria anomala (Wilkinson); Chelonus* sp.*; Chelonus munakatae Munakata; Coelinius* sp.*; Hormius* sp.*; Macrocentrus* sp.*; M. philippinensis; Ashmead; M. thoracicus (Nees); Meteorus bacoorensis Ashmead; Microplitis* sp.*; Opius* sp.*; Orgilus* sp.*; Rhysipolis* sp.*; Spathius* sp.*; Brachymeria* spp.*; B. excarinata Gahan; B. lasus (Walker); B. obscurata (Walker); Podagrion ahlonei Mani; Elasmus* spp.*; E. brevicornis Gahan; E. corbetti Ferriere; E. philippinensis Ashmead; Tetrastichus ayyari Rohwer; T.israeli (Mani & Kurian); T. schoenobii Ferriere; T. howardii; Tetrastichus* sp.*; Barylypa apicala (Cameron); Brachycoryphus nursei (Cameron); Charops brachypterum Gupta &Maheswary; C. nigrita Gupta and Maheswary; Chorinaeus facialis Chao; Diatora lissonata (Viereck); Eriborus argenteopilosus (Cameron); E. ryukyuensis (Momoi); E. sinicus Holmgren; E. vulgaris (Morley); Ischnojoppa luteator (Fabricius); Copidosomopsis nacoleiae (Eady); Stenomesius*

	japonicus Ashmead; Itoplectis narangae (Ashmead); Melcha maculiceps Cameron; Temelucha sp.*; T. philippinesis (Ashmead); T. stangli (Ashmead); Trathala flavorbitalis (Cameron); Trichomma cnaphalocrosis Uchida; Xanthopimpla* sp.*; X. flavolineata Cameron; X. lepcha (Cameron); X. punctata (Fabricius); X. trifasciata (Smith); Trichomalopsis apanteloctena (Crawford); Trichospilus pupivora Ferriere; T. nrshirakii (Crawford)*
Predators	*Casnoidea interstitialis Schroeter; Chlaenius* spp.*; C. bioculatus Chaudoir; C. posticalis (Motschulsky); Cicindela lacrymosa (Fabricius); Drypta geniculata (Klug); Drypta japonica Bates; Ophionea ishii ishii Habu; O. indica; O. nigrofasciata Schmidt — Goebel; Coccinella* sp.*; C. repanda (Thunberg); Harmonia octomaculata (Fabricius); Micraspis* sp.*; M. crocea Mulsant; Menochilus sexmaculatus (Fabricius); Paederus fuscipes Curtis; Euborellia stali (Dohrn); Proreus simulans (Stal.); Paederus fuscipes Curtis; Proboscidocoris* sp.*; Andrallus spinidens (Fabricius); Polytoxus fuscovittatus (Stal.); Scipinia horrida (Stal.); Staccia diluta (Stal.); Odontomacus* sp.*; Odontoponera transversa (F. Smith); Solenopsis geminata (Fabricius); Chalybion bengalense (Dahlbom); Sceliphron madraspatanum conspicillatum (Costa); Monomorium pharaonis (Linnaeus); Eumenes campaniformes (Fabricius); Agriocnemis pygmaea Rambur; A. femina femina Brauer; Ischnura senegalensis Rambur; Pseudagrion pilidorsum Brauer; Macrodiplax cora Brauer; Pantala flavescens (Fabricius); Trithemis aurora (Burmeister); Conocephalus indicus; C. maculates; Anaxipha longipennis (Serville); Euscyrtus concinnus (de Hann); Metioche vittaticollis (Stal.); Conocephalus longipennis (de Hann); Argiope aemula (Walckenaer); A. catenulata (Doleschall); Neoscona theisi (Walckenaer); Nephila maculata (Fabricius); Clubiona japonicola Boesenberg & Strand; Lycosa pseudoannulata (Boesenberg & Strand); Oxyopes javanus Thorell; Marpissa mandali; Plexippus paykulli (Audouin); Tetragnatha javana (Thorell); T. mandibulata Walckenaer; T. maxillosa Thorell*

Source: Khan *et al.* (1988, Sarao (2007)

Abundant natural enemies in irrigated rice usually prevent significant insect pest problems. IPM extension education of depth and quality is required to discourage unnecessary insecticide use that upsets this natural balance, and to empower farmers as expert managers of a healthy paddy ecosystem. We have to advice the farmers that "When we kill off the natural enemies of a pest we inherit their work".

18.2.4. Behaviour modifying compounds

Semiochemicals have emerged as promising and safer alternatives in pest control, which can be integrated in an IPM package. Semiochemicals (Gk. *Semeon*, a signal) are chemicals that mediate interactions between organisms and are subdivided into allelochemicals (interspecific) and pheromones (intraspecific). Allelochemicals are subdivided into several groups *viz.*, Allomones-when the response of the receiver is adaptively favourable to the emitter but to the receiver; Kairomones-favourable to the receiver but not the emitter or Synomones-favourable to both emitter and receiver. Pheromones (Gk. *phereum*, to carry; *horman*, to excite or stimulate) are released by one member of a species and cause a specific interaction with another member of the same species. They are further classified as alarm, aggregation or sex pheromones. It is the sex pheromones of insects that are important for IPM in the crops and can be utilized in many ways such as:

Indirect control (Monitoring). Surveying and monitoring for early warning, timimg of control measures, knowing the population dynamics, mapping of pest distribution (dispersal & migrating behaviour), quarantine work, risk assessment, knowing effect of control measures, enhancing natural enemy effectiveness, detecting insecticide resistant insects

Direct control. Mass trapping, mating disruption, lure and kill (Table 18.6).

Table 18.6. Studies with sex pheromones in rice ecosystem

Technique/ Pest	Crop	Reference
Mass trapping		
Scirpophaga incertulas	Rice	Varma, 1999, Krishnaiah *et al.,* 2000, Katti *et al.,* 2001, Pushpakumari and Tewari, 2005
Mating disruption		
S. incertulas	Rice	Ganeswara Rao *et al.,* 1994, Cork and Basu, 1996, Cork *et al.,* 1996, Cork *et al.,* 1998
Monitoring		
S. incertulas	Rice	Krishnaiah *et al.,* 1998a
Auto-confusion technique		
S. incertulas	Rice	Murthy, 2008
Integration of pheromones with other strategies		
S. incertulas + *C. medinalis*	Rice	Katti *et al.,* 2001
Synergizing interaction through plant extract		
S. incertulas	Rice	Krishnaiah *et al.,* 1998b, Varma *et al.,* 2002

Extensive investigation on all the three aspects of sex pheromones of yellow stem borer in rice are given below:

Monitoring. The principle use of sex pheromones is to attract insects to traps, for detection and determination of temporal distribution. Efficient monitoring is the foundation of sound IPM programmes, without which proper pest management decisions cannot be made. Krishnaiah (1995) reported several parameters that would enable efficient pheromone based monitoring for insect pests of cereal crops. Krishnaiah *et al.* (1998a) optimized three traps per acre as trap density for monitoring of rice yellow stem borer.

Varma *et al.* (2000) conducted monitoring studies on rice yellow stem borer (*S. incertulas*) using natural blend (3:1) of (Z) -11-hexadecenal and (Z) -9-hexadecenal and noticed three peak periods of moth activity during July, August and September in *Kharif* and two during January and March or April during *Rabi*. It is evident from the temporal relationship between peak periods of pheromone trap catches and field infestation that peak infestation occurs 2 to 3 weeks after peak/ sudden increase in moth captures in pheromone traps. Three sleeve traps baited with rubber dispenser (impregnated with pheromone @ 5mg/ dispenser) should be installed at inter trap distance of 60 m in a triangular pattern in rice field of about one acre. The trap height should be maintained at 0.5 m at vegetative stage and just above the crop canopy subsequently. The dispensers have to be replaced at three weekly intervals.

Mating Disruption. Mating disruption is accomplished in two principle ways, false trail following or confusion. False trail following results from placing many point sources of pheromone (hollow fibers, flakes or other point sources) per acre than the anticipated numbers of females in the crop. Thereby chances of males finding females, at the end of the pheromone trail will be greatly reduced.

Male confusion is achieved through high ambient pheromone concentrations permeating the field and hiding the trails of calling females (large doses from diffuse sources such as microcapsules or larger doses of pheromone in point source dispensers such as tie-on polyethylene ropes). Added to the effect is the adaptation of antennal receptor sites and/or habituation of the insect's central nervous system. Ganeswara Rao *et al.* (1994) reported that the pheromone applied @ 40 g a.i. /ha through 625 point sources disrupted communication for mating during *Rabi* 1992. Typical dose of the pheromone is 10,000-40,000 mgs/ha/season.

Auto-confusion technique. Auto-confusion technique employs electro-statically charged, wax powder impregnated with pheromone that attract male moths, on close approach male moths, and their antennae in particular are coated with powder that prevents them orienting to other pheromone source and causes them to act as decoys for untreated males. Auto-confusion is thought to work in part by incapacitating the male moths' ability to orientate to pheromone sources (Howse, 2004). This results in reduction of dead heart and white ear heads by yellow stem borer. In this individual Exosex YSB pellets have the appearance of a small cylinder of compressed powder that should be applied by hand into the internodal ridge of bamboo stick or in the hangers. The pellets should be pushed into the hangers until 3-4 mm remains exposed at the front of the hanger. Typical dose of the pheromone is generally 480 mgs/ha/ season and there is transfer of pheromone from one male insect to another in the secondary transfer of the pheromone containing powder.

Mass trapping or male annihilation. The rationale of this technique is to place enough traps to concentrate pest insects in to a restricted space (catch enough males) and leave the females of the species without mates. Mass trapping technique offers great promise against monophagous pests like yellow stem borer on rice with crop often cultivated extensively and contiguously. Mass trapping can be integrated with biological control (Katti *et al.,* 2001).

The adoption of this technology is low mainly because of:

1. Lack of ready availability of traps and pheromone lures at affordable prices
2. Pheromone lures require frequent replacements (3-4 weeks) during crop period
3. Lack of awareness among farmers and extension workers
4. Poor quality of dispensers / lures affecting faith of the farmers

18.2.5. Chemical control

Insecticides use is an indispensable evil in any ecosystem when any body thinks in terms of ecological conservation. Their use becomes inevitable when population increase above economic threshold levels (ETLs) (Table 18.7). Insecticides recommended at national level (Table 18.8) and state level (Table 18.9) has been enlisted below. For efficient and economical pest control only need based chemicals should be used in the form as given below:

Economic threshold level. Calendar schedule or blanket application of insecticides cause many ecological problems. Insecticides should be used only when pest population reach ETL. The ETL relates to pest population level at which control measures must be started so that the population may not reach up to economic injury level (Table 18.7).

Seed treatment. Krishnaiah and Kalode (1983), reported that soaking of sprouted rice seed in 0.2% emulsion of chlorpyriphos or suspension of isofenphos for 3 hours before sowing reduce the damage by gall midge during nursery period. Shi *et al.,* 2001 oserved that mixing 1 kg of dry seed with 12-20 g of carosulfan 35 DC gave significant control of thrips upto 24 days after sowing.

Table 18.7. ETLs of some insect-pests of rice

Insect-pest	ETL
Stemborers	2% dead hearts on basmati 5% dead hearts on non-basmati
Leaf folder	10% damaged leaves (1/3rd dam aged leaf)
Planthoppers (BPH & WBPH)	>5 insects / hill
Green leafhopper	10-20 insects/hill (2 insects/hill in RTD endemic areas)
Hispa	1 adult or 1-2 damaged leaves
Gall midge	5 % silver shoots
Gundhi bug	1-2 bugs/hill
Rice ear-cutting caterpillar	1 larva/hill

Source: Anonymous, 2001 and 2010.

Table 18.8. Insecticide used in rice at national level

Group	Effective against	LD50 (mg/kg)	
		Oral	Dermal
Organophospates			
Quinalphos	Sucking and lepidopterous insects	62-137	1250-400
Phosalone	Acaricide & wide range of pest species	120-170	1500
Monocrotophos	Wide range of pests including mites, sucking insects , leaf eating beetles & bollworms	14-23	336
Chlorpyriphos	Soil & foliar pests, mosquitoes, flies, household pests	135-163	-
Acephate	Aphids, thrips, leaf miners, saw flies and lepidopterous larvae	866-945	2000
Fenitrothion	Selective acaricide but of low ovicidal activity	250	3000
Dichlorvos (DDVP)	Quick knock down effect, lepidopterous larvae, sucking pests	80 male rat, 56 Female	107 male rat, 75 Female
Fenthion	A wide variety of pests particularly fruit flies, mango nut weevil etc.	215-245	320-330
Phorate	Sorghum shoot fly and rice gall midge	16-37	2.5-6.2
Phosphamidon	Sucking pests, leaf miners & mites	17.9-30	374-530

Ethoprophos	Nematicide and soil insecticide	62 (rat)	26 (rabbit)
Carbamates			
Carbaryl	Wide spectrum of insect pests	-	-
Carbofuran	Sucking and soil inhabiting pests	8.8-14.1	10,200
Carbosulfan	Wide spectrum of pest species	185-250	2000
BPMC or Fenobucarb	Sucking pests	410	4200
MIPC or Isoprocarb	Rice pests, leaf hoppers	403-485	500
Insect Growth Regulator			
Buprofezin	Plant hoppers, Mealy bugs, whiteflies, scales, etc	2000	5000
Phenyl Pyrazoles			
Fipronil	Suking and lepidopterous insects	100	2000
Ethiprole	Sucking pests		
Neonicotinoids			
Imidacloprid	Sucking pests	450	5000
Thiamethoxam	Rice hoppers, aphids, whiteflies, thrips, rice bugs, leaf miners	1563	2000
Thiacloprid	Molluscicide & Insecticide	444	-
Neristoxins			
Cartap	Rice stem borer, leaf folder, sugarcane shoot borer, cabbage diamond back moth	345 male, 325 female	1000
Thiocyclam	Lepidopterous and Coleopterous pests	310 male	-
Ether Derivatives			
Ethofenprox	Pests of rice particularly BPH & leafhoppers	-	2140
Fermentation Products			
Spinosad	Active against *Helicoverpa armigera*, leafhoppers, aphids & whiteflies on cotton	-	-
Diamides			
Flubendiamide	Activity against a board spectrum of lepidopterous insect pests including resistant strains	2000	2000

Table 18.9. Insecticide used in rice at state level

Insect-pest/ Formulation	Insecticide	Brand	Dose/ acre
Stem borers			
Non-Basmati			
Spray	Triazophos	Hostathion 40 EC	350 ml
Monocrotophos	Nuvacron/Monocil 36 SL	560 ml	
Chlorpyriphos	Coroban/Dursban/Lethal/ Chlor-guard/ Durmet/ Classic/ Force 20EC	1000 ml	
Basmati			
Spray	Monocrotophos	Nuvacron/ Monocil 36 SL	560 ml
Chlorpyriphos	Coroban/Durban/Lethal/ Chor-guard/ Durmet/ Classic/Force 20EC	1000 ml	
Fipronil	Fipronil 80% WG	15 g	
Granular	Cartaphydrochloride	Padan/Caldan/ Kritap/ Sanvex/ Nidan/ Marktap 4G	10 kg
Fipronil	Regent/Mortel 0.3 G	6 kg	
Chlorpyriphos	Dursban 10G	4 kg	
Phorate	Foratox 10G	5kg	
Leaf folder			
Spray	Triazophos	Hostathion 40 EC	350 ml
Monocrotophos	Nuvacron/Monocil 36 SL	560 ml	
Chlorpyriphos	Coroban/Durmet/Force 20EC	1000 ml	
Triazophos	Hostathion 40 EC	350 ml	
Planthoppers			
Spray	Imidacloprid	Confidor 200 SL	40 ml
Endosulfan	Thiodan/Endocel 35 EC	560 ml	
Monocrotophos	Nuvacron 36 SL	560 ml	
Quinalphos	Ekalux/Quinguard 25 EC	800 ml	
Chlorpyriphos	Coroban/Dursban 20EC	1000 ml	
Rice hispa			
Spray	Methyl Parathion	Methyl Parathion 50 EC	120 ml

Monocrotophos	Monocil 36 SL	560 ml	
Quinalphos	Ekalux 25 EC	800 ml	
Chlorpyriphos	Dursban 20EC	1000 ml	
Rice-ear-cutting-caterpillar			
Spray	Quinalphos	Ekalux/ Quinguard 25 EC	400 ml
Monocrotophos	Nuvacron 36 SL	560 ml	
Root weevil			
Granule	Phorate	Thimet/Foratox 10G	3 kg

Seedling root dip. This practice ward off the transplanted crop from early stage pests like gall midge, stem borers and whorl maggots. Chlorpyriphos as 0.02% emulsion with normal dipping period of 12 hours could effectively reduce the damage by these pests. However, for hoppers carbosulfan 0.02% emulsion was very effective. This practices will also help to conserve natural enemies like spiders, which colonize the rice fields early in the crop season (Krishnaiah and Kalode, 1986).

Root zone application. When insecticide was placed in the root zone, the persistence of insecticides was increased due to minimum losses by oxidation, vaporization, run-off and exposure to sunlight. Attempts were also made to coat urea super granules with wettable powders of carbofuran either with coal tar or with neem oil and placing coated material in the root zone. This will save labour cost and results in supply of both insecticide and N to crop efficiently.

Pest surveillance. Insect-pest damage can be prevented if the fields are monitored regularly and effectively. Planthoppers, stem borer and leaf folder manifest visual symptoms of attack rather late and in this case surveillance information is of immense practical utility for planning crop protection measures.

Selective insecticides. Selectivity means that a chemical may be highly toxic to one organism but not or less toxic to other organism and this difference in toxicity is called selectivity. Granular insecticide cartap hydrochloride is highly toxic to lepidopterous pests but less toxic to spiders. Endosulfan and quinalphos have been found to be safe to the mirid bug and the egg parasitoid of stem borer, respectively.

Spray application technique. In case of hoppers for better effectiveness, direct the spray towards the base of the plants. If damage noticed at hopper burn stage, treat the affected spots along with their 3-4 meter periphery immediately as these spots harbour high population of the insects. Whereas in case of rice ear-cutting caterpillar which is a nocturnal pest in behaviour, the spray should be done in evening hours for getting better results.

18.2.6. Botanicals

Settling and Orientation behaviour. Saxena *et al.,* 1980a reported that neem oil possesses a strong repellent action to brown planthopper (BPH), very low number of BPH settled on treated plants. At 3% concentration of oil, the spray treated plants attracted not more than 36% of BPH. When neem oil treated leaves at 12% or higher concentration were offered to leaf folder in a choice test, it significantly lowered the number of larvae that arrived on treated leaves (Saxena *et al.,* 1980b). However, similar effects were not observed in case of green leafhopper females (Heyde *et al.,* 1983). Similarly, neem cake applied to soil did not restrain BPH settling on TN 1 in a choice test (Saxena *et al.,* 1984).

Feeding behaviour. Feeding deterrent effect of neem oil was studied in many insect pests. Saxena *et al.,* 1980a reported that BPH adults were restless, took longer time to locate feeding sites and ingested smaller quantities of food on treated plants. Such effects were not observed in vegetable oil treated plants. Significant reduction in quantity of food ingested by BPH, WBPH and GLH occurred even at 1% concentration of neem oil (Heyde *et al.,* 1983). Saxena and Khan (1985) observed reduced phloem feeding increased xylem feeding, repeated probing and profuse salivation of GLH on plants sprayed with 10% neem oil. Custard apple oil was relatively more effective as antifeedant than neem oil against BPH, WBPH and GLH (Saxena *et al.,* 1983).

Growth and metamorphosis. Saxena *et al.,* 1980a reported that at 3 per cent concentration of neem oil only 3-9 per cent BPH nymphs became adults. Similarly, Heyde *et al.,* 1983 also observed the effect of neem oil on nymphal development and found that at 3% concentration of neem oil, only 12 per cent of WBPH and 2 per cent of GLH nymphs became adults.

18.2.7. Biopesticides

Bt formulations were evaluated for their effectiveness against lepidopterous insect-pests like, leaf folder and stem borers. Rath (1999) studied the effectiveness of four B.t. formulations against rice yellow stem borer and observed that BTK II and Dipel 3.5% were more effective than the other two formulations *viz.,* Deflin 85% and Biolep. Panda *et al.* (1999) from Orissa observed that Bioasp, Biolep, Biotox, Dipel, and Delfin @ 2000 g of formulation/ha were highly effective against leaf folder and moderately effective against yellow stem borer and comparable to the insecticide monocrotophos @ 500 ml of formulation/ha. Lal (2001) also evaluated four B.t. formulations of var. *kurstaki viz.,* BTK II, Delfin, Dipel and Biolep at 2000g of formulation/ha against leaf folder and yellow stem borer under field conditions and reported that these were as effective as chlorpyriphos @ 250 g a.i./ha.

18.2.8. Future strategies for Insecticide use in rice IPM

- New insecticides with better mode of action and low residual life should be evaluated at as many locations as possible to assess their entire spectrum of effectiveness against the whole range of insect-pests infesting the rice crop.
- Insecticide induced resurgence should be monitored for hoppers (BPH, WBPH and GLH) and lepidopterons (stem borers, leaf folder) at national level.
- Natural enemies in rice ecosystem play an important role in reducing the insect-pest populations. So, the insecticides that are recommended against insect-pests must be screened for its toxicity to the natural enemies.
- In case of organic rice the use of botanical and bio-pesticides should be assessed for better insect-pest management. Exploratory attempts to get effective molecules/ strains should be encouraged.
- Studies on compatibility of insecticides with fungicides should be under taken to form the ready made recommendation strategies against insect pest and disease situations.
- Soil ph, Soil EC and organic carbon has great effect on the granular insecticides, so these factors should also be studied for effectiveness of granular insecticides before recommendation under diverse edaphic situations.
- Rainfall frequently washed off the insecticide sprays applied to crop. Studies should be conducted to find the minimum time interval required between spray and rain fall for getting greater effectiveness of the compound.

These insecticides should be applied alternately. Spray formulations should be applied using 100 litres of water per acre in Knapsack sprayer, whereas granular insecticides should be broadcasted in standing water, with person applying these granules should wear glove. Further application of any of these insecticides may be repeated as and when damage reaches economic threshold level. Padan/Caldan/Kritap/Sanvex/ Nidan/ Marktap 4G/ Regent/Mortel 0.3G/ Dursban 10G/ Fipronil 80% WG also control leaf folder.

18.3. CONCLUSIONS

Successful IPM requires research at all levels of the system. Breakthroughs made by research efforts enabled early programmes to be built on a sound scientific foundation. Implementation must be an integral part of IPM research and development programmes. Creative ways must be found to involve farmers more fully in rice IPM activities and to promote and support farmer-to-farmer training. Scientists must ensure that new technology must promote, rather than endangers, the ecological balance in rice. Varieties with stable resistance and tolerance to insect-pests continue be valuable complements to natural control. The conservation and enhancement of natural controls is a key research area. A better understanding of when and why pest outbreaks are triggered is necessary for creating ecologically sound control strategies, and for planning and improving of present levels of natural control.

REFERENCES

Anonymous (2008) Global rice production and area. www.faostat.com

Anonymous (2001) *Integrated Pest Management for Rice.* IPM package no. 1. Govt. of India, DPPQS, Faridabad, pp. 1-30.

Anonymous (2010) *Package of Practices for Kharif Crops.* Punjab Agricultural Univ., Ludhiana pp.1-23.

Cork, A. and Basu, S.K. (1996) Control of the yellow stem borer, *Scirpophaga incertulas* by mating disruption with PVC resin formulation of the sex pheromone of *Chilo suppressalis* (Lepidoptera : Pyralide) in India. *Bull. Ent. Res.* **86**: 1-9.

Cork, A., De Souza, K., Krishnaiah, K., Kumar, D.V.S.S.R., Reddy, A.A., Casagrande, E. and De Souza, K. (1996) Control of yellow stem borer, *Scirpophaga incertulas* (Lepidoptera : Pyralide) by mating disruption on rice in India: effect of unnatural pheromone blends and application time on efficacy. *Bull. Ent. Res.* **86**: 515-524.

Cork, A., De Souza, K., Krishnaiah, K., Reddy, A.A. and Zainulabuddin, S. (1998) Season long control of yellow stem borer, *Scirpophaga incertulas* (Lepidoptera : Pyralide) by mating disruption with the natural ratio of pheromone components. *Bull. Ent. Res.* **88**: 109-116.

Dhaliwal, G.S., Arora, R. and Dhawan, A.K. (2004) Crop losses due to insect pests in Indian Agriculture: An update. *Indian J. Ecol.* **31**:1-7.

Dupo, A.L.B. and Barrion, A.T. (2009) Taxonomy and general biology of delphacid planthoppers in rice agroecosytems. In: Heong, K.L. and Hardy, B. (edt) *Planthoppers: New Threats to the Sustainability of Intensive Rice Production Systems in Asia*, International Rice Research Institute. pp. 3-155.

FAO (Food and Agricultural organization) (1967) Report of the first session of the FAO Panel of Experts on Integrated Pest Control, Rome

Ganeswara Rao, A., Krishnaiah, K., Cork, A. and Hall, D.R. (1994) Disruption of sex pheromone mediated communication in yellow stem borer, *Scirpophaga incertulas* Walker with a controlled release pheromone formulation. *J. Ent. Res.* **18**: 287-288.

Geier, P.W. (1966) Management of insect pests. *A. Rev. Ent.* **11**: 471-490.

Geier, P.W. and Clark, L.R. (1961) An ecological approach to pest control. In: *Proc. Eight tech. Meeting. International Union for Conservation of Nature and Natural Resources.* Pp. 10-18.

Gurr, G.M., Wratten, S.D. and Altieri, M.A. (2004) *Ecological Engineering: Advances in Habitat Manipulation for Arthropods.* Collingwood, CSIRO Publishing. Pp. 232.

Heyde, J.V.D., Saxena, R.C. and Schmutterer, H. (1983) Neem oil and neem extracts as potential insecticides for control of hemipterous rice pests. *Proc. 2nd International Neem Conference, Rauischolzhausen,* pp. 377-390.

Howse, P. (2004) Mechanism of the Exosex autocofusion system of mating disruption. *Int. Pest Cont.* **45**: 321-322.

Jena, K.K. and Kim, S.M. (2010) Current status of brown planthopper (BPH) resistance and genetics. *Rice* **3**: 161-171.

Kalode, M.B., Sain, M., Bentur, J.S. and Kondal Rao, Y. (1983) Investigations on host -plant resistance in rice to insect pests. In: *Proceedings of Rice Pest Managament Seminar.* Tamil Nadu Agricultural University, Coimbatore. pp. 182-191.

Katti, G., Pasalu, I.C., Varma, N.R.G. and Krishnaiah, K. (2001) Integration of pheromone mass trapping and management of yellow stem borer and leaf folder in rice. In: *Proc. 88th Indian Science Congress*, New Delhi pp. 38.

Khan, Z.R., Barrion, A.T., Litsinger, J.A., Castilla, N.P. and Joshi, R.C. (1988) A bibliography of rice leaf folders (Lepidoptera: Pyralidae). *Insect Sci. Applic.* **9**: 129-174.

Khanna, H.K. and Raina, S.K. (2002) Elite Indica transgenic rice plants expressing modified Cry1Ac endotoxin of *Bacillus thuringiensis* show rehanced resistance to yellow stem borer (*Scirpophaga incertulas*). *Transgenic Res.* **11**: 411-423.

Kogan, M. and Ortman, E.F. (1978) Antixenosis- A new term proposed to define Painter's 'Non-prefence' modality of resistance. *Bull ent Soc Am* **24**: 175-176.

Krishnaiah, K. (1995) Insect pheromones in integrated pest management of cereals. In: *Pheromones in Integrated Pest Management.* (Eds. Krishnaiah, K. and Diwakar, B.J.). Pl. Prot. Assoc. of India, pp. 33-39.

Krishnaiah, K., Varma, R.G., Pasalu, I.C. and Zainulabuddin, S. (1998b) Improvement in efficacy of sex pheromone of rice yellow stem borer, *Scirpophaga incertulas* Walker with co-attractants and plant extracts. *Indian J. Pl. Prot.* **26**: 131-137.

Krishnaiah, K., Varma, R.G., Pasalu, I.C., Zainulabuddin, S. and Yadav, J.S. (2000) Development of mass trapping technique for management of yellow stem borer (*Scirpophaga incertulas* Walker) in rice. In: *Innovative pest and disease management in Horticultural and Plantation crops* (Eds. Narasimhan, S., Suresh, G. and Daniel, W.), Allied Publication, Chennai, pp. 33-40.

Krishnaiah, K., Zainulabuddin, S., Ganeswara Rao, A., Kumar, D.V.S.S.R. and Varma, R.G. (1998a). Pheromone monitoring system of rice yellow stem borer, *Scirpophaga incertulas* Walker. *Indian J. Pl. Prot.* **26**: 99-106.

Krishnaiah, N.V. and Kalode, M.B. (1983) Effect of simulated rainfall on insecticide spray effectiveness. IRRN, **8**: 12.

Krishnaiah, N.V. and Kalode, M.B. (1986) Toxicological investigations against rice green leafhopper, *Nephotettix virescens* (Distant). *Trop. Pest Mgmt.* **32**: 44-48.

Lal, R. (2000) Evaluation of some granular insecticides against whitebacked and brown planthopper in rice. *Pesticide Res. J.* **12**: 96-98.

Mitsch, W.J. and Jorgensen, S.E. (2005) *Ecological Engineering and Ecosystem Restoration.* John Wiley, New York. Pp. 424.

Murthy, K.S.R.K. (2008) Auto confusion pheromone-A new technique for the management of leaf folder and yellow stem borer of rice based on their biology. *Pestology* **32**: 58-61.

Murthy, K.S.R.K. and Krishnaiah, K. (2008) Auto confusion Pheromones-A new approach for the management of rice yellow stem borer, *Scirpophaga incertulas. Indian J. Pl. Prot.* **36**: 158-162.

Nagadhara, D., Ramesh, S., Pasalu, I.C., Kondal Rao, Y., Krishnaiah, N.V., Sharma, N.P., Bown, D.P., Gatehouse, J.A., Reddy, V.D. and Rao, K.V. (2003) Transgenic indica rice plant resistant to sap sucking insects. *Pl. Biotech.* **1**: 239-250.

Ooi, P.A.C and Shepard, B.M. (1994) Predators and Parasitoids of Rice Insect Pests In: Heinrichs, E.A. (ed.) *Biology and Management of Rice Insects*, Wiley Eastern Limited pp. 585-612.

Padmavathi G., Deen R., Ram T., Kondal Rao and Pasalu I.C. (2006) Genetics analysis of resistance to brown plant hopper (*Nilaparvata lugens* Stal.) in rice cultivars. In: 2nd International Rice Conference, Oct 9-13, N. Delhi pp. 261-262.

Panda, S.K., Nayak, S.K. and Behera, U.K. (1999) Field efficacy of five commercial formulations of *Bacillus thuringiensis* var. *kurstaki* against the rice stem borer and leaf folder. *Pest Mgmt. Eco. Zoo.* **7**: 143-146.

Painter, R.H. (1951) *Insect Resistance in Crop Plants.* The McMillan Co., New York. 520 p.

Pathak, M.D. and Dhaliwal, G.S. (1981) *Trends and Strategies for Rice Insect Problems in Tropical Asia.* IRRI Research paper Series No. 64, The International Rice Research Institute, Los Banos, Phillippines.

Prasada Rao, U., Rama Prasad, A.S. and Mishra, B. (2001) *High Yielding Rice Varieties of India-2000.* Directorate of Rice Research, Hyderabad. pp. 102.

Pushpakumari, A.S. and Tewari, S.N. (2005) Management of yellow stem borer, *Scirpophaga incertulas* (Walker) (Lepidoptera : Pyralide) in rice through sex pheromone mediated mass trapping of male. *Pestology* **29**: 11-17.

Ranjekar, P.K., Patankar, A., Gupta, V., Bhatnagar, R., Bentur, J.S. and Anand Kumar, P. (2003) Genetic engineering of crop plants for insect resistance. *Curr. Sci.* **84**: 321-329.

Rath, P.C. (1999) Evaluation of some *Bacillus thuringiensis* and neem formulations against yellow stem borer on rice under rainfed low lands. *Oryza* **36**: 398-399.

Reddy, A.P.K. and Bentur, J.S. (1999) Insect and disease resistance in rice. In: *Rice Breeding and Genetics* (ed.) J.S. Nanda. Oxford & IBH Publishing, New Delhi. Pp. 143-167.

Santhanalakshmi, S. and Kadirvel, P. (2007) Reaction of introgression lines for rice to a BPH population from India. IRRN **32**: 23-24.

Sarao, P.S. (2007) Eco-friendly approaches for the management of leaf folder, *Cnaphalocrocis medinalis* (Guenee) on Rice. Ph.D. Dissertation, Punjab Agricultural University, pp. 152.

Saxena, R.C. and Khan, Z.R. (1985) Effect of neem oil on survival of *Nilaparvata lugens* (Homoptera: Delphacidae) and on grassy stunt and ragged stunt virus transmission. *J. econ. Ent.* **78**: 647-651.

Saxena, R.C., Epino, P.B., Tee, C.W. and Puma, B.C. (1983) Neem, chinaberry and custard apple: Antifeedant and insecticidal effects of seed oils on leaf hopper and plant hopper pests of rice. Proc. 2nd International Neem Conference, Rauischolzhausen, pp. 403-412.

Saxena, R.C, Justo, H.D. Jr and Epino, P.B. (1984) Evaluation and utilization of neem cake against brown plant hopper *Nilaparvata lugens* (Homoptera: Delphacidae). *J. econ. Ent.* **77**: 502-507.

Saxena, R.C., Liquido, N.J. and Justo, H.D. (1980a) Neem seed oil, a potential antifeedant for the control of the rice brown plant hopper, *Nilaparvata lugens*. Proc. 1st Int. Neem Conf., Rottach Egem, pp. 171-188.

Saxena, R.C., Waldbauer, G.P., Liquido, N.J. and Puma, B.C. (1980b) Effect of neem seed on rice leaf folder, *Cnaphalocrocis medinalis*. Proc. 1st Int. Neem Conf., Rottach Egem, pp. 171-188.

Shi Ji Cheng, Peng Zheng Wen, Huang Xiao Ming, Cai Sheng Bin, Xu Zheng Wen and Huang Shi Wen (2001) Control of insect pests of late rice with carbosulfan DS. *Pl. Prot.* **27**: 40-41.

Siddiq, E.A., Reddy, A.P.K., Krishnaiah, K. and Bentur, J.S. (1998) Genetic improvement in rice for resistance to diseases and insects. Vol. **3**: pp. 75-103. In: *IPM Systems in Agriculture* (ed.) R.K. Upadhyay, K.G. Mukerji and R.L. Rajak. Aditya Books Pvt. Ltd. New Delhi.

Varma, N.R.G. (1999) Management of yellow stem borer, Scirpophaga incertulas on rice by sex pheromone. Ph.D. Thesis, ANGRA Univ., Hyderabad.

Varma, N.R.G., Krishnaiah, K., Pasalu, I.C. and Reddy, D.D.R. (2000) Monitoring of rice yellow stem borer, *Scirpophaga incertulas* Walker using pheromone traps and thermal summation. *Indian J. Pl. Prot.* **28**: 84-93.

Varma, N.R.G., Zainulabuddin, S., Bhavani, B. and Rao, P.R.M. (2002) Efficacy of certain new insecticides against rice planthoppers under field conditions. *Indian J. Pl. Prot.* **31**: 31-33.

Theory and Practice of Integrated Pest Management
R. Arora, B. Singh and A.K. Dhawan (Eds)
Scientific Publishers (India), Jodhpur 2011, pp. 321 - 337.

19

EMERGING TECHNOLOGIES FOR INTEGRATED PEST MANAGEMENT IN FORAGE CROPS

Ramesh Arora

Department of Plant Breeding & Genetics
Punjab Agricultural University, Ludhiana - 141 004, India

Integrated pest management (IPM) involves the use of a variety of cultural, mechanical, biological and chemical control measures for minimizing the damage caused by insect pests. As per one of the earlier and popular definitions IPM refers to an ecological approach in PM in which all the available necessary techniques are consolidated in a unified program, so that pest populations can be managed in such a way that economic damage is avoided and adverse side effects are minimized' (NAS, 1969). The concept of IPM has evolved over the years and as per a newer definition, "IPM is a decision support system for the selection and use of pest control tactics, singly or harmoniously coordinated into a management strategy, based on cost/benefit analyses that take into account the interests of and impacts on producers, society and the environment" (Kogan, 1998).

IPM is of critical importance in forage crops as the green fodder is fed to animals and the presence of residues of pesticides applied indiscriminately may harm farm animals. Moreover, fodder crops are selected/bred for absence of anti-nutritional factors. This may indirectly increase their susceptibility to insect pests. Therefore, cultural control methods along with limited application of safer pesticides are recommended for management of insect pests on these crops.

19.1. LOSSES CAUSED BY INSECT PESTS

Insect pests cause enormous losses in most of the fodder crops (Table 19.1). The crops grown for seed production are especially susceptible to the damage caused by insect pests. The gains in seed yield, due to gram caterpillar, *Helicoverpa armigera* management in *Rabi* forage legumes have been reported. These were lowest in Persian clover (Shaftal) genotype S69 (39.9%), intermediate in alfalfa (Lucerne) genotype LLC 5 (51%), and highest in Egyptian clover (Berseem) genotype BL-10 (295%). Among the Berseem genotypes, gains in seed yield were 123, 126, 156, 226 and 295 per cent in BL-22, BL-180, BL-1, BL-42 and BL-10, respectively. The important insect pests of fodder crops alongwith their life histories and management strategies are discussed herein (Sandhu, 1976; Dhaliwal, 1998; Atwal and Dhaliwal, 2005).

Table 19.1. Losses caused by important insect-pests of forage crops

Crop	Insect-pests	Period of activity damage	Estimated loss (%) under unprotected conditions
Egyptian clover or Berseem (*Trifolium alexandrinum*)	Bihar hairy caterpillar	Sept-Oct and Mar-Apr	30-80
	Green semilooper	Mar-Apr	25-50
	Gram caterpillar	Mar-Apr	40-90
Alfalfa or Lucerne (*Medicago sativa*)	Thrips	Apr-June	20-35
	Lucerne Weevil	Nov-Mar	25-35
	Bihar hairy caterpillar	Sept-Oct and Mar-Apr	30-40
	Gram caterpillar	Apr-June	45-55
	Green semilooper	Mar-Apr	20-55
	Spotted alfalfa aphid	Feb-Apr	15-25
Persian clover or Shaftal (*Trifolium resupinatum*)	Green semilooper	Mar-Apr	45-65
	Gram caterpillar	April	35-45
Oats (*Avena sativa*)	Oat bird cherry aphid	Feb-Mar	30-52
	Oat thrips		30-52
Maize (*Zea mays*)	Maize Jassid	Mar-Apr	15-52
	Maize borer	Mar-Nov	20-30
Sorghum (*Sorghum vulgare*)	Sorghum shoot fly	Apr-Sept	45-85

	Maize borer	Mar-Nov	20-25
	Sugarcane pyrilla	Sept-Oct	10-30
	Pollen eating beetle	Sept-Oct	24-30
	Green aphid	Sept-Oct	10-25
Cowpea (*Vigna unguiculata*)	Bihar hairy caterpillar	Sept-Oct	15-35
	Cowpea pod borers		
	(i) Blue butterfly	Oct-Nov	12-20
	(ii) Maruca	Oct-Nov	10-15
	(iii) Gram Caterpillar	Oct-Nov	30-35
	Black aphid	Sept-Oct	10-18
	Blister beetle	Sept-Oct	65-80

Source : Dhaliwal (1998)

19.2. GRAM CATERPILLAR (*Helicoverpa armigera* (Hubner) (Noctuidae: Lepidoptera)

Gram caterpillar (*Helicoverpa armigera* (Hubner) (Noctuidae Lepidoptera) is a major threat to agriculture in Asia, Africa, Oceania and parts of Europe. A diverse range of crops including grain crops, forage and vegetable legumes, fibre crops, oilseeds, horticultural and ornamental plants are among the economically important hosts attack by the pest (Dhaliwal and Arora, 2006). The pest has a preference for feeding on the floral parts of its host plants although young foliage is also attacked frequently. *H. armigera* is the key pest of Egyptian clover (berseem) seed crop in Northern India and losses upto 70-80 per cent in seed yield have been reported (Arora and Dhawan, 2010). The larvae prefer to feed on the flowers and the developing seeds inside pods. The pest also attacks the seed crops of lucerne and shaftal during the *Rabi* season and cowpea during the *kharif* season.

Life-cycle. The females lay eggs singly on tender parts of the plants especially on buds and flowers. High humidity tremendously increases the fecundity of the females and a single female may lay as many as 741 eggs in 4 days. The eggs are shining greenish yellow and are round. They hatch in 2-4 days in April to October and 6 days in February and the young larvae feed on the foliage for some time and later bore into the pods and feed on the developing grains, with their bodies hanging outside. They move from pod to pod and are full-fed in 13-19 days and measure 35 mm in the last instar. The full-grown larvae come out of the pod and pupate in the soil. The pupa is dark brown and has a sharp spine at the posterior end. In the active season, the pupal period lasts 8-15 days, but in winter the duration is prolonged, particularly in northern India. There may be as many as 8 generations in a year.

Campoletis chlorideae Uchida (Ichneumonidae) is a larval parasite of major importance especially on berseem and chickpea. During the peak of its activity, it may parasitize 50-60 per cent of *H. armigera* larvae.

Management. (i) Avoid raising berseem crop adjoining to tomato, gram, late sown wheat, *sathi* moong, *sathe* mash and sunflower on which the pest multiplies and later shifts to berseem, lucerne and shaftal (Table 19.2). In case this is not possible, the pest must be properly managed on these crops. (ii) During the vegetative phase, the pest is kept in check through timely cuttings and feeding by insectivorous birds especially at the time of cutting. (iii) At the initiation of flowers, the crop must be sprayed with 500 ml of indoxacarb 15 EC or 150 ml spinosad 48 SC or 2.5 litres of endosulfan 35 EC in 200-250 litres of water per hectare using a manually operated knapsack sprayer. If the attack persists, spray should be repeated after 10 days.

Table 19.2. Effect of adjoining crop on *H. armigera* incidence in berseem

Adjoining crop	Caterpillars/m^2	Pod damage (%)	Seed yield (q/ha)
Berseem	3.87	22	1.85
Gram	8.73	41.5	1.20
Sathe mash	8.1	74.0	0.33
Late wheat	9.08	37.1	1.68
Tomato	5.12	55.4	1.09
Sathe mung	9.9	82.0	0.29

Source : Dhaliwal and Sidhu (1988)

19.3. STEM BORER *Chilo partellus* (Swinhoe) (Lepidoptera : Pyralidae)

It is the most destructive pest of maize and sorghum in Sri Lanka, India, Pakistan, Afghanistan, Uganda, Central and East Africa. It is found throughout India. This insect has also been recorded on pearl millet, sugarcane, Sudan grass, sarkanda (*Saccharum munja*) and some other grasses. Its caterpillars damage maize and sorghum by boring into the stems, cobs or ears. The grown up caterpillars are about 20-25 mm long and dirty grayish white, with black head and four brownish longitudinal stripes on the back. The adults are yellowish-grey moths, about 25 mm across the wings when spread.

Life-cycle. The insect breeds actively from March-April to October and for the rest of the year it remains in hibernation as a full-grown larva in maize and sorghum stubble, stalks or unshelled cobs. The larvae pupate sometime in March and emerge as moths in the end of that month or in early April. They are active at night, when they mate and lay eggs on the underside of the leaves of various host plants, particularly the early crops of maize and sorghum grown as fodder. The eggs are flat, oval, yellowish and are laid in overlapping clusters each containing up to 20 eggs. A female lays over 300 eggs during its life-span of 2-12 days and the eggs hatch in 4-5 days in summer. The young larvae first feed on the leaves, making a few shot holes and then bore their way downwards through the central whorl as it opens. More shot holes become visible, indicating an earlier attack and the plant also shows dead-hearts. In sorghum, the midribs are attacked by the young larvae, more than one plant being attacked by a larva. The larva becomes full-fed in 14-28 days, passing through six stages and after making a hole in the stem pupates inside it. The life-cycle is completed in about 3 weeks and there are probably 5 generations in a year. The full-grown caterpillars of the last generation hibernate in stubble, stalks, etc., and remain there till the next spring.

Management (i) The potential for carry over may be reduced by destroying the subble, weeds and other alternate hosts of the stem borer by ploughing the field after harvest, (ii) Removal and destruction of dead-hearts and destruction of infested plants showing early pin-hole damage has been found to be successful practice in reducing the pest incidence. (iii) Destruction of crop residues and chopping of stems harboouring diapausing larvae could be very effective in reducing borer population (iv) Release laboratory bred adult populations of *Trichogramma* spp. (v) (a) Spray the crop 2-3 weeks after sowing or as soon as borer injury to the leaves is noticed with fenvalerate 20EC @ 100 ml/ha, cypermethrin 10EC @ 100 ml/ha or deltamethrin 2.8 EC @ 200 ml/ha. (b) Alternatively, the crop should be sprayed with any of the following insecticides; endosulfan 35EC @ 250 ml/ha or carbaryl 50WP @ 250 g/ha.

Push-Pull IPM

The term was first conceived by Pyke *et al.* in 1987 in Australia for the use of repellent and attractant stimuli, deployed in tandem, to manipulate the distribution *Helicoverpa* spp. in cotton. Later Miller and Cowles called it the Stimulo-deterrent diversion for the management of onion maggot, *Delia antigua*. Z.R. Khan *et al.* at the International Centre for Insect Physiology and Ecology, Nairobi, developed the concept for the management of cereal sem borers of food and forage crops (Khan *et al.*, 1997, Khan and Pickett, 2004). The technology has been especially successful for the management of *Chilo partellus* in maize and sorghum. Molasses grass, *Melinis minutiflora* and silverleaf, *Desmodium uncinatum* are used as push crops for repel ovipositing borer. Napier grass, *Pennisetum purpureum* and Sudan grass, Sorghum *vulgare sudanense* are grown as pull crops (trap plants and reservoir for NEs).

Arora (2008) evaluated the potential of Napier grass hybrid PBN-233 as border rows for management of stem borer in fodder maize. There were three treatments maize crop alone (unprotected), maize crop surrounded by two border rows of pearl millet-Napier grass hybrid and maize crop alone protected by application of two sprays (15 and 22 days after sowing) of 375 g ha^{-1} of Sevin 50 WP for protection against maize borer. It was recorded that mean incidence of stem borer was lowest (3%) in maize crop protected by border rows of pearl millet-Napier grass hybrid as compared to chemically protected crop (5%) and unprotected check (9%). Thus, cultural control of maize stem borer by planting border rows of pearl millet-Napier hybrid was found to be quit effective and economical.

19.4. SORGHUM SHOOT FLY, *Atherigona* spp. (Diptera: Muscidae)

Sixteen species of shoot fly have so far been recorded from Africa and Asia, out of these, only six species viz. *A. soccata, A.bidens, A.falcata, A.punctata, A.orientalis* and *A. nequii* are represented in India.

The sorghum shootfly, also known as the sorghum stemfly, is a widely distributed pest in Europe, Africa and Asia. In India, it is more serious in southern parts. Besides sorghum, it infests maize, wheat, broom corn, small millets (*Panicum* spp.) and grasses. It causes damage to the seedlings as well as to the early stages of the crop. The maggots bore into the stem and cut the main shoot. The high-yielding hybrids are more susceptible to the attack of this fly.

Life-cycle. The female fly lays approximately 40 eggs singly on the underside of the leaves during its life span of about one month. The eggs are elongate, flattened and somewhat boatshaped and are provided with two wing like lateral projections. The eggs hatch in 1-2 days and the tiny maggots creep out and reach in between the sheath and the axis, and bore into the stem. They feed inside the main shoot for 6-10 days and, when full grown, they may pupate either inside the stem or come out and pupate in the soil. The pupal period in the summer lasts about a week. Several generations are completed in a year. In northern India, the pest over-winters in the pupal stage.

Damage. The insect attacks the young crop when it is in the six leaf stage. Six weeks after planting, the crop is seldom attacked. As the maggots feed on the main shoot, the growing point is destroyed and by the time they pupate, the plant is almost dead. The young plants show typical dead-heart symptoms. When the attacked plants are somewhat older, tillers are produced, which mature later than the main crop. The total loss in yield is sometimes as high as 60 per cent. Cloudy weather favours the multiplication of this insect and it is believed that infestation is also higher in irrigated fields.

Management (i) The crop sown from early June to the second week of July normally escapes the attack of shoot fly. (ii) Grow resistant varieties like CHS-7, CHS-8, Indian sorghum types IS-5566, 5285 and

5613. (iii) Seed treatment with imidacloprid 70 WS or thiometham 70 WS @ 5 g/kg seed. (iv) Spray 250 g carbaryl 50WP in 125 litres of water/ha.

19.5. CABBAGE SEMI-LOOPERS, *Thysanoplusia orichalcea* (Fabricius) and *Autographa nigrisigna* (Walker) (Lepidoptera : Noctuidae)

These two polyphagous pests are widely distributed in north-western India and cause considerable damage to berseem and shaftal crops. They also attack cabbage, cauliflower and other winter vegetables. The caterpillars are plump and palish green. They cause damage by biting round holes into cabbage leaves. On walking, they form characteristic half-loops and are often seen mixed with cabbage caterpillars.

The adults of *T. orichalcea* are light palish brown with a large golden patch on each fore wing. They measure about 42 mm across the spread wings. The adults of *A. nigrisigna* are darker and have dark-brown and dirty-white patches on the fore wings.

Life-cycle. These insects are active during the winter and it is not known how they survive the heat of summer. During the active jperiod, they lay eggs on leaves of host plants and the caterpillars feed individually, biting holes of varying size according to the stage of their development. When full-grown, they pupate in the debris lying on the ground. The moths are very active at dust on flowers in gardens and public parks, where they are seen in hundreds during the spring season.

Management

(i) Timely cutting (30 days intervals) exposes the larvae to the predatory birds which feed on it. The pest is more serious in lodged crop where it multiplies rapidly (Table 19.3). A delay of 10 days in cutting more than doubles the larval attack while doubling the cutting interval to 60 days nearly quadruples the pest incidence.

Table 19.3. Harvesting stage and *T. orichalcea* in berseem

Cutting interval (DAS)	Larvae/m^2	Forage yield (t/ha)
30	2.05	21.9
35	3.55	22.0
40	5.45	19.9
45	7.50	15.3
60	7.90	11.1

Source : Dhaliwal (1992 a)

Table 19.4. Effect of intercropping on *T. orichalcea* attack in berseem

Crop	Caterpillar/m^2
Berseem sole	2.8
Berseem + ryegrass	0.5
Shaftal sole	3.5
Shaftal + ryegrass	1.0

Source: Dhaliwal (1992a)

19.6. BIHAR HAIRY CATERPILLAR, *Spilosoma obliqua* Walker (Lepidoptera : Arctiidae)

The Bihar hairy caterpillar is a sporadic pest and is widely distributed in the Orient. In India, it is very serious in Bihar, Madhya Pradesh, Uttar Pradesh and the Punjab as a polyphagous pest, particularly of berseem and shaftal crops. It also attacks cowpea, sesamum, *mash, mung,* linseed, mustard, sunflower and some vegetables.

Damage is caused by full-grown caterpillars which measure 40-45 mm in length and are profusely covered with long grayish hair. The moth measures about 50 mm across the wing spread. The head, thorax and under side of the body are dull yellow. The antennae and eyes are black.

Life-cycle. The pest breeds from March to April and again from July to November. It passes the hottest part of the summer (May-June) and winter (December to February) in the pupal stage amidst plant debris. Adults emerge from the over-wintering larvae in March. The moths are nocturnal and they mate during the night. The female lays 400 to more than 1200 light-green, spherical eggs in clusters on the underside of leaves. The eggs hatch in 8-13 days and during the first two stages, the tiny caterpillars feed gregariously, but afterwards they disperse widely in search of food. They grow to maturity through 7 stages, within 4-8 weeks. When full-grown, larva spins a loose silken cocoon in which pupation takes place in plant debris or in the soil. The pupal stage lasts 1-2 weeks in the active period and the moths live for about a week. The life-cycle is completed in 6-12 weeks and the pest passes through 3 or 4 broods in year.

Management

(i) Destroy weeds like gutputna (*Xanthium strumarium*), jangli palak (*Rumex aspra*) and wild castor (*Ricinus* sp.) adjoining to berseem and shaftal fields before sowing. These weeds act as alternate host plants of the pest which migrates from these weeds to the berseem crop (Table 19.5).

(ii) Sow the berseem and shaftal crops away from the fields of moong, mash, maize, groundnut and sesamum which harbour this pest during *kharif* season. The full grown larvae later on migrate to young Berseem and shaftal crops in September-October. Growing of radish or turnip for vegetable purpose on bunds of berseem and shaftal fields should be avoid to minimize the incidence of Bihar hairy caterpillar.

Table 19.5. Effect of weeds on *S. obliqua* in berseem

Treatment	Caterpillars/plant
Weeds eradicated	0.4
Weeds not eradicated	8.17
Population on weeds	
Xanthium strumarium	536.0
Ricinus sp.	473.0
Canabis sativa	3.0
Chenopodium album	2.4
Rumex aspra	1.4

Source : Dhaliwal (1992b)

19.7. LUCERNE WEEVIL *Hypera postica* (Gyln.) (Curculionidae: Coleoptera)

Young larvae first feed on the opening leaf buds at the tips of the plants in a concealed manner. The attacked plants show skeletonizing or shredding of the upper leaves. Grown up larvae eat up the upper foliage and the growing tips and make the growth stunted.

Life-cycle. Eggs are oval, less than 1 mm long, bright yellow when freshly laid, subsequently turns darker yellow, brown and black just prior to hatching. Newly hatched larva yellowish white with black head. Later on it becomes green with prominent, white mid-dorsal stripe. Full grown larva 9 mm long, legless but with well-developed ridges on the underside of the body which takes the place of legs. Larva pupates in loosely woven spherical cocoon. Adultiare brown with distinct snout projecting downward. It is 4-5 mm long with a darker stripe extending down its back more than half the length of its body.

19.8. GRASSHOPPERS

Various species of grasshoppers are widely distributed in India. They are polyphagous and feed on leaves of fodder crops in addition to rice, maize, millets, sugarcane, grasses, sunhemp, *arhar*, etc.

19.8.1. Rice Grasshoppers *Hieroglyphus banian* (Fabricius) and *H. nigrorepletus* Bolivar (Orthoptera : Acrididae)

Rice grasshoppers are somewhat like locusts but are smaller. The adults are 40-50 mm long and are shining greenish yellow, having three black lines running across the pronotum. Nymphs are yellowish, with many reddish brown spots in the early stages, but become greenish as they grow older. The nymphs emerge from the eggs after the first heavy rainfall during July-August. Initially they feed on the wild plants on or near the field bunds. Later they migrate to main crop and both the nymphs and the adults cause extensive defoliation. The insects prefer to hide in the central whorl and due to presence of excreta in the central whorl, the fodder gives a foul smell and is not eaten by the animals . Napier grass and Napier millet crops are preferred by the pest but maize and sorghum are also sometimes seriously damaged.(Arora and Shera, 2011).

Life-cycle. *Hieroglyphus* spp. are univoltine (one generation in a year) and pass the winter and dry part of summer in the egg stage. The eggs are found in the soil and they hatch in June or in early July, a few days after the first shower of the monsoon. On emergence, the nymphs start feeding actively and complete their development in seven stages, within 3 weeks. The adults are seen feeding voraciously during August and September. When they are two months old, they mate. The female starts laying eggs by inserting her abdomen in the soil. The eggs are laid 5-8 cm deep, in pods, each containing 30-40 eggs. The egg-laying continues from September to November and the adults die soon after, sometime in the winter.

19.8.2. Sporadic Grasshopper, *Oxya nitidula* Walker (Orthoptera : Acrididae)

Sporadic grasshopper is widely distributed in the Orient and is a pest of sporadic occurrence. In Punjab, it is serious in the low-lying areas of Amritsar, Gurdaspur and Hoshiarpur which are flooded occasionally. Damage is done by nymphs and adults which feed on various *kharif* crops, such as maize, sorghum, rice, sugarcane, etc. The nymphs are shiny green and have a broad, black band extending from the base of the eyes to the wing pads. The legs are greenish-yellow and the full-grown nymphs measure about 15-25 mm in length. The adults are similar in body colour but are larger (about 25-30 mm) and have green wings.

Life cycle. The pest is active from March to December and passes winter in egg stage in the soil. The over-wintered eggs hatch in March and the emerging nymphs feed on berseem, *khabbal* grass (*Cynodon dactylon*) and other available crops. They moult 6 or 7 times by June and are transformed into winged adults in 60-64 days. The adults also feed on various *kharif* crops and live for 40-49 days. They lay brown eggs in clusters on the underside of leaves of various food plants, particularly sugarcane. A female may lay 29-143 eggs, which hatch in 16-24 days. The nymphs of the second generation developing during July-August are transformed into winged adults in 46-57 days. These adults live for about 60-71 days and lay eggs in the soil instead of on plants. Some of these eggs hatch in October and give rise to third-generation nymphs which never complete their development and eventually die of cold in December. A majority of eggs, however, do not hatch and remain in the soil till next March. Thus, the pest complete two broods in a year.

19.8.3. Surface Grasshopper, *Chrotogonus trachypterus* (Blanchard) (Orthoptera : Acrididae)

This species of surface grasshoppers is widely distributed in the Orient and Africa. In India, it is common in the north whereas *C. oxypterus* occurs in the southern regions. The pests are very common in the field and assume serious proportions in certain years. The former causes considerable damage to the germinating cotton in April-May and the germinating wheat in November. The grasshoppers are polyphagous and, besides cotton and wheat, they feed on a number of other cultivated crops such as *berseem* sugarcane, *kharif* fodders, barley, etc.

Damage is done by the nymphs as well as the adults. The adults are stoutly build, about 20 mm long and 8 mm broad, with a rough brownish appearance. There are two longitudinal rows of black dots on the underside of the white body. The nymphs are similar in body form but are smaller and have wing pads instead of wings.

Life-cycle. *C. trachypterus* remains active throughout the year but its activity is considerably reduced during winter when only eggs and adults are found. A female may lay 36-434 eggs in 2-15 egg pods, at a depth of 3-6 cm in the soil. There are generally 16-46 eggs in a pod. In summer, the eggs hatch in 19-33 days whereas in winter they hatch in about 5 months. The developing nymphs feed on grasses or other soft foods. They moult 5-7 times and are full-fed in 40-70 days. Adults are long-lived, the female for about three months and the male about 1.5 months. There are two complete generations in a year and sometimes the eggs laid in October also hatch giving rise to a third partial brood, which dies in the nymphal stage during the winter.

Management

(i) Spores of *Nosema locustae,* a microsporidian are impregnated in wheat bran flakes and applied on field. Grasshoppers get infected in 1-3 weeks. Nymphs are most susceptible and succumb to the disease easily. The adults become lethargic and unsteady, and are easy prey for birds as well as healthy grasshoppers.

(ii) *Metarhizium acridum* has been developed as a fungal biopesticide against locusts and grasshoppers. However, both these products are not available in the Indian market. Deltamehrin 2.8 EC and acephate 75 SP at 200 ml and 1 kg /ha have been found effective against both the nymphs and the adults. Destruction of field bunds after termination of the crop reduces the incidence of *Hieroglyphus* spp. during the next year.

19.9. SORGHUM MIDGE, *Contarinia sorghicola* (Coquillett) (Diptera : Cecidomyidae)

This insect is distributed in all the sorghum growing tracts of the country and causes considerable losses. The adult fly is a very small (2 mm) fragile mosquito-like insect with a bright orange abdomen and a pair of transparent wings. The damage is caused by maggots which feed on the ovaries and thus preventing the formation of grains. The infestation is dependent on humidity (more sever in low lying humid areas), initial infestation and duration of flowering of sorghum.

Life-cycle. The female inserts the eggs singly into developing florets at the time pollen is being shed and lays about 30-100 eggs. The eggs hatch in 2-3 days and the maggot feeds inside the developing grain and pupates there itself. The larva has four instars with a total duration of 10-11 days. It emerges between the tip of the glumes leaving the white pupal case attached to the tip of the floret which is very characteristic. The pupal stage lasts about a week and the total life-cycle is completed in 14-22 days.

Management. (i) Both early and late maturing varieties of sorghum should not be grown in the same area as it would provide the pest a continuous supply of flowers. The varieties having the same flowering and maturity time would reduce midge damage considerably. (ii) Since the damaged and aborted seeds are he main source of carry over of the pest from one season to another, collect and burn the panicle and post-harvest trash. (iii) Spray 1 litre of malathion 50EC or endosulfan 35EC or 1.25 litres of lindane 20EC or 2.0 kg of carbaryl 50WP in 500 litres of water/ha. Only the earheads should be treated at 90 per cent panicle emergence followed by second spray after 4-5 days. (iv) Dusting the earheads with endosulfan 4% or carbaryl 5% or lindane 2% @ 12 kg/ha has also proved effective.

19.10. APHIDS

Several species of aphids attack fodder crops. Black aphid *Aphis craccivora* Koch (*Aphididae : Hemiptera*) has been recorded on lucerne, shaftal, senji, cowpea and cluster-bean, spotted alfalfa aphid *Therioophis trifolii* (Monell) and pea aphid *Acyrthosiphon pisum* (Harris) on lucerne and cereal aphids *Rhopalosiphum padi* (Linn.), *R. maidis* (Fitch), *R. rufiabdominalis* (Sasaki) and *Schizaphis graminum* (Rondani) on oats. *R. maidis* and *Aphis sacchari* Zehnt. are important on sorghum.

Damage. In case of legumes, the colonies of aphids suck the sap from the underside of leaves and from terminal shoots. Severely attacked shoots get withered and blighted. Cereal aphid populations usually began their development on the lower leaves and in the leaf sheaths of oats but often move up the plants as the population increased. Large populations also fed on the inflorescence especially on the branches of the panicles at the glume bases.

Management. (i) Aphids are kept in check by predators like coccinellid beetles and syrphid flies on fodder crops. (ii) Heavy rainfall also washes down the aphids from the crop. (iii) In case of oats early sown upto (Oct. 10) crop suffers less aphid attack than the late sown crop. (iv) In case of severe attack, spray the crop with 100 ml/ha of imidachloprid 200SL or 50 g/ha of acetamiprid 20SP or 100 g/ha of thiomethoxam 25 WG.

19.11. BLISTER BEETLE *Mylabris pustulata* Thunb. (Coleoptera: Meloidae)

It is an important pest on sorghum and cowpea crops. Adult blister beetles are conspicuous by their large size and bright colours ranging from metallic blue, green, black and red yellow or brown. Their feeding on inflorescence results in poor seed set (Taneja and Nwanze, 1994).

Management. (i) Raise the crops away from cow-dung pits which act as reservoirs of the pest. (ii) Collection and destination of adults in kerosenized water is effective in reducing damage. (iii) Carbaryl and endosulfan sprays have been found effective for its control.

19.12. CHAFER BEETLES

The important species of chafer beetles which feed on sorghum and pearl millet in India are *Chiloloba acuta* Walker and *Oxycetonia versicolor* Fab. The adults feed on pollen and stigma resulting in poor seed set in the seed crop (Taneja and Nwanze, 1994).

Management. Collection and destruction of the adult beetles by putting them in kerosenized water minimizes the damage of the pest in seed crop of sorghum and pearl millet.

19.13. SUGARCANE LEAFHOPPER, *Pyrilla perpusilla* Walker (Lophopidae : Hemiptera)

The sugarcane leafhopper, *Pyrilla perpusilla* Walker, sometimes damages the fodder crops of sorghum, maize and millets. The nymphs and adults suck sap from the lower surface of the leaves leading to withering of affected plants. A female lays about 600-800 eggs in batches in the leaf sheath or on the undersurface of leaves along the midrib. The egg batches are covered with fine whitish hairs produced from the female's, anal pads. Incubation period is 7-10 days and the nymphal duration lasts for 34-52 days. There are 3-5 overlapping generations in a year.

Management. Conservation, augmentation and mass release of *Epiricania melanoleuca* Fletcher effectively suppressed *Pyrilla* without the need for insecticide application.

(ii) Hand collection of egg masses reduces pest build-up.

19.14. JASSIDS

Cereal Jassid, *Zygnidia manaliensis* (Singh) (Hemipera: Cicadellidae)

In recent years, the so called 'cereal jassid' has attained the status of sporadic pest in northern India on fodder maize sown during spring season. The pest has also been reported on sorghum, pearl millet and sudan grass. Both nymphs and adults suck the sap from lower side of the leaves where characteristic silver stipplings appear in due course of time. With the extensive feeding by the insects, the size of the stipplings increases resulting in the loss of chlorophyll and the leaves turn white with scortched margins. The insects generally attack the lower leaves of the plants or the newly emerged leaves of the growing plant. Heavily infested plant gives a sickly appearance and at times the growth is completely stunted. There is a loss of 30-70 per cent in the protein content and the plants loose their fodder value (Mathur, 1994).

Empoasia motti **Pruthi** damages cluster bean, berseem, lucerne and pearl millet. *Distantasca terminalis* (Distant) attacks cowpea, lucerne, berseem, maize, etc. Both the species feed on the lower side of leaves which became yellow followed by curling and crumpling.

Management. In case of severe attack, spray the insecticides mentioned for control of aphids.

19.15. THRIPS

Thrips attacking fodder crops include lucerne thrips *Thrips hawaiiensis* (Morgan) (Thripidae : Thysanoptera), Oats thrips *Haplothrips ganglbaueri* (Schmutz.) (Thripidae : Thysanoptera), cowpea thrips *Megalurothrips distalis* Karny (Thripidae : Thysanoptera) and sorghum thrips *Anaphothrips sudanensis* Trybom (Thripidae : Thysanoptera) (Sandhu, 1976). All of them except the sorghum thrips attacked the flowers which dried up and were shed prematurely in lucerne and cowpea. Sorghum thrips desapped the young leaves resulting in their whitening.

Management. In case of severe attack spray the crop with 100 ml/ha of imidacloprid 200 SL or 375 ml/ha of novaluron.

19.16. LEAF-MINERS

The leaf-miners attacking fodder crops include pea leaf-miner *Chromatomyia horticola* (Goureau) (Agromyzidae : Diptera) on berseem and lucerne, cowpea leaf-miner *Acrocerops* sp. (Gracillariidae : Lepidoptera) on cowpea and groundnut leaf-miner *Aproaerema modicella* Deventer (Gelechiidae : Lepidoptera) on lucerne. All three species mined the leaves and cowpea leaf-miner caused blister-like blotches on the leaves. Groundnut leaf-miner also webbed the leaves together (Sandhu, 1976).

Management. In early stage, remove and destroy the leaf-miner infested leaves to check the build up of the pest.

19.17. ALFALFA SEED CHALCID, *Bruchophagus roddi* Grussakovsky (Eurytomidae : Hymenoptera)

The larvae of this insect feed inside the developing seeds of lucerne. Adults emerge from the seeds when the crop is near maturity. Sometimes seed crop is badly damaged.

Management. Insecticides recommended for management of gram caterpillar in alfalfa seed crop also control this pest.

19.18. FUTURE RESEARCH NEEDS

Insect pests cause huge losses in tonnage as well as quality of fodder crops. Pesticides have only limited acceptability for management of these pests. Moreover, a waiting period of 2-3 weeks is usually required before the treated crop can be harvested for use as fodder. Many times such a long waiting period is not practicable or acceptable to farmers. Therefore, there is a need to strengthen non-chemical approaches and recommend biopesticides for management of insect pests in fodder crops. Some of the newer and safer pesticides with short waiting periods may also fit into the IPM programmes in these crops.

REFERENCES

Arora, R. (2008) Evaluation of pearl millet - Napier grass hybrid as a trap crop for management of *Chilo partellus* in fodder maize. *Proc. Second Insect Sci. Congr: Contributed Papers*. Feb. 21-22, 2008, ISAIS, Punjab Agric. Univ., Ludhiana, pp. 174-75.

Arora, R. and Dhawan, A.K. (2010) Efficacy of selected insecticides for the management of *Helicoverpa armigera* (Hubner) on Egyptian clover seed crop. *J. Insect Sci.* **23:** 81-86.

Arora, R. and Shera, P.S. (2011) First record of rice grasshopper *Hieroglyphus nigrorepletus* damaging Napier grass and Napier millet. *J. Insect Sci.* (In Press).

Atwal, A.S. and Dhaliwal, G.S. (2005) *Agricultural Pests of South Asia and their Management.* 5th Revised ed., Kalyani Publishers, New Delhi.

Dhaliwal, G.S. and Arora, R. (2006) Integrated Pest Management: Concepts and Approaches. 2nd Revised ed., Kalyani Publishers, New Delhi.

Dhaliwal, J.S. (1992a) Effect of cultural practices on cabbage semilooper *Thysanoplusia orichalcea* (Fab.) infesting Berseem. *Proc. Natn. Symp. Recent Advances in Integrated Pest Management.* Oct. 12-15, 1992, ISAIS, Punjab Agric. Univ., Ludhiana, pp. 30-31.

Dhaliwal, J.S. (1992b) Role of some weeds in the carry over of Bihar hairy caterpillar, *Spilosoma obligua* Walker to Egyptian clover. *J. Res. Punjab Agric. Univ.* **30:** 168-70.

Dhaliwal, J.S. (1998) Recent advances in integrated pest management of forage crops. In: A.K. Mukherjee, S. Maiti and M.K. Nanda (eds). *Strategy for Maximization of Forage Production.* BCKV, Kalyani, pp. 169-73.

Dhaliwal, J.S. and Sidhu, D.S. (1988) Effect of adjoining host crops, last cut and natural enemies on *Heliothis armigera* (Hubner) infesting berseem (*Trifolium alexandrinum* L.) seed crop. *Indian J. agric. Sci.* **58:** 832-86.

Hassanali, A., Herren, H., Khan, Z.R., Pickett, J.A. and Woodock, C.M. (2008) Integrated pest management : the push-pull approach for controlling insect pests and weeds of cereals, and its potential for other agricultural systems including animal husbandry. *Phil. Trans. R. Soc.* **363:** 611-21.

Khan, Z.R. and Pickett, J.A. (2004) The 'push-pull' strategy for stem borer management: a case study in exploiting biodiversity and chemical ecology. In: G.M. Gura, S.D. Wratten and M.A. Altieri (eds). *Ecological Engineering for Pest Management: Advances in Habitat Manipulation for Arthropods.* CAB International, Wallingford, pp. 155-164.

Khan, Z.R., Chiliswa, P., Ampong-Nyarko, K., Smart, L.E., Polasjek, A., Wandera, J. and Mulaa, M.A. (1997) Utilization of wild gramineous plants for management of cereal stem borers in Africa. *Insect Sci. Appl.* **17**: 143-150.

Kogan, M. (1998) Integrated pest management : Historical perspectives and contemporary development. *A. Rev. Ent.* **43:** 243-70.

Mathur, L.M.L. (1994) Advances in insect pest management in maize In : G.S. Dhaliwal and R. Arora (eds). *Trends in Agricultural Insect Pest Management.* Commonwealth Publishers, New Delhi, pp. 113-160.

Miller, J.R. and Cowles, R.S. (1990) Stimulo-deterrent diversionary cropping : a concept and its possible application to onion maggot control. *J. Chem. Ecol.* **16**: 3197-3212.

NAS, (1969) Principles of Plant and Animal Pest Control. Vol. 3. Insect Management and Control. National Academy of Sciences, Washington, USA.

Sandhu, G.S. (1976) Some insects recorded as pests of fodder crops at Ludhiana (Punjab). *J. Res. Punjab Agric. Univ.* **19**: 449-459.

Taneja, S.L. and Nwanze, K.F. (1994) Current status and strategies for management of insect pests in sorghum and millets. In : G.S. Dhaliwal and R. Arora (eds). *Trends in Agricultural Insect Pest Management.* Commonwealth Publishers, New Delhi, pp. 161-85.

Theory and Practice of Integrated Pest Management
R. Arora, B. Singh and A.K. Dhawan (Eds)
Scientific Publishers (India), Jodhpur 2011, pp. 338 - 369.

20

MANAGEMENT OF INSECT PEST COMPLEX IN MAIZE

Vijay Kumar and Uma Kanta
Department of Entomology
Punjab Agricultural University, Ludhiana - 141 004, India

20.1. INTRODUCTION

Maize or Corn (*Zea mays* L.) originated in America and is now principle cereal crop in tropics and subtropics. In world it ranks third among the cereals after wheat and rice in terms of area and production. In India it is extensively grown for food grains, feed and fodder purposes and is cultivated in an area of 70 lakh hectares with a production of 140 lakh metric tonnes (Anonymous, 2004a). The average yield of the crop in India is 20.0 quintal per hectare, which is very low as compared to average yield of 49.06 quintal per hectare in the world (Anonymous, 2004b). Maize is traditionally grown for grain and fodder production in India during the *kharif* (monsoon) season under rainfed conditions. However, during the last decade better yields have been realized from the *rabi* (winter) crop grown under irrigation. Maize is also grown as fodder in the Indo-Gangetic plains and summer maize in Punjab during the spring season.

The availability of maize at one or more stages throughout the year added new dimensions to the pestilence front. Insects hitherto unknown to attack the maize have become problems. A number of insects cause minor to serious damage to maize crop. Sarup *et al.* (1987) listed 130 such insect species. However, the total number of insect and mite pests exceeds 250 (Mathur, 1991). Of the various insect species, only about dozen of these are quite serious and require control measures. Right from sowing till harvest, maize crop suffers from depredation of these pests. All parts of maize plant, i.e. root, stem, leaf, grain (milky and ripe) and seeds are attacked and damaged by these pests resulting in either partial or total yield loss. The insects namely *Chilo partellus* (Swinhoe), *Sesamia inferens* (Walker) and *Atherigona naqvii* (Steyskal) and *A. soccata* (Rondani) have been recorded as most serious pests of this crop and wide spread in India causing substantial reduction in crop yield. Amongst the most serious pests are the tissue borers. The insect pests of economic importance and their distribution in the country are given in Table 20.1. The multiple pest complex of maize crop posses a serious limitations in the intensification of maize cultivation in different agro-climatic regions of India.

Various methods of pest control, namely mechanical, cultural, biological and chemical, have been developed to check the damage due to different insects in maize. There has been more emphasis on chemical control. Several insecticides have been identified that provided fairly successful control. The chemical control measures however, are often not adopted by the farmers to the desired extent due to various reasons. The farmers are poor, illiterates, not well informed of the plant problems and have small holdings. Moreover, maize crop has low economic return and insecticide uses cause environmental pollution, residue problems, development of resistance, pest resurgence, secondary pest outbreak and destruction of natural enemies of pests. The insecticide application needs to be discouraged, as the grains and the fodder are directly consumed by human beings and milch animals, respectively. So, there is need to develop Integrated Pest Management (IPM) strategies for the control of economically important insect pests of maize.

20.2. STEM BORER, *Chilo partellus* (swinhoe) (Crambidae: Lepidoptera)

The maize stem borer, *C. partellus* is the key pest throughout India.

Distribution. The maize stem borer, *C. partellus* is most serious and widely distributed pest of maize throughout the subcontinent and throughout South East Asia to Indonesia and Taiwan. The pest first appear in East Africa in the Early 1950 and has now spread as for as in Northern Sudan, Botswana, Zaire, Sudan to South Africa (Harris, 1985).

Host plants. It is a major pest of *Jowar* and maize and at times also reported from sugarcane, *Bajra* and rice. In addition to these, it has been recorded from Sudan grass (*Sorghum vulgare* var. *sudanensis* Stapf.), Nachini (*Eleusine coracana* Guratu), Baru-Johnson grass (*Sorghum halepense* Pers), Job's tears (*Coix lachryme*- Jobi L.), Kawadia or Burgur (*Polytoca barbata* Stapf), pearl millet, finger millet and wild grasses etc.

Table 20.1. Important insect pests of maize in India

Common name	Scientific name	Distribution
Major pests		
Maize stem borer	*Chilo partellus* (Swinhoe)	Throughout the country
Pink stem borer	*Sesamia inferens* (Walker)	South India and Punjab
Shoot fly	*Atherigona naqvii* Steyskal; *A. soccata* Rondani and *A. orientalis* Schiner	Uttar Pradesh (Tarai area), Andhra Pradesh, Rajasthan, Punjab and Himachal Pradesh North India
Occasional pests		
Armyworm	*Mythimna loreyi* (Duponchal) *M. separata* (Walker)	South India and Punjab Himachal Pradesh, Punjab and Karnataka
Corn earworm	*Helicoverpa armigera* (Hubner)	Throughout the country
Silk cutter	*Euproctis virguncula* (Walker)	Throughout the country
Cut worm	*Agrotis* spp.	Kashmir, Tarai area of Uttar Pradesh and Bihar.
Climbing cutworm	*Rhyacia herculea* (Corti and Draudt)	North Bihar
Rice yellow hairy caterpillar	*Psalis pennatula* (Fabricius)	Throughout the country
Sugarcane leaf hopper	*Pyrilla perpusilla* Walker	Punjab, Uttar Pradesh, Bihar and Maharashtra
Maize jassid	*Zyginidia manaliensis* (Singh)	Punjab and Himachal Pradesh
Cereal thrips	*Anaphothrips sudanensis* Trybom	Uttar Pradesh, Rajasthan and Punjab
Cotton grey weevil	*Myllocerus undecimpustulatus maculosis* Desbrochers	Throughout the country
Termite	*Microtermes obesi* Holmgren, *Odontotermes obesus* Ramb. & other *Odontotermes* spp.	Uttar Pradesh, Bihar and Assam Throughout the country
Pentatomid bug	*Dolycoris indicus* Stall	Uttar Pradesh, Punjab and South India
White grub	*Holotrichia consanguinea* Blanchard	Karnataka, Gujarat, Rajasthan, Uttar Pradesh and Bihar
Toka	*Chrotogonus* sp.	Throughout the plains of North India except Punjab

Phadka grasshopper	*Hieroglyphus nigrorepletus* Bol.	Rajasthan, Maharashtra, Madhya Pradesh, Uttar Pradesh, Bihar and Punjab
Rice grasshopper	*H. banian* F.	Rajasthan, Maharashtra, Madhya Pradesh, Uttar Pradesh, Bihar and Punjab
Paddy grasshopper	*Oxya velox* F.	Andhra Pradesh, Maharashtra, Madhya Pradesh, Tamil Nadu, Orissa, Punjab and Rajasthan
Minor pests		
Corn aphid	*Rhopalosiphum maidis* Fitch	Throughout the country
Leaf roller	*Marasmia trapezalis* Guence	Uttar Pradesh and Punjab

Identification. The adult moth is dull brown and is about 2.5 cm in wing expanse. The forewings are pale brown and the hind wings are white. The eggs are small, flat oval, and creamy-white arranged in rows overlapping one another in a mass. An egg mass contains about 30 eggs. These are laid generally on the lower surface of the leaves. Morethan one egg mass may be found on a single leaf or plant. The freshly hatched larvae are tiny, about 2 mm long. The full-grown caterpillar is dirty brown with black head and is about 2.5 cm long. It carries 4 longitudinal stripes on its body. Pupa is brown and is found usually in the stem tunnel.

Life-cycle. Various workers have studied the biology of this pest. According to Butani (1955) egg, larva and pupal stage lasts for 4-5, 18-20, 6-8 days, respectively, and life cycle is completed in 29-33 days in case of non-hibernating stage during summer while corresponding figures are 8-9, 28-93, 9-24 and 83-210 days, respectively, for hibernating stages in winter in India. According to Singh (1963) incubation, larval and pupal period of this pest is 4-8, 19-35 and 7-9 days, respectively. The average fecundity per female is 87 to 110 eggs.

It is active throughout the year in the southern region of the country. But in the northern states, it passes winter as hibernating larvae from October to February. The larvae hibernate in the stalks and stubbles of maize, *jowar, Bajra* and Baru and in cobs. The percentage of hibernating larvae is 76.7, 13.5 and 9.8% in stubbles, stems and cobs of maize, respectively. As the season warm up after winter, the larvae change to pupae. By the end of march, the moth starts emerging from the pupae. They mate generally during the early hours of the day. The female lays eggs during evening hours on the maize and bajra sown for fodder. A female lays about 300 eggs in its lifetime. These eggs hatch in 3-4 days to produce the tiny caterpillars. These larvae scrap on the tender whorl leaves and shortly thereafter enter into stem through the whorl and feed there for about 3-4 weeks. The full-grown larvae transform into pupae in the stem tunnels. After about week, the moth emerges from the pupae again starts the life cycle. After completing about two generations on the spring sown fodder crops (Maize and *Bajra*), this pest shifts to main *kharif* maize crop by the middle of June or so and remains active there up to September-October. Some of population hibernates but some larvae, pupae and the moths shift to the late sown fodder to pass another generation.

Field behavior of maize stem borer. It prefers to lay more eggs on younger crop of about one to four weeks old than the older one. There is negative correlation with the age of the plant. The application of insecticides should, therefore, be restricted on about to 10 to 30 days old crop. Use of chemicals on older crop may not be of much use.

About one third of the infested plants show dead hearts in about 3-4 days after appearance of infestation symptoms. The first application of insecticides must, therefore be done within 2-3 days of appearance of symptoms of attack to avoid dead heart (drying of central whorl of the plant) formation.

In young crop about 80% of the infested plants receive primary or direct attack by small larvae, newly hatched from the eggs, laid on plants and such infested plants carry heavy larval population (10-20 larvae per plant). So from the young crop the infested plants should be removed and destroyed at hoeing to kill the larvae from the primary infestation.

The attack usually occurs in patches. Hence, the insecticides must be used as 'spot treatment' only in the whorl of the infested plants in a patch.

Since the pest passes winter as hibernating larvae in the stalks and stubbles of maize, *jowar* and *Bajra* and in the maize cobs, the practice of ploughing of maize and jowar fields soon after the harvest of these crops, disposing of their stalks, cobs and use of healthy cobs for seed would reduce the carry over of the pest and its further multiplication to a great extent.

Economic threshold. The severity of damage by the maize stalk borer is significantly more in 10-day old crop as compared to the later stages of the crop. Any control measures after 30 days of germination may not prove economical. The threshold of economic injury level for maize stem borer on spring maize crop was above 4 per cent infestation (Ghouri *et al.,* 1978) and in Kharif maize was 2.0 per cent dead heart formation (Singh and Uma Kanta, 2002).

Nature of damage. It attacks every part of the maize plant except roots. Newly hatched larvae first scrap the central leaves of the whorl of the maize plant and then tunnel into the stem throughout the whorl. The affected leaves shows small pin holes and scraped leaf injury. Grown up larvae produce bigger holes or slit holes in the whorl leaves. They may pass tunnel through the stem directly. The severe attack results in stunted growth, deadhearts and stem breakage and consequently the yield loss. The larvae also damage the emerging tassels and developing grains in the cobs.

Losses. In fact, it is a limiting factor in the successful cultivation of this crop. This borer is reported to cause 24 to 83 percent loss in India (Chatterje *et al.,* 1969; Sarup, 1973; Mathur and Rawat, 1981). The losses due to *C. partellus* ranged from 27.6 to 80.4 percent (Sarup *et al.,* 1979) and combined avoidable losses in maize yield due to stem borer and shoot fly has been estimated to be 20 to 87 per cent (Mathur, 1992; Pathak *et al.,* 1971).

Management

Keeping in view the studies on the field behavior of maize borer, many comprehensive schedules of management were developed against this pest.

Resistant germplasm / Sources of resistance. The development and use of variety resistant to *C. partellus* is an ideal approach to manage this pest. Extensive evaluation of diverse germplasm has been carried out under artificial infestation of maize stem borer. Most of the work has been done at Punjab Agricultural University (PAU), Ludhiana and Indian Agricultural Research Institute (IARI), New Delhi and some work has been done at G.B. Pant University of Agriculture and Technology (GBPUAT), Pantnagar, Rajasthan Agricultural University (RAU), Udaipur and Andhra Pradesh Agricultural University (APAU), Hyderabad. These studies have led to the identification of some relatively resistant germplasm (Table 20.2). Most outstanding example of host plant resistance is Antigua Gr 1 (CM 500), a Caribbean introduction. Many introductions from CIMMYT showed promise and have contributed to the maize breeding programmes in India (Table 20.2).

Table 20.2. Germplasm showing relatively resistant reaction to maize stem borer.

Germplasm	Locations	Reference
Comp. A 53 (SA) x Comp A 54 (EV) RU 21, EH 2230, EH 3136, J 22, Opaque B-15	Ludhiana	Anonymous (1973)
A 6, A 21, Amarillo Cristalino-1, Antigua Gr 1, Antigua Gr 2 Sel, Blanco, Antigua 7 D, Antigua 8 D, British Virgin Island 117, Caribbean Flint Comp, Cuba 9, Cuba 12, Cuba 40, Gautemala 257, M 512, MCPD 6, Mezcla Amarillo Baja,	New Delhi	Sarup *et al.* (1978)

Serie SB, R2CII, Thai DMR Comp 17, V520CA, CISTRON, EA1712 (late), FV 147x BUP116, FV147x ZP2077, K 10 x 2 PR 588, LP 1712 x ZPR 588, MR 21x R 588, MR 21x SD 10, O 118 ax BUP 43, SD 10 x BUP 116, Syn 60j, T 146 x BUP 116, T 146x SD 10, T 166x ZP2077/54, T 116x ZPR 588, T 169, T 341x WF 9, U 221x ZPR 588, VTR 116x ZPR 588, YUCP 2077/54		
Amber, Deccan 103, Sona, Vikram	Pantnagar	Sarup *et al.* (1979)
Harnampur Local, Kesri Local, BS 20, Iowa Long Ear Syn	New Delhi	Sarup *et al.* (1981)
Honey June, NC 59663, Pool 15, Pool 16, Pool 17, Pool 19, Pool 24, Pool 25, Pool 26, Pool 27, Pool 28, Pool 29, Pool 30, Pool 32, Pool 33, Pool 33 QPM, Tuxperate x Tropical QPM (Dent), Tuxpeno QPM	Ludhiana	Annonymous (1984, 1985)
Tuxpeno QPM, Antigua Gr 1, Mex 17	New Delhi	Durbey and Sarup (1985)
Hunius, BS7, BS 8, BS 14, Cooks Early Yellow Dent	New Delhi	Panwar and Sarup (1985)
Ganga 5, Antigua Gr. 1, J 22, J 605	Ludhiana	Sekhon (1985)
Ageti 76, Deccan Hybrid, Ganga 5, Ganga 2, Jawahar, Kisan, Tarun, Vijay, Vikram Amarillo Pak, Caribbean Flint Comp, Cuba 11 J, D 818, Golden Crystal, H 207, Hybrid Vaznyl, Mo x 117, Mo x 57, N 21, N 22, Pop 31, PR 7921, Suwan 7528	Pantnagar	Sharma (1987)
Chandan, Deccan 103, Ganga 5, Jawahar Comp, Kundan	New Delhi	Sarup *et al.* (1987)
Bulandshahar Local, Meerut Yellow Local, Saharanpur Local, Monghia Local, Gore Local	Pantnagar	Sharma (1987)
Ganga 4, Dhawal, Hunius, Jawahar x Thai	New Delhi	Singh (1988)
Arun (A 68), Kiran (J 660), Tarun, D 791, EH 5041, J 3022, Composite, Pool 15, Pool 17, Pool 27, (J 54x Mo 17 B.57)-17-1-2-2-1-1-1 #, SW1 (S) C6-40-1-1-1-2 #, SW1 (S) C6-53-1-1-1-2-2 #	Ludhiana	Uma Kanta and Sekhon (1989)
(J 54 x Mo 17. B 57) -17-1-2-2-1-1#, Suwan 1(s) C6-40-1-1-1-2-1#, Suwan 1(s) C6-53-1-1-1-2-2#, Arun, D 791, Kiran, Pool 17, Pool 27	Ludhiana	Uma Kanta and Sekhon (1994)

Cultural control

- It is considered as first line of defence against the pest that includes the manipulation of the date of sowing (Sarup *et al.*,

1978 and Goyal, 2005), higher plant population (Panwar and Sarup, 1980; Sharma and Sarup, 1980) and use of sorghum as a trap crop for the control of *C. partellus* in maize (Sarup *et al.,* 1977), use of two rows of teosinte and *bajra* as a barrier crop (Goyal, 2005) two rows of napier bajra hybrid (Kumar & Kanta, 2008) can be effective means for achieving the targeted yield of maize. Early sowing (mid june) coupled with higher plant population (66,000/ha) under assured irrigation significantly increased the maize yield (Sarup *et al.,* 1978). But the early sown maize crop got relatively more attack of the maize borer. So the farmers try to sow the crop late to avoid attack of this pest.

- Intercropping of maize with pulses, viz. soyabean, cowpea and green gram reduce the incidence of *C. partellus* than pure crop of maize. The yield was also higher in intercropping of maize with all the pulses (Srinivas and Panwar, 2003).
- Plough up the field soon after harvesting the maize, jowar and *Bajra* crops. Collect and destroy the stubbles. Use the stalk by the end of February and chop the remaining stalks for subsequent use. The cores of maize cobs may harbour the borer larvae. These should also be destroyed by the end of February. For seed keep only healthy cobs free from borer.
- Use the recommended seed rate.
- Remove and destroy the plants showing severe borer injury while hoeing the crop.
- The manipulation of soil moisture with the help of controlled irrigation may influence crop damage by insects through its effect on plant vigour and growth. However, in rainfed agriculture, there is little scope of manipulation of soil moisture.
- The uses of fertilizers enhance the plant nutrition, which influence the susceptibility of the crop to the insect.
- Crop rotation with non-host crops reduces the pest infestation by interrupting the continuity of the food chain.

Natural enemies. *Cotesia* sp. (*Apanteles* sp.) is a dominant larval parasite of *C. partellus* throughout India. It is known to govern the population density of this pest to some extent. High fecundity strain of *Trichgramma exigua* caused 98 per cent mortality of the eggs of *C. partellus*. Verma and Saxena (1989) reported the possibility of effectiveness of an exotic parasite, *Allorphogas pyralophagus* in managing population of *C. partellus* in field condition. Some predator viz., *Menochilus sexmaculata, Coccinella undecimpunctata* and different species of spiders are known to predate upon the larvae of this pest but their potential in the biological control has not been utilized (Sarup *et al.,*

1987). Recently Jalali and Singh (2003) reported that in India, *Trichogramma chilonis* Ishii as an egg parasitoid was found most promising for field studies and *Cotesia flavipes* (Cameron) was the promising larval parasitoid in fodder maize ecosystem. The release of *T. chilonis* at 3-day interval was the best treatment to obtain higher yield in *kharif* fodder maize as compared to 5-day interval releases @ 1.0 lakh per hectare. Singh *et al.,* and Uma Kanta *et al.* (2004) also reported that releases of *Trichogramma chilonis* @ 1.0 Lakh/ha were effectively controlling the incidence of *C. partellus* under Punjab conditions. *Bacillus thuringiensis* strains particularly HD-73 and HD-263 were highly toxic to *C. partellus* larvae and can be exploited for control in field conditions (Khanna *et al.,* 1995).

Chemical control. Spraying the crop with Sevin (Carbaryl 50 WP) @ 1.68 kg a.i./ha four times at weekly interval, starting from 15 days after sowing, reduced the dead hearts infestation (Young, 1962). Atwal *et al.,* (1966) found that four sprays of carbaryl 0.2 per cent at 10-day interval after germination reduced the maize borer incidence. Treating the seed of maize with BHC (13% gamma isomer) @ 5, and Lindane 1% (granules) @ 20 g/100 gm protected the crop upto 11 and 15 days, respectively. The higher doses of BHC were phytotoxic (Lal *et al.,* 1961). For the control of the borer, mephosfolan 5 G @ 0.375 a.i. /ha or carbafuron 3 G @ 0.225 Kg @ 0.225 kg a.i./ha should be applied in foliar whorl in 15-25 days old plants. Carbfuron 3 G @ 5 kg/ha after 15 days of germination was also effective against borers. If necessary, foliar whorl application may be followed by Mephosfolan 0.25 kg a.i./ha after a month. Endosulfan (35EC + 4G) and Carbafuron (3G or 40 WP) also effectively controlled the stalk borers. Treatment with endosulfan 35 EC 0.1% @ 450 l/ha and 4G @ 5 kg/ha in the form of either spray+ granules or granules, 10-15 and 25-30 days after germination were effective (Bharadwaj *et al.,* 1980).

Integrated pest management

- Kill the borer larvae hibernating in plant remnants like stubbles, stalks, cobs and cores. Plough up the fields after harvesting the maize crop. Collect the stubbles and burn them. Use maize stalks, cobs and cores by the end of February. Chop the remaining stalks, if any, for subsequent use. For seed, keep only healthy cobs, free from borer attack.
- Remove and destroy the plants showing severe borer injury, while hoeing the crop.
- Spray the crop 2-3 week after sowing or as soon as borer injury to the leaves is noticed using 60-litre water per acre with knapsack sprayer with any of the following synthetic pyrethroids:

Sumicidin 20 EC (fenvalerate)	40 ml
Ripcord 10 EC (cypermethrin)	40 ml
Decis 2.8 EC (deltamethrin)	80 ml

Normally no additional spray or granular application is required after the spray with pyrethroids.

OR

Spray as above with any of the following insecticides:

Sevin 50 WP (carbaryl)	100 g
Lindane 20 EC (gamma BHC)	200 ml
Thiodan 35 EC (endosulfan)	100 ml
Nuvacron 36 SL (monocrotophos)	110 ml

After spray apply twice, (at 7 to 10 days interval) Lindane 6 g @ 200-500 g per acre in the whorl of only infested plants that show fresh injury in the central leaves through bottle with few holes in its cap (Anonymous, 2005).

20.3. PINK STEM BORER, *Sesamia inferens* Walker (Noctuidae : Lepidoptera)

It is the second major tissue borer of maize. *S. inferens* causes serious damage to maize in southern India. It remains active throughout the year, but damage is more serious during the *rabi* season. The incidence of pink stem borer was noticed for the first time at PAU Ludhiana during the *rabi* season of 1981-82 when its average incidence was hardly 0.5 per cent.

Distribution. This species occurs in Burma, China, India Indonesia, Malasia, Srilanka, Taiwan and Phillipines (Teetes *et al.,* 1983). In India it is distributed in Andra Pradesh, Bihar, Madhya Pradesh, Maharashtra, Karnataka, Orrissa, Punjab, Uttar Pradesh, Harayana, and West Bengal states (Chundurwar, 1989).

Host plant. Besides maize, it is known to infest sugarcane, sorghum, bajra rice, wheat finger millet and ragi (Jepson, 1954).

Identification. The adults are medium sized 14 17 mm long, stout and fawn coloured moths. They have straw coloured fore wings with a marginal dark brown streak and hyaline white hind wings. Eggs are round, creamy white and pearl like which changes to brown near hatching. These are laid in the lower most leaf sheath, preferably of young plants. These are also laid on soil near the plant base. Full-grown larvae are 25-30 mm long, smooth and cylindrical. These are purple pink dorsally and white ventrally with orange -red head. Pupae are stout, 16-20 mm long and reddish brown with purple tinge near the head. The pest complete life cycle in 1-2 months in summer and 3-5 months during winter.

Life-cycle. The female lays bead like eggs in 2 to 3 longitudional rows within the lower most leaf sheath preferably of young plants of maize. The incubation period is 6 to 7 days. The larval period varies from 24 to 36 days. The pupal period is about 10 days. The total life cycle varies from 40 to 80 days. The larval period is prolonged in severe winter. The adult lives for 4 to 6 days (Rao, 1983). One generation from egg to adult stage takes six to seven weeks. Thus 6 to 8 generations are found in a year (Ayyar, 1940).

Nature of damage. The larvae of pink stem borer are noctuid. The young larvae avoid light and feed concealed in the leaf sheath in the early stage of growth. The pest was found to attack very young seedlings. As a result of feeding, the tips of tender leaves of the whorl started wilting. Later the larvae eat their way in central shoot causing dead hearts in young plants. They also form tunnels inside the stem with exit holes lined with excreta. The maize crop in the early stages of growth was found to be vulnerable to the attack of the pink borer. Maximum number of dead hearts were formed when artificially infestation was done on 7 to 10 days old plants, and the plants showed greater tolerance when infested 15 days after germination and the injury due to the borer was found to be negligible when infested 20 days after germination.

Losses. It also cause serious damage to maize in winter season particularly in peninsular India where the average yield loss in maize may vary from 25.7 to 78.9 per cent (Sarup *et al.,* 1979).

Management

Resistant germplasm/sources of resistance. The germplasm relatively more resistant to pink stem borer were Dhawal (VC 80xEto.Tuxp br 20, Ganga 11, Antigua 4 D, Comp D1 x J 236, EH 424472, EH 439573, J 26, H 2, J 12, J 22, J 603, ZFS 3, M 13, MCU 204, MCU 507, MCU 607, Pool 17, Puerto Rico Gr 1x TAD, Syn A22, Syn A 226, Syn B 19, Syn B 41, Syn B 13 and Thai DMR (Rao, 1983; Mathur, 1987).

Cultural control

- Removal and destruction of dead hearted plant is recommended.
- The crop rotation of Wheat-paddy and maize-potato-wheat would be more suitable than maize-paddy, wheat.
- Sex Pheromone trap with (Z)-11-hexadecenyl acetate and (Z)-11-Hexadecen-1-0 L in the ratio of 4:1 proved an effective method for fore-casting the occurrence of *S. inferens* in China (Chen *et al.,* 1987).

Natural enemies. Pink stem borer is parasitized by *Telenomus* sp. and eggs of *Trichogramma australicum*, larva of *Apanteles flavipes*, *Bracon chinensis* and *Winthemia semiberbis* in larval stage and *Tetrastichus ayyari* Rohw and *Xanthopimpla* sp. in pupal stage (Anonymous, 1981).

Chemical control. The control practices recommended for maize stem borer should be followed in this case also.

20.4. SHOOT FLY OR STEM FLY *Atherigona* spp. (Anthomyiidae: Diptera)

In northern India, shoot fly species (*A. naqvii* Steyskal and *A. soccata* Rondani) cause extensive damage in spring-sown maize. The infestation of shoot fly in maize was first recorded in 1925. Later in 1968, at Pantnagar it was observed to cause serious damage to spring planted crop. Four different species of shoot fly injuring maize seedlings were recorded by Sandhu and Kaushal (1976) in spring and autumn seasons. While according to Mathur (1989) six species belonging to genus *Atherigona* have been reported to attack maize. The principal species identified are as *Atherigona naqvii* Steyskal, *A. soccata* Rondani, *A. fulcata* (Thomson), *A. indica* Malloch, *A. orientalis* Schiner and *A. punctata* Karl have been reported to attack maize in India. The fly has been identified as *Atherigona orientalis* from Pantnagar (Rathore and Young, 1975) and *A. naqvii* from Ludhiana (Sandhu and Kaushal, 1976). The maize crop sown at the normal time is generally not infested by shoot fly, but the spring-sown maize is heavily affected by this insect in Uttar Pradesh (Rathore *et al.,* 1969), Delhi (Chatterji *et al.,* 1973), Punjab (Sandhu and Kaushal, 1976) and Andhra Pradesh (Seshu Reddy and Davies, 1977). *A. naqvii* is a most serious pest of Sathi maize grown during summer season in Punjab.

Distribution. Shoot fly complex prevalent in India (especially northern parts), Ceylon, Nigeria, Pakistan, Ethopia, Saudi Arabia and South Africa. *Atherigona* spp. are serious pests of spring crop in the tarai region and western Uttar Pradesh, Andhra Pradesh and Punjab (Rathore *et al.,* 1969; Sandhu and Kaushal, 1976; Sesha Reddy and Davies, 1977).

Host plant. Maize (in spring season), wheat, barley, sorghum.

Extent of damage. Studies on the incidence of shoot fly spp. in relation to date of sowing revealed that as the sowing time delayed from early February to end of February the per cent incidence of shoot fly increased from 21.4 to 35.9. It declined from 21.1 to 4.2 in sowing of March to first week of April and reduced to 3.0 per cent by mid April sowing at and around Ludhiana (Sajjan *et al.,* 1983). In tarai areas of Uttar Pradesh the maize crop sown during the first and second week of February escaped the shoot fly incidence. The delay in the sowing beyond this period resulted in more incidence with consequential loss in yield. In H.P. the incidence has been reported to be 9.6 per cent in May and 26.1 in July (Kumar and Chander, 1987). Studies on the incidence conducted on winter maize in the Punjab revealed that the maize crop sown during the last week of October to mid November remained almost free or had very low incidence of shoot fly (0-4.2%) while the January sown crop sustained the highest damage from shoot fly (53.1 to 56.2%) followed by December sown crop (Sekhon *et al.,* 1986).

Life-cycle. The fly lays white minute (0.72 x 0.16 mm) rice grain shaped eggs sculptured with horizontal longitudional ridges singly or in small batches on the stem above the ground or just below the ground level and also on the under surface of cotyledonary or first leaf of young seedlings (Sandhu and Kaushal, 1976). *A. naqvii* is known to lay eggs mostly in cracks and crevices of soil. According to Sandhu and Kaushal (1976) the egg, larval and pupal stage of this pest lasted for 1-3, 15-18 and 8-9 days, respectively. While adult flies survived for 1-2 weeks on yeast glucose dry food. According to Kumar and Chander (1987) total life cycle was completed in 17-33 days.

Nature of damage. The larvae feed on leaf surface in the funnel primarily as “whorl maggot”. Irregular slits or short holes and deformed leaves are formed as a result of attack. Dead hearts are also produced in young and slow growing plants within 5-7 days of egg laying. Such plants give rise to side tillers, which do not grow much and give a bushy appearance. Seedlings that grow fast recover from the damage to some extent. In most of the severely attacked plants tillering is initiated at an early stage. A small circular cut sometimes appears on the plant at the point of entrance, which become more prominent on the lamina of the emerging leaf 6-7 days after infestation. Such leaves turn yellowish green, look crumpled and folded inward, from the margins. Severely infested seedlings wilt, terminal growth is arrested and remain stunted. Infestation has been reported to be more on plants in shady places than in sunny sites (Kanaujia and Sharma, 1977).

Losses: The avoidable losses in maize yield due to shoot fly have been estimated to be 69 to 97 per cent respectively (Mathur 1992) and upto 20 per cent loss in grain yield (Pathak *et al.,* 1971).

Management

Resistant germplasm. In India, uptill now about 2000 maize germplasm lines have been screened against *Athergona* species and number of resistant sources has been identified by various workers from different locations. The resistant germplasm are given below in the Table 20.3.

Table 20.3. Germplasm showing relatively resistant reaction to shoot fly

Germplasm	Location	Reference
Ganga 3, Ganga 7 and Kisan	Pantnagar	Rathore *et al.* (1969)
Vikram, Vijay, H 1 and Jawahar x Antigua Gr 1	New Delhi	Anonymous (1972)
Comp A 53 (SA) x Comp A 54 (EV) RU 21, EH 2230 EH 3136, J 22, Opaque B-15	Ludhiana	Anonymous (1973)
Vijay, CM 202, D 1, Syn P 529 x Kisan	Pantnagar	Choudhry and Sharma (1975)
(CM 201), CM 202 x CM 111 x (C B C), D 37 (Deotox G C C) br2 # (Ver63 # EB) br2 # # White semi flint br2 Pop, Yellow Tuxpeno Comp x Costal Trop Flint f (DMS-1) # #	Pantnagar	Sharma and Singh (1975)
Ganga 5, Ganga 7 (CM 202 x CM 111) x Puerto Rico 22 D x Pureto Rico 17 D), J 1 x Antigua Gr II, Yellow Tuxpeno Comp x Coastal Trop Flint (F2)	Pantnagar	Pandey and Sharma (1980)

Cultural control

- Removal and destruction of the dead hearted plants in the area of low infestation.
- In the areas where it is of regular occurrence and the incidence is not severe, increased seed rate should be practiced. The extra-infested plant should be uprooted when the plants are about 25-day old.
- The early planting of short duration varieties in the month of February may escape the pest incidence.
- Destruction of alternate hosts like wild sorghum, sarkanda etc. should be done regularly as they serve as the source of primary infestation and the pest may survive during off-season on these hosts.

Natural enemies. Three parasites viz., *Tetrastichus nyemitawus* Rohwer, *Tetrastichus sp* and *Neochrysocharis* spp. (Enlophidae) of *A. naqvii* have been reported from Udaipur by Kundu and Kishore (1972).

Chemical control. In the area where shoot fly is of regular occurrence, prophylactic use of Furadon 3G @ 5 kg/acre or Thimet 10G @ 4 kg/acre in the furrows at the time of sowing. Seed treatment with carbafuron 2 to 4% before planting in the area where it is a regular pest. Phosphamidon 85 WSC @ 150 ml/ha, Metasystox (oxydematon methyl 25 EC) @ 1250 ml, Malathion 50 EC @ 1000 ml and Rogor (dimethoate 30 EC) @ 1250 ml and Sevin (carbaryl 50 WP) @ 375 g a.i./ha provide good control of shoot fly, seven days after germination (Singh and Uma Kanta,

2004). Sajjan and Sekhon (1985) observed that seed treatment with Furadon @ 10 kg per 100 gm seed and soil application of Furadon 3 G @ 12.5 kg/ha and Thimet 10 G (phorate) @ 10 kg a.i./ha at sowing time were effectively controlling the shoot fly. Shoot fly can be controlled with furrow application of phorate 10 G @ 120 ml (0.03%) or Malathion 50 EC @ 200 ml (0.05%) a.i./ha (Bharadwaj *et al.,* 1980). Application of Furadon 3G (carbafuron) @ 5 kg/acre or Thimet 10G (phorate) @ 4 kg/acre in the furrows at the time of sowing (Anonymous, 2005).

20.5. ARMYWORM, *Mythimna separata* (Walker) (Noctuidae: Lepidoptera)

The armyworm, *M. separata,* is a polyphagous graminaceous pest. In Punjab, it has been reported as a major pest of wheat, oats and barley. It has also been found attacking *rabi* maize and causing considerable damage to this crop in eastern Uttar Pradesh (Singh *et al.,* 1981). In northern India, the armyworm, *M. separata* has been recorded feeding extensively in the plant whorls of winter season crop. Four prominent species of the genus *Mythimna, M. separata* (Walker), *M. unipunctata* (Haworth), *M. venalba* (Moore) and *M. loreyi* (Duponchal) cause economic losses to a number of crops including maize, throughout the year (Srivastava and Rathore, 1961). The outbreak of this pest occurs in Godhra (Gujarat) during *kharif,* 1978. It is a polyphagous graminaceous insect pest in Punjab. Srivastava and Khan (1961) observed that the armyworm cause heavy losses to maize at two stages of its growth, viz., immediately after germination up to the development of nodes and internodes and secondly one month after sowing, when whorls are formed. Early infestation gives the appearance of a grazed crop, while late infestation results in complete defoliation due to larval feeding from leaf edge toward the midrib. No attempt has been made to evaluate maize germplasm for resistance to this armyworm.

Distribution. *Mythimna loreyi* (Duponchal) is available in South India and Punjab but *M. separata* (Walker) is recorded in Himachal Pradesh, Punjab and Karnataka

Host plant. It has been reported as a major pest of wheat, oats, rice barley rye, tobacco and also attacks *jowar* and certain wild grasses.

Identification. The adults are pale brown and stout moths. These are attracted to light. The eggs are oval, round, light green when freshly laid and turn pale yellow and finally black. Eggs are laid singly in rows or in clusters on dry and fresh plants or on soil. Freshly hatched larvae are very active and dull white and later on changes to pale brown. Full-grown larvae are 35-40 mm long, green to light pink and with brown to reddish brown head. The larvae have four dark stripes on either side of their body.

Life-cycle. The female moth lays eggs in rows or in clusters on dry or fresh plants or on the soil. The eggs are round, light green, when freshly laid and turn pale yellow later and finally black. The incubation period is recorded from 4-11 days from March to May and 19 days in December-January, in Punjab. In spring, larval period is 13-14 days but it may be prolonged to 88-100 days in winter months. Pupation takes place in the soil and also under dry leaves among the stubbles or fresh tillers. The pupal period is 9-13 days in May and 36-48 days during winter months. The pupation takes place near the water channels. The longevity of moth is 1-9 days.

Nature of damage. The young larva attacks the tender leaves of the whorl of the plants at night. They produce scraped injury and small holes in the leaves. The grown up larvae feed on foliage voraciously. In case of heavy incidence the whole leaf, including the mid rib is consumed by the caterpillars and attacked plants look like grazed by the cattle, but the plants are seldom killed. The larvae throw up the faecal pellets on the leaves or in the whorl from which their presence is easily detected. The pest attacks the winter maize crop throughout the season, generally with low incidence. But its incidence increases during March-April. The border rows of the crops, particularly adjoining the wheat fields, suffer more damage, probably due to migration of the larvae of this pest from the ripening wheat crop. The larvae also attack maize silk in April-May. The populations build up starts in the beginning of March and increases with the rise of temperature in the spring. The life cycle is completed in about one month in summer and in five months during winter.

Management

Natural enemies. Armyworm has large number of natural enemies of which *Apanteles ruficrus* (Holiday), *Eurgtoma* sp. and *Ooencyrtus* sp. are major larval parasites. *Megaselia* spp. are major larval parasites. *Megaselia* sp. and *Anthrenus flavipes* Le Conte are major larval-pupal parasites during September-October (Bharadwaj *et al.,* 1980).

Chemical control. This pest can be controlled by spraying 250 g Sevin 50 WP or 250 ml Thiodan 35 EC or 35 ml Cymbush 25 EC (Cypermethrin) or 40 ml Sumicidin 20 EC (fenvalarate) in 60 litres of water per acre (Anonymous, 2005). Use of BHC 10% dust @ 25 kg/acre is also very effective (Bharadwaj *et al.,* 1980).

20.6. CORN EARWORM, *Heliothis armigera* (Hubner) (Noctuidae : Lepidoptera)

The corn earworm is a polyphagous pest and attacks some vegetables and cultivated crops. This has been found attacking *rabi* maize regularly in eastern Uttar Pradesh. A slight incidence of this pest has been noticed in summer as well as in winter season on maize in Punjab.

Distribution. The insect has a worldwide distribution.

Host plant. It is a polyphagus pest and feed on vegetables and cultivated crops. Cotton, sorghum, gram, soyabean, pea, safflower, chillies, groundnut, tobacco, okra, many legumes and other plants.

Identification. The eggs are shining, greenish yellow and are round. The larvae are striped and may be yellow, green or brown. The full-grown larvae measure 35 mm in length. The moth is large and brown with a V-shaped specks and dull black border on the hind wings. The pupae are dark brown and have a sharp spine at the posterior end. The colour of the larvae, found feeding on leaves was greenish and of those found on tassel was light brown.

Life-cycle. The spherical yellowish eggs are laid singly on tender parts. A single female may lay as many as 741 eggs in four days. The incubation period is 2-4 days in April to October and 6 days in February. The young larvae feed on the foliage for sometimes and later bore into the pods and feed on developing grains. Cannibalism among larvae is common. The full-grown larvae come out of the pods and pupate in the earthen cell. The pupa is dark brown and has sharp spines at the posterior end. The pupal period lasts 8-15 days in an active season, which may be prolonged in winter. There may be as many as 8 generations in a year.

Nature of damage. The young larvae start damaging tassel and move downward, eat silk and may tunnel in the ears and damage grains, in the milky stage. The damage in the ear may form a channel like structure, which is seen after exposure of the grains. The attacked cobs show masses of moist excreta on them.

Management

Cultural control. Chop and dispose off the drying maize tassels after pollination to remove eggs and reduce incidence.

20.7. HAIRY CATERPILLAR/SILK CUTTER, *Euproctis virguncula* Walker (Noctuidae : Lepidoptera)

Among the hairy caterpillars, attacking maize, the white moth hairy caterpillar, *E. virguncula* is more common. It is a polyphagus pest. It has been observed as minor pest of the *kharif* maize upon which it is primarily a silk cutter. Its larvae feed on the freshly matured silk during September-October, but do not effect the pollination (Sandhu and Kaushal, 1972; Sandhu *et al.*, 1974).

Distribution. Throughout the country

Host plant. It feeds on rice, pearl millet, barley and sorghum

Identification: The adult is white moth with small black dots on the wings and red line along the anterior margins. Body is bright orange with black bands and spots. A female lays bright yellow spherical eggs in batches, covered with hairs from anal tuft, on underside of leaves. Freshly hatched larvae congregate on the leaves and later on disperse and feed green matter /vegetation available in the area. As they grow, they become deep orange in colour and are covered profusely with thick tufts of hairs. The full-grown caterpillar is 6-7 cm long. The pupae are found in the soil 10-15 cm deep in earthen cells.

Nature of damage. It appears in an epidemic form, cause serious damage by feeding on the leaves and the tender stems, when young feed gregariously. The grown up caterpillar may migrate from one field to another. Its larvae feed on the freshly matured brown silk and reduce it to the level of cob husk during September-October and did not effect the pollination (Sandhu *et al.,* 1974). The infested plants can be easily distinguished from a distance.

Management

Cultural control

- The light traps set on as soon as the moths are visible after the rain start. One petromax for 4 hectares and one lantern for one hectare would be sufficient. The petromax or the lantern should be placed in kerosinized or insecticide mixed water trough to kill the moths.
- The plants or twigs with large number of first instar larvae feeding on them can also be collected and destroyed.
- During the summer season and after the harvest of *kharif* crops, the field should be ploughed to expose the pupae to birds and bright sun.

Integrated pest management

- Use light-trap for the destruction of moths.
- Young larvae are gregarious. They can be destroyed by crushing them under feet or by picking and putting them into kerosenized water. If the population is high, control it by spraying 500 ml of Thiodan 35 EC or Ekalux 25 EC or 200 ml of Nuvan 100 EC (dichlorovos) in 100-200 litres of water with a manually operated sprayer, per acre.

20.8. CUTWORM, *Agrotis* Spp. (Noctuidae : Lepidoptera)

Two species belonging to *Agrotis* have been reported to attack maize. They were identified as *A. ipsilon* Hufnagal and *A. spinifera* (Hubn.). Cutworm, though is not regular pest of maize, it occurs in epidemic in some seasons. The insect belonging to genus *Agrotis* are subterranean in habit. The pest remains active throughout the year but minor pest during August to October and is major limiting factor in certain regions (Bharadwaj *et al.,* 1980).

Distribution. India, Burma, Sri Lanka, China, Indonesia, Australia, America, Europe and North Africa. It has been reported from Kashmir, Tarai area of Uttar Pradesh and Bihar.

Host plant. Maize, okra, tomato, cabbage, lucern and chillies were severely damaged by *A. ipsilon* and *A. spinifera* (Kushwaha *et al.,* 1972a).

Identification. The full-grown larva is dark brown with a plump and greasy body and measures 42-45 mm in length. The adult has dull brownish fore wing with numerous wavy lines and spots and fawn coloured hind wings.

Life-cycle. The pest is active from October to April. The female lays from 199 to 344 creamy-white, dome shaped eggs in clusters of about 30 each, either on the under surface of the leaves of food plants or in the soil. The incubation period varies 2 days in summer to 8-13 days in winter. The larvae period lasts for 30-34 days in February to April. The full-grown larva makes earthen chambers in the soil and pupate underground. The pupal stage is 10 days and 30 days in summer and winter, respectively. The total life cycle is completed in 48-77 days. There is three generation in a year.

Nature of damage. The caterpillar, which hides during day time in cracks in the soil, becomes active at dusk, feed on leaves and also cut seedlings plants and tender shoots of the young plants. The larvae cause damage to the maize seedlings growing during winter season. Initial damage is done by feeding on leaves by the younger larvae, while the older larvae of 4th, 5th and 6th instars kill the plant by cutting it just above the soil surface (Jarvis *et al.,* 1981). The larvae damage the developing root and stem.

Losses. In 1972 spring, it appears in epidemic form at Pantnagar University Farm, as high as 18 larvae per meter row length were recorded (Sharma and Singh, 1972).

Management

Resistant germplasm. According to Kushwaha *et al.* (1972b) antibiosis existed in some maize varieties. Bassi local proved to be resistant, as it reduced the longevity and weight of each larval instar, pupa and increased the mortality of *A. spinifera* as compared to Ganga 2, Ganga 4, Malan Comp and Deoto x TAD. Jarvis *et al.* (1981) screened about 2378 maize germplasm lines against cutworm. They suggested that resistance in maize against 5th instars larvae of cutworm was more important than early instars larvae. Tseng *et al.* (1984) evaluated maize germplasm for resistance to leaf feeding by 3rd instars larvae of black cutworm. Out of 109 single cross hybrid and 100 inbred lines, none was found resistant to this pest.

Cultural control

- Pick the larvae from the base of the plants, where they take shelter under the clodes during the daytime.
- Flooding of fields could control the pest effectively. The flooding of the field helped in two ways. First it causes the larvae to float on the surface and expose them to predator, birds. Secondly the wet and compact soil caused due to water reduces the free movement of larvae and ultimately its activity.

Chemical control: Give prophylactic treatment of Thimet 10G (phorate), or Furadan 3G @ 15 kg per ha. Lindane @ 1.5 kg and 2.0 kg a.i./ ha also showed promise for the cutworm control (Rathore and young, 1975).

20.9. CLIMBING CUTWORM, *Rhyacia herculea* Corti and Draudt (Noctuidae: Lepidoptera)

It is also a serious pest of winter sown maize crop. It has been reported to cause 20 to 40 per cent loss (Verma *et al.,* 1979). Climbing cutworm is noctuid in habit. An epidemic of climbing cutworm *Rhyacia herculea* Corti and Draudt was recorded in the winter season of 1962-63 in North Bihar (Singh and Sinha, 1965). With the introduction of winter maize during recent years, the insect has become a limiting factor to its cultivation in the state.

Distribution. North Bihar.

Nature of damage. This pest starts its attacks on 20 to 30 day old plants. The caterpillar feed voraciously on the apical portions and the margins of the leaves. It feed from tip to edges and leave behind mid rib only. The damage to the leaves adversely affects the photosynthesis and plant growth.

Losses. The losses caused by this pest vary from 18 to 34 per cent (Verma *et al.* 1979).

Management

Resistant germplasm. Early maturing varieties are more preferred than late maturing ones. Further the infestation of the pest showed significantly negative correlation with the trichome density, fibres and silica concentration of the leaves.

20.10. SUGARCANE LEAF HOPPER OR PYRILLA, *Pyrilla perpusilla* Walker (Fulgoridae (Lophopidae) : Hemiptera)

It is very active insect and cause serious loss to the Sathi maize crop during March-May. It also attacks the normal *kharif* crop. The sugarcane Pyrilla appeared as a very serious pest on maize and sorghum during *kharif* 1975 around Ludhiana (Punjab).

Distribution. India (Punjab, Uttar Pradesh, Bihar, Haryana, Rajasthan and Maharashtra), Pakistan, Sri Lanka, Burma, and Thailand

Host plant. Sugarcane, maize, wheat, millet, rice, barley, oats, sorghum, *bajra, baru*, Guinea grass, Swank, Sudan grass, Napier grass, wild grasses and some dicotyledons.

Identification. The young ones (nymphs) are white having a pair of small white tails at the end of the body. The full-grown nymphs is pale yellow, 10-15 mm long and has two white prominent feather like filament at the tail end of its body. The adult is about 20 mm long and has straw-coloured body with dark patches or spots on the wings. At the front end it has a snout like prolongation and prominent red eyes.

Life-cycle. The female hopper lays 300 to 536 eggs in clusters on the underside of the leaves during summer and within the leaf sheath during winter. The clusters are covered with a white, fluffy material from the anal tuft of the females. As this fluff is removed, oval pale white eggs are seen in three to five longitudinal rows of 35-50 eggs. The eggs hatch in about three to four weeks during November or December and eight to ten days in the summer. The nymphs grow to maturity through five stages within eight weeks in summer and 5-6 months in winter. The adults live 27-52 days in the summer and 18-20 weeks in the winter. In all, 3-4 generations are completed in a year.

Nature of damage. Both the nymphs and the adults keep sucking plant sap from the underside of the leaves. The infested leaves turn pale and the plants cannot synthesize the food material and thus they do not grow well. In case of severe attack, the crop may dry up completely. There also develops black fungus on the leaves which hinders photosynthesis and consequently the development and growth of the plants.

Losses. During the years of outbreak on sugarcane, it is also known to attack maize, pearl millet, sorghum, paddy, barley and wheat, causing considerable damage to these crops (Sarup *et al.*, 1969)

Management

Natural enemies. *Ageniaspis pyrillae* Mani, *Tetrastichus pyrillae* Crawford, and *Cheiloneurus pyrillae* Mani are egg parasites of the insect.

Chemical control. Spray Metasystox 25 EC or Rogor 30 EC @ 200 ml/ acre in 50 litres of water with manually operated knap-sack sprayer or in 20 litres of water with a low volume sprayer (Anonymous, 2005). Monocrotophos 40 SC @ 0.3 kg, Carbaryl 50 WP @ 0.5 kg, Endosulfan 35 EC @ 0.263 kg and Trichlorofon 50 EC @ 0.19 kg a.i./ha proved effective upto 5 days after application and reduced the nymphs and adults of Pyrilla (Sajjan *et al.,* 1985).

20.11. MAIZE JASSID, *Zyginidia manaliensis* (Singh) (Cicadellidae : Hemiptera)

Maize jassid is a serious pest of spring crop of fodder maize in Punjab (Sandhu *et al.,* 1975) has been found infesting fodder maize crop during spring season in Northern India.

Distribution. Punjab and Himachal Pradesh.

Host plant. Apart from maize it has been seen breeding on barley, oat, crab grass, goose grass, pearl millet, sorghum, sudan grass, love gra-ss, guinea grass, rice, sarkanda and wheat in the Punjab and Himachal Pradesh (Singh, 1969; Sandhu and Deol, 1972; Sandhu *et al.*, 1975).

Identification. It is small, active insect. The adults are dirty white and measures 3.1 mm in body length. The eggs are creamy white and oblong and are laid inside the veins of the plants. The newly hatched nymphs are dirty white and the older ones are yellow. These nymphs are seen jumping on the leaves and the adults fly away actively on slight disturbance of the plant.

Nature of damage. Both nymph and adult cause economic damage by sucking the plant sap from the under surface of the leaves and produce whitish dotted or broken streaks or result in typical stippling leaf injury. In case of severe attack, the leaves turn totally white. This whitening and severe stippling impair photosynthesis, give bad look to the fodder and the quality of the fodder is also adversely affected due to this pest in terms of low protein concentrations.

Losses: Economic losses of 32 per cent in fodder yield during spring season and 15 per cent loss in grain yield of winter crop (Singh, 1991).

Management

Resistant germplasm. Host plant resistance for maize jassid has been evaluated only at PAU, Ludhiana under natural infestation. Singh *et al.,* (1978) evaluated germplasm for leaf injury during spring season. They reported Antigua Gr1, CM 300, Comp A 51 (II) x Comp A 54, Comp H 3, EH 2340, EH 3136, EH 3597, EH 4018, Ganga 5 (Path), Opaque and Syn B 15 to be relatively resistant. Singh (1991) on the basis of jassid population and per cent plant infested on winter crop, found CM 202, CM 600 and V 188 to be relatively resistant. Sekhon and Sodhi (1994) found that the varieties VL 88, CM 600 and CM 202 proved relatively more resistant to maize jassid.

Chemical control. Spray Metasystox 25 EC or Rogor 30 EC @ 200 ml/ acre in 50 litres of water with manually operated knap-sack sprayer or in 20 litres of water with a low volume sprayer (Anonymous, 2005). Carbafuron 3 G @ 6 kg a.i./ha, Fenitrothion 50 EC @ 0.125 kg and Disulfoton 5 G @ 1.0 kg a.i./ha proved most effective in controlling jassid (Sajjan *et al.,* 1982).

20.12. CEREAL THRIPS, *Anaphothrips sudanensis* (Thripidae : Thysanoptera)

It is also very small insect about one mm long. Adults are brown and nymphs are pale. As many as 40-50 thrips are seen in a single whorl.

Distribution. This pest has been reported from Rajasthan and Punjab but it is more prevalent in the former state

Nature of damage. It causes damage to seedlings, tassel and ear of maize crop grown in the winter and spring seasons (Verma and Ramzan, 1965; Sandhu *et al.,* 1974; Noor *et al.,* 1973). Both the nymphs and adults suck plant sap from top of the young leaves and gradually congregate and form colonies within the leaf sheath. The feeding result in small silvery specks which enlarge to form brown patches and further coalesce resulting in complete drying of leaves. Heavy infestation at the seedling stage may cause complete loss of the crop.

Losses. The infestation has been reported to range between 29 to 86 per cent (Noor and Kushwaha, 1976; Mathur, 1983).

Management

Resistant germplasm. Noor and Khushwaha (1976) and Mathur (1983) reported Ageti 76 (J 603), Malan, Renuka (G 25), B 21, Diara EV MF S 4, Diarazs EV ZFS 1, E 215-216, G 24, G 26, G 27, Hunius EV (PFS) 3 F 5, Hunius EV (ZFS 3), H 216, MCU 78-U-1025, MCU 314 and MCU 501 to be relatively resistant.

Chemical control. Spray Metasystox 25 EC (Oxydemeton methyl) or Rogor 30 EC (dimethoate) @ 200 ml/ acre in 50 litres of water with manually operated knap-sack sprayer or in 20 litres of water with a low volume sprayer (Anonymous, 2005). Carbafuron 3 G @ 6 kg, Fenitrothion 50 EC @ 0.125 kg and Disulfoton 5 G @ 1.0 kg a.i./ha proved most effective in controlling thrips (Sajjan *et al.,* 1982). Spraying the crop with Phosphamidon 100 EC @ 120 ml (0.03%) or Malathion 50 EC @ 200 ml (0.05 %) a.i./ha (Bharadwaj *et al.,* 1980).

20.13. COTTON GREY WEEVIL, *Mylocerus undecimpustula-tus maculosus* Desbrocheres (Curculionidae : Coleoptera)

It causes serious loss to the Sathi maize crop during March-May. They also attack the normal *kharif* crop.

Distribution. The insect is distributed throughout India.

Host plant. It is a pest of cotton crop but sometimes it damages maize plants also. It also attacks sorghum, sugarcane, *bajra, bhindi* and pigeon pea, mango, apple, pear, peach, pomegranate, guava and groundnut.

Identification. The grubs are white legless, cylindrical and measures about 8 mm in length. The adult is small grey weevil and is 3-6 mm long.

Life-cycle. The weevil lays avoid, light yellow eggs in the soil and as many as 361 eggs are laid at a depth of about 8 cm. The eggs hatch in about 3-5 days. The young grub feeds on the roots and complete their development in 1-2 months. They pupate inside the earthen cells, forming creamy white pupae, which changes into adults in about one week. During the active period, the life cycle is completed in 6-8 weeks. The pest probably breeds three to four times a year.

Nature of damage. The grubs remain in the soil feeding on the roots of the seedlings and destroy them and adult beetles feed on leaves from edges.

Management

Chemical control. Spray Metasystox 25 EC or Rogor 30 EC @ 200 ml/ acre in 50 litres of water with manually operated knap-sack sprayer or in 20 litres of water with a low volume sprayer (Anonymous, 2005).

20.14. PENTATOMID BUG, *Dolycoris indicus* Stal. (Pentatomidae : Hemiptera)

Distribution. It was observed infesting ear heads of maize during 1971-72 in Ludhiana (Anonymous, 1975) and all growth stages of paddy in Uttar Pradesh (Garg and Tandon, 1982).

Host plant. The yellowish pentatomid bug was recorded as a minor pest of sorghum, pear-millet, finger millet, wheat, sunflower and safflower from southern India (Fletcher, 1914).

Life-cycle. Four clusters of eggs (14-28 per cluster) laid on the upper surface of maize leaves in the field were used for studying the life history of the bug. The pinkish barrel shaped eggs developed a distinct circular cap while hatching. The egg stage lasted for 6 days. The distal egg cap was pushed up slowly up by the emerging nymphs and they rested for 2 days on the respective egg shells before moving in different directions. One group of nymphs was provided tender maize grains in the milky stage of development. The nymphs passed through 5 instars, the average duration being 5.5, 16.5, 8.5, 8.0 and 6 days respectively. The total nymphal period was 44.5 days. The adult survived for 36 days. The total life cycle was completed in about 3 months.

Nature of damage. The adult bugs were observed in the field on maize ears with naked tip. Theses did not affect the leaves. The damage to the grain was only one per cent. The population of the bug was at its peak during September to November when these were found on the earheads of maize, sorghum, pearl millet, rice and sunflower. These were also noticed on wheat and barley during March-April and on spring-sown sorghum and pearl millet during May-June.

20.15. *PHADKA* GRASSHOPPER, *Hieroglyphus nigrorepletus* Bolivier (Acrididae : Orthoptera)

Grasshoppers generally do not cause much damage to maize crop. However in favourable season, it may prove very harmful and leave nothing on the plant except stem and midrib of leaves. It is a sporadic pest during September to October (Bharadwaj *et al.,* 1980).

Distribution. Widely distributed in India, Pakistan, Ceylon, Vietnam, Indonesia, Thailand and South China. In India it is distributed in Rajasthan, Maharashtra, Madhya Pradesh, Uttar Pradesh, Bihar and Punjab.

Host plant. Rice, maize, sorghum, *bajra,* millets, sugarcane, Arhar, sunhemp and grasses.

Identification. The adults are 40-50 mm long and are shining greenish yellow, having three black lines running across the pronotum. Nymphs are yellowish, with many reddish brown spots in the early stages, but become greenish, as they grow older.

Life-cycle. The pest passes the winter and dry parts of the summer in the egg stage. The eggs are found in the soil, hatch in few days after the first shower in June or early July. The newly emerged nymphs start feeding actively and are fully developed after passing through seven stages, in about 3 weeks time. The adults feed voraciously during August and September. Two-month-old adults mate and the females start laying eggs by inserting her abdomon in the soil. The eggs are laid in the pods 5-8 cm deep in the soil and are covered by a gelatinous substance. Each pod contains 30-40 eggs. The egg laying continues from September to November. There is only one brood in a year.

Nature of damage. Both nymphs and adult generally feed on grasses before attacking the crop. The leaves are completely eaten, leaving the mid rib. An individual grasshopper during its lifetime eats about 42 gms (green weight) of maize leaves.

Management

Regular summer ploughing and destruction of bunds, which expose the insect eggs to bright sun and natural enemies.

The newly hatched nymphs, which are gregarious in nature and feed on the grasses growing on the bunds, can very easily be controlled by crushing them mechanically.

Spray the crop with 1 kg acephate 75 SP or 200 ml deltamethrin 2.8 EC per hectare (Arora and Shera, 2011).

REFERENCES

Anonymous (1972) *Proceedings of the XV Annual Maize Workshop.* IARI, New Delhi.

Anonymous (1973) *Investigation of the Major Insect Pest of Maize with Special Reference to Insect-Plant Relationship.* Final Technical Report PL-480 Project. Department of Entomology, Punjab Agricultural University, Ludhiana.

Anonymous (1981) *Proceedings of the All India workshop on Pest Management, Recommendations and future Guidelines of Research.* Entomological Society of India. 108 pp.

Anonymous (1984) *Twenty- seventh Annual Progress Report.* All India Coordinated Maize Improvement Project, Indian Agricultural Research Institute, New Delhi.

Anonymous (1985) *Twenty-eighth Annual Progress Report.* All India Coordinated Maize Improvement Project, Indian Agricultural Research Institute, New Delhi.

Anonymous (2004a) *FAO Production Year Book.* Food and Agriculture Organization of United Nations, Rome.

Anonymous (2004b) *Statistical Abstract of Punjab,* Govt. of Punjab.

Anonymous (2005) *Package of Practices for Crops of Punjab: Kharif.* Punjab Agricultural University, Ludhiana.

Arora, R. and Shera, P.S. (2011) First record of rice grasshopper as a pest of Napier grass and Napier millet. *J. Insect Sci.* (In Press)

Atwal, A.S., Chahal, B.S. and Sandhu, G.S. (1966) Control of maize borer for greater returns. *Prog. Fmg.* **2** (6): 17-18.

Ayyar, T.V.R. (1940) Handbook of Economic Entomology for South India, Govt. Printing Press, Madras.

Bharadwaj, S.C., Singh, S.P. and Gupta, R.S. (1980) Proceeding of the All India Workshop on Pest Management Recommendations and Future Guidelines of Research. Entomological Society of India, New Delhi. pp 14-17.

Bing, J.W. and Guthrie, W.D. and Dicke, F.F. (1992) Genetics of resistance in maize to the corn leaf aphid (Homoptera: Aphididae) colonization to DIMBOA content in maize inbred lines. *J. econ. Ent.* **84** (4): 1626-1632.

Butani, D.K. (1955) The influence of temperature on the development of insects with special reference to *Chilo zonellus* Swinhoe. *Indian J. Ent.* **17** (2): 280-282.

Chatterji, S.M., Young, W.R., Sharma, G.C., Sayi, I.V., Chahal, B.S., Khare, B.P., Rathore, Y.S., Panwar, V.P.S. and Siddiqui, K.H. (1969) Estimation of loss in yield of maize due to insect pests with special reference to borers. *Indian J. Ent.* **31**: 109-51.

Chatterji, S.M., Bhamburkar, M.W., Marwaha, K.K, Panwar, V.P.S. and Siddhiqui, K.H. (1973) Evaluation of some systemic insecticides for the control of shoot fly infesting spring maize. *Indian J. Ent.* **35** (2): 161-163.

Chaudhary, R.N. and Sharma, V.K. (1975) Note on the comparative resistance of some elite maize germplasm to shoot fly in spring. *Indian J. agric. Sci.* **45**:561.

Chen, I.W., Ouyang, X.Y. and Lu, X.L. (1987) Application of the sex pheromone of *Sesamia inferens* (Walker) to forecasting the pest. *Insect Knowledge* **24** (1): 8-10.

Chundurwar, R.D. (1989) Sorghum stem borers in India and Southeast Asia. In: *International Workshop on Sorghum Stem Borers*, Nov. 17-20, 1987, ICRISAT, India, pp. 19-26.

Durbey, S.L. and Sarup, P. (1985) Antibiosis due to powdered dry plant material of maize varieties incorporated in artificial diet for rearing of the stalk borer, *Chilo partellus* (Swinhoe). *J. ent. Res.* **9**: 201-206.

Fletcher, T.B. (1914) *Some South Indian Insects and other Animals of Importance.* Govt. Press, Madras.

Garg, D.K. and Tandon, J.P. (1982) Major insect pest of rice on hilly tracts of Uttar Pradesh, India. *Int. Rice Res. Newsl.* **7** (1): 11-12.

Ghouri, A.S.K., Irshad, M., Akram, M. and Ismail Shah (1978) Studies on the economic significance of spring maize pests and their control. *Int. Pest Contr.* **20** (5): 14-15.

Goyal, G. (2005) Management of *Chilo partellus* (Swinhoe) (Lepidoptera: Pyralidae) with cultural practices in *kharif* maize. M.Sc. Thesis, Punjab Agricultural University, Ludhiana.

Harris, K.M. (1985) Lepidopterous stem borers of sorghum. In: Proc. *Int. Sorghum Ent. Workshop*, 15-21 July, 1984, College Station, Texas, USA. International Crops Research Institute for the Semi-Arid Tropics, Pattancheru, Hyderabad, India, pp. 161-167.

Jalali, S.K. and Singh, S.P. (2003) Biological control of *C. partellus* (Swinhoe) in fodder maize by inundative releases of parasitoids. *Indian J. Pl. Prot.* **31**: 93-95.

Jarvis, J.L., Guthrie, W.D. and Robbins, J.C. (1981) Evaluation of maize plant introductions for resistance to black cutworm larvae. *Mydica* **26** (4): 219-225.

Jepson, W.F. (1954) *A Critical Review of World Literature on the Lepidopterous Stalk Borers of Tropical Graminaceous Crops.* Commonwealth Institute of Entomology, London, UK, pp. 127.

Kanaujia, K.R. and Sharma, V.K. (1977) *Indian Farm. Digest.* **10** (6): 29-37.

Khanna, V., Gupta, V.K. and Kanta, U., Dhaliwal, H.S. and Sekhon, S.S. (1995) Control of *Chilo partellus* (Swinhoe) by *Bacillus thuringiensis* based bio-insecticides. *J. ent. Res.* **17** (2): 101-105.

Kumar, J. and Chander, R. (1987) Occurrences of shoot fly *Atherigona naqvii* Steyskal (Diptera: Muscidae) on maize. *Himachal J. agric. Res.* **10** (1): 64.

Kundu, G.G. and Kishore, P. (1972) New host record of *Atherigona naqvii* Steyskal (Anthomyiidae: Diptera) from India together with new record of its three hymenopterous parasites. *Indian J. Ent.* **34** (1): 80-81.

Kushwaha, K.S., Noor, A. and Pareek, B.L. (1972a) Cutworm damage and its control in Rajasthan. *Univ. Udaipur Res. Stud.* **9**: 69-71.

Kushwaha, K.S., Pareek, B.L. and Noor, A. (1972b) Varietal resistance of maize to the cutworm *Agrotis spinifera* Hubner (Noctuidae: Lepidoptera). *Univ. Udaipur Res. Stud.* **10**: 75-77.

Lal, R., Rai, B.K. and Shashi, K. (1961) Preliminary investigation on seed treatment for the protection of young maize plants from insect infestation. *Indian J. Ent.* **23** (2): 155-57.

Majumdar, N. and Teotia, T.P.S. (1965) *Indian J.Ent.* **27** (2): 153-159.

Mathur, L.M.L. (1989) Maize pest problems in India and future scope of host plant resistance. Towards Insect resistant maize for the third world. *Proceedings of the International Symposium on Methodologies for Developing Host Plant Resistance to Maize Insects, Mexico, D.F. CIMMYT.* pp. 291-294.

Mathur, L.M.L. (1992) Insect pest management and its future in Indian maize programme. *XIX International Congress of Entomology,* June 27 to July 4, 1992, Beijing, China.

Mathur, L.M.L. and Rawat, H.S. (1981) Studies on maize pests with certain observations on the survival of the hibernating larvae of *Chilo partellus* (Swinhoe) and its incidence in relation to sowing date. *Rajasthan J. Pestic.* **8**:17-31.

Mathur, L.M.L. (1983) *Research Findings in Maize Entomology.* Mohanlal Sukhadia University, Udaipur.

Mathur, L.M.L. (1987) Maize pest problems in India and future scope of host plant resistance. In: *Toward Insect Resistant Maize for the Third World.* Proc. Int. Symp. Methodologies Developing Host Plant Res. Maize Insects CIMMYT, Mexico. Pp. 291-294.

Mathur, L.M.L. (1991) Genetic of insect resistance in maize. In: *Maize Genetics Perspectives.* K.R. Sarkar, N.N. Singh and J.K.S. Sachan (Eds.). Indian Society of Genetics and Plant Breeding, New Delhi. Pp. 238-250.

Noor, A. and Kushwaha, K.S. (1976) Relative susceptibility of maize varieties to the thrips, *Anaphothrips sudanensis* Trybom. In: *All India Symp. Modern Concepts Pl. Prot.*, University of Udaipur, Udaipur. Pp. 70-71.

Noor, A., Kushwaha, K.S. and Rawat, H.S. (1973) Damage of wheat thrips, *Anaphothrips sudanensis* Trybom in maize at Udaipur Univ. *Udaipur Res. J.* **11**: 40-41.

Pandey, K.C. and Sharma, V.K. (1980) Comparative susceptibility of some released and elite germplasm against shoot fly, *Atherigona* spp. *Indian J. Ent.* **42**: 21-23.

Panwar, V.P.S. and Sarup, P. (1980) Exploring possibility of varied plant population levels of maize for the control of *Chilo partellus* (Swinhoe). *J. ent. Res.* **5** (2): 123-128.

Panwar, V.P.S. and Sarup, P. (1985) Significance of extreme conditions for screening of maize germplasm for resistance to stalk borer *Chilo partellus* (Swinhoe). *J. ent. Res.* **9**: 237-240.

Pathak, P.K., Sharma, V.K. and Singh, J.M. (1971) Effect of date of planting of spring sown maize on the incidence of shoot fly, *Atherigona sp.* and loss in yield due to its attack. *Annual Report* 1970-71. Experimental Station, U.P. Agricultural University, Pantnagar.

Rai, M.K. and Sharma, V.K. (1999) Relative susceptibility of some maize varieties to aphid, *Rhopalosiphum maidis. Indian J. Ent.* **61** (2): 181-183.

Rao, A.B. (1983) Techniques of scoring for resistance to maize stalk borer (*Sesamia inferens*). In: Joginder Singh (ed.). *Techniques of scoring for Resistance to the Major Insect Pests of Maize.* All India Coordinated Maize Improvement Project, Indian Agricultural Research Institute, New Delhi. Pp. 16-26.

Rathore, Y.S. and Young, W.R. (1975) *Indian J. Ent.* **37**(4): 406-407.

Rathore, Y.S., Chatterji, S.M. and Asnani, V.L. (1969) A note on the screening of maize varieties for resistance to shoot fly. *Atherigona* spp. *Indian J. Ent.***31** (3): 277-278.

Sajjan, S.S. and Sekhon S.S. (1985) Control of shoot fly, *Atherigona naqvii* Steyskal (Diptera: Anthomyiidae) on maize through soil and seed treatment. *Indian J. Ent.* **47** (4): 371-375.

Sajjan, S.S., Sekhon, S.S. and Kanta, U. (1985) Efficacy of some important ingredients against *Pyrilla perpusilla* Walker (Lophopidae: Homoptera) infesting maize. *J. ent. Res.* **9** (1): 43-45.

Sajjan, S.S., Sekhon, S.S. and Kanta, U. (1982) Control of jassid, *Zygnidia manaliensis* (Singh) and thrips *Anaphothrips sudanensis* Trybom, infesting maize. *J. ent. Res.* **6** (1): 18-21.

Sajjan, S.S., Sekhon, S.S. and Uma Kanta (1983) Incidence and control of shoot fly, *Atherigona naqvii* Steyskal on spring-sown maize in Punjab. *J. Res. Punjab agric. Univ.* **20** (3): 302-306.

Sandhu, G.S. and Deol, G.S. (1972) New records of insect pests of wheat in the Punjab. Paper presented at the 11th All India Wheat Research Workers Workshop, Aug. 24-28, Hissar.

Sandhu, G.S. and Kaushal K.K. (1976) Occurrence and biology of maize shoot fly in India. *Ent. Res.* **88** (1): 27.

Sandhu, G.S. and Kaushal, K.K. (1972) Incidence of yellow hairy caterpillar, *Euproctis virguncula* Walker in Punjab. *Entomologists Newsl.* **2** (7): 43-44.

Sandhu, G.S., Chahal, B.S. and Kanta, U. (1975) Chemical control of cereal jassid on maize in Punjab. *Indian J. Ent.* **37**: 243 246.

Sandhu, G.S., Singh, B. and Kanta, U. (1974) Control of cereal thrips *Anaphothrips sudanensis* Trybom, on maize with systemic insecticides. *J. Res. Punjab Agric. Univ.* **11**: 403-404.

Sandhu, G.S., Singh, J., Kaushal, K.K. and Sajjan, S.S. (1974) The biology of *Euproctis virguncula* Walker (Lymentridae: Lepidoptera) on cereals in the Punjab. *J. Res. Punjab Agric. Univ.* **11** (2): 187-190.

Sarup, P. (1973) Strategy for the control of insect pests of maize during the fifth plan. Paper Presented at the All India Coordinated Maize Improvement Workshop, Pantnagar.

Sarup, P., Marwaha, KK., Siddqui, K.H. and Panwar V.P.S. (1979) Investigations on major insect pests of maize with special reference to Insectrelationship. *Res. Bull. Div. Ent. IARI* No.3: 183.

Sarup, P., Marwaha, K.K., Panwar, V.P.S. and Siddiqui, K.H. (1977) Studies on insect plant relationship - evaluation of `Introductory Nursery' for resistance to maize stalk borer, *Chilo partellus* (Swinhoe) under artificial infestation. *J. ent. Res.* **1**(2): 151-157.

Sarup, P., Marwaha, K.K., Panwar, V.P.S. and Siddiqui, K.H. (1978) Identification of sources of resistance to maize stalk borer, *Chilo partellus* (Swinhoe) amongst world maize germplasm comprising 'International Nursery'. *J. ent. Res.* **2** (2): 154-159.

Sarup, P., Marwaha, K.K., Panwar, V.P.S. and Siddiqui, K.H. (1978) Evaluation of some exotic and indigenous maize germplasms for resistance to *Chilo partellus* (Swinhoe) under artificial infestation. *J. ent. Res.* **2** (1): 98-105.

Sarup, P., Siddiqui, K.H. and Marwaha, K.K. (1987) Trends in maize pest management research in India together with bibliography. *J.ent. Res.* **11**: 19-68.

Sarup, P., Siddiqui, K.H., Panwar, V.P.S. and Marwaha, K.K. (1981) Response of diverse maize germplasms to artificial infestation of the stalk borer, *Chilo partellus* (Swinhoe). *J. ent. Res.* **5**: 70-75.

Sarup, P., Siddqui, K.H. and Marwaha, K.K. (1987) Trends in maize pest management research in India together with bibliography. *J. ent. Res.***11**: 19-68.

Sarup, P., Singh, D.S., Srivastava, V.S., Lal, R. and Peswani, K.M. (1969) Testing of different pesticides against the adults of sugarcane leafhopper, *Pyrilla perpusilla* Walker. *Indian J. Ent.* **31** (2): 161-165.

Sekhon, S.S. and Sajjan, S.S. (1985) Antixenosis (non-preference) mechanism of resistance in maize against oviposition by maize borer, *Chilo partellus* (Swinhoe). *Indian J. Ent.* **47**(4): 427-432.

Sekhon, S.S. and Sodhi, R.S. (1994) Comparative resistance of *Zea mays* L. Varieties to maize jassid *Zygindia manaliensis* (Singh). *J. ent. Res.***18** (1): 93-94.

Sekhon, S.S., Sajjan, S.S. and Bhinder, S.S. (1986) Influence of cultural practices on the incidence of shoot fly, *Atherigona naqvii* Steyskal on winter maize in the Punjab. *Indian J. Ecol.* **13** (1): 105-108.

Seshu Reddy, K.V. and Davies, J.C. (1977) Species of *Atherigona* in Andhra Pradesh. PANS **23** (4): 379-383.

Sharma, V. K. (1987) Maize insect pest problems, present status and future of host plant resistance in India. In: Towards *Insect Resistant Maize for the Third World.* Proceeding of the International Symposium on Methodologies for Developing Host Plant Resistance to Maize Insects. Mexico *D.F. CIMMYT.* pp. 281-85.

Sharma, V.K. and Singh, J.M. 1972. Chemical control of maize borer, chilo zonellus (Swinhoe). Annual Report 1971-72. Directorate Experiment station, G.B. Pant University of Agric. and Tech., Pantnagar, 337-339.

Sharma, V.K. and Singh, J.M. (1975) Screening for resistance to *Atherigona* spp. in spring-sown maize. *Indian J. Ent.* **37**: 3

Singh, B., Kanta, U. and Sandhu, G.S. (1978) Sources of resistance to the cereal jassid, *Zygindia manaliensis* (Singh) in maize germplasms. *J. Res. Punjab agric. Univ.* **15**: 350-352.

Singh, D.P., Brar, K.S., Uma Kanta, and Gourav, G. (2004) Development of bio-intensive IPM module for the management of maize borer, *C. partellus* (Swinhoe). Paper presented at National Conference on role of Bio-Pesticide, Bio-agents and Bio-fertilizers for sustainable Agriculture and Horticulture, Lukhnow.

Singh, H. and Uma, Kanta. (2004) Evaluation of insecticides as foliar spray and time of application against the shoot fly, *Atherigona naqvii* Steyskal on summer maize. *Pest Mgt. Econ. Zoo.* **12** (2): 203-207.

Singh, L. and Uma, Kanta (2002) Determination of damage threshold for economic control of *Chilo partellus* (Swinhoe) in maize. *J. Insect Sci.* **15** (1 & 2): 75-76.

Singh, M.P. and Sinha, M.M. (1965) Some cutworm pests new to Bihar. *Indian J. Ent.* **22**: 113-114.

Singh, N.N. (1988) Maize improvement in India-Problems and Prospects. pp. 46-51. In: *Proc. Third Asian Regional Maize Workshop*, Chinese Academy of Agricultural Sciences and the CIMMYT- Asian Regional Maize Program, Kunming and Nanning, China, June 8-15, 1988.

Singh, R. (1991) Studies the damage and losses caused by jassid, *Zygnidia manaliensis* (Singh) to winter and spring maize and its control. M.Sc. Thesis, Punjab Agricultural University, Ludhiana.

Singh, S. (1969) Fifteen new species of jassids (Cicadellidae) from Himachal Pradesh and Chandigarh. *Res. Bull. (N.S.) Punjab Univ.* **20** (3 & 4): 339-361.

Singh, S.P. (1963) Studies on the development, oviposition and appearance of different broods of maize borer, *Chilo zonellus* (Swinhoe) with a view to timing the insecticidal application, Unpubl. *M.Sc. Thesis, Punjab Agric. Univ. Ludhiana.*

Snehdeep, Kaur and Kanta, U. (2003) New Records of Pyralid, *Loxostege massalis* Walker on Kharif Maize. *Insect Environ.* **9** (1): 16.

Srinivas, P.S. and Panwar, V.P.S. (2003) Combined effects of intercropping maize with pulses and potash fertilizer on stem borer, *Chilo partellus* (Swinhoe) infestation in maize. *Ann. Agric. Res.* **24**: 461-65.

Srivastava, B.K. and Khan, R.M. (1961) Field tests with newer insecticides for control of the maize armyworms. *Indian Agric.* **5** (1) 113-114.

Srivastava, B.K. and Rathore, Y.K. (1961) Armyworms-a menance to maize in Rajasthan. *Indian Fmg.* **11** (6) 25-31.

Teetes, G.L., Seshu Reddy, K.V., Leuschher, K. and House, L.R. (1983) *Sorghum Insect Identification Handbook.* Information Bulletin No. 12. International Crop Research Institute for Semi-Arid Tropics, Hyderabad, India, pp. 124.

Tseng, C.T., Tollefson, J.J. and Guthrie, W.D. (1984) Evaluation of maize single cross hybrids and inbred lines for resistance to third instar black cutworm larvae (Lepidoptera: Noctuidae). *J. econ. Ent.* **77** (3): 565-568.

Uma Kanta and Sekhon, S.S. (1989) Location of sources of resistance amongst different varieties and inbred of *Zea mays* to *Chilo partellus.* Paper presented in Int. Cong. on Biology and Control of Pests of Agricultural and Medical Importance, Madurai, India.

Uma, Kanta and Sekhon, S.S. (1994) Location of sources of resistance amongst different varieties and inbreds of *Zea mays* to *Chilo partellus. J. ent. Res.* **18**: 1-6.

Verma, A. and Saxena, H. (1989) Field recovery of an exotic parasite, *Allorphogas pyralophagus* Marsh against sorghum borer, *Chilo partellus* (Swinhoe). *Indian J. Pl. Prot.* **17** (1): 101-102.

Verma, A.N. and Ramzan, M. (1965) *Annual Report* (1964-65). Department of Entomology, Punjab Agricultural University, Ludhiana.

Verma, G.D., Singh, J. and Singh, I.P. (1979) Field evaluation of some insecticidal treatments for control of *Rhyacia herculea* Corti and Draudt on *rabi* maize in Bihar. *Entomon* **4**: 129-131.

Young, W.R. (1962) A suggestion for insect control in experimental maize. Proceedings of maize conference part 1. Conference of hybrid Maize seed Producers, New Delhi.

Theory and Practice of Integrated Pest Management
R. Arora, B. Singh and A.K. Dhawan (Eds)
Scientific Publishers (India), Jodhpur 2011, pp. 370 - 376.

SCIENTIFIC PUBLISHERS

21

CHANGING SCENARIO OF INSECT PESTS OF WHEAT AND THEIR MANAGEMENT

Beant Singh

Department of Plant Breeding & Genetics
Punjab Agricultural University, Ludhiana - 141 004, India

Wheat is major *rabi* cereal crop of Punjab state and was grown over 35.26 lac hectares with annual production of 157.33 lac tonnes of food grains during 2008-09 (Anonymous, 2010). The annual monetary loss of Rs 41368 millions has been reported due to insect pests in wheat from India. (Dhaliwal *et al.,* 2010). They further reported that the losses due to insect pests in wheat have increased from 3 to 5 per cent after green revolution. This could be ascribed to change in the pest scenario in wheat. Till late sixties, barring *Gujhia* weevil and *Toka* (grasshopper) there was hardly any serious pest of wheat, but with the introduction of high yielding semi dwarf varieties, changed cropping system and use of new agro techniques, some new insects attain the status pest. The termites, aphids, armyworm, American pod borer and brown mite are the major pests of wheat (Deol, 2002).

Presently, the pest scenario of wheat is again undergoing a change with the changes in climate and tillage systems and less important insect pests like pink stem borer *Sesamia inferens* Walker and root aphid *Raphalosiphum rufiabdominalis* are becoming important pests of wheat. Dhaliwal *et al.* (2010) has included pink stem borer in the list of emerging insect pest of different crops. Summarized description about their damage and management in wheat is described in this article.

21.1. TERMITES OR WHITE ANTS, *Odontotermis obesus, Microtermis obesi*

Distribution. Mainly in the northern and central India, but also in some pockets of peninsular India. In Punjab, Termites that were previously known to be serious pests of rainfed crop (sub mountainous regions comprising parts of Ropar, Hoshiarpur and some parts of Gurdaspur) are now a day commonly infesting irrigated crop also.

Symptoms of damage. They are social insects and live in colonies known as termitaria. Termites feed on roots and underground portion of stem causing wilting and withering of the affected plants. Patchy gaps are formed due to the death of seedlings resulting in poor or uneven plant stand. The workers of the termite may be invariably found in soil or at the base of freshly damaged or wilting seedlings. At the earing stage when plant is attacked the roots are completely or partially eaten. Such plants dry up without producing any grains (white ears) and can be easily pulled out.

Management

(i) **Seed treatment.** Before sowing, treat the seed with Regent 5% SC (fipronil) @ 6 ml or Dursban/Ruban/Durmet 20 EC (chlorpyriphos) @ 4 ml or Thiodan 35 EC (endosulfan) @ 7 ml/kg seed. For this, dilute 240 ml of Regent or 160 ml of Dursban/Durmet/Ruban or 280 ml of Thiodan in one litre of water and spray over 40 kg seed uniformly spread on the *pucca* floor, tarpaulin or polythene sheet. The seed should be turned over for proper mixing. The treated seed must be kept overnight for drying before sowing.

(ii) **Broadcasting treated soil.** If the termite damage is noticed in the standing crop, dilute one litre of Thiodan 35 EC (endosulfan) in 2 litres of water, mix in 20 kg soil and broadcast evenly in one acre. This is the only effective method to control termites in standing crop under rainfed area. If the crop is irrigated, apply light irrigated, after broadcasting.

Seed treatment is the most effective and economical method of termite control as compared to broadcasting of treated soil. The insecticides applied for termite control will also control root aphid and *Gujhia* weevil. Seed treated with chlorpyriphos is less attacked by birds.

Biocontrol of termites. Termites can managed with a potentially lethal fungus, such as (*Metarhizium anisopliae*) or (*Beauveria bassiana*), using an attractant, such as the fungus (*Gloeophyllum trabeum*) or its volatile products. Seed treatment with *Metarhizium anisopliae* also provide promosing results.

21.2. APHIDS, *Sitobion avenae, S. miscanthi, Rhopalosiphum padi* AND *R. maidis*

Distribution and Development. All wheat growing areas, especially in NWPZ and Peninsular India. The aphids exist in different stages, viz., winged (alates), wingless (apterous) sexual and asexual forms. The rapid spread takes place through asexual reproduction where females give rise directly to nymphs rather than eggs. Infestation usually occurs during second fortnight of January till crop maturity.

Symptom of damage. Wheat aphids pose a serious threat to wheat cultivation in Punjab. These are soft bodied, green to blackish green louse like insects found in small colonies on leaves and earhead. Aphid develops colonizing on leaves, leaf sheath, flag leaf and in the spikelets of the ear heads. Both nymphs and adults suck the sap from the leaves and maturing grains. Due to their rapid multiplication within a few days, they usually cover the entire surface of the shoot causing yellowing, curling and subsequent drying of leaves, which ultimately leads to reduction in number, and size of earheads. The aphids also excrete honeydew on which black sooty mould develops that hinders photosynthetic activity of the leaves. The high populations of aphids at the earhead stage can cause substantial losses in yield. The damage of the aphids first starts from the periphery of the field, particularly from the side on which trees are present. If sprayed there, its further spread in field can be checked. Some natural enemies like coccinellids, syrphids and chrysopids prey upon the aphids in field and keeps their population below economic threshold level. But in case of high population, chemical control should be practiced.

Management. As the colonizing of aphids starts, spray the crop with 40 ml of imidacloprid 200 SL or 20 gm of thiamethoxam 25 WG or 12g of clothianidin 50 WDG or 150 ml dimethoate 30 EC or oxydemeton methyl 25 EC in 80-100 litres of water per acre with knap-sack sprayer or in 30 litres of water with power sprayer. If necessary repeat the spray after 15 days.

Note

1. Control aphids at earhead stage at economic threshold level of 5 aphids per earhead as recorded from 10 randomly selected earheads in each of the 4 quarters of one acre.
2. Since the aphids appear first on borders of the crop. Spray only the infected strip to check their further spread.

Biocontrol of wheat aphids. wheat aphids are attacked by a variety of predators and parasites. These include the parasitic wasps, several species of lady beetles and green lacewings. These beneficials help keep greenbug populations from increasing to damaging levels in many years.

(i) Lady Beetles. Several species of lady beetles can be found in wheat out of which the seven-spotted lady beetle *Coccinella septempuncutata* areI most common.

As a group, lady beetles provide some control of aphids in small grains every season. But since lady beetles consume the aphids they leave no evidence of their control. Sometimes the only indication is the lack of significant aphid problems later in the year. No one lady beetle species can completely reduce an aphid population, and research has shown that augmentation by releasing lady beetles is of little benefit. The most significant action a farmer can take to help lady beetles control aphids is conservation. Conservation primarily means avoiding broad-spectrum insecticides and using the lowest rate of insecticide that achieves adequate control of the target pest. It is also helpful to apply insecticides when the lady beetle population consists of egg or pupal stages rather than young larvae.

(ii) Parasitic wasps. *Lysiphlebus testaceipes*, *Diaeretiella rapae* and *Aphelinus varipes* are tiny, black wasps which parasitize wheat aphids wherever this pest is found. Although the adult wasps are not commonly seen, wheat producers should be able to recognize the distinctive parasitized wheat aphids mummy which remains attached to the wheat leaf. Greenbug mummies parasitized by *L. testaceipes* and *D. rapae* are beige or tan and are round and swollen.

(iii) Lacewings. Green lacewings are found in many crops, including wheat. However, there is little information on the importance of lacewings in the biological control of greenbugs. The larval stages of all lacewings are predators which feed primarily on aphids but also on insect eggs and small caterpillars. The adults of some lacewing species, such as *Chrysoperla carnea*, feed only on honeydew or pollen. Lacewing larvae are not common in wheat but compliment other predators.

21.3. ARMYWORM, *Mythimna separata*

Distribution and Development. Mostly in the warmer climates of central India and to some extent in northern plains. The larvae are found in the cracks of soil and hide during the day but feed during night or early morning. In wet and humid weather, they may feed during day time also.

Symptom of damage. The pest is destructive in the larval stage only. The newly hatched larvae feed on tender leaves. The grown up larvae feed on leaves, flag leaf, awns, glumes and on the developing grains. The larvae are shy of sun and very rarely found on the plants during daytime. However, they remain hidden in tillers, feed at night in the shade or at dawn and dusk. But its attack can be located from the presence of larvae and blackish green faecal pallets on the ground close to the stem of the plant. At higher populations it defoliates the plants completely and feed on the grains in the milky stages. Under such situations it can cause yield losses upto 40 per cent. Late sown, high yielding, heavily fertilized and well irrigated crop suffers more losses due to attack of armyworm.

Management. To control this pest, spray the crop with Sevin/ Hexavin 50 WP (carbaryl) @ 1.2 kg/acre or Nuvan 85 SL (dichlorvos) @ 200 ml/acre or Ekalux 25 EC (quinalphos) @ 400 ml /acre in 80-100 litres of water with a knap-sack sprayer. To obtain better control always spray the crop in the evening hours because larvae are more active at this time.

21.4. AMERICAN POD BORER, *Helicoverpa armigera*

Distribution and Development. This is a sporadic pest of wheat and do not come in pest form every year. In the years when its population goes high it damages by feeding on the grains in the ear heads. The incidence is high in fields where wheat follows cotton. The young larvae feed on the leaves, whereas the grown larvae feed directly on the grains in the earheads. It generally appears in the month of March-April. One can identify this pest by the presence of whitish faecal pallets on the ground.

Management. Spray the crop with Sevin 50 WP (carbaryl) @ 1.2 kg/acre or Ekalux 25 EC (quinalphos) @ 800 ml/acre by mixing in 100 litres of water. In some fields mixed infestation of American pod borer and armyworm has also been observed, In such situations insecticides recommended against American pod borer should be used, as these will also control armyworm.

21.5. PINK STEM BORER, *Sesamia inferens* Walker

Symptoms of damage. This pest is mainly observed in fields where rice-wheat cropping pattern is practiced. It generally attacks the wheat crop at seedling stage. The larva bore into the stem of young plant and kills the central shoot causing 'dead heart'. The infested tillers first looks pale brown and ultimately dry up. At the ear emergence due to its attack 'white ears' are produced which have little or chaffy grains.

Management. Spray the crop with quinalphos 25 EC @ 800 ml/acre by mixing in 80-100 litres of water. Removal or destruction of the stubbles at the time of first ploughing after harvest of rice reduces the carry over to wheat. Ploughing and flooding field is also effective in killing the larvae.

21.6. BROWN WHEAT MITE, *Petrobia latens*

Distribution. In most of the wheat growing areas, under rainfed conditions, especially in the states of Rajasthan, Haryana and M.P. Sometimes, it is a pest in humid and warm conditions of irrigated areas also.

Symptoms of damage. The brown wheat mite is cosmopolitan pest of wheat. It is more serious on unirrigated than the irrigated crop. The nymphs and adults of mite suck sap from leaves, leaf sheath, green stem and spikes. Some times it develops serious population from tillering till harvest. The damage appears first on the lower leaves with slowly traveling upward in the form of whitish minute specks on leaf lamina. The infected plants loose their cell sap, become chlorotic and remain stunted. Severely damaged leaves starts drying from tip backwards and highly infested crop gives palish appearance. The grains get shrivelled and loose boldness. The mite can be seen on the plants during day time when sun shine in bright and wind velocity is low.

Management. Heavy rains and hailstorm reduce its population temporarily. But when it appears in serious form spray any one of the following insecticides:

Metasystox 25 EC (oxydemeton methyl) or Rogor 30 EC (dimethoate) or Nuvacron 36 SL (monocrotophos) @ 150 ml per acre in 80-100 litres of water with knap-sack sprayer or in 30 litres of water with power sprayer. If necessary repeat the spray after 15 days.

21.7. ROOT APHID, *Raphalosiphum rufiabdominalis*

In recent years, the incidence of root aphid is also observed in wheat. It is sporadic pest of wheat and its infestation is seen in isolated fields.

Symptoms of damage. The symptom of damage of root aphids appears on 3-4 week old crop. It such sap from the roots and collar region of wheat plant near soil surface. The effected tillers shows signs of wilting and sometimes its damages confuses with the damage of termite as both these pest attack at same time buts its damage distinguished from termite by the presence of insect in the collar or root zone of wheat plant and also it does not destroy the roots as did termites.

Management. The root aphids can be managed by applying irrigation to the effected field. Apart from this some of the insecticidal seed treatments are also giving promising results for root aphid control.

REFERENCES

Anonymous (2010) *Package of Practices for Crops of Punjab - Rabi.* Punjab Agricultural University, Ludhiana. pp. 1-17.

Atwal, A.S. and Dhaliwal, G.S. (1997) *Agricultural Pests of South Asia and their Management.* pp 179-200. Kalyani Publishers, New Delhi

DeBach, P. 1974. *Biological Control by Natural Enemies.* Cambridge University Press 323 pp.

Deol, G.S. (2002) Latest trends for insectmanagement in wheat. Paper presented in specialized workshop on *"Identification and management of weeds, insect-pests and diseases in wheat"*, held at CETWPT, P.A.U., Ludhiana.

Dhaliwal, G.S., Jindal, V. and Dhawan, A.K. (2010) Insect Pest Problems and Crop Losses: Changing Trends. *Indian J. Ecol.* 37 : 1-7

Knutson Allen, Michels, G.J. Jr and Gilstrap Frank (1997) *Biological Control of Insect Pests in Wheat.* Texas A & M University.

Parker, B.L., El-Bouhssini, M. and Skinner, M. (2001) *Field Guide: Insect Pests of Wheat and Barley in North Africa, West and Central Asia.* International Center for Agricultural Research in the Dry Areas. Aleppo, Syria. 120 pp.

Singh, V.S. (1986) Management of insect and mite pests. In: *Twenty Years of Coordinated Wheat Research 1961-86* (Ed.) *J.P. Tandon and A.P. Sethi)* Wheat Project Directorate, All India Coordinated Wheat Improvement Project, ICAR. pp. 158-88.

Theory and Practice of Integrated Pest Management
R. Arora, B. Singh and A.K. Dhawan (Eds)
Scientific Publishers (India), Jodhpur 2011, pp. 377 - 409.

22

MANAGEMENT OF INSECT PESTS IN CITRUS

P.K. Arora and D.R. Sharma*
Regional Research Station, Abohar - 152 116
*Department of Horticulture
Punjab Agricultural University, Ludhiana - 141 004, India

There is a significant growth in production and area expansion under citrus for the last few years but our productivity and fruit quality is far below to those of advanced countries. The major reason for this is the inadequate insect-pest management practices resulting in short productive life of plants and decline. Citrus suffers extensively on account of the varied insect-pest problems. A brief account of the insect-pests infesting citrus and their management is presented here:

22.1. CITRUS PSYLLA, *Diaphorina citri* Kuwayama

Citrus psylla is most destructive and the most important of all the insect pests of citrus in Punjab, Haryana, Himachal Pradesh, Maharashtra and Delhi but is of minor importance in Southern Peninsula. Damage is caused by the nymphs and adults. The adult is grey coloured, small, 3 mm in length, active, and rests on the leaf surface with closed wings, the tail end of the body being turned upwards. The wings are membraneous, semitransparent, with brown band in the apical half of the forewings. The hind wings are shorter and thinner than the forewings. The nymphs are flat, louse-like and orange-yellow in colour. The pest is active throughout the year, but only adults are to be found in winter months when it is very cold (Arora *et al.,* 1997). With the onset of new vegetative flush in February-March in citrus the overwintering females lay almond-shaped, orange and stalked eggs on unopened tender leaves and shoots of citrus trees. The lower end of the egg stalk is embedded in the plant tissue. The light-yellow nymphs emerging from the eggs have a tendency to stick close to the egg-shell and these can be seen congregating on the young half-opened leaves, but later instars can move to older leaves also. There are five nymphal instars and there are 8-9 overlapping generations in a year under Punjab conditions. The psyllid nymphs and adults suck the cell sap from the floral buds, young leaves and tender shoots. The infested twigs die-off from tip backwards. The adjacent branches also dry-up due to the toxins injected by the psyllids. Sooty mould developed on the honey dew excreted by the nymphs also affects the plants by interfering with the photosynthetic activity. The loss caused ranges from 83 to 95%. Dhara *et al.* (1989) conducted ecological studies on citrus psylla on limes and recommended sampling sizes for eggs, nymphs and adults as 40, 38 and 19 shoots per tree, respectively.

Association with greening disease. Psylla is also known as a potential vector of the greening disease of citrus. Fifth instar nymph is the vector of greening disease, but only adult transmits the causal organism which accelerates the decline (Capoor *et al.,* 1967). Hence control of psylla at early instars is very essential.

Other Hosts. Apart from citrus, it also breeds on Curry leaf (*Murraya koenigi*). Chakravarthy *et al.* (1998a) studied the biology of psylla on Acid lime, Sathgudi sweet orange and Curry leaf and reported that curry leaf was the most preferable host with high rate of fecundity and short duration of life cycle. So do not grow *Murraya* in the vicinity of citrus orchards since it acts as breeding ground for psylla.

Management. Control measures like insecticides, botanicals, mineral oils, insect growth regulators, biocontrol and cultural control have been tried by different workers against citrus psylla with encouraging results. A brief account of these measures is discussed here.

Chemical Control. This pest cannot be controlled easily because of greater adult longevity, high fecundity and more number of overlapping generations. But if timely efforts are made to control it during February-April and September when there is restricted movement of nymphs on tender foliage, we can manage its population by spraying 1250 ml Rogor (dimethoate) 30EC or 1000 ml Metasystox (oxydemeton-methyl) 25EC or 200 ml Confidor (imidacloprid) 17.8 SL or 160g Actara (thiamethoxam) 25 WG in 500 litres of water per acre at ten days intervals. For its effective control, start spraying at the time of on-set of new flush because over wintering adults oviposit their eggs in the half opened leaves and repeat the spray after ten days interval if needed (Anonymous, 2010; Aulakh *et al.,* 2008). The insecticides like difenthiuron, dimethoate, imidacloprid, monocrotophos, oxydemeton-methyl, phosalone, phosphamidon, profenophos, quinalphos, thiamethoxam and triazophos are tried successfully against citrus psylla under different climatic conditions (Arora *et al.,*

2005a; Batra *et al.,* 1990; Chakravarthy *et al.,* 1998b; Patel *et al.,* 1998). Shivankar *et al.* (2000) also mentioned that two to three sprays of monocrotophos / dimethoate / thiodemeton at 10-15 day interval, soil application of aldicarb/dimethoate at 16 ml a.i. per tree or trunk treatment with Azodrin 40 were effective against citrus psylla. Buitendag and Bronkhrost (1980a) tried new technique i.e. trunk injection of undiluted monocrotophos and obtained good control of psylla. Trunk application products, methamidophos (citrimet), with relatively short control period (16-18 days) and imidacloprid, reduce the likelihood of psylla developing resistance to the trunk applied products (Buitendag, 1993).

Mineral oils / Plant / Animal products. Apart from insecticides different workers have tried plant/animal products, Insect growth regulators (IGR's) and mineral oils for the control of citrus psylla. Beattie *et al.* (1999) has shown that petroleum spray oils (PSOs) can be used to control pests simultaneously by drowning them and/or affecting behaviour. The old, broad-range PSOs were sometimes phytotoxic but with the development of oil formulations, these have become more effective against pests and safer to plants (Cen *et al.,* 1999).

Chakravarthy *et al.* (1998b) achieved >90% control of citrus psylla on Sathgudi sweet orange after spraying with azadirachtin (0.03%) and neem (*Azadirachta indica*) oil. However, insecticides were more effective and persistent in action when compared to botanicals like neem oil (1.0%), pongamia oil (1.0%) and animal origin insecticide, fish oil rosin soap (0.2%) which were only effective for a short period. Spray of neem and naffatia (*Ipomoea fistulosa / I. carnea*) leaf extracts was found effective against citrus psylla by Borad *et al.* (2001). The extracts of botanicals like *Vitex nigundo, Acorus calamus* and Nemax (*A. indica*) gave more than 90% mortality of psylla nymphs after 20 hours of treatment (Shivankar and Singh, 1997). Crude plant extracts of Garlic (*Allium sativum*) and Datura (*Datura stramonium*) were also found to be promising against citrus psylla nymphs (Arora *et al.,* 2008a).

Biocontrol

Predator. Many species of predators viz., *Coccinella septempunctata* L., *C. rependa, Menocheillus sexmaculatus, Chilocoris nigritus, Brumoides suturalis* (Coccinellidae : Coleoptera), *Apertochrysa crassinervis* (Chrysopidae : Neuroptera) and Syrphids have been reported on psylla nymphs (Shivankar and Rao, 2001b). Gupta and Bhatia (2000) observed four species of predatory coccinelids, viz., *Leis dimidiata (Harmonia dimidiate), Menochilus sexmaculatus (Cheilomenes sexmaculata), Micraspis cardoni* and *Coccinella septempunctata,* and a chrysopid predator, feeding on nymphs. Liu (1989) reported that psylla population in field is effected by *Chrysopa boninensis (Mallada boninensis*) density.

Parasitoids. *Tamarixia radiata causes upto 95% parasitism in D. citri* in North India (Atwal, 1962). The parasitoid, *T. radiata,* is one of the promising parasites for regulating the psylla population throughout the world (Shivankar *et al.,* 2000). It is the primary ecto-parasite. Eggs are laid on third, fourth and fifth instar nymphs. The larvae remain on the ventral surface of nymph and feed on its body contents. Adults pupate inside mummified nymphs and emerge out in 9-14 days by chewing a hole through the thorax of the host. In China, population of *Tetrastichus (Tamarixia)* along with some environmental factors determines the population density of psylla in field (Liu, 1989b).

Entomopathogenic Fungi. Pathogenic fungi, *Fusarium coccophilum (Nectria flammea), Cladosporium oxysporum* and one undetermined fungus species AcH1611 are known to attack nymphs of *Trioza erytreae* in South Africa where as adults are attacked by *N. flammea* only (Berg and Berg, 1996). Role of these fungi in biological control of psylla has not been exploited fully yet.

Cultural Control. Prune the affected and dried shoots and give ventilators in the inner dense canopy to maximize the light interception.

IPM Module. IPM Module consisting of Clean cultivation + *Mallada boninensis* @ 30 larvae/tree + Sweet Flag (3%) + *Verticillium lecanii* @ 5 g/ litre + trap crop- *Murraya* on border lines, proved effective against psylla (Shivankar and Rao, 2008). Berg (1999) advocated some methods to check the spread of greening disease in South Africa. These methods include, restricting the area of citrus production; growing cultivars with coinciding flushing periods, growing a single cultivar in an orchard; avoiding pruning and skirting; eradicating infected plant material; inspecting and monitoring orchards for psylla outbreaks and chemical control.

22.2. CITRUS LEAFMINER, *Phyllocnistis citrella*

The pest has originated in South-East Asia and has now spread across all the citrus tracts of the world. In India Citrus leafminer (CLM) was first recorded by Stainton in 1856 and is now one of the most destructive pests of citrus especially under nursery conditions (Atwal, 1964). Under Punjab conditions, the leafminer remains active from March to mid - December and it has two critical periods of infestation i.e. last week of April to mid-June and last week of July to mid-October. The leafminer larvae feed on tender leaves making silvery serpentine mines. It may also mine the epidermis of tender branches and very rarely fruits. The full grown larva measures 5.1 mm in length and is pale yellow or pale green with light brown well developed mandibles. Adult is a tiny moth, measuring 4.2 mm across the wings. On the front wings there are brown stripes and black spots along the tips. The hind wings are pure white and both pairs are fringed with hairs. The female lays minute, flattened, transparent eggs on young leaves or tender shoots, usually on the lower surface near the midrib. The legless larvae emerging from these eggs mine into the leaf tissue and form zigzag galleries. Mature larvae settle down in the galleries near the margin. They spin cocoons for pupation and due to this the leaves get distorted and crumpled. The affected foliage eventually dry up and fall down. This also encourages the incidence and severity of citrus canker.

Losses and ETL. According to Atwal (1964) CLM causes the loss of 5-10% to citrus industry in Punjab. Boughdad *et al.* (1999) studied the pest status and biology of leafminer in Morocco and reported that each leafminer larva destroyed from 4 to 100% of leaf surface, before reaching the pupal stage. This study concluded that only young citrus plantations require control measures against CLM. Wang *et al.* (1999) showed that after damage by the leafminer, the leaves had certain compensatory abilities in photosynthesis and tissue repair. Pena *et al.* (2000) studied the effect of CLM density on yield of 'Tahiti' lime and calculated injury levels indicated that 16-23% and 18-85% of leaf area damage caused significant yield reductions in 15 and 5 year old trees, respectively. Huang and Li (1989) studied the damage and economic threshold of CLM on citrus and concluded that when <20% of leaf area was damaged there was no influence on growth, development or yield. The economic threshold was estimated at 0.74 larva per leaf. On the same lines, Wang *et al.* (1999) estimated the damage of CLM to 4 citrus varieties and have set economic threshold on Xuegan variety at 1.21 larvae of first instar per tender leaf.

Association with other pests. Folds of the leafminer affected leaves may harbour other insects such as mealy bugs, dusky cotton bugs and citrus psylla (Atwal, 1964). This is also known to encourage the incidence and severity of citrus canker. Kacic *et al.* (1997) observed that there is a relationship between the injury by this miner and citrus canker caused by bacterium, *Xanthomonas compestris* pv. *Citri*. Rodrigues *et al.* (1998) also suggested that CLM can disseminate and facilitate infection by canker. In Brazil, Chagas *et al.* (2001) reported that damage caused by this insect was responsible for the increase in citrus canker infestation. An 11-fold lower infection rate was observed compared to the leaves injured by the insect 7 days after inoculation. The significance of the damages caused by CLM in terms of the increase of citrus canker is noted since the favourable microclimatic conditions of temperature and relative humidity inside the mines built by the larvae account for an improved development of the bacterium.

Management

Chemical Control. The recent trends in controlling CLM include the use of chemicals which are ecofriedly. It includes insecticides, plant/animal products, Insect growth regulators (IGR's) and mineral oils.

Time of application of insecticide is as important as the choice of insecticide. Perovic and Malidzan (2001) reported that beginning of shoot growth and leaf flush are the optimal period for insecticide application whereas according to Wu (2000) the best times for spraying were when the young leaves were 1 cm in length and a week later. Su (2000) while working on variety Ponggan reported that the most suitable period for spraying the shoots was when they were 2-3 cm long. Spraying the nursery and full grown plants with 1875 ml Durmet (chlorpyriphos) 20EC or 1250 ml Hostathion (triazophos) 40 EC or 1000 ml Ripcord (cypermethrin) 10EC or 500 ml Sumicidin (fenvalerate) 20 EC or 200 ml Confidor (imidacloprid) 17.8 SL in 500 litres of water per acre at fortnightly intervals during flushing periods is recommended. It is further advised to alternate the spray of conventional group of pesticides with pyrethroid group to overcome the problem of resistance and resurgence of other pests (Anonymous, 2010). Various insecticides like abamectin, acetamiprid; carbaryl, deltamethrin, fenvalerate, fenpropathrin, lambda cyhalothrin, imidacloprid and permethrin are recommended for the control of CLM by different workers (Arora *et al.*, 2008b; Bei *et al.*, 1999a,b; Radke and Thakare, 1989; Seraj, 1999a,b; Sharma *et al.*, 2005b; Su, 2000; Wu, 2000; Zhang *et al.*, 2001). Rao and Shivankar (2008) reported that (i) abamectin @0.32 ml per litre of water followed by novaluron @0.87 ml per litre of water and petroleum spray oil @ 3.72 ml per litre of water and (ii) abamectin @0.32 ml per litre of water followed by spinosad @0.34 ml per litre of water and novaluron @0.87ml per litre of water; were very effective for a period of 15 days. But, Vargas (1999) observed that foliar applications of tebufenozide (0.24% a.i.) and abamectin (0.018%) were not so effective.

Mansant *et al.* (1999) discussed the role of Confidor 200SL in developing IPM strategies in Spain and mentioned that that it has become one of the fore most products for foliar treatment. Because of its good systemic properties it became a very important product in practice for application via the irrigation system or undiluted application to the bark. Soil applied imidacloprid (Sharma *et al.*, 2001a,b; Cruz and Dale, 1999) and thiamethoxam had more persistence compared to acetamiprid (4 g per plant) as well as foliar application of imidacloprid @0.008% (Bhatti and Sharma, 2010). This mode of application also meets the requirement that beneficial organisms in the sensitive citrus ecosystem should come to no harm. However, soil application of granular insecticides viz. cartap hydrochloride 4G, chlorpyriphos 10G, fipronil 0.3G and phorate 10G each @20kg/acre proved inferior to foliar application of decis and fenvalerate @ 0.01% a.i. (Sharma *et al.*, 2001c). Seraj (1999b) injected demeton-s-methyl into lime tree stems by trunk injection technique and obtained good control of CLM.

Perovic and Malidzan (2001) reported that Imidacloprid, abamectin + mineral oil and mineral oil manifested high efficacy in CLM control in citrus plantations. Seraj (1999c) mentioned that oils are an essential component of integrated pest management programme for citrus based on the use of natural enemies. He has evaluated the efficacy of 2 and 0.5% (v/v) petroleum spray oils and found that these suppressed CLM population significantly. Combining the oil with copper fungicides did not reduce its efficacy against the CLM. Petroleum oils are known to act as an ovipositional deterrent to the pest.

Biocontrol. Many species of parasitoids viz., *Tetrastichus phyllocnistoides, Elamus claripennis, Cirrospilus quadristriatus, Bracon phyllocnistoides, Tetrastichus* spp and *Simpieses purpurea* have been recorded causing 30-45% parasitisation (Thakare *et al.,* 1984-85). According to Atwal (1964), the larval predator blackchalcid (*Ageniaspis* spp) and chrysopid hold good promise for CLM control. *Ageniaspis* spp. alone causes more than 80% larval parasitism. In France, Malausa and Vacante (1997) imported *Ageniaspis citricola* from Florida and used it in biological control programme against CLM. Similarly, Willink *et al.* (1998) described the biological control of CLM using inundative releases of *A. citricola* in citrus orchards in Argentina. Vercher *et al.* (2000) imported different parasitoid species for the biological control of CLM in Spain and these six species viz. *Ageniaspis citricola, Quadrastichus* sp., *Semielacher petiolatus, Galeopsomyia fausta, Cirrospilus ingenuus* and *Citrostichus phyllocnistoides*, were released in field. Out of these, *Q. sp.* and *C. phyllocnistoides* became established in many orchards. It was observed that parasitism increased while the egg laying density, per cent foliar surface destroyed by the CLM and adult CLM density decreased significantly. Viggiani (2001) reported that *S. petiolatus* is quite successful against CLM in Italy while Mineo & Mineo (1999) described it as the most abundant parasitoid of CLM in Italy. It was observed that 4th instar larvae were parasitized by *S. petiolatus*. Similarly in Algiers, *S. petiolatus* yielded some satisfactory control results on CLM (Doumandji *et al.,* 1999).

Plant/Animal products. Muthiah *et al.* (1998) reported that 3% neem oil treatment resulted in a decrease in damage by CLM but not significantly. On the other hand, spray of neem and naffatia (*Ipomoea fistulosa / I. carnea*) leaf extracts was found effective against CLM by Borad *et al.* (2001). Fish oil, resin soap, pongamia, mahua and neem oils were found effective against CLM by Katole *et al.* (1993). Neem seed cake extract (2%) was more effective than neem, mahua, pongamia oils (2 & 4%) and pongamia seed (2%) against CLM on Lime (Nair *et al.,* 1975; Dhara *et al.,* 1990, 1993b). Nestor *et al.* (1998) obtained very good control of CLM with Citrolina and neem each @5%. Pawar *et al.* (1999) found that neem based insecticide, achook (1%) and emulsified neem oil (1%) are very effective against CLM.

Pheromone Traps. In China, traps baited with the sex attractant (7Z, 11Z) — 7,11-hexadecadienal caught an average of 103.9 adults/day over a 6-day period and traps containing 1 mg attractants were effective over a distance of 280 meters and for 20 days.

Cultural Control. Heavy pruning and burning of the infested leaves is also found effective at smaller scale. But avoid the pruning during active growth periods as it may induce further new flush and thereby allow the pest to have more number of generations. Prune the heavily infested shoots during winter only (Ayyar, 1963). In Bermuda, the Department of Agriculture has requested that the public should prune and destroy the very new growth from their citrus trees, so as to reduce the amount of potential feeding sites for newly introduced CLM (Jesssey, 1999).

22.3. WHITEFLY COMPLEX

Of the 30 species of Aleurodidae on citrus throughout the world, 12 are found in India (Pruthi and Mani, 1945). Out of these, Citrus whitefly (*Dialeurodes citri* Ashmead) and black fly (*Aleurocanthus woglumi* Ashby) are the serious pests of citrus in Punjab. These are active from April to November, but cause more damage during April-May and September-October. The nymphs and adults suck the cell sap from the leaves thereby reducing the plant vigour. Severely infested foliage turns yellowish green, curl and finally shed. Sooty mould grows on the voluminous honey dew excreted by the nymphs and adults (Arora *et al.,* 2005b) which gives tree a black look and is known to interfere with the photosynthesis. Affected trees produce few flowers and the fruits are insipid. In the sooty mould infested leaves, the contents of total chlorophylls, proteins, amino acids, total phenols, orthodihydroxyphenols, nitrogen, phosphorus and potassium decreased, whereas total sugar increased (Arora *et al.,* 2000). The adult of citrus whitefly (CWF) is a minute insect, measuring 1.02-1.52 mm. The eyes are transparent, red and kidney-shaped, with the lower half covered over with bristles. The head is somewhat pointed. The wings are more than twice the length of body, extending beyond the tip of the abdomen. Both the wings and body are completely covered with a white waxy powder that is why it is called whitefly. Female lays eggs singly on the underside of soft young leaves. The eggs are oval, pale yellow, and rest on small stalks. Nymph of whitefly is pale yellow, with purple eyes and its body is marginally fringed with bristles. It may be distinguished from the pupa which is broadly oval, pale yellow, with an orange or yellow area in the middle of the body. The whitefly passes through three broods under Punjab conditions whereas in case of blackfly there are two broods in a year. Citrus Blackfly (CBF) lays eggs in spiral rings on the lower side of new leaves. Blackfly nymphs are black in colour. Freshly emerged adults are reddish but later on get covered with a heavy pulverulence of bluish look.

Management

Chemical Control. According to Shivankar and Rao (2001a) application of insecticides at the appropriate developmental stage of the insect-pest is more important than the choice of insecticide for pest management. In case of CBF, pesticides should be applied at 50% egg hatching stage when 1st and 2nd larval instars predominate. Under Punjab conditions, spraying with 1140 ml Thiodan (endosulfan) 35EC or 1000 ml Phosmite (ethion) 50 EC or 1250 ml Hostathion (triazophos) 40EC in 500 litres of water per acre during April-May and September-October is recommended (Anonymous, 2010). A large number of insecticides viz. acephate, buprofezin, butocarboxim, dichlorvos, dimethoate, ethion, flucythrinate, imidacloprid, mexacarbate, monocrotophos, omethoate, phosalone, phosphamidon, pyriproxyfen and suprocide have been tried against whitefly complex by different workers in India, China and Israel (Arora *et al.,* 1998; Batra *et al.,* 1989, 2001a&b; Katole *et al.,* 1999; Sharma *et al.,* 2005a; Shivankar and Rao, 2001a; Zade *et al.,* 1986; Zhang and Wang, 2001; Zhou *et al.,* 2001). In Texas, use of aldicarb, chlorpyriphos, ethion, fenpropathrin, imidacloprid, methidathion, petroleum oil, pyridaben and pyriproxyfen against CBF is recommended (Holloway and Smith, 2010).

Nimbalkar *et al.* (1989) determined the relative toxicity of different insecticides against pupae of CBF. On the basis of the LC50s, cypermethrin was 55.07 times, permethrin 53.12 times, acephate 2.97 times, dimethoate 1.75 times, malathion 1.4 times, quinalphos 1.36 times and phosalone 1.19 times as toxic as phosphamidon.

Plant/Animal Products. Emulsified neem oil @1% (Jadhav and Pujari, 1999a), neemark@1% and fish oil rosin soap @3.33% sprayed during adult peaks reduced egg laying (Katole *et al.,* 1994). Jadhav and Pujari (1999a) found that neem oil is quite effective. But summer oil was found less effective than insecticides (Gokkes *et al.,* 1990).

Use of Traps. Yellow traps painted with castor or mobil oil and placed in association with low wattage light bulbs were found effective for monitoring adults of the CBF in Maharashtra (Kuchanwar *et al.,* 1989).

Cultural Control. The incidence of whitefly and black fly can be reduced by avoiding close plantation. Pruning and destruction of affected twigs if attack is located on smaller scale is also quite effective (Aulakh *et al.,* 2008). Keeping orchards clean from weeds and pruning diseased and dried tree branches is recommended in Texas (Holloway and Smith, 2010).

Biocontrol. The larvae and adults of several species of parasitoids (*Eretmocerus* spp., *Prospaltella* sp. and *P lahorensis* Howard) and predators (*Brumus suturalis* F., *Scymnus* sp., *Chrysopa* sp.) feed on whitefly eggs and nymphs; *Lasius* sp., a black ant, feeds on eggs and crawlers actively; *Amystus* sp., a red spider mite and an unidentified spider feed upon adult whiteflies (Saini, 1964). *Mallada boninensis* (Neuroptera : Chrysopidae), *Brumus suturalis, Scymnus* sp. are the predominant predators of CBF and CWF. These feed on eggs and nymphs. But these alone have not proved so effective as the natural fecundity of these predators was not high enough to compensate the natural fecundity of black flies (Shivankar and Rao, 2001b). In Nagpur region, the predators, *Catana parcesetosa (Serangium parcesetosum), Anisochrysa boninensis (Mallada boninensis) and an unidentified mite were found feeding on CBF* (Satpute *et al.,* 1986). Among the parasitoids, *Amitus hesperdium, Encarsia opulent* and *E. clypeali* are found quite effective against CBF in Florida and Texas (Ross, 1988). These three species along with *E. smithi* were sent to Mexico from India and Pakistan during 1948-50 for suppression of CBF (Clausen, 1958; Rao *et al.,* 1971). Shivankar and Rao (2001b) described the mass production and utilization of CBF natural enemies viz. *M. boninensis, Encarsia* spp., *A. hesprdium* and *Eretmocerus* sp. in detail.

Entomapathogens. Fungi namely, *Aschrsonia aleyrodis and A. webberi* were found attacking all the stages of CBF (McCoy, 1978). Hussain and Khan (1945) found a parasitic fungi (*Cladosporium* sp.) destroying a large population of the pest. Microbial sprays of *A. papillate* (Bose, 1953), *Cladosporium* sp. (Saini, 1964) and *A. aleyrodis* (Dharamraju and Reddy, 1975) are reported effective against nymphal stage and later attack pseudopupae under high humidity.

IPM Module. IPM Module consisting of Clean cultivation + *Mallada boninensis* @ 30 larvae/tree+Sweet Flag (3%) + *Verticillium lecanii* @ 5 g/litre +trap crop on border lines, proved effective against blackfly (Shivankar and Rao, 2008).

22.4. MEALYBUGS

Three species of mealybug viz. *Planococcus citri* Risso; *P. lilacinus* (Cockerell) and *Nipaecoccus viridis* (Newstead) have been found infesting citrus in Punjab (Sharma and Arora, 2009). But out of these *P. citri* is the only species which has been reported as economically important pest of citrus. Mealybug has a soft, oval, flattened, distinctly segmented body which is covered with white powdery wax. Young mealybugs have less wax and are yellowish in colour. The mealybugs can be observed damaging new growth, cut ends and fruits in form of fluffy cottony mass. They often settle near the stem-end of the fruit resulting in heavy fruit drop (Bindra, 1970). The nymphs and adult females inflict injury by extracting the juice from the cells whereas the male is devoid of mouthparts and does not feed. Due to heavy sucking the plant turn pale, wilt down and consequently dry up. Besides they also excrete honey dew on which sooty mould develops.

Management

Chemical Control. Spraying with chlorpyriphos (0.075%), profenophos (0.075%), acephate (0.075%) and thiamethoxam (0.008%) is highly effective against mealy bugs (Sharma and Arora, 2009; Aulakh *et al.*, 2008) whereas Jadhav and Pujari (1999b) obtained good control of mealybug with chlorpyriphos at lower dosage of 0.05%. Spraying with 2ml dichlorvos + 25g of fish oil rosin soap per litre of water was found effective in reducing mealybug population. Since the mealybugs harbour under the loose bark, debarking followed by pasting with chlorpyriphos and methyl parathion both @ 4ml/litre of water helps in minimizing the mealybug population (Mani and Krishnamoorthy, 1996). Complete drenching of tree with methyl parathion (0.05%) gave good control of mealybug. Use of Dimethoate (1.5 ml) + Kerosene oil (2.5 ml) or carbaryl (1 ml) + Kerosene oil (1.0 ml) or malathion (2.0 ml) per litre of water is recommended for the control of mealybugs (Shivankar, 1999). Earlier, Atwal (1976) mentioned that this pest can be controlled by spraying the trees with diazinon / parathion / phosphamidon @0.03% or malathion/dimethoate @ 0.05%.

Insect Growth Regulators. Diofenolan showed excellent activity at low rates against *P. citri* (Streibert *et al.,* 1994). Exposing *P. citri* nymphs to aqueous suspension of 25% wettable buprofezin powder (0.008 t0 0.005% a.i.) resulted in 91-99% nymphal mortality.

Botanicals. Jadhav and Pujari (1999b) obtained moderate control of *P. citri* with Pongamia oil (1%) and Repelin (1%).

Biocontrol. *Coccidoxenoides peregrinus* (Timberlake), *Tetracnemoidea indica* (Ayyar) and *Cryptolaemus montrouzieri* (Muls) are natural enemies of of mealybugs found in South India. Population of *C. montrouzieri* @ 20/plant is sufficient enough to check the mealybug population (Mani and Krishnamoorthy, 1996). Previously, Krishnamoorthy and Mani (1989) mentioned that the chrysopids, *Mallada boninensis, Chrsopa lacciperda* (*Plesiochrysa lacciperda*), *Anisochrysa basalis (M. basalis*) and *Chrysoperla carnea* can be used to control mealybugs. Two encyrtid parasitoids, *Leptomastix dactylopii* and *C. peregrinus* usually attack mealy bugs but out of these, *C. peregrinus* is more abundant and is responsible for decline of *P. citri* causing 10-30% parasitism (Mani, 1994). Release of the parasitoid, *L. dactylopii* at rate of 5000-7000 adults per hac gives complete control of *P. citri* within 3-4 months (Krishnamoorthy and Mani, 1994).

Cultural methods. Aulakh *et al.* (2008) mentioned different cultural measure for the management of mealybugs in citrus orchards. These measures are :

(i) Do not grow crops like cotton, chari, bajra, maize, barseem, bhindi, guara and vegetables in the orchards.

(ii) Survey the orchards regularly to monitor mealy bug population.

(iii) Sanitation of the orchards is extremely important. Uproot and destroy the weeds in the vicinity of orchards.

(iv) Do not allow the branches of trees to touch the ground.

(v) Destroy ant nests in the orchards as they are the carriers of mealy bugs to their feeding sites. Ants share mutualistic relationships with mealybug. The ants receive honeydew, while the mealybug are protected from attack by parasitoids, cleaned of excreta and have suitable conditions for reproduction.

(vi) Removing and destroying the heavily infested branches and fruits has also proved very effective.

22.5. CITRUS APHIDS

Three species of aphids, *Toxoptera aurantii* (Boyer de Fonscolombe); *Aphis gossypii* Glover and *Myzus persicae (Sulzer)* in general are sporadic pests, but now these are becoming regular pests of citrus. *T. aurantii* and *A. gossypii* are reported as most efficient vectors of Citrus Tristeza Virus (CTV). Under Punjab conditions, these cause more damage during February-April and September-October. The nymphs and adults suck the cell sap of young leaves, flower buds and tender twigs. This impairs the vitality of the tree and causes severe curling and deformation of young leaves resulting into stunted growth. The honeydew excreted by the aphids also provides a good substrate for the growth of sooty mould, which affects the photosynthetic activity of plant. Citrus aphids are parthenogenetic and viviparous in nature. The nymphs moult four times. Winged forms commonly develop when colony of aphids become crowded or when leaves become old and hard, whereas no winged forms are present when aphids feed on very tender leaves and are not large in number. The entire life cycle is completed in three weeks and there are 12 generations in a year.

Vector. Various citrus species found on citrus are vectors of Citrus tristeza virus (CTV).

T. citricida is the most efficient of all the aphid vectors of CTV causing stem splitting and debilitation of grapefruits and oranges.

Management

Chemical Control. For the effective control of aphids, spraying with systemic insecticides viz. 1250ml Rogor (dimethoate) 30EC or 1000ml Metasystox (oxydemeton-methyl) 25EC or 200ml Confidor (imidacloprid) 17.8 SL or 160g Actara (thiamethoxam) 25 WG in 500 litres of water per acre in the second week of March and again in the first week of September is recommended (Aulakh *et al.,* 2008). Foliar sprays of acetamiprid, carbosulfan, cypermethrin, fenvalerate, imidacloprid, methomyl, mevinphos monocrotophos, naled and omethoate have been found effective for the control of aphids by different workers under different climatic conditions (Arora *et al.,* 2009; Buitendag and Bronkhorst, 1980b; De-Biase and Russo, 1996; Deshmukh *et al.,* 1989; Dikshit and Lal, 2002; Lei *et al.,* 2000; Ye *et al.,* 1996). Melia and Blasco (1980) also obtained best results against aphids with dimethoate, ethiofencarb, phosphamidon, menazon, oxydemeton-methyl and pirimicarb. But according to Nucifora (1998) pirimicarb had a very low activity against citrus aphids whereas imidacloprid activity was very satisfactory and long lasting. Methomyl had an excellent immediate action against aphids but failed to ensure long term efficacy.

Some other methods of insecticide application has also been used successfully by different workers. Gokkes *et al.* (1988) tried ground air blast or aerial spraying methods in Israel and achieved good control of aphids in citrus for 7-16 days with ethiofencarb 50EC @ 2-4litres per hac. Buitendag and Bronkhrost (1980a) tried new technique i.e. trunk injection of undiluted monocrotophos and obtained good control of aphids. Galli *et al.* (1995) obtained effective control of *T. citricidus* with Aldicarb (19.5g a.i./tree) and Aldicarb (19.5g a.i./tree) + ethion (75g a.i./100 litres water) in Brazil.

Sprays of clay suspension containing kaolinite alone or mixed with montmorillonite (sticker) markedly reduced the rate of colonization by winged aphids on trees of *Citrus aurantifolia* (Bar and Frankel, 1979).

Botanicals. Mahua and pongamia oils @1% and neem and pongamia seed extracts @2%are recommended for the control of citrus aphid (Dhara *et al.,* 1993a).

Biocontrol. Many predators and parasitoids were found attacking citrus aphids. The major predators of aphids were Coccinellids, Chrysopids Syrphids and Spiders, Coccinellids being the most abundant (Radhakrishnan and Muraleedharan, 1993). In Turkey, Yumruktepe and Uygun, (1994) reported that *A. citricola (A. spiraecola), A. gossypii, A. craccivora, T. aurantii* and *Myzus persicae* were the dominant aphid species alongwith their natural enemies belonging to the Coccinellidae (22 species), Syrphidae (9 species), Braconidae (8 species), Lygaeidae (2 species) and Cecidomyiidae, Miridae, Nabidae, Anthocoridae and Ascidae (one species each). In Lebanon, Hussein and Kawar (1985) achieved 95% control of *Toxoptera aurantii* with braconid parasites *Aphidius matricariae, Lysiphlebus fabarum, Trioxys angelicae* and *Praon volucre. Lysiphlebus testaceipes* was the main parasitoid found on citrus aphids, and its rate of parasitism was significantly high on some aphids, such as *T. aurantii* in Spain (Michelena *et al.*, 1994). Larvae of *Allothrombium pulvinium* are parasitic on citrus aphids in Iran (Zhang and Faraji, 1994).

Entomopathogenic Fungi. A pathogenic fungi, *Fusarium lateritium* isolated from dead bodies of citrus aphids gave very good results when applied in the field in China to control citrus aphids (Song, 2001)

Traps. Seif (1987) recommended the use of yellow water traps for monitoring of aphid populations in sweet orange orchards in the Kenya coast. Yellow coloured stickey traps are found quite successful against aphids under Punjab conditions also. Melia (1989) used yellow traps for the control of citrus aphids and found that yellow traps situated at the ground level were more effective for monitoring the aphid populations than those situated at the higher level.

22.6. CITRUS THRIPS (*Scirtothrips citri*)

Thrips are minute insects but can be seen with naked eyes. Males are practically unknown and the reproduction is by parthenogenesis. The female inserts eggs in the incisions made in leaf tissues, which hatch in 2 to 7 days. It passes through 3-4 nymphal instars before attaining maturity. There are several generations in a year. The nymphs and adults rasp and suck the cell sap from fully developed flowers, leaves and also from young and grown-up fruits. Leaves become cup shaped, leathery, distorted and wrinkled. Two white lines parallel to leaf mid-rib and whitish silvery ring around the fruit neck are the typical symptoms. Thrip damage in nursery causes stunted growth.

Management

Chemical Control. Under Punjab conditions, spray the recommended insecticides viz., 1140 ml Thiodan (endosulfan) 35EC or 1000ml Phosmite (ethion) 50 EC or 1000 ml Metasystox (oxydemeton-methyl) 25EC or 1250 ml Hostathion (triazophos) 40 EC in 500 litres of water per acre in mid March and April if thrip attack is observed (Aulakh *et al.*, 2008). Dimethoate or phosphamidon or monocrotophos @ 1 ml per litre of water should be sprayed at bud burst stage (Shivankar and Rao, 2001a). Use of soft insecticides like soaps, botanicals such as Sabadilla sugar and even simply washing with clean water should be done to check the thrips in nurseries (Shivankar *et al.*, 2008).

Biocontrol. Out of the four species of predaceous mites viz., *Iphiseius degenerans, Euseius tularennsis, Neoseiulus cucumeris* and *N.*

barkeri evaluated against citrus thrips, *I. degenerans* showed promise for commercial releases and control of thrips in citrus nurseries (Grafton-Cardwell *et al.,* 1999).

Wind breaks. Grout and Richards (1990) showed that oranges growing adjacent to *Grevillea robusta* windbreaks had more severe scarring due to thrips than fruits on trees in the middle of the orchards or on trees in adjacent orchards with different windbreak species. Higher number of adult thrips were recorded in citrus near *G. robusta* windbreaks than in the centre of the orchards.

MITES. According to Ebeling (1959) at least 30 species of mites infest citrus. But in India, out of these only three species viz., *Eutetranychus orientalis* Klein., *E. banksi* and *Tetranychus fijiensis,* are recorded as major pests of citrus. Mite incidence is maximum during dry and hot period (May-June) or sometimes become serious in August-September. Both nymphs and adults abrade the surface and suck the cell contents from the upper surface of the leaves, tender shoots, flowers and fruits. Affected surface loose normal colour, become yellowish brown and tend to collect lot of dust which impairs photosynthetic activity. Mite feeding on fruits results in irregular areas of scarred tissues on fruit. Mite symptoms may in some instances resemble either wind scaring or thrip symptoms. However in case of thrip, a whitish silvery ring around the fruit neck are the typical symptoms where as in case of mite and wind scarring there is presence of irregular areas of scarred tissues all over the fruit.

Management

Chemical Control. Different workers have tried various chemicals for the control of mites on citrus with encouraging results. Mites can effectively be controlled by spraying 1250 ml Rogor (dimethoate) 30EC/ 1000ml Phosmite (ethion) 50EC / 1000ml Ekalux (quinalphos) 25EC in 500 litres of water per acre. Repeat the spray after 15 days in case of severe infestation (Aulakh *et al.,* 2008). Acaricides like tetradifon or dicofol @ 0.025% and sulphur @0.3% should be sprayed to get effective control of mites (Shivankar and Rao, 2001a). Propargite (Bei and Lang, 2000), bifenthrin, lambda cyhalothrin, fenpropathrin, dicofol and abamectin (Helal *et al.,* 1995) were also found effective for the control of mites on citrus. Seraj (1999b) injected demeton-s-methyl into lime tree stems and obtained good control of mite. In Jordan, Sharaf (1989) obtained effective control of mites on lemons with flubenzimine and omethoate. The best level of mite control was obtained with mixture of lime sulphur liquid (calcium polysulfids) + hexythiazox that reduced mite numbers by 100% in seven days after spraying (Zhan and Peng, 1990). Previously, malathion, parathion and carbophenothion all @0.02% as foliar spray have been found effective by Sethi (1967). Bindra *et al.* (1970)

reported the reduction of mite population when systemic insecticides,Viz., dimethoate (12 &16 ml a.i./tree), formothion (20 ml a.i./tree), phorate (20g a.i./tree) and cytrolane (20 ml a.i./tree) were applied in the basins of citrus trees.

Biocontrol. Mites, *Euseius hibisci* and *Ambylesius hibisci* are the most important natural enemies of citrus mites. The predatory mite *E. scutalis* fed voraciously on the adults of the mites *E. orientalis.*

Release of 750 individuals/branch would suppress the pest population (Donia *et al.,* 1995) Other species like *Scymus gracilis, Arthrocnodax* sp., *Chrysopa* sp. and *Hemerobius* sp., have also been reported to be predacious (Pruthi and Mani, 1945).

Cultural Methods. Water stress aggravates the mite problem. So citrus trees should be well irrigated, particularly during the stress in the summer months.

Entomopathogenic Fungi. The entomopathogenic fungi, *Vertcillium lecanii* (Sewify and Mabrouk, 1990) and *Hirsutela thompsoni* (Garsen *et al.,* 1979) were reported parasitising mites.

22.7. LEMON CATERPILLAR (*Papilio demoleus*)

The lemon caterpillar is most destructive to citrus nurseries and young plantations. The caterpillars are voracious feeders and generally start feeding from margin inwards reaching the midrib. They cause complete defoliation and result in serious setback. The butterflies lay eggs on tender shoots and fresh leaves, mostly on undersurface. The eggs are placed singly or in groups of 2-5. The eggs are glued firmly on to the surface of the leaf and are pale or greenish yellow, when freshly laid, but turn brown later on, becoming dark grey just before hatching. The young caterpillar is brown to black with white markings and resembles to bird droppings. When full grown, it changes to yellowish green and possesses a horn like structure on the dorsal side of abdomen, with a hump like appearance in front. The mature larva spins a supporting girdle around its body and pupates on a twig, a dry stick or any other raised structure. The adult is a beautiful butterfly with small black and yellow hairs on its body. There are three to four generations per year.

Management

Chemical Control. Spraying with 1140 ml Endosulfan 35EC or 1000 ml Ethion 50 EC or 1250 ml Triazophos 40EC or 1875ml Chlorpyriphos 20EC in 500 litres per acre gives very good control of lemon caterpillar (Aulakh *et al.,* 2008). Foliar sprays of any of the contact or systemic insecticides viz., dimehoate@1.5ml or fenitrothion@1ml or carbaryl @2g or endosulfan 1.4ml per litre of water at an early larval stage can effectively check the pest infestation (Anonymous, 1997). Earlier, BHC

@0.15%, DDT @0.15%, diazinon @0.04%, malathion @0.03% and parathion @0.02% were fond quite effective (Brar, 1959; Sethi, 1965).

Solunke and Deshpande (1991) fed the *P. demoleus* larvae on citrus leaves dipped in permethrin, aqueous extracts of *Ficus religiosa, Monchoria hastactolia, Parthenium hysterophorus, Azadirachta indica, Calotropis gigantean* and *Datura stramonium*. After 72 hours of feeding permethrin recorded highest mortality of 85.4% followed by *P. hysterophorus* extract with 52.8% larval mortality. Among the plant products, azadirachtin (0.03%) was found significantly superior over neem seed kernel extract and neem leaves extract both @5% (Chauke *et al.,* 1999).

Hand-Picking. At small scale, hand-picking and destruction of larvae as a control measure in nursery plants is found quite effective. According to Brar (1959) picking with forceps is much quicker, and in less than 4 hours 1 man can destroy the larvae from one acre of nursery.

Biocontrol. Yellow wasp (*Polistes hebereus*), preying mentid (*Creobractor gemmatus*) and spiders are reported as the bioagents of citrus pests (Singh and Singh, 1998). Atwal (1964) observed that *Pteromalus* spp. parasitized pupae of lemon butterfly. A braconid, *Apanteles flavipes* and ichneumonid, *Melalophacharaps* sp. are the predominant larval parasitoids, and a chalcid, *P. puparium* and a ichineumonid, *Holcojoppa coelopyga* are the major pupal parasitoids (Anonymous, 1997).

Microbial. 'Dipel', a commercial formulation of the bacterium *Bacillus thuringiensis* has been found to be very effective when used at 0.05% concentration (Shivankar and Rao, 2001a). Entomopathogens like bacterium, *Serratia marcescens* and fungus, *Fusarium* sp. also kill the lemon butterfly population substantially (Singh and Singh, 1998). *Neoaplectana carpocapsae* Dutky applied at 600,000 nematodes per tree resulted in 63.33% kill of third instar and 67.67% mortality of fourth and fifth instar caterpillars (Srivastava, 1978).

22.8. FRUIT SUCKING MOTH (*Eudocima fullonia*)

Fruit sucking moth is a sporadic pest of citrus orchards especially in the sub-montane areas. It is polyphagous in nature. This pest remains active from July to October. The moths get attracted towards the ripening fruits during dusk and dawn. The moth punctures the rind of fruit. Infested fruits invariably drop-off prematurely perhaps due to the toxins injected by the moths while feeding. A circular spot appears at the site of feeding which gives frothy jet of fermented juice when squeezed.

Management

Chemical Control. Partial control can be achieved dusting 2% Follidol dust (methyl -parathion) during evening time (Batra, 1990). Zhou *et al.* (2001) recommended spraying with dichlorvos and trichlorofon for the control of fruit piercing moth in China.

Cultural Methods. For its management, destroy the breeding hosts especially belonging to family *Manispormaceae.* Disposal of fallen and decaying fruits is found useful. Creating smoke in the orchards after sunset is quite effective.

Poison Bait. Use of poison bait is recommended. Prepare the poison bait by mixing 2 litres citrus fruit juice, 250 ml Malathion 50EC and 1 kg Gur/molasses in 10 litres of water and hung the pitcher pots containing poison bait at 2-3 places around the tree. Treat all the border rows one month before fruit ripens. The bait should be renewed at weekly intervals (Pruthi and Batra, 1960).

22.9. FRUIT FLY, *Bactrocera* spp.

The damage is more prevalent in sub-mountaneous zone. The female adult fly punctures the ripening fruits by penetrating its ovipositor and lays the eggs inside. Larva on hatching feeds on the pulp inside. Infested fruit shows many dark green depressions due to punctures caused by the insertion of ovipositor by female. On squeezing the infested fruit, a number of jets of juice come out, as there are many holes on a single fruit. Fungal and bacterial infections take place through the puncture hole due to which rottening of fruit occurs resulting in fruit drop.

Management

Cultural methods. Collection of the infested fruits and their destruction by burning or burying them deep in pits checks the fruit fly effectively.

Traps. For the control of fruit fly, Shivankar (1999) recommended male attracting trap containing 0.1% methyl eugenol and 0.05% malathion@ 25traps per hac starting from 60 days before fruit set. Male attracting traps containing methyl eugenol are highly specific, low cost, environmentally safe and effective (Alyokhin *et al.,* 2001).

22.10. CALIFORNIA RED SCALE, *Aonidiella aurantii*

The nymphs and adults suck sap from leaves, twigs and fruits and devitalize the plants. Scales also inject a toxic substance into the plant sap. Yellow spots appear at the feeding sites. Each mark indicates the point at which the scale have thrusted its stylets. The branches soon loose chlorophyll, turn scurffy and start drying. It also secretes waxy covering over its dorsal surface.

Management

Chemical Control. Control of armored scale is very difficult as they have a protective cover over their body which does not allow the insecticides to penetrate. However, frequent spraying with phosphamidon or parathion each @ 0.03% afforded good control of this pest. The mixing of sticker (Tenac) with parathion enhanced its toxicity (Ghose, 1965). Application of chlorpyriphos @0.15% (Rawhy *et al.,* 1980) or dimethoate @0.125% or methadathion (Supracide) @0.125% (Ali *et al.,* 1978) gave the good control of the scale insects in Egypt. Salama and Amin (1983) showed that 0.15-0.2% sprays of malathion, dimethoate, parathion-methyl or dichrotophos alone or in combination with 2% mineral oils were very effective against scales. Buitendag and Bronkhrost (1980a) tried new technique i.e. trunk injection of undiluted monocrotophos and obtained good control of scales.

Biocontrol. The *Aphelinus* sp. parasitises a high percentage of scales. The coccinellids, *Chilocorus nigritus* Fabr. and *Scymnus quadrillum* are predators on *A. aurantii* throughout India. Release of *C. nigrata* at the the rate of 10 and 15 adults per tree offered complete control of scale in 70 and 50 days, respectively (Krishnamoorthy and Rajagopal, 1999).

Cultural Methods. Removing and destroying the heavily infested branches and fruits has been proved very effective.

Insect Growth Regulators. Diofenolan showed excellent activity at low rates against *A. aurantii* (Streibert *et al.,* 1994). The emergence of males of *A. aurantii* was inhibited, when 1st and newly moulted 2nd instar nymphs were sprayed with RO 13-5223 (ethyl(2-(4-phenoxyphenoxy) ethyl) carbamate) @ 0.0125-0.1%, but the effect on the females was much less marked (Peleg, 1982).

Sex Pheromones. Tremblay and Rotundo (1980) mentioned that traps baited with virgin females proved very effective for monitoring populations. Of the two components of the sex pheromone of *A. aurantii*, one ((3Z,6R)-3-methyl-6-(1-methylethenly)-3,9-decadienyl acetate) is more attractive.

Entomapathogens. Gao and Quyang (1981) found a fungus (*Fusarium coccophilum*) parasitising the *A. aurantii*. in China.

22.11. CITRUS LEAF-FOLDER, *Psorosticha zizyphi*

Citrus leaf-folder is a sporadic pest of citrus and it has gained importance in Punjab for the last few years. Young caterpillars feed in congregation on tender and succulent leaves or sprouting buds. Later on, they secrete silken strands to roll up the leaves longitudinally and start feeding from tip downwards. It is active from April to October.

Management

Cultural Control. Clipping off of the rolled leaves is also effective at smaller scale.

Chemical Control. Spraying with 625 ml Nuvacron (monocrotophos) 36SL or 1875 ml Dursban (chlorpyriphos) 20EC or 1000ml Ekalux (quinalphos) 25 EC in 500 litres of water per acre is recommended (Aulakh *et al.,* 2008).

22.12. BARKCATERPILLAR, *Indarbela quadrinitata*

Bark-eating caterpillar is a pest of neglected orchards. It is polyphagous pest infesting many fruit trees, road side trees, forest trees and ornamental bushes. It causes damage by boring holes into the stem and branches and feeds on the bark under the cover of its excreta.

Management

For its effective management remove the webbing and inject kerosene oil into the holes during September-October and again in January-February. Also treat all the alternate hosts in the vicinity (Aulakh *et al.,* 2008). According to Patel *et al.* (1964) bark eating caterpillars can easily be controlled by inserting kerosene-soaked cotton swabs into their holes followed by closing of these holes with mud. Injecting a few ml of monocrotophos or dichlorvos (1%) in each tunnel and spraying the affected portion with the same solution after clearing the larval excreta and frass is quite effective (Shivankar, 1999).

Bioagents. Larval parasitisation of *Indarbela quadrinitata* by *Apanteles* sp. was reported. For its control use of fungus, *Aspergillus candidus* has also been recommended (Singh and Singh, 1984).

22.13. AMERICAN BOLLWORM, *Helicoverpa armigera*

The larvae feed on the leaves, flower buds and young fruits of Kinnow mandarin. They damage flower buds and young fruits by making holes. The infested fruits turn black, dry-up and drop pre-maturely. American bollworm causes more damage to citrus during month of April (Arora *et al.,* 1996).

Management

(i) Do not grow cotton in the orchards.

(ii) Spraying with 625 ml Nuvacron (monocrotophos) 36SL or 1875 ml Dursban (chlorpyriphos) 20EC or 1000 ml Ekalux (quinalphos) 25 EC in 500 litres of water per acre is quite effective against early instars of American bollworm.

22.14. CONCLUSIONS

At present in order to protect their fruit plants from insect pests, growers use a variety of recommended and un-recommended pesticides without knowledge about their impact on the natural enemy complex, resurgence of secondary pests and residue implications of the insecticides. Frequent application of these insecticides may result in persistence of their residues in fruits at the time of harvest, which may eventually reach the consumers resulting in health hazards. Secondly, under the strict legislations of foreign countries, the presence of residues above the prescribed maximum residue limits (MRL's) in fruits meant for export purpose rendered them unfit for export. As consumers demand for high quality and safe food is constantly increasing, the challenge for growers is to keep up with the standards of quality. In Abohar, the major citrus belt of Punjab, the situation is more complex because Kinnow is grown in Cotton based ecosystem where pesticide load on Kinnow crop is increased due to increased common pest problems like Mealy bugs, *Spodoptera, Helicoverpa, aphids* etc. So the rationalization of current spray schedules and development of IPM strategies is regarded as an immediate priority. The introduction of the system of Good Agricultural Practices (GAP) to harmonise and upgrade the Kinnow farming system is the need of time. This will include shifting to environment friendly pesticides (green chemistry), more selective insecticides, more accurate time and precise/ efficient method of pesticide application, introducing pest scouting, encouraging sanitation, biological and physical control methods.

REFERENCES

Ali, A.M., M.F. Abou-Ghadir; A.G.A. Salman, AMKEl-Sayed, S. H. Mannaa, M.F. Abou-Ghadir and A.M.K. El-Sayed (1978) Evaluation of insecticides and time of application for the control of the red and black scale insects on orange trees in Assiut, Upper Egypt. *Bull. Entomol. Soc. Egypt Econ.* 11 : 173-180.

Alyokhin, A.V., R.H. Messing and J.J. Duan (2001) Abundance and mating behavior of Oriental fruit flies (Diptera : Tephritidae) near methyleugenol-baited traps. *Pan-Pacific Entomol.* 77(3) : 161-167.

Anonymous (1997) *A decade of NRCC,* NRC for Citrus, Nagpur (Maharashtra). pp. 97.

Anonymous (2010) *Package of Practices for the Cultivation of Fruit Crops.* PAU, Ludhiana.

Arora, P.K., S.K. Thind, S.K. Thatai, J.S. Josan, N.K. Mehrotra and R.C. Batra. (1996) Outbreak of *Helicoverpa armigera* (Hubner) on citrus in Punjab. *Indian J. Ecol.* 23(2) : 145-147.

Arora, P.K., S.K. Thind, N Kaur and I.S. Grewal (1997) Population dynamics of citrus psylla, *Diaphorina citri* Kuwayama, on Kinnow mandaruin in relation to temperature and relative humidity. *Indian J. Ecol.* 24(2): 179-183.

Arora, P.K., R.C. Batra., S.K. Thind., N.K. Mehrotra and J.S. Josan (1998). Relative efficacy of some organophosphatic insecticides against citrus whitefly. *Indian J. Appl. Ent.,* 12 : 51-54.

Arora, P.K., N. Kaur., S.K. Thind, P.K. Monga and J.K. Arora (2000) Biochemical changes induced by sooty mould in leaves of some citrus cultivars. *Indian. J. Pl. Pathol.* 18 (1&2) : 73-74.

Arora, P.K. and D.R. Sharma (2005) Efficacy of insecticides against nymphs of citrus whitefly, *Dialeurodes citri* Ashmead. In: *1st Congress on Insect Science.* PAU, Ludhiana Dec. 15-17, 2005. p. 193.

Arora, P.K. R.S. Battu and B. Singh (2005a) Bioefficacy of some insecticides against citrus psylla viz-a-viz determination of quinalphos residues in kinnow mandarin fruits. *Pest Management Hortic. Ecosys.* 11 (1): 33-38.

Arora, P.K., S.K. Thind, and P.S. Aulakh (2005b) Correlation of citrus whitefly population with sooty mould development and its control. *Pest Management Hort. Ecosys.* 11 (1) : 61-64.

Arora, P.K., S. Raheja., P.K. Monga, N. Kaur, P.S. Aulakh and J.S. Josan (2008a) Bioefficacy of some botanicals against citrus psylla on Kinnow mandarin. In: *11th Punjab Science Congress on Science & Technology : Opportunities & Challenges,* February 7-9, 2008, Thapar University, Patiala. AP 59. p. 84.

Arora, P.K., S.K. Thind, P.S. Aulakh and Kamaldeep Singh. (2008b) Population dynamics of citrus leafminer, *Phyllocnistis citrella* Stainton and its control on Kinnow mandarin. *Environ. Ecol. 26 (4A):1841-1844.*

Arora, P.K., D.R. Sharma., P.K. Monga and P.S. Aulakh (2009) Bioefficacy of imidacloprid against sucking pests of kinnow mandarin. In: *12th Punjab Science Congress* held on February 7-9, 2009 at PAU, Ludhiana. AP-60. p. 60.

Atwal A.S. (1962) Insect pests of citrus in Punjab. I. Biology and control of citrus psylla, *Diaphonia citri* Kuwayama (Hemi., psyllidae). *Punjab Hort. J.*, 2(2) : 104-108.

Atwal A.S. (1964) Insect pests of citrus in Punjab. VI. Biology and control of citrus leaf miner *Phyllocnistis citrella* Stainton (Lep., phyllocnistidae). *Punjab Hort. J.*, 4(2) : 100-103.

Atwal, A.S. (1976) Pests of Citrus. In : *Agricultural Pests of India and South-East Asia.* Kalyani Publishers, Delhi, Ludhiana. Pp.195-213.

Aulakh, P.S., S.K. Thind and P.K. Arora (2008) *"Kinnow"*. Published by Additional Director of Communication, Punjab Agricultural University, Ludhiana. Pp. 48.

Ayyar, T.V.R. (1963) *Handbook of Economic Entomology for South India,* pp 516. Govt. of Madras.

Batra, R.C., D.R. Sharma and Y.R. Chanana. (1989) Relative efficacy of different insecticides for the control of whitefly, *Dialeurodes citri. Indian J. Hort.* 46 : 122-125.

Batra, R.C. (1990) Role of insect pests in citrus decline. In : *Proceedings of National Seminar-cum-Workshop on Nursery and Orchard Management.* January 15-27, 1990. PAU, Ludhiana.: 207-217.

Batra R.C., G.S. Sandhu and D.R. Sharma (1990) Relative efficacy of different organophosphorus insecticides for the control of psylla, *Diaphorina citri* Kuwayama. *Res. Devel. Reptr.* 7 (172) : 53-58.

Batra, R.C., D.R. Sharma and P.K. Arora (2001a) Impact of abiotic factors on *Dialeurodes citri* Ashmead and its chemical control. *J. Res. PAU.,* 38 (3-4) : 212-217.

Batra, R.C., P.K. Arora, D.R. Sharma and R. Kumar (2001b) Biological activity of some insecticides against *Dialeurodes citri* Ashmead. In: *IPM in Horticultural Crops : Emerging Trends in the New Millennium (Proceedings of the Second National Symposium on Integrated Pest Management (IPM) in Horticultural Crops: New Molecules, Biopesticides & Environment).* Abraham Verghese & P. Parvatha Reddy (Eds.) October 17-19, 2001. Bangalore. p. 31.

Bar, J.M. and H. Frankel (1979) Spraying lime trees with clay suspension prevents their colonization by aphids. *Hassadeh* 60 (1) : 65-67.

Beattie GAC; D.J. Rae, D.M. Watson, M.D. Huang and P. Broadbent (1999) Simultaneous control of *Diaphorina citri* Kuwayama other citrus pests and some diseases with petroleum spray oils. *Proceedings of the citrus germplasm conservation workshop,* Brisbane, Australia, 6-7 October, 1999 : 40-42.

Bei, X.F. and G.L. Lang (2000) Experiment of control of citrus mite by using Comite insecticide. *South China Fruits* : 29 (2) : 20-21.

Bei, X.J., L.Y. Jiang, X.L. Chi and Q.W. He (1999a) Experiment of spraying imidan for control of citrus leaf miner. *South China Fruits* : 28(5) : 23.

Bei, X.J., L.Y. Jiang, X.L. Chi and Q.W. He (1999b) Control of citrus leaf miner with imdacloprid WP. *Pl. Prot.* 25(4) : 52-53.

Berg, M Vanden (1999) Measures to reduce citrus psylla population and the spread of greening disease. *Neltropika Bull.* 303 : 5-6.

Berg M.A. and Berg M.A. (1996) Pathogenic fungi attacking the citrus psylla, *Trioza erytreae* (Hem : triozidae), in two field populations in Mpumalanga, South Africa. *J. S. Afr. Soc. Hortic. Sci.* 6(2) : 97-98.

Bhatti, I.S. and D.R. Sharma (2010) Bio-efficacy of soil applied neo-nicotinoid insecticides against *Phyllocnistis citrella* Stainton on drip irrigated Kinnow plants. In : *National Seminar on Impact of climate change on fruit crops,* October 6-8,2010 held at PAU, Ludhiana.p.144.

Bindra, O.S. (1970) Insects. In : *Citrus Decline in India.* Jt. Publication of PAU/OSU/USAID, Ludhiana, Punjab. Pp. 64-78.

Bindra, O.S., G.C. Verma, G.S. Sandhu, H.K. Chhabra (1970) Studies on the control of citrus pests by soil application of systemic insecticides. *J. Res. PAU,* 7 : 197-202.

Borad P.K., M.J. Patel, N.M. Waghela, B.H. Patel, M.G. Patel, V.H. Patel and J.R. Patel (2001) Evaluation of some botanicals against citrus leafminer

(*Phylocnistis citrella*) and psylla (*Diaphorina citri*) on kagzi lime (*Citrus aurantifolia*). *Indian J. Agric. Sci.* 71(3) : 177-179.

Bose, S.K. (1953) An entomogenous fungus of citrus whitefly. *Proceedings 40th Indian Science Congress* . pp79.

Boughdad, A; Y Bouazzaouo and L. Abdelkhalak (1999) Pest status and biology of the citrus leaf miner, *Phyllocnistis citrella* Stainton (Lepidiptera : Phyllocnistidae), in Morocco. *Proceedings of the fifth International Conference in Pests in Agriculture,* Part 2, France, 7-9 december, 1999 : 251-259.

Brar, T.S. (1959) Bionomics and control of *Papilio demoleus* Linnaeus (Papilionidae : Lepidoptera) . *M. Sc. Thesis Punjab Univ., Chandigarh.*

Buitendag, C.H. and G.J. Bronkhrost (1980a) Injection of insecticides into tree trunks- a possible new method for the control of citrus pests.. *Citrus Subtrop. Fruit J., 556* :5-7.

Buitendag, C.H. and G.J. Bronkhrost (1980b) Phosdrin 2EC for controlling American bollworm and aphids in citrus orchards with biological pest control. *Citrus Subtrop. Fruit J.* 603 :16.

Buitendag, C.H. (1993) Living with citrus greening in South Africa. *Citrus J.* 3(1) : 1018.

Capoor, S.P., D.G. Rao and S.M. Wiswanath (1967) *Diaphorina citri* Kuwayama, a vector of the greening disease of citrus in India. *Indian J. Agric. Sci.* 37 : 572-576.

Cen, Y.J., C.V. Xu and M.Y. Tian (1999) Advances in the use of petroleum spray oils in control of pests in citrus. *J. S. China Agric. Uni.* 20(2) : 118-122.

Chagas, M.C.M., J.R.P. Parra, J.S. Namekata, Hartung and P.T. Yamamoto (2001) *Phyllocnistis citrella* Stainton (Lapidoptera : Gracillariidae) and its relatio-nship with the citrus canker bacterium Xanthomonas axonopodis pv citrii in Brazil. *Neotrop. Ent.* 30 (1) : 55-59.

Chakravarthi, V.P., P. Savithri, P.R. Prasad and N.V. Rao (1998a) Biology and seasonal abundance of citrus psylla, *Diaphona citri*, Kuwayama (Homoptera : psyllideae). In : *Advance in IPM for Horticultural crops* (Reddy PP, NKK Kumar and A Verghese- Eds.). Proceedings of the First National Symposium on Pest Management in Horticultural Crops : Environmental implications and thrusts, Bangalore, 15-17 Oct., 1997, p 26-29.

Chakravarthi, V.P., P. Savithri, P.R. Prasad and V.J. Naidu (1998b) Efficacy of various insecticides against citrus psylla, *Diaphona citri*, Kuwayama (Homoptera : psyllideae). In : *Advances in IPM for Horticultural crops* (Reddy PP, N.K.K. Kumar and A. Verghese Eds.). Proceedings of the First National Symposium on Pest Management in Horticultural Crops: Environmental implications and thrusts, Bangalore, 15-17 Oct., 1997, p 32-33.

Chauke, R.P., A.N. Pasalwar and S.R. Dalal (1999) Efficacy of various insecticides and plant products against the larvae of lemon butterfly (*Papillio demoleus* L.). *Hi-Tech Citrus Management* (eds. Shyam Singh and SP Ghosh). *Proceedings of International Symposium on Citriculture.* Nov. 23-27, 1999) held at NRC for Citrus, Nagpur. Pp. 928-930

Clausen (1958) Biological control of insect pests. *Ann. Rev. Ent.*, 3 : 291-310.

Cruz, R. and W.E. Dale (1999) Control of Citrus leafminer by drench treatment with imidacloprid on desert soils in Peru. Special issue : Imidacloprid. *Pflanzen. Nachrich. Bayer* 52 (3) : 310-319.

De-Biase, L.M. and L.F. Russo (1996) Control trials against citrus aphids in the Pignataro Maggiore area (Caserta). *Informatore Fitipatologico* 46(11): 60-61.

Deshmukh, P.B., K.S. Pillai and A.T. Mathai (1989) Bioefficacy studies of some insecticides against black citrus aphid, *Toxoptera aurantii. Pestology* 13(6): 18.

Dhara, J.B., A. Verghese and P.L. Tandon (1989) Ecological studies on citrus psylla, *Diaphorins citri* Kuwayama (Hemiptera : Psyllidae) with special reference to its spatial distribution and sampling plan. *Entomon* 14(3 & 4) : 319-324.

Dhara, J.B., P.L. Tandon and A. Verghese (1990) Evaluation of different plant oils and extracts against citrus leaf miner, *Phyllocnis citrella* Stainton. In : *Proceedings National Symposium on problems and prospects of Botanical pesticides in IPM* held at Rajamundry, AP, India, pp .26.

Dhara, J.B., A. Verghese. and P.L. Tandon (1993a) Evaluation of different plant oils and extracts against citrus aphid, *Toxoptera citricidus* (Kirkaldy). *Indian J. Pl. Prot.* 18(2) : 251-254.

Dhara, J.B., P.L. Tandon and A. Verghese (1993b) Evaluation of different plant oils and extracts against citrus leaf miner, *Phyllocnis citrella* Stainton. In: *Botanical Pesticides in Integrated pest management.* Indian Society of Tabacco Sciences, Rajamundry, 340- 342 pp.

Dharamraju, E. and C.S. Reddy (1975) Insect-pests associated with citrus decline. *Citrus Decline in Andhra Pradesh. APAU,* Rajendra Nagar, Hyderabad, India.

Dikshit, A.K. and O.P. Lal (2002). Safety evaluation and persistence of imidacloprid on acid lime (*Citrus aurantifolia* Swingle). *Bull. Envir. Contam. Toxicol.* 68(4) : 495-501.

Donia, A.R.A., E.M. Helal and M.M.A. El-Hamid (1995) Mass rearing and field evaluation of released predaceous mite. *Alexandria J. Agric. Res.* 40(3) : 209-219.

Doumandji, M.B., N. Chahbar and L. Saharaoui (1999) Survey of the population dynamics and the parasitic complex of *Phyllocnistis citrella Stainton* 1856 (Lapidoptera : Gracillariidae) on two species of citrus in the region of Rouiba (Algiers). In *: Proceedings 51st International Symposium on Crop Protection, Gent, Belgium,* 4 May, 1999, 155-162.

Ebeling, W. (1959) *Subtropical Fruit Pests.* University of California Press. Pp.436.

Galli, J.C., F.D. Albuquerque, R.V. Gallao and R.A. Pinto (1995). Strategies for IPM programs to control scale, false rust mite and aphid on citrus and the effect on predators. *Revista de Agricultura Piracicaba* , 70 (3) : 249-268.

Gao , R.X. and Z. Quyang (1983) A preliminary investigation on parasitic fungi on citrus pests. *Weishengwuxue Tongbao,* 8(2) : 57-58.

Garsen, V., R. Kenneth and I.A. Muttath (1979). *Hirsutella thompsonii*, a fungal pathogen of mites. *Ann. Appl. Biol.* 91(1):29-40.

Ghose, S.K. (1965). Control of orange coccid, *Labioproctus polei Green. Indian J. Agric. Sci.* 35 : 32-38.

Gokkes, M., Y. Bar, I. Gal, Y. Oren, G. Eshel, I. Ofer and S. Zarka (1988) Aphid Control in mature citrus groves. *Hassadeh* 68(7): 1329-1330.

Gokkes, M., U Tadmor, D. Carmon and R. Shneider (1990). An uncorroborated field study for the control of citrus whitefly, *Dialeurodes citri* (Ashmead). *Alon Hanotea* 44(8): 759-761.

Grafton-Cardwell, E.E., Y. Ouyang and R.A. Striggow (1999) Predacious mites for control of citrus thrips, *Scirtothrips citri* (Thysanoptera : Thripidae) in nursery citrus. *Biol. Cont.* 14(1) : 29-36.

Grout, T.G. and G.I. Richards (1990) The influence of windbreak species on citrus thrips, *Scirtothrips citri* (Thysanoptera : Thripidae) populations and their damage to South African citrus orchards. *J. Entomol. Soc. S. Afr.* 53(2) : 151-157.

Gupta D and R. Bhatia (2000) Population dynamics of the citrus psylla, *Diaphorina citri* Kuwayama in lower hills of Himachal Pradesh. *Pest Mgmt. Econ. Zool.* 8 (1) : 41-45.

Helal, E.M., M.M.A. E.l-Hamid; A.R.A. Donia and E.A. Zakzouk (1995) Toxicity of some pesticides to the citrus brown mite, *Eutetranuchus orientalis* Kelein and its predatory mite, *Euseius scutalis* (Athias-Henriot). *Alexandria J. Agri. Res.* 40(3) : 221-230.

Holloway, R.L. and D. Smith (2010) A Texas citrus Pest Management Strategy. Pp46. Internet (*http://www.ipmcenters.org/pmsp/pdf/TXcitrus.pdf*)

Huang, M.D. and S.X. Li (1989) The damage and economic threshold of Citrus leafminer, *Phyllocnistis citrella Stainton* to Citrus. In : *Studies on the Integrated Management of Citrus Insect-pests.* pp. 84-89.

Hussain, M.A. and A.W. Khan (1945) The citrus aleyrodidae (Homoptera) in the Punjab and their control. *Mem. Ent. Soc. India* 1 :41.

Hussein, M.K. and N.S. Kawar (1985). The role of aphid natural enemies in regulating population densities of *Toxoptera aurantii* (Homoptera : Aphididae) on citrus trees in Lebanon. *Arab J. Pl. Prot.* 3 (1) : 11-17.

Jadhav, V.B. and C.V. Pujari (1999a). Management of citrus blackfly (*Aleurocanthus woglumi* Ashby)with conventional and botanicals. In : *Hi-Tech Citrus Management* (eds. Shyam Singh and SP Ghosh). *Proceedings of International Symposium on Citriculture.* Nov. 23-27, 1999) held at NRC for Citrus, Nagpur. Pp. 800-802.

Jadhav, V.B. and C.V. Pujari (1999b) Field evaluation of some insecticides and plant oils against citrus mealybug (*Planococcus citri* Risso). In : *Hi-Tech Citrus Management* (eds. Shyam Singh and S.P. Ghosh). *Proceedings of International Symposium on Citriculture.* Nov. 23-27, 1999) held at NRC for Citrus, Nagpur. Pp. 902-904.

Jessey, A.C. (1999) Citrus Leaf miner (CLM). Monthly Bulletin, Department of Agriculture, Fisheries, Bermuda. 70 : 2, 9-11.

Kacic S., K. Zanic and M. Catalinic (1997) Citrus leaf miner (CLM) *Phyllocnistis citrella* Stainton (Lapidoptera : Gracillariidae) in Croatia. *Fragmenta : Phytomedica-et- Herbologica*, 25 (1-2) : 15-25.

Katole S.R., H.S. Takare and R.K. Mahajen (1993) Effects of some plant products and insecticides on the infestation of citrus leaf miner on Nagpur mandarin. *J. of Maharashtra Agric. Uni.* 18(1) : 67-68.

Katole, S.R., A.V. Kolhe; A. Muqueem; K.B. Kale and S.D. Khiratkar (1994) Pesticidal management of sooty mould syndrome on citrus. *Crop Res.* 8(1) : 16-20.

Katole, S.R., R.G. Ughade; H.V. Ingle and U.S. Satpute (1997) Effect of weather parameters on the incidence of citrus leafminer (*Phyllocnistis citrella*) Stainton. *PKV Res. J.* 21 (2) : 252-253.

Katole, S.R., T.V. Kurhatkar and S.A. Nimbalkar (1999) Comparative efficacy of some insecticides against nymphal and pupal stages of citrus blackfly, Aleurocanthus woglumi Ashby. In: *Hi-Tech Citrus Management* (eds. Shyam Singh and SP Ghosh). *Proceedings of International Symposium on Citriculture.* Nov. 23-27, 1999) held at NRC for Citrus, Nagpur. Pp. 909-913.

Krishnamoorthy, A. and M Mani (1989) Recovery of an exotic parasitoid, *Leptomastrix dactylopii* How. From *Planococcus citri* (Risso) infesting some horticultural crops. *J. Biol. Cont.* 3 : 125.

Krishnamoorthy, A. and M Mani (1994) Manage mealybug in citrus orchards. *Indian Hortic.* 39 (1) :12-13.

Krishnamoorthy, A. and D Rajagopal (1999) Biocontrol of California red scale, *Aonidiella aurantii* (Mascale) on citrus using Coccinellid predator, *Chilocorus nigrita* (F.) in India. *Abstracts International Symposium on Ciriculture* at NRC for Citrus, Nagpur, pp.148.

Kuchanwar D.B., M.G. Hardas, H.S. Thakare and R.B. Gawande (1989) Seasonal incidence and management of citrus blackfly, *Aleurocanthus woglumi* Ashby on citrus. *Pestology* 13(11) : 21-29.

Lei, H.D., B.M. Lin, H.Y. Li; C Ran, Q.V. Zhang, K.M. Qian and W.H. Tian (2000). Experiment on the control of citrus aphids and leaf miner by carbosulfan insecticide. *S. China Fruits* 29 (5) : 15.

Liu, Z.M. (1989a) Studies on the interaction system of Citrus psylla, *Diaphorina citri* Kuwayama and its natural enemies, *Tetrastichus sp.* and *Chrysopa boninensis* Okamoto.In : *Studies on the Integrated Management of citrus insect pests*, pp. 144-164.

Liu, Z.M. (1989b) Population dynamics of citrus psylla and its ectoparasite. In : *Studies on the Integrated Management of citrus insect pests*, pp. 165-175.

Malausa, J.C. and V. Vacante (1997) Statement on the advancement of classical biological control of citrus leaf miner of citrus fruits *Phyllocnistis citrella* in France. *Integrated control in citrus fruit crop* (Proceedings of the meeting held at Florence, Italy, 29 August, 1996). *Bulletin- OILB SROP*, 20(7): 78-80.

Mani M. (1994). Effectiveness of the exotic encyrtid parasitoid, *Leptomastix dactylopii* HOW in control of mealybug, *Planococcus citri (*Risso) in guava orchards. *J. Entomol. Res.* 18(4) : 351-355.

Mani M. (1995) Studies on the natural enemies of oriental mealybug, *Planococcus lilacinus* (Ckll.) (Homoptera psuedococcidae) in India. *J. Ent. Res.* 19(1) : 61-70.

Mani M and A Krishnamoorthy (1996) Mealy bug problem on fruit crops. *Indian Hortic.* 41(3) : 43-45.

Mansant V, J.V. Sanz., J.I. Izquierdo and J.M. Puiggross (1999) Imidacloprid : New strategy for controlling citrus leaf miner (*Phyllocnistis citrella*) in Spain. *Special issue : Imidacloprid. Pflanzen. Nachrich. Bayer* 52(3) : 360-373.

Massa, B and M.C. Rizzo (2000) *Agromyza hiemalic* Becker (Diptera: Agromyzidae) leaf miner of nettle (*Urtica species*) : Phenology and Parasitoides in Italy. *Phytophaga Palermo* 10 : 53-67.

McCoy, C.W. (1978) Entomopathogens in arthropod pest control programs for citrus. In : *Microbial control of insect pests : Future strategies in Pest Management System* (Allen, G.E., C.M. Ignoffo and R.D. Jaques, Eds). Gainesville FL, pp. 211-223.

Melia, A. and J. Blasco (1980). Citrus aphids. The results of various tests of products to determine their effectiveness against different species. *Boletin del servicio de defense contra Plagas e Inspection Fitipatologica* 6(1) : 67-73.

Melia, A. (1989). Use of yellow traps on the control of citrus aphids (Homoptera : Aphididae). *Boletin de Sanidad Vegetal Plagas* 15(2) : 175-185.

Michelena, J.M., A. Sanchis and P. Gonzalez (1994) Aphidiinae on greenflies in fruit orchards in the Valencia community. *Boletin de Sanidad Vegetal Plagas* 20(2):465-470.

Mineo, G. and N. Mineo (1999) Further data on the establishment of *Semielacher ptiolatus* (Girault) (Hym., Eulophidae) in Sicily. *Bollettino de zoologia agraria de Bachicoltura* 31 (2) : 235-239.

Muthiah, C; G. Logiswaran and R.S. Azhakiamanavalan (1998). Effect of certain insecticides and plant products against leafminer (*Phyllocnistis citrella*) on sweet orange (*Citrus sinensis*). *S. Indian Hortic.* 46(1-2) : 68-70.

Nair, M.R.G.K. (1975) *Insects and Mites of Crops in India.* ICAR, New Delhi. pp. 180.

Nestor, B, M. Oscar; G. Morales, S.J.L. Carrillo and B.M. Hiram (1998) Mortality of *Phyllocnistis citrella* with mineral oil and neem. *Manejo Integredo de Plagas* 50 : 29-33.

Nimbalkar S.A., A.V. Morey; S.S. Narkhade and Y.M. Taley (1989) Relative toxicities of different insecticides against pupae of citus blackfly, *Aleurocanthus woglumi* Ashby. *Magazine College of Agriculture*, Nagpur, 56-59 : 25-27.

Nucifora, S. (1998) Imidacloprid against aphids on citrus. *Proceedings of Giornate Fitipatologiche* held at Scicli, Italy, from May 3-7, 1998. pp. 185-190.

Patel, V.C., H.K. Patel and R.C. Patel (1964) Effect of various recent insecticides on the bark borer (*Indarbela* sp.) of road side trees in Gujrat. *Indian J. Ent.* 26 :103-106.

Patel J.J., J.R. Patel, V.M. Valand, B.H. Patel and M.J. Patel (1998) Bioefficacy of some of the new insecticides against leaf miner, *Phyllocnistis citrella* and psylla, *Diaphorina citri* infesting citrus. *Indian J. Ent.* 60(1) : 101-103.

Pawar, N.P., S.A. Nimbalkar, S.M. Dadmal and S.R. Katole (1999). Non pesticidal management of citrus leaf miner, *Phyllocnistis citrella* Stainton. In : *Hi-Tech Citrus Management* (eds. Shyam Singh and SP Ghosh). *Proceedings of International Symposium on Citriculture.* Nov. 23-27, 1999) held at NRC for Citrus, Nagpur. Pp. 895-898.

Peleg, B.A. (1982) Effect of a new growth regulator, RO 13-5223, on scale insects. *Phytoparasitica* 10 (1) :27-31.

Pena, J.E., R. Duncan; E. Klema and A. Hunsberger (1999). Evaluation of direct and indirect action of insecticides and acaricides for the control of lime and avocado pests. In : *Proc. Fl. State Hortic. Soc.* No.112 : 213-217.

Perovic, T and S Malidzam (2001) Chemical control of citrus leafminer, *Phyllocnistis citrella* (Lapidoptera : Gracillariidae). *Jugoslovensko Vocarstvo* 35(3-4) : 175-182.

Pruthi, H.S. and M.S. Mani (1945) Our knowledge of the insects and mite pests of citrus in India and their control. *Imp. Conc. Agric Res Sci Monger,* 16 : 42.

Pruthi, H.S. and H.N. Batra (1960) Important fruit pests of North West India. *ICAR Bull. No.*80 :133.

Radhakrishnan, B. and N. Muraleedharan (1993) Bio-ecology of six species of syrphid predators of the tea aphid, *Toxoptera aurantii* (Boyer de fanscolombe) in Southern India. *Entomon* 18(3&4) : 173-180.

Radke, S.G. and A.Y. Thakare (1989) Chemical control of citrus leafminer. *PKV Res. J.* 13(1) : 44-47.

Rao, V.P., M.A. Chani., T. Sankaran and K.C. Mathur (1971) A review of the biological control of insects and other pests in South-East Asia and the Pacific region. *Commonw. Inst. Biol. Control Tech. Common.* 6 : 1-49.

Rao, C.N. and V.J. Shivankar (1999) Natural parasitism of *Phyllocnistis citrella* Stainton in Central India. In : *Hi-Tech Citrus Management* (eds. Shyam Singh and SP Ghosh). *Proceedings of International Symposium on Citriculture.* Nov. 23-27, 1999) held at NRC for Citrus, Nagpur. Pp. 885-888.

Rao, C.N. and V.J. Shivankar (2008) Field appraisal of certain bio-rational insecticides against citrus leafminer, *Phyllocnistis citrella* Stainton under nursery conditions. In : *Proceedings National Symposium on Citriculture : Emerging trends.* July 24-26,2008. NRC for Citrus, Nagpur. Pp. 230-233.

Rawhy, S.H., F Rofail, M.A. Hanafi, M.M. Abou-Setta, S.F. Mahmoud, M. Ghabbour and M.M. Abou-Setta (1980) Relation between the summer spray for controlling certain scale insects and the percentage of sound citrus fruits. *Agri. Res. Rev.* 58(1) : 259-263.

Rodrigues J.C.V., V. Rossetti, M.A. Machado, J.T. Sovrinho, N.N. Lima and N.N. de-Lima (1998) Citrus leaf miner : A factor for increase of pests and citrus canker. *Laranja* 19(1) : 49-60.

Ross, M. (1988) Twenty years of biological success on citrus in United States. In : *Proceedings of 6th International Citrus Congress. Tel Aviv, Israel.*

Saini, B.S. (1964) Bionomics and control of *Dialeurodes citri* (Riley and Howard) (Homoptera : Aleurodidae). *M.Sc. Thesis, PAU, Ludhiana.*

Salama, H.S. and A.H. Amin (1983) Chemical control of scale insects (Homoptera : Coccidae) infesting citrus trees in Egypt. *Crop Prot.* 2(3) : 317-324.

Satpute, P.P., Y.M. Taley, S.A. Nimbalkar and N.R. Kadu (1986) Natural enemies of citrus blackfly, *Aleurocanthus woglumi* Ashby, in Nagpur region. *Bulletin of Entomology* 27(2) :128-131.

Seif, A.A. (1987) Comparison of green and yellow water traps for sampling citrus aphids at the Kenya coast. *E. Afr. Agric. Forestry J.* 53 (3): 159-161.

Seraj, A.A. (1999a). Effects of some insecticides and petroleum spray oils on the control of citrus leafminer in Dezful district and review of its IPM. *J. Agricu. Sci., Islamic Azad Univ.* 5 : 17-18, 17-42, 98.

Seraj, A.A. (1999b) Injection of insecticides into tree trunks : A new method for the control of citrus pests. Iranian *J. Agric. Sci.* 30(1): 121-128.

Seraj, A.A. (1999c) Effects of petroleum spray oils without and with copper oxychloride on the control of citrus leaf miner. *Iranian J. Agric. Sci.* 30(3): 551-562.

Sethi, S.L. (1965) Insecticides for the control of citrus caterpillar, *Papilio demoleus* Linn. *J. Res. PAU* 2: 205-207.

Sethi, S.L. (1967) Control of Texas citrus mite, *Eutetranuchus banksi* in India with new pesticides. *J. Econ. Ent.* 60 :180-181.

Sewify, G.H. and A.M. Mabrouk (1990) The susceptibility of different stages of citrus brown mite, *Eutetranychus orientalis* Oudemans (family : Tetranychidae) to entomopathogenic fungus *Verticillium lecanii. Proceedings and Abstracts, 5th International Colloquium on Invertebrate Pathology and Microbial Control,* Adelaide, Australia, 20-24 August, 1990. p. 395.

Sharaf, N.S. (1989) Monitoring spider mite populations on lemon for effective control with flubenzimine and omethoate. *Dirasat* 16(7) : 53-64.

Sharma, D.R., R.C. Batra., P.K. Arora and S. Kumar (2001a) Bioactivity of insecticides against *Phyllocnistis citrella* Stainton. In: : Abrham Verghese & . P. Varvtha Reddy (Eds.) *IPM in Horticultural Crops : Emerging Trends in the New Millennium* (Proceedings of the Second National Symposium on Intergrated Pest Management (IPM) in Horticultural Crops: New Molecules, Biopesticides & Environment). October 17-19, 2001, Bangalore, p. 86.

Sharma, D.R., R.C. Batra, S. Kumar and P.K. *Arora* (2001b) Toxicity of Imidacloprid against *Phyllocnistis citrella* Stainton in Kinnow nursery. In: *International Conference on Pesticides, Environment, Food Security* 19-23, Nov. 2001. IARI, New Delhi p 199-200.

Sharma, D.R., S. Kumar, R.C. Batra and P.K. Arora (2001c) Residual toxicity of some insecticides against *Phyllocnistis citrella* Stainton in Kinnow nursery. In: *International Conference on Pesticides, Environment, Food Security* 19-23, Nov. 2001. IARI, New Delhi. p. 196-198

Sharma, D.R., S. Singh and P.K. Arora (2005a) Population fluctuation and control of citrus blackfly, *Aleurocanthus woglumi* Ashby in Punjab. In: *1st Congress on Insect Science*. PAU, Ludhiana Dec. 15-17, 2005. p. 179.

Sharma, D.R., S Kumar., R.C. Batra and P.K. Arora (2005b) Toxicity of insecticides against different populations of *Phyllocnistis citrella* Stainton in Punjab. *J. Insect Sci.* 18 (1): 68-71.

Sharma, D.R. and P.K. Arora (2009) Status and control of mealy bugs on fruit crops in Punjab. *J. Insect Sci.* 22 : 156-161.

Shivanker, V.J. and S. Singh (1997) Insect pests of Nagpur mandarin and their management. *Extension Bulletin No 12.* NRC for citrus, Nagpur.

Shivankar, V.J. (1999) Recent trends in insect pest management of citrus. In : *Hi-Tech Citrus Management* (eds. Shyam Singh and SP Ghosh). *Proceedings of International Symposium on Citriculture.* Nov. 23-27, 1999) held at NRC for Citrus, Nagpur. Pp. 762-784.

Shivankar, V.J., CN Rao and S. Singh (2000). Studies on citrus psylla *Diaphorina citri* Kuwayama : A Review. *Agricult. Rev.* 21(3) : 199-204.

Shivankar, V.J. and C.N. Rao (2001a) Insect pests of citrus and their management. In : Citrus (eds. Shyam Singh and SAMH Naqvi). International Book Distributing Co. Lucknow, India. Pp.327-344.

Shivankar, V.J. and C.N. Rao (2001b) Biological control of Citrus insect pests. In : Citrus (eds. Shyam Singh and SAMH Naqvi). International Book Distributing Co. Lucknow, India. Pp.347-372.

Shivankar, V.J. and C.N. Rao (2008) Evaluation of bio-intensive pest management modules against citrus blackfly, *Aleurocathus woglumi* Ashby and psylla, *Diaphorina citri* Kuwayama. In: *Proceedings National Symposium on Citriculture : Emerging trends.* July 24-26,2008. NRC for Citrus, Nagpur. Pp. 238-241.

Shivankar, V.J., D.R. Sharma and C.N. Rao (2008). New horizons in insect pest control in citrus. In : *Proceedings National Symposium on Citriculture : Emerging trends.* July 24-26, 2008. NRC for Citrus, Nagpur. Pp. 201-214.

Singh, R and J Singh (1984) New record of *Aspergillus candidus* Link. A potential entomogenous fungus on *Indarbela* spp. *Sci. and Cult.* 48(8) :282-283.

Singh, S.P. (1985) Biological Control of insect pests of Horticulture. *Proc. Natl. Sem. Entomoph. Inst.*, Calicut. Pp.221-231.

Singh, S.B. and H.M. Singh (1998) Natural enemies associated with insect pests of lemon, *Citrus limon* Burm. *Shashpa* 5 (1) : 109-110.

Solunke, B.R. and S.V. Deshpande (1991) Studies on use of plant products for the control of lemon butterfly larvae. *J. Maharash. Agric. Univ.* 16(2): 302-303.

Song, Z. (2001) A species of entomogenous fungus, *Fusarium lateritium* isolated from citrus aphid. *Scientia Silvae Sinicae* 37(6): 66-70.

Srivastava, R.P. (1978) Studies on the feasibility of the use of microbial pesticides for the integrated control of *Papilio demoleus. X International Horticultural Congress,* Sydney, Australia, 15-23 Aug., 1978; International Society for Horticultural Science. Abs. No.1624.

Streibert, H.P., M.L. Frischknecht and F Karrer (1994) Diofenolan - A new insect growth regulator for the control of scale insects and important Lepidopterous pests in deciduous fruit and citrus. *Proceedings- Brighton Crop Protection Conference, Pests and Diseases,* 1 : 23-30.

Su, WW. (2000) Experiment on the control of citrus leafminer. *S. China Fruits* 29 (3) :19.

Thakare K.R., S.J. Radke, M.N. Borle and H.T. Ghughuskar (1984-85). Two decades of research on citrus pest management in Maharashtra. *Brochure Pub. by DOR PKV,* Akola.

Tremblay, E. and G. Rotundo. (1980) The use of sex pheromones in the control of citrus scale insects. *Standardization of Biotechnical metods of Integrated Pest Control in Citrus Orchards.* pp.59-66.

Vargas O.H.; D. Bobadilla; R..M. Jimenez and C.H. Vergas (1999) A preliminary trial about larval susceptibility of *Phyllocnistis citrella* Stainton (Lapidoptera : Gracillariidae) to insecticides sprayed on the foliage. *IDESIA,* 16 : 23-27.

Vercher R; M.F. Garcia; C.J. Costa, C Marzal and C Granda (2000) Importation and establishment of parasitoides of the citrus leafminer, *Phyllocnistis citrella* (Lapidoptera : Gracillariidae). *Boletin —D —Sanidad —Vegetal, Plagas* 26 (4) : 577-591.

Viggiani. G. (2001) Turning point in the biological control of citrus leaf miner. *Informatore —Agrario* 57 (4) : 53-55.

Wang L.D., Y Myinsheng and W. Qiung (1999) Damage of citrus leaf miner to citrus and its economic threshold. *Chinese J. Appl. Ecol.* 10(4) : 457-460.

Willink E, D Figueroa, P Zamudio and H. Salas (1998) Biological control of the citrus leaf miner. *Avance Agroindustrial* 18(73) : 14-18.

Wu X.F. (2000) Control of the citrus leafminer for Sijiju mandarin variety. *S. China Fruits* 29(4) : 19-20.

Ye, Q.M., H.Y. Bao, Z Li, Y.H. Zhu; X.X. Ye; L.F. Jin and Y. Zheng (1996) Population dynamics and chemical control of citrus aphids in Zhejiang. *Acta Horticulturae, Zhejiangensis* 8(5) : 281-286.

Yumruktepe, R and N Uygun (1994) Determination of aphid species (Homoptera : Aphididae) and their natural enemies in citrus orchards in the Eastern Mediterranean region. *Turkiya III. Biyolojik Mucadele Kongresi Bildirileeri. 25-28 Ock, Ege Universitesi Ziraat Fakultesi, Bitki Koruma Bolumu, Izmir.* pp. 1-12.

Zade V.J., Y.M. Taley; S.A. Nimbalkar and N.R. Kadu (1986). Efficacy of different insecticides against the nymphs of citrus blackfly, *Aleurocanthus woglumi* Ashby. *Indian J. Ent.* 48(2) : 218-219.

Zhan, Z.X. and W.F. Peng (1990) Study on compound preparation and application of acaricides used on citrus. *Fujian Agril. Sci. Tech.* 4:23-24.

Zhang, G.C. and J.Q. Ma (1998) Effect of using Nurelle-D505 for controlling citrus leafminer. *S. China Fruits* 27 : 3,22.

Zhang, Z.Q. and F. Faraji (1994) Notes on *Allothrombium pulvinum* Ewing (Acari : Trombidiidae) new to the fauna of Iran, *Acarologia* 35(4) :357-360.

Zhang, G.B. (2000) The causes of rampancy of citrus dirty mould disease and its control. *South China Fruits.* 29 (1) : 10.

Zhang, Z.H. and G.G. Wang (2001) The biological characters of citrus whitefly and its control by chemicals. *S. China Fruits* 30 (2) : 7-9.

Zhang, Q.V., H.D. Lei; B.M. Lin; C. Ran; H.Y. Li; K.M. Qian and W.H. Tian (2001) Experimental control of citrus leaf miner. *S. China Fruits* 30 (4) : 17.

Zhou, YS; T.M. Ye and Q.W. Huang (2001) The integrated control of citrus diseases and pests. *S. China Fruits* 30 (5) :19-20.

Theory and Practice of Integrated Pest Management
R. Arora, B. Singh and A.K. Dhawan (Eds)
Scientific Publishers (India), Jodhpur 2011, pp. 410 - 425.

23

MANAGEMENT OF INSECT PESTS IN FRUIT CROPS OTHER THAN CITRUS

D.R. Sharma, P.K. Arora and Sandeep Singh
Department of Horticulture
Punjab Agricultural University, Ludhiana - 141 004, India

23.1. INTRODUCTION

Based on the prevailing climatic conditions, there are six agro-climatic zones in India, wherein the fruit crops of one regional environmental requirement are grown in another region. This type of crop husbandry has some typical pest problems as the insect and mite pests have easy access to any of these regions depending upon the fruit crop. It is estimated from the available information in the literature that there are about 1000 insect and mite species on about 100 major and minor fruit crops in India. This number would certainly be affected with the collection and evaluation of germplasm and improvement in production technology coupled with the climatic changes as all these factors significant influence the nefarious activities of the pest species. The wide geographical diversity and host range of perennial fruit crops also provide almost a permanent niche for many insect pests. Thus, a comprehensive approach of management of the existing or new pests is needed. An overview of literature on the management of fruit pests in India indicated that persistence efforts were made to develop various non-chemical and chemical technologies based on the most vulnerable stage and seasonal activity of the pest in relation to host plants, precise knowledge of orchard ecosystem, identifying principal factors and regular assessment of pest and natural enemy's populations. These management technologies include cultural, mechanical and physical, chemotherapeautic, regulatory, biological, plant resistance and rcproductive methods either to curb pest outbreaks or create conditions unfavourable for outbreaks of the pest, are discussed with some pertinent examples (Deol and Sharma, 2005; Sharma 2005a,b; 2006; 2007a,b; Sharma and Batra, 2007; Sharma and Singh 2006; Sharma *et al.* 2005).

23.2. CULTURAL CONTROL

Crop husbandry of fruit crops is radically different to that of field crops and under many situations the pest problems are of some permanent nature. Several cultural practices if performed regularly in orchard ecosystem would be helpful in either elimination or atleast reducing the carryover of pest population.

Clean culture. The practices which create conditions unfavourable for pest development, are very important in regulating the pest populations. Clean culture, which eliminates the weeds and other secondary hosts, is useful in avoiding the carry over of some pests. It is useful in either reducing the opportunity of pest to lay eggs or feeding in the off-season. It has been found promising in controlling banana leaf and scaring beetle; scales and mealy bugs on pineapple and mango; thrips and shot hole borer on grapes and apple; pomegranate butterfly [*Deudorix isocrates* (Fabricius)]; San Jose scale [*Quadraspidiotus perniciosus* (Comstock)] on apple, peach leaf curl aphid (*Myzus persicae* Sulzer) on peach and hoppers and mealy bug on mango.

Tillage. It has been proved useful in controlling the dormant stages of pests either by burying them deep in the soil so that they are unable to emerge or brought to the soil surface for exposing them to inclement weather conditions and/or natural enemies. Many insects like mango mealy bug [*Drosicha mangiferae* (Green)], gall makers, leaf-cutting weevil, litchi nut borer (*Blastobasis* sp. and *Gatesclarkeana* sp.), fruit fly of guava, pear and peach [*Bactrocera dorsalis* (Hendel)] and *ber* (*Carpomyia vesuviana* Costa) are found to be controlled by plouging the soil. Mango mealy bug (*D. mangiferae)* can effectively be controlled by ploughing the soil during May-June as the gorged females lay eggs in the soil. The eggs are exposed to either intense heat or natural enemies. Similarly, the pre-pupal and pupal populations of *B. dorsalis* occurred in the soil and shallow ploughing with cultivator immediately after harvest is effective in exposing and killing the pupating larvae/pupae which are mostly present at 4-6 cm depth. Ploughing also destroys the hibernating mango stone weevil [*Sternochaetus mangiferae* (Fabricius)] while raking the soil during October-November and again in March can manage the pest in mango orchards.

Plant density. Plant densities change the micro-climate habitat e.g. dense banana planting provides congenial conditions for banana aphid

(*Pentalonia nigronervosa* Coquillett) population as closer plantation develops canopy favourable for multiplication of aphid in shady humid climate. The infestation of mango hopper [*Amritodus atkinsoni* (Lethierry)], scales, mealy bugs, citrus whitefly [*Dialeurodes citri* (Ashmead)] and blackfly (*Aleurocanthus woglumi* Ashby) is facilitated in close plantations.

Good agricultural practices. Good cultural or agronomical practices namely irrigation, fertilization and pruning have been found to stimulate the rapid plant growth and maturity. Vigorous growth thus tolerates the pest attack of aphids, mealy bugs, leaf folders, borers etc. due to the fact that out-growth withstands the insect injury. The neglected and poorly maintained plants are severely attacked. Encouraging the plants to put intermittent flushes by regular irrigation and split dosages of nitrogenous fertilizers enhances the attack of mango hoppers. Heavy irrigation of orchards during October, however, kills the eggs of *D. mangiferae* and pupae of mango midges.

Orchard sanitation. Orchard sanitation i.e. keeping plantations clear of any plant debris will be helpful in reducing the pest incidence or found effective in managing many insect pests e.g. cuttings of banana pseudostems after harvesting remain in the field, which otherwise provide suitable site for banana rhizome and pseudostem weevil to lay eggs. The regular removal of infested twigs and fruits are also effective against mealy bugs, fruit flies and litchi nut borer.

Pruning. This horticultural practice is used to maintain maximum yield of high quality fruit as well as to control tree size and shape. In fruit trees, which are perennial in nature, proper pruning of undesirable plant parts often removes some active and overwintering stages of insect pests like mealy bugs and scales on several fruit crops and shoot gall psylla and leaf webber on mango. This may also facilitate better spray coverage or light penetration in the canopy of tree. Pruning or removing of some of the mango branches enhances the light perception and is effective against mango hoppers. Pruning of infested leaves, twigs and branches and destroying the same, reduces the incidence of mango scale (*Aspidiotus destructor* Signoret).

Mosaic planting. This practice has been found to have direct impact on the survival of pest, parasitoids and predators. For example, planting of blackberry in grapevine orchards is useful for the control of grapevine leaf hopper.

Removing of sick plants. Fruit trees heavily infested by some insect pests should be uprooted and destroyed. The practice is useful against shot hole borers, trunk borer, bark eating caterpillar on various fruit crops and papaya plants infected with virus which is being transmitted either by aphid (*Aphis gossypii* Glover) or whitefly [*Bemisia tabaci* (Gennadius)] to healthy plants.

23.3. MECHANICAL AND PHYSICAL CONTROL

In fruit crops the mechanical and physical methods are largely practiced against a number of insect pests. It is easy, quick and effective but labour intensive.

Removal or destruction. These practices generally involve the removal and destruction of pests mostly by using manual labour. At small scale, it is very effective and profitable especially to homestead fruit growers. Collection and destruction of insect stages is a predominating and effective method in checking the outbreak of many pests e.g. brushing of scales of mango; dropped guava, *ber*, pear, peach fruits infested by fruit fly maggots (*Bactrocera* sp. and *C. vesuviana)*; dropped mango leaves mined by grubs of leaf cutting weevil; mummies of litchi for litchi nut borer; rolled leaves by leaf roller and fruit borers on many fruit crops, mealy bugs and scales on various fruits, gall makers on fig and, bud and fruit borer leaves on sapota. Removal and destruction of infested plant parts is also effective in regulating the pest populations of mango stem borer [*Batocera rufomaculata* (de Geer)] and guava (*Microcolona technographa* Meyrick) and mango (*Chlumetia transversa* Walker) shoot borers. Regular removal and burying of fallen fruits at least at 60 cm depth has been proved to be very effective against fruit fly, *B. dorsalis* in guava, peach and pear orchards. The collection and destruction of leaves (litchi mite, mango midges and sapota leaf webber) and fruits (mango stone weevil) reduces the source of pest carryover.

Bagging. Bagging the fruits of pomegranate with butter paper or cloth immediately after fruit formation prevents the damage of butterfly *(D. isocrate).*

Banding. Slippery alkathene sheet or sticky Ostico band, encircling the trunks of trees are used to prevent certain stages of insects from climbing to the foliage from the ground. This method is widely used to control insects that spend a part of their lives in the soil and move to the tree tops by creeping up the trunks. By using slippery alkathene sheet, swarming population of newly hatched nymphs of mango mealy bug (*D. mangiferae)* is barred to climb the tree due to slippery nature of sheet. The crawlers die due to starvation as the suitable site of feeding i.e. tender apical parts for feeding are not available. The slippery band consists of 15-20 cm wide alkathene sheet and is applied at the basal end of the stem of the mango tree. The top or upper part of the sheet is secured with 1-3 nails (2 cm) while lower end should be covered with compact soil so as to prevent the crawlers to climb. Occasional wiping of the band is desirable during rains. Fixing of sticky bands on the main trunk prevents the movement of ants and thus the dispersal of scales and mealy bugs in sapota orchards is reduced.

Spiking. The adults of palm rhinoceros, [*Oryctes rhinoceros* (Linnaeus)] which feed within bored hole just below the crown, and cause the death of tree could be spiked out with the help of iron hook. It is very economic when practiced after detecting the pest in the hole. The beetle can be detected either by fresh chewed fibrous material coming out from the bored hole or by audible gnawing sound of the beetles. This method not only controls the pest population but also reduces its carryover.

Shaking and beating. It can be practiced to control the sluggish adults of *ber* beetle, which cause heavy damage to the foliage during night hours. The adults can be dislodged by shaking or beating the trees with a stick and are collected on a cloth spread below the tree in the early morning and are destroyed.

Traps. Various trapping devices, viz. light, bait, colour traps etc. are being used to attract the insects. Yellow sticky or pan water trap is used in daytime to catch aphids on various fruit crops. Traps baited with some attractive material namely fermented sugar, molasses etc. and poisoned with chemicals have also been found effective in capturing fruit fly adults. Likewise, freshly chopped pseudostems of banana when spread in the banana orchard during night, are found an effective means of catching the adult weevils.

23.4. BIOLOGICAL CONTROL

In fruit crops, several predator and parasitoids of different insect pests have been identified and appeared to be ideal for the control of pests, but actually there are few instances where the biocontrol agents are being used to reduce the pest population below the economic injury level. The bioagents are generally used indirectly, except few instances where they are used directly for the immediate control of pest species e.g. predatory coccinellid; *Cryptolaemus montrouzieri* Mulsant has been used against mealy bugs and scales on citrus, grapes, pomegranate, sapota, *ber,* custard apple, guava etc. Among different bioagents, parasitoids have received increasingly more attention than predatory insects. Hitherto many entomophagous spiders, mites, fungi, bacterial, protozoan, virus, nematodes and other vertebrates have been found to play a significant role in balancing the pest population of the fruit crops. The biological pest suppression is not only an important tool for sustainable agriculture but also eco-friendly and safe for mankind. A complex of natural enemies affecting all life stages of insect pest would be desirable to promote population regulation below damaging level rather than reliance on a single parasitoid species to function on its own. A number of natural enemies have been reported on fruit crops in India but only the successful examples are discussed in this text.

Mango. Several natural enemies are found to attack mango hoppers in India. Among them, the reduviid, *Irantha armiper* Stal preys voraciously on *A. atkinsoni*. In nature, *Verticillium lecanii* (Zimmemlan) @ 1 x 10^8 spores/ml is observed to wipe out mango hoppers around Lucknow. Among the different natural enemies, *Thompsonisca desantiellus* Shafee is commonly found on *A.destructor*. *Aulacaspis tubercularis* (Newstead) is found predated by *Chilocorus nigrita* (Fabricius), *Chilocorus circumdatus* (Schonh) and *Scymnus* sp. An aphelinid, *Pteropteryx koebeli* (Howard) is found to cause about 90 percent parasitism in nature on *A. tubercularis*. *C. nigrita* is rec-ommended for the suppression of both *A. tubercularis* and *A. destructor*.

Grapevine. The pink mealy bug, *Maconellicoccus hirsutus* (Green) is a major pest of grapevine. Among the local natural enemies, the encyrtid parasitoid, *Anagyrus dactylopii* (Howard) and the coccinellid predator, *Scymnus coccivora* (Ramakrishana Ayyar) are of considerable importance. An exotic parasitoid, *Leptomastix dactylopii* Howard causes up to 70 per cent parasitism in nature while the coccinellid predator, *C. montrouzieri* feeds about 1000 eggs or 300-500 mealy bug nymphs. Adult beetles (10 days old) at the rate of 5000/ha are recommended for release in the evening hours to control the grape mealy bug and is giving good control in Maharashtra, Andhra Pradesh, Kamataka and Tamil Nadu. *Nipaecoccus viridis* (Maskell) and *Planococcus citri* (Risso) also appear in serious form on grapes in certain locations and the release of *C. montrouzieri* helps in controlling these two mealy bugs. Both *Spodoptera litura* (Fabricius) and *Helicoverpa armigera* (Hubner) cause severe damage to the leaves and bunches of grapevine and application of nuclear polyhedrosis virus controls these two pests on grapevine.

Guava. Green shield scale, *Chloropulvinaria psidii* (Maskell) occurs on the fruits, leaves, shoots and trunk. They excrete honeydew resulting in development of sooty mould on the leaves and thereby hindering the photosynthetic activity. Several natural enemies have been found to attack scale, however, *C. montrouzieri* is known to feed about 4000 eggs during its development and the release @ 20 adult beetles/tree is found very effective against C. *psidii*. Wax scale, *Drepanococcus (=Ceroplastodes) chiton* (Green) is also very serious in some orchards around Bangalore. Parasitoids, *Anicetus ceylonensis* Howard and *Cephaleta brunniventris* Motschulsky are highly effective in suppressing *D. chiton* population. Local predator, *C. nigrita* observed in large numbers, also helps in suppressing its population and in the subsequent years, the scale population does not appear at all. Among different mealy bugs, *Ferrisia virgata* (Cockrelle) is very serious in many guava orchards in South India. It occurs on shoots and fruits. Fruits infested with the mealy bugs are covered with sooty mould and loose the market value. In guava orchards, several local parasitoids and predators in nature very well regulate the mealy bugs.

Among the different natural enemies, the parasitoids, *Aenasius advena* Compere, *Blepyrus insularis* (Campere) and the predators, *Scymnus coccivora* (Ramakrishana Ayyar), *Mallada boninensis* (Okomato), *Brumus suturalis* (Fabricius) and *Spalgis epius* (Westwood) play a significant role in suppressing the *F. virgata* population. *A. advena* causes up to 50 per cent parasitism in nature while *B. insularis* also causes about 32 per cent parasitism. The parasitoid can be bred easily on 5-10 days old mealy bugs in large numbers. Release of *C. montrouzieri* @ 20-30/tree gives excellent control of *F. virgata* on guava.

In guava orchards, parasitoids, *Coccidoxenoides peregrinus* (Timberlake), *Allotropa citri* Muesebeck and the predators, *S. epius* and *C. montrouzieri* regulate *P. citri* in nature. The release of exotic paras-itoid, *L. dactylopii* also found effective and in many orchards, *P. citri* disappeared subsequently due to its activity. The predators, *S. epius* and *C. montrouzieri* are commonly associated with the mealy bugs, *Planococcus lilacinus* (Cockrelle) and *M. hirsutus*. Release of *C. montrouzieri* helps in managing these two mealy bug species on guava.

Sapota. The green scale, *Coccus viridis* (Green) and the green shield scale, *Chloropulvinaria psidii* are found to infest leaves and tender shoots of young sapota plants. The green scale is suppressed by *C. nigrita* and C. *montrouzieri* while the green shield scale is controlled by *C. montrouzieri*. Mealy bugs, *P. lilacinus* and *P. citri* are parasitized by two encyrtids, *L. dactylopii* and *C. peregrinus* and are able to suppress the populations of *P. citri* while *C. montrouzieri* is highly effective in checking the populations of *P. lilacinus*.

Pomegranate. Ash whitefly, *Siphoninus sphillyreae* (Haliday) is serious on the leaves and the entire plant becomes sick due to the development of black sooty mould on the honeydew excreted by the whiteflies. The aphelinid, *Encarsia azimi* Hayat cause as high as 90 per cent parasitism in nature. Mealy bugs, namely *P. lilacinus, P. citri, N. viridis* and *M. hirsutus* are checked by *S. epius* and *C. montrouzieri* efficiently. The parasitoid, *L.dactylopii* is recommended for the suppression of *P. citri* while *C. montrouzieri* for all other mealy bug species in pomegranate orchards.

Anar butterfly, *Deudorix epijarbas* Moore is a serious pest in north India especially Himachal Pradesh and Jammu & Kashmir. The eggs of *D. epijarbas* are found parasitised upto 62 per cent and the larvae upto 15 per cent in nature by various parasitoids. Release of *Trichogramma* spp. also reduces the pomegranate fruit borer infestation. *Deudorix isocrates* (Fabricius) another butterfly is parasitized by the egg parasitoids namely, *Telenomus* sp., *Ooencyrtus papilionis* Ashmead and *Trichogramma chilotraeae* Nagaraja and Nagarkatti upto 60 percent in south India. Release of *T. chilonis* @ 2.5 lakhs/ha four times at 10 days interval reduces the fruit borer infestation.

Ber. Wax scale, *D. chiton* is a serious pest on the shoots and leaves. Two parasitoids, namely *Cephaleta brunniventris* Motschulsky and *Anicetus ceylonensis* (Howard) cause heavy parasitism on the scales and thus disappeared subsequently in the orchards. The local parasitoids and the predator *S. epius* play a significant role in reducing the pest population of mealy bugs, *N.viridis, M. hirsutus, P. lilacinus* and *P. citri*. Release of *C. montrouzieri* is also found to be effective against *N. viridis* and *P. citri* whereas *L. dactylopii* is being used against *P. citri* in *ber* orchards.

Custard apple. The predators namely *S. epius,* C. *montrouzieri* and *S. coccivora* are commonly found feeding on the mealy bugs,viz. *N. viridis, M. hirsutus, P. lilacinus, Planococcus pacificus* Cox and *P. citri*. Release of *C. montrouzieri* helps in clearing the mealy bugs quickly on the fruits. *Aphytis* sp is known to cause up to 30 per cent parasitism on scale insects namely, *Aonidiella orientalis* (Newstead) and *Parasaissetia nigra* (Nietner) while *Scutellista cyanea* Motschulsky causes about 42 per cent parasitism on *P.nigra*. Release of *C. nigrita* also helps in bringing down the scale population.

Apple. A large number of predators and parasites are ecorded on San Jose scale, *Q. perniciosus*. Among them, the parasitoids namely *Encarsia* (= *Prospaltella*) *perniciosi* Tower, *Aphytis* sp. *proclia* group, *Aphytis diaspidus* Howard, *Aphytis mystilaspidus* (Leb), *Chiloneurus* sp. and predators such as *Chilocorus bijugus* Mulsant, *C. circumdatus*, and *C. septumpuncta* are causing significant control of scale. The Russian, Chinese and Japanese strains of *E. perniciosi* introduced in 1959 in Himachal Pradesh caused 8-89 per cent parasitism. Predators namely *C. montrouzieri*, *Chilocorus kuwanae* Silvestri and *Cybacephalus gibbulus* Errick were also released for scale control. *Aphytis proclia* group Azotus, *E. perniciosi* and *Teleterebratus perversus* Compere & Zinna and predators like *C. bijugus. Coccinella septempunctata* (Linnaeus) and *Sticholotis marginalis* Kapur are effective natural enemies of San Jose scale in Jammu and Kashmir, while in Himachal Pradesh *A. proclia. C. bijugus. C. septempunctata* and *P. flexibilis* are reported attacking the pest. In Kashmir, the release of *E. perniciosi* caused upto 9.1 per cent parasitism after its establishment in apple orchards. The release of predator, *C. bijugus* in Himachal Pradesh had 10-39.7 per cent reduction of scale population after 3 years of release.

In 1937, attempts were made to introduce and establish exotic parasitoid, *Aphelinus mali* Haldmann from U K in Kullu valley. It established successfully on woolly apple aphid, *E.lanigerum* within a short time and provided up to 98 per cent control. Afterwards, the parasitoid was also introduced against *E.lanigerum* at Mashobra and Kotgarh in Shimla hills, Chaubattia (Uttaranchal). Coimbatore (Tamil Nadu) and Shilong (Meghalaya) and provided significant control of aphid at the initial stages but later on failed to establish due to varying climatic conditions. Various predators have also been reported to exercise fairly good control of woolly apple aphid but none is released at commercial scale. However, the combined affect of *A. mali* and predators, *Chrysopa scelestes* Banks and *Syrphus confrater* Weid on woolly apple aphid population caused 23.8-34.8 per cent reduction. *A. mali* is well established in valley where insecticides are used at a minimal level. It can withstand lower winter temperatures than its host. Parasitization reaches high levels in the late summers, when other factors have slowed down the aphid increase. The best to conserve *A. mali* is to minimize insecticidal sprayers.

23.5. PLANT RESISTANCE

Despite the great achievement in plant resistance in field crops, this aspect has gained little attention in fruit crops. Baring the historical examples of plant resistance in apple variety Winter Majetin against wooly apple aphid *(Eriosoma lanigerum* Hausmann*)* and wild American grapes *(Viteus viniferata)* against *Phylloxera vitifolia* Fitch, much has not been achieved in fruit crops and these two examples are still legendary. An overview of host plant resistance of fruit crops in India (Sharma *et al.,* 2004) revealed that information on sources of resistance based on the insect damage is available in mango, apple, guava, peach, plum, banana, grapes, date palm, *ber*, sapota, etc. but little efforts have been made to incorporate the resistant/tolerant (Table 23.1) germplasm in breeding programme. The bases and mechanisms of host plant resistance in some fruit crops are discussed.

Ber. The preference of egg laying and subsequent larval development on a cultivar also depicts its level of tolerance. The reaction of cultivars were presumed to vary according to the antibiosis and the percent fruit infestation, number of eggs or egg punctures per fruit, number of larvae/fruit and percent larval hatching in fruits of fruitfly, *C. vesuviana*. The fruitfly preferred egg laying on Dandan, Gola, Nazuk and Rashmi. The preference varied from 6.7 to 7.3 percent on these cultivars. The percent hatching on different cultivars also varied from 6.1 to 83.2. In cultivars Tikadi, Illaichi and Umran, only a small percentage of eggs developed into larvae and, therefore, were less susceptible to the fruitfly. *C. vesuviana* infestation was positively correlated with fruit weight, pulp stone ratio, total soluble solids and total sugars, while it was negatively correlated to acidity, Vitamin C and total phenols. The varieties high in pulp, TSS, total sugars, low acidity, vitamin C and total phenols were highly susceptible to the pest.

Guava. The effect of guava cultivars viz., AC 10, Chittidar, Lucknow 46 and Lucknow 49 on larval survival, adult occlusion and size of *Bactrocera correcta* Bezzi were investigated and significant variations in larval and pupal periods were observed between cultivars, but there was no discernible difference in pupation rates (Fig. 23.1). Adult emergence was higher on Chittidar (90.6%) and minimum on Lucknow 46 (59%). Chittidar yielded the largest adult (>2.17 mm head width). The role played by shape of the guava fruit in the ovipositional behaviour of *B. correcta* revealed that Lucknow 49 (round big) was the most resistant cultivars with 0.07 punctures/fruit, 0.71 maggots/fruit and 4.8 per cent fruit damage. Smooth skinned *cv.* Chittidar was most susceptible with 1.28 punctures/fruit, 2.18 maggots/fruit and 49.38 per cent fruit damage. Fruits of Lucknow 46 (bell shaped) and AC 10 (oblong) were intermediate in reaction.

Table 23.1. Resistant/tolerant stock of different fruit crops

Fruit crop	**Insect/mite pest**	**Resistant/tolerant stock**
(1)	(2)	(3)
Datepalm	Stone beetle	Late ripening Medjool, Thoory
	Greater date moth	Barhee
Chikoo	Chikoo moth	Cricket ball, PKM-I, Guthi, Oval, Kirthabarthi
	Bud borer	PKM-I, Kirthabarthi, Cricket ball, Guthi, Oval
Banana	Rhizome weevil	Locatan, Cavendish, Cavendish (R), Poovan, Chakkia, Pedohapack, AAB clone
	Rhinoceros beetle	Chowghat dwarf orange x West coast tall
	Thrips	ABBB clone
	Lacewing bug	Jurmony, Malakali, Padathi, Kali, Krishan vazahai, Agneswar, Malbhog, Chenichampa, Bhimkal, ABBB clone
Ber	Fruit fly	Seb, Jogia, Tikadi, Illaichi, Nazuk
	Fruit borer	Illaichi, Jogia, Dandan
	Leaf roller	ZG-2, Golar, Sanaur-1, Nazuk, Ladoo, Kala Gola
	Mites	Chomuocal
Guava	Fruit fly	Nasik, China Surkha, Behat Coconut, Pear shaped, Red flesh, Smooth Green hybrid
	Tea mosquito big	Bangalore Round, Bapatala, AC 10, Seedless
	Shoot borer	Red flush
Pomegranate	Butter fly	Sindh, Swarupkathi
	Castor capsule borer	Guinea, Red flesh

(1)	(2)	(3)
	Hoppers	Pulhora, Kala Hapus, Keshar Basti, Annanas
Mango	Gall insect	Annanas, Anain, Delicious, Gulabkhas, KO7, KO11, Maharaja of Mysore, Salem, Banglora, Vellakachi
	Fruit fly	Toranjo, Monteiro, Manjurad
	Leaf gall midge	Deshi Malgoba, Alam Baneshan
	Mites	Mohandas, Malgoa, Pulliadi
Apple	Wooly apple aphid	MM_9, M_{21}, M_{997}, M_{793}, MM_{111}, MM_{115}, M_9, M_{16}, MM_{101}, MM_{106}, Crab apple, Yellow Delicious, Isreal type 2, 4, 5
	San Jose scale	Benoni, Green sweet, Pecks Pleasant, Summer Golden Pippin, Pippin, Cox's orange, Mcintosh, Cortlant, Tompkins

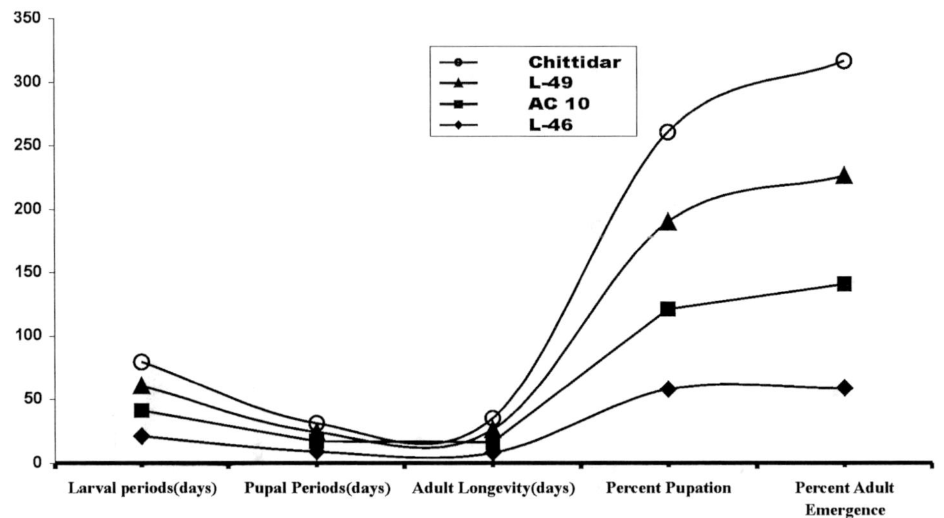

Fig. 23. 1. Development of guava fruit fly *Bactrocera correcta* in different cultivars

Fruits of Red Flesh and Seedless guava with rough and wrinkled skin had very low fruit fly, *B. dorsalis* infestation whereas Pink Flesh (Gulabi) and Strawberry cultivars with rough and gritty skin, respectively, exhibited low infestation. Baramasi and Guinea varieties with intermediate mellow skin had moderate infestation while smooth skinned Allahabad Sufeda, Apple colour and Lucknow 49 were severally infested. Fruit morphological characters like skin surface and flesh colour significantly influenced the preference of fruitfly. Fruits with red flesh and rough skin had less damage compared to those with white flesh and smooth skin. However, fruit shape did not show any effect on the incidence. EC 147037, EC 147039, Kamsari, Spear Acid and Superior Sour Lucidum were the least affected (<10% damage) against different insect and mite pests in India while Bangalore Local, Seedless Smooth Green and White Flesh were susceptible with 41.0, 40.66, 37.67 and 35.33 per cent fruit damage, respectively.

High levels of vitamin C, total soluble solids and total phenols in the fruits of different guava cultivars were found to contribute the resistance to *B. correcta*. The percent infestation of *B. dorsalis* was positively correlated to total sugars, TSS and total proteins but was negatively correlated to total phenols, orthodihydroxy phenols and flavonols. Acidity and vitamin C in fruits showed a very weak but negative impact on fruit fly incidence.

Mango. Chemical analysis of inflorescence showed that the varieties such as Baneshan Bangalora, Chinnarasam and Khander resistant to leaf hoppers had less nitrogen, more phosphorus, potassium, calcium, total phenol and orthodihydroxy phenols and less of reducing sugars. Magnesium and non-reducing sugar content were not related to resistance. A moderately resistant cultivar Gaddemar also had similar composition as that of a resistant cultivar but had more reducing sugars. Moreover, another moderately resistant cultivar Rumani contained more nitrogen than that in resistant one but had less than that in a susceptible cultivar. Susceptible cultivar such as Himayudin had more of nitrogen but less of potassium.

23.6. REPRODUCTIVE OR AUTOCIDAL CONTROL

The reduction of insect population by using substances that cause sterility by altering sexual behaviour or otherwise disrupt the normal reproduction in insects is found effective in many insect pests. Male sterility has successfully been used on oriental fruit fly, melon fruit fly and codling moth. It involves the artificial rearing, sterilization with direct application of -radiation or by using some chemosterilants like TEPA, METAPA and apholate and releasing otherwise healthy male insects in numbers that will compete with natural males. The females thus mated with such sterile males, produce unfertilized and nonviable eggs and ultimately the suppression of insect population. No work on autocidal control on fruit crops is done in India. Using of sex attractants or pheromone is also found effective in monitoring the pest populations of codling moth and Mediterranean fruit fly. Methyl eugenol traps are being used for monitoring of fruit fly on several fruit crops but its success for management is yet to be explored.

23.7. REGULATORY CONTROL

It is a lawful regulation of eradication or to prevent or to control infestation or to reduce damage by insects including some other problems arising out of pest control programme. Initially plants and plant products of all kinds could be moved freely both between and within the countries of the world. As a result, a number of insect and mite pests introduced in areas where they were not present e.g. San Jose scale and woolly apple aphid. They reproduce and establish at alarming rate and become as major pests. To check this spread, regulations can be either through quarantine (foreign and/or domestic) and embargos to prevent invasions of new pests or inspection and certification to ensure that fruit plants and products including other plant materials are not infested with noxious insects. By enforcing such laws, tremendous benefits can be obtained in the elimination of infestations. The successful campaigns have been conducted against major fruit pests like Oriental fruit fly, citrus black fly, mango stone weevil, apple leaf roller, cottony cushion scale and San Jose scale.

23.8. CHEMICAL CONTROL

Chemical control of insect and mite pests of fruit crops has a phenomenal increase in fruit production but their indiscriminate use against a wide range of fruit pests, is causing serious problems like health hazards, environmental pollution, rapid development of resistance and resurgence of target or non-target insect species. Furthermore, toxicity of insecticides as such and their residues in/on fruits pose a serious threat in fruit crops, wherein the crop products are normally consumed without allowing reasonable time for degradation of the toxic insecticides applied near the harvest. Keeping in view, the deleterious effects of insecticide residues, an ideal orchard insecticide is never possible. However, the application made at economic threshold levels (ETL) or otherwise providing a long pre-harvest interval would produce table fruits free from the insecticide residues. The information in literature showed that little efforts are made on determining the ETL on fruit crops (Table 23.2) and there is a need to develop the strategies to avoid the insecticides residues on fruits. Moreover, to obtain high efficacy of insecticides on fruit trees the following factors must be taken into consideration:

- Right stage, site and activity period of the pest.
- Proper selection of insecticide, dosage and application equipment.
- Coverage of the tree canopy.

It has been seen that due to heavy foliage of fruit trees, the spray fluid does not reach at the site of attack of the pest e.g. the attack is underside the leaves or outer/inner canopy of the tree or terminal portion of shoots or top of the trees. Depending upon these factors, the Entomologists have evaluated a large number of insecticides and their dosages against different insect and mite pests of fruit crop, only some practical aspects of insecticidal usage are elaborated.

Table 23.2. Economic threshold levels of different insect and mite pests

Insect pests	**Economic thresholdlevels**
Aphis spiraecola	5-10% infested shoots
Aphis gossypii	25% infested shoots
Toxoptera aurantii	25% infested shoots
Dialeurodes citri	5-10 nymphs/leaf on mandarin
Aleurocanthus spp.	First colonies occurrence
Parabemisia myricae	First colonies occurrence
Planococcus citri	5-10% infested fruits
Saissetia oleae (soft scale)	3-5 nymphs /leaf
Aonidiella aurantii (armoured scale)	3-5 nymphs /leaf
Prays citri (flower moth)	50% infested flowers.
Ceratitis capitata	20 adults/trap/week (Clementine)
Eriophyes sheldoni	50-70% infested buds
Aculops pelekassi	2-3% infested fruits.
Panonychus citri	3 specimens/leaf or 50% infested leaves
Tetranychus urticae	2% infested fruits or 10% infested leaves.
Odoiporus longicollis	5% infested plants
Cosmopolites sordidus	4 weevils/trap

Besides dusting or foliar spraying as normally done on other field crops, some other methods of application of insecticides in fruit crops are applied due to the fact that the fruit trees are perennial in nature. Thus insecticidal application is also made to roots, trunk, bark, stems, branches, twigs, leaves, flowers, fruits and seeds. The different methods of application found effective in many cases are discussed as under:

Dipping. Certain pests are carried over through nursery plant from one place to another. Prior to planting, these plants may be treated by dips or washes with some safer insecticides, which may not be harmful to the tender nursery plants. Aqueous solutions are often used in planting stocks of banana. Before shipment in and abroad, nursery stocks or harvested fruits are also treated by dipping to disinfect the stock of any pest.

Tree injection/root feeding. Injection of insecticides into sap stream by removing a band of bark around the tree trunk at a proper place and applied through root system, kill insects infesting trunks. It also protects the tree against subsequent infestation by sap or boring insects and infection by rotting fungi. Aqueous solution of monocrotophos is applied through roots for grubs of red palm weevil or solution of methyl parathion is injected for the control of mango stem borer, *B. rufomaculata*. Insertion of cotton soaked in carbon disulphide, chloroform, formalin or petrol and sealing the holes with mud is effective against bark eating caterpillar, *Indarbela* spp.

Baiting. Pests like fruit flies, Bactrocera spp. are attracted to and feed upon various baits. By addition of desired amount of suitable toxic insecticides like malathion to the attractive mixtures (pheromones or other insecticides), the insects can be killed. For this purpose dry (saw dust, wheat bran etc.) and wet ingredient along with sweet juices (mostly molasses or gur or sugar) are conveniently blended with poisonous insecticides and placed at the target sites. This practice is being done in guava, peach and other fruit crops.

Painting. In the protection of fruit trees, one of the most efficacious barriers against insect attacks is a coat of paint containing low percentage of active ingredient of insecticide together with filler often used over the surface of susceptible region. The substances like oil solution of creosote, lindane, chlorpyriphos, pentachlorophenol, coal tar, lime sulfur etc. having repellency and residual toxic effects against bark eating insects that deposit their eggs and/or feed on the bark are being used. Swabbing the trunk after removing the all loose bark with IIHR swab mixture (carbaryl 6 gm + copper oxychloride 10 gm + neem oil 1 ml + kerosene oil 1 ml + sticker 1 ml per litre of water) after pruning of grapevine mealy bug *M. hirsutus* and grapevine shot hole borer, *Xyleborus crassinsculus* (Motschulsky).

Soil application. Soil inhabiting pests like white grubs, white ants, root eaters or other dormant stages of insects may be killed or the soil may be made unsuitable for insect species by mixing insecticides directly in the top layer of soil or applied through irrigation water. Soil application of insecticides like chlorpyriphos or imidacloprid may provide the fruit grower a powerful weapon against some soil arthropod like termites. Broadcasting of chlorpyriphos or imidacloprid in sand followed by irrigation or in standing water or application of these insecticides to basal part of the tree has been found very effective against termites.

23.9. CONCLUSION

Tactics commonly adopted for the control of fruit pests are classified as curative/direct and preventive/indirect methods. These methods largely include non-chemical (cultural, mechanical and physical, biological and plant resistance) or chemical approaches or in few cases regulatory enforcement or reproductive suppression of pest population. However, an overview of the literature showed that fruit crops are more amenable for most of the non-chemical approaches, but the farmers are mostly adopting the chemical control measures to combat the pest problems on fruit crops.

Biological control and plant resistance have yet to gain its significance and concerted efforts are needed on these important management components.

REFERENCES

Deol, G.S. and Sharma, D.R. (2005) Advances in integrated pest management of fruit crops. In: *Advances in Integrated Pest Management of Horticultural, Spices and Plantation Crops.* B.S. Chillar, V.K. Kalra, S.S. Sharma and Ram Singh (eds). Centre of Advanced Studies, Dept. of Entomology, CCS HAU, Hisar pp 1-8.

Shamra, D.R., Batra, R.C., Gill, H.S. and Arora, P.K. (2004) Advances in fruit crop resistance to insect pests, pp 439-491. In *Host Plant Resistance to Insects: Concepts and Application.* G.S. Dhaliwal and Ram Singh (eds). Panima Publishing Corporation, New Delhi.

Sharma, D.R. (2005a) Advances in integrated pest management of litchi, papaya and pomegranate. In: *Advances in Integrated Pest Management of Horticultural, Spices and Plantation Crops.* B.S. Chillar, V.K. Kalra, S.S. Sharma and Ram Singh (eds). Centre of Advanced Studies, Dept. of Entomology, CCS HAU, Hisar pp 18-25.

Sharma, D.R. (2005b) Advances in integrated pest management of mango, sapota and grapevines. In: *Advances in Integrated Pest Management of Horticultural, Spices and Plantation Crops.* B.S. Chillar, V.K. Kalra, S.S. Sharma and Ram Singh (eds). Centre of Advanced Studies, Dept. of Entomology, CCS HAU, Hisar pp 9-17.

Sharma, D.R. (2006) Management of insect pests of tropical and sub-tropical fruit crops. pp.123-35. In: *Emerging Trends in Economic Entomology.* B.S.Chillar, R.K.Saini and Roshan Lal (eds). Centre of Advance Studies, Dept. of Entomology, CCS, HAU, Hisar. pp 280.

Sharma, D.R. (2007a) Eco-friendly Approaches for the Management of Fruit Pests. pp 137-155. *Sustainable Pest Management* D. Prasad (ed.). Daya Publishers, New Delhi, pp 520.

Sharma, D.R. (2007b) Management of polyphagous mealy bugs on fruit crops. Pp. 213-221. *Recent Trends in Biology and Management of Polyphagous Pests of Agricultural Importance.* J.P. Bhanot, H.R. Rohilla and V.K. Kalra (eds). Centre of Advance Studies, Dept. of Entomology, CCS, HAU, Hisar. pp 264.

Sharma, D.R. and Batra, R.C. (2007) Recent Advances in Management of Insect Pests of Fruit Crops. pp. 333-351. In: *Entomology: Novel Approaches.* P.C. Jain and M.C. Bhargava (eds). New India Publishing Agency, New Delhi.

Sharma, D.R. and Sandeep, Singh (2006) Management of insect pests of temperate fruits. pp. 111-122. In: *Emerging Trends in Economic Entomology.* B.S. Chillar, R.K. Saini and Roshan Lal (eds). Centre of Advance Studies, Dept. of Entomology, CCS, HAU, Hisar. pp 280.

Sharma, D.R., Singh, S. and Arora, P.K. (2005) Insect Pests. *40 Years of Research on fruit crops.* Y.R. Chanana, J.S. Bal, V.K. Mehan, J.K. Sharma and M.I.S. Gill (eds). PAU, Ludhiana. pp 81-93.

Theory and Practice of Integrated Pest Management
R. Arora, B. Singh and A.K. Dhawan (Eds)
Scientific Publishers (India), Jodhpur 2011, pp. 426 - 442.

24

MANAGEMENT OF INSECT PESTS OF VEGETABLE CROPS IN NET HOUSE

Sandeep Kaur

Department of Vegetable Crops
Punjab Agricultural University, Ludhiana - 141 004, India

24.1. INTRODUCTION

Day by day, the demand for insecticide residue free vegetables has increased with the increasing awareness of consumers. The development of insecticide resistance, pest resurgence and out breaks of secondary pests are the other areas of concern. Different methods for the production of quality vegetables either have been designed or being designed all over the world to meet the demand of good quality vegetables in market throughout the year. The insect pest problems and their severity levels in vegetables are also variable under protected cultivation structures and open field conditions. The micro environment of net house / Poly house / low tunnel grown plants is generally found congenial for the growth and development of sucking insect pests. Moderate temperatures, high humidity levels and diffused sun light under protected structures result in more succulent plant growth to which sucking insect pests like aphids, whiteflies, leaf miners and some times even tobacco caterpillars often get attracted. These pests have more preference for such environments. During hot months, the pests like thrips and mites multiply very fast in net/ poly-houses. These insects grow very fast once they get a chance to enter in net house/ poly-house due to the favourable conditions. Generally, the different stages of insects over winter in soils of net-house/ polyhouse / low tunnels. On coming of the favourable season, they resume their activity on plants. The different stages of sucking pests: nymphs and adults being of very small in size are generally wind borne especially at the times of high speed winds or cyclones. Sometimes, nets are also not completely insect proof. Further, wearing and tearing of nets also invites many insect pests in nets-houses. So, proper identification of these pests, their damage symptoms and timely management actions against these insect pests help to prevent their further spread. Some commonly occurring pest problems in our conditions (Singh *et al.,* 2002) in the following crops are mentioned below.

Crop	Insect pest	Period of Peak activity
Capsicum	Aphids, *Aphis gossypii* Glover and *Myzius persicae* (Sulzer)	February-April
	Thrips, *Scritothrips dorsalis* Hood, *Caliothrips indicus* (Bagnall), *Frenkliniella schultzi* (Trybom) and *F. occidetalis* (Pergande)	March-June
	Broad mite, *Polyphagotarsonemus latus (Banks)*	March -August
	Tobacco caterpillar, *Spodoptera litura* (Fabricius)	March -August
Tomato	Aphid *Aphis gossypii* Glover and *Myzius persicae* (Sulzer)	September-December and February-March
	Whitefly, *Bemisia tabaci* (Gennadius)	September-December
	Tobacco caterpillar, *Spodoptera litura* (Fabricius)	March -August
Brinjal	Aphids, *Aphis gossypii* Glover and *Myzius persicae* (Sulzer)	February-April
	Thrips, *Thrip tabaci* (Lindeman)	March-June
	Red spider mites, *Tetranychus* spp.	April-June
	White fly, *Bemisia tabaci* (Gennadius)	March -October
	Tobacco caterpillar, *Spodoptera litura* (Fabricius)	March -August

In the present scenario, the sole reliance on the insecticidal sprays is not enough to tackle the pest problems n net/poly houses. Thus, there is urgent need to develop and follow the integrated pest management practices against vegetable pests. The IPM practices includes physical methods of control, monitoring of insect pest, Economic Threshold Levels (ETL), cultural control, mechanical control, biocontrol, botanical control, microbial control etc. The work done on the different IPM components at national and international level is reviewed as under:

24.2. PHYSICAL CONTROL

Nylon net. Cultivation of capsicum and tomato is recommended under both under net house and polyhouse while for brinjal, it is recommended under net house by Punjab Agricultural University, Ludhiana for the higher yield, good quality with minimum use of insecticides (Anon., 2008). The net of 40 mesh size is recommended for net house and for nursery raising of vegetable crops to prevent insect infestation by large size insects such as brinjal shoot and fruit borer, okra soot and fruit borer, tomato fruit borer, fruit fly and other foliage feeders like beetles etc. Some other precautions to prevent insect entry in net/polyhouse are also suggested such as: (1) Use double door system in net/polyhouse. (2) Plug all holes in doors and walls of net house. (3) All the ventilators in poly houses must be covered with insect proof nets. (4) Use under ground pipe system for irrigation. (5) Inspect net house regularly for wearing and tearing of net/polyhouse.

However, fifty mesh size nylon net is recommendation for insect free vegetable seedling production by Indian Institute of Horticultural Crops, Banglore (www.iihr.res.in/technologies%20developed.HTM). Thiruvudainambi *et al.* 2002 reported that when seedlings were raised in beds by covering with white nylon net to avoid sucking pests and seedling were given root dip treatment in imidacloprid (0.2 per cent) for three hours before transplanting followed by three sprays of imidacloprid (0.1 per cent), recorded least incidence of chilli mosaic disease which is vectored by insects like white fly.

Harmanto *et al.* (2006) suggested that based on internal microclimate and ability to exclude some major insects, the 52 mesh net seems to be selected as a compromise size of nets appropriate for tropical greenhouse as compared the 40 and 78 mesh size nets.

OptiNet 50 is a patented technology of USA. It blocks penetration of green house thrips and other insects in two ways: (a) optic protection: insects are repelled by screen's optical additives. (b) physical protection: fine 50 mesh screen provides a second line of protection. The dual protection significantly reduces the number of pests entering the structure, while maintaining adequate airflow. As a result fewer pesticide treatments are needed, complying with European standards and saving costs (http://www.Signaturesupply.com , www.polysack.com).

Talekar *et al.* (2003) reported that good quality, uniformly woven 32-mesh size netting does not allow diamond back moth or striped flea beetle to pass through. It will allow a few aphids and thrips to penetrate but there numbers are miniscule and will not damage leafy vegetables harvested with in four to six weeks. Common army worm, *Spodoptera litura* is a big moth and cannot enter through 32 mesh size net. But female moth of this pest lay eggs in small batches on the top of the net and tiny larvae hatching from these eggs drop through the holes over the crop inside the net house. This starts a new infestation.

Plastic Sheets. Prabhat and Poehling (2006) reported that fewer whitefly, aphids and thrips were recorded in UV blocked greenhouses. However, thrips, *Ceratothripoides claratris* were most abundant in greenhouse with low UV intensity. During open gates experiments, infection level reached up to 96-100 per cent under UV non- blocking greenhouses compared with 6-10 per cent infection levels in green houses where UV radiations were blocked. Apparently virus symptoms were delayed in the greenhouse where UV block roof material was used.

Costa *et al.* (2002) found that crops grown in greenhouses under a plastic that blocked UV light at wavelengths of 380 nm and below had lower numbers of aphids and thrips compared with a plastic that blocked UV light at wavelengths of 360 nm and below. There was no reduction in greenhouse whitefly, *Trialeurodes vaporariorum* Westwood, populations under the <380 UV-absorbing plastic compared to <360 nm UV-absorbing plastic in these field trials. The type of greenhouse plastic used in a structure can affect population levels of some insect species, and may be useful tools in developing integrated pest management programs for insect management.

Sticky traps. Gu *et al.* (2008) observed that yellow sticky cards trapped *B. tabaci* adults when placed parallel to tomato rows @ 1 per 3m row and significantly reduced whitefly populations on tomato.

Mulches. Reflective mulch proved effective against not only aphids but also against white flies, thrips and leaf hoppers in vegetable crops. Reflective mulches caese to repel insect when the plant canopy covers more than about 60 per cent of the soil surface. Polyethylene plastic film can be used for mulch. Silver or grey colour is the most effective colour for mulch. Aluminium metalized polyethylene and silver embossed polyethylene are also effective as much but are expensive (Dreistadt *et al.* 2007).

Pan-traps, yellow sticky traps and foliar counts showed that reflective (Synthetis) and buckwheat (living) mulches consistently had fewer numbers of adult whiteflies and aphids compared with the standard white (Syntheic) mulch treatments. Living mulch treatments had higher natural enemy populations than synthetic mulch and bare-ground treatments (Frank and Liburd 2005).

Monitoring of insect pests. Monitoring of insect pests in vegetable is an integral part of IPM practices for the timely start of protective measures against the insect pests. There are two ways of monitoring insect population on crop: (a) by observing insect stages manually on plant parts (b) by using different types of traps to monitor different insect pests in vegetable crops. Dreistadt *et al.* (2007) explained different ways of monitoring of thrip adults and larvae by branch beating or shaking of foliage or flower on to sheet paper or a beating tray. Adult thrips can be monitored by using yellow sticky traps. Blue sticky traps are most effective in capturing western flower thrips. Monitoring of crop in net / polyhouse for insect pest incidence should be done on every alternate day as recommended by Punjab Agricultural University (Anon. 2008)

Sticky traps/cards. Chen *et al.* (2004)[a] and Chen *et al.* (2004)[b] reported that Blue LEDs (light emitting diodes) increased the efficacy of blue sticky cards 2.0-2.5 times for catching thrips and may be useful in greenhouses for monitoring and thrips control. The higher trap catches and reduced costs indicated that blue sticky cups would be the better choice for thrips monitoring and control.

Hoelmer and Simmons (2008) also found the yellow sticky traps very effective for observing trends in whitefly *Bemisia tabaci* (Gennadius) biotype B and its exotic parasitoid *Eretmocerus emiratus* Zolnerowich and Rose populations at a particular site and for detecting parasitoids at specific locations in the organic fields of spring cantaloupe and watermelon in the Imperial Valley, CA.

Kim *et al.* (2001) suggested that to obtain an unbiased mean density of white fly, *Trialeurodes vaporariorum* per greenhouse, sticky traps should be placed at least >12.5 m apart in greenhouse to ensure that they are spatially independent.

The yellow traps are also used for monitoring of green peach aphid population as reviewed by John (2005).

Water Pan traps. McDonald *et al.* (2003) monitored seasonal dispersal and diversity of aphid species by using water pans on pepper farms in St. Catherine, Jamaica throughout 1998 and 1999 to identify the most likely vectors of tobacco etch virus (TEV) in pepper fields. The water pan traps are used for the monitoring of green peach aphid's population (John, 2005).

Economic threshold level (ETH). Some of workers have suggested the economic threshold levels for different insect pests. Shipp *et al.* (1998) suggested 10 to 26 adult thrips per sticky trap per day as economic injury levels depending on the seasonal stage of sweet pepper production and market prices in Ontario, Canada. Whereas, Shipp *et al.* 2000 reported the economic injury levels for *Frenkliniella occidentalis* 20 to 50 adults per sticky trap per day or 3 to 7.5 per flower for cucumber crop under average greenhouse production conditions in Ontario, Canada. In India, two thrips per leaf as ETH level for chilli thrip has been reviewed by Dhandapani *et al.* (2003).

John (2005) reviewed that the action threshold of 4°C can be used to predict egg hatching and immigration of alates of green peach aphid since Green peach aphid is a pest of cold weather.

One mite per leaf has been reviewed as economic threshold levels for Chilli mites (*Polyphagotarsonemus latus*) in India (Dhandapani *et al.* 2003).

24.3. CULTURAL CONTROL

Some workers reported different cultural practices to prevent insect pest build up in net house/polyhouses.

Sterilization of Structures. Lethal exposures to CO2 or N2 were determined by Held *et al.* (2001) for common greenhouse pests including fungus gnat larvae, Bradysia sp.; green peach aphid, *Myzus persicae* (Sulzer); sweetpotato whitefly, *Bemisia* sp.; twospotted spider mite, *Tetranychus urticae* Koch; and western flower thrips, *F. occidentalis* (Pergande). Exposure for 12-18 h to >99% N_2 or CO_2 caused complete mortality of aphids, mites, thrips, and whiteflies. However, there was variation in response among plant species and cultivars, with effects ranging from delayed flowering to mortality.

Soil Sterilization. Soil sterilization should be done before the sowing/ transplanting of any vegetable crop in net/ polyhouse to disinfest the soil. Sterilize the soil in net/ polyhouse with 2 per cent formalin solution. Mix 20 ml of formalin in one litre of water and drench in the soil @ 4-5 litre/ m^2. Cover the soil with the polythene sheet for 48-72 hours. Then remove the sheet and turn the soil for 3-4 days for complete elimination of formalin before transplanting (Anon. 2008).

Sanitary Measures. Some sanitary measures have been suggested by Punjab Agricultural University to check insect pests infestation in net /polyhouse (Anon. 2008). Remove dry fallen leaves and other plants parts at frequent intervals to check insect pest carryover in net/polyhouse as different stages of insects generally overwinter / hibernate / pupate in soil.

The timely removal of plant and growing medium debris from greenhouses or to place debris into refuse containers with tight-sealing lids to prevent the further spreading of insect pests in green house is suggested to the greenhouse producers by Hogendorp and Cloyd (2006). Talekar *et al.* (2003) also suggested that after harvest of crop, plant debris should be cleared. It will reduce the carryover of insect pests and diseases to the subsequent crops. Clean the net if it is soiled. Use the pressurized water to clean the net. The net should be cleaned from inside as it will reduce the accumulation of water in the net house and will enable planting of crops without delay. Such cleaning will removes the dust and insect eggs from the net and increase the transmission of light. Further, manure to be applied in net house should be fully decomposed and free of insect pests specially grubs. The manure should be finely ground and sieved through 8 mesh size sieve. Key points for the nets houses suggested by AVRDC are as under: (a) Two double doors at right angle to each other is the better arrangement for prevention of insect pest entry in net house. (b) Keep net house tightly sealed all the time. (c) Maintain cleanliness and integrity of netting. (d) For weeding operations in low tunnels, remove net and uproot weeds quickly and close the net promptly to avoid insect entry. If there are several beds of crop, then open only one bed at a time. Dreistadt *et al.* 2007 suggested to remove weeds as they are the alternate hosts for thrips. Avoid excess application of nitrogen. Remove old and spent flowers.

Effect of Irrigation and Fertilizer. Dreistadt *et al.* 2007 reported that vigorous plants generally out grow thrips damage. So keep plants well irrigated.

Staking of plants. Tomato variety Punjab Tropic and hybrid Naveen were when staked suffered low fruit infestation by *H. armigera* (Kaur and Singh 1997).

24.4. MECHANICAL CONTROL

In case of infestation of borers (okra shoot and fruit borer, brinjal shoot and fruit borer, tobacco caterpillar) and leaf miner which may appear in net/poly house either by negligence or incidentally, remove and destroy infested leaves, shoots and fruits by burying them deep in soil out side the net/ poly house. Destroy the egg masses and tobacco caterpillers mannualy (Anon., 2008). Kaur *et al.,* 2004 and Chiranjeevi *et al.,* 2002 suggested that clipping of infested shoots of brinjal shoot and fruit borer contributed less damage in brinjal crop. Similarly, destruction of tomato fruit borer infested tomato fruits at green stage at regular intervals not only prevents the carryover of the pest in the field, but also results in lowest fruit infestation as compared to the control treatment or when the infested fruits were destroyed at red ripe stage (Jitendar, 2009).

24.5. BIOLOGICAL CONTROL

Predators. Omkar *et al.* (2006) reported that the neonates of aphidophagous ladybird beetles, *Propylea dissecta, Coccinella transversalis* and *Coelophora saucia* preferentially consumed conspecific eggs in presence of essential aphid prey, *Aphis gossypii* and this preference was strongest in *P. dissecta.* The optimum temperature for the mass rearing of of *P. dissecta* is suggested 27°C (Pervez and Omkar, 2004)[b]. The order of suitability of prey species for lady bird beetle, *Cheilomenes sexmaculata* (Fabricius) was *Aphis. craccivora* Koch > *A. gossypii* Glover > *Rhopalosiphum. maidis* (Fitch) > *Myzus persicae* (Sulzer) > *Uroleucon compositae* (Theobald) > *Lypaphis erysimi* (Kaltenbech) > *A. nerii* Boyer de Fonscolombe as reported by Omkar and Bind (2004). Pervez and Omkar (2004)[a] studied the predation potential, development, immature survival and reproduction of an aphidophagous ladybeetle, *P. dissecta* (Mulsant) when fed on seven aphid prey, viz. *A. gossypii*, *A. craccivora*, *L. erysimi*, *U. compositae*, *B. brassicae*, *R. maidis* and *M. persicae*. *A. gossypii* was most suitable and consumed by the larvae and adults of *P. dissecta*, while *M. persicae*, the least.

Xu *et al.* (2006) studied the efficiency of *Orius insidiosus* for bio control of either *Frankliniella occidentalis* or *Tetranychus urticae* alone or for the two pests in combinations on beans. When two pests were simultaneously offered, the efficiency of *O. insidiosus* in controlling *T. urticae* markedly decreased. Furthermore, mite control decreased with increasing *T. urticae* densities and was also affected by the density of *O. insidiosus*. The presence of mites at initial densities of 20-80 females did not significantly influence thrips control by *O. insidiosus*. The presence of *F. occidentalis* resulted in higher oviposition by *O. insidiosus* females than the presence of mites, indicating that thrips are a more suitable resource than *T. urticae* for *O. insidiosus*.

The two other polyphagous predators *Orius laevigatus* and *O. majusculus* have the potential as biocontrol agents for *Bemisia tabaci* when this whitefly coexists in a vegetable crop with the thrips *Frankliniella occidentalis* (Arno *et al.,* 2008).

Weintraub *et al.* (2003) reported that the predatory mite *Neoseiulus cucumeris* (Oudemans) (Acarina: Phytoseiidae) successfully controlled the broad mite *Polyphagotarsonemus latus* (Banks) (Acarina: Tarsonemidae) on two varieties of greenhouse-grown sweet peppers (*Capsicum annuum* L.). Releasing *N. cucumeris* on each or every second plant was as effective in controlling broad mites as sulfur treatments in terms of plant height, dry mass and yield. Plants treated with sulfur, however, had significantly higher thrip population and fruit damage.

The predator *Macrolophus pygmaeus* develops well on the aphid *M. persicae* or on the whitefly *T. vaporariorum*, both of which are important pests of vegetable crops. This predator is also well adapted to the temperatures that occur both in greenhouses and in the open field in the Mediterranean region. Compared to other natural enemies of whiteflies, such as *Encarsia formosa* Gahan (Hymenoptera: Aphelinidae), *Macrolophus pygmaeus* can increase at relatively low temperatures (Perdikis and Lykouressis, 2002).

Parasitoid. Under field conditions, *Trichogramma pretiosum* (Riley) 1.0 lakh ha^{-1} six times from flower initiation stage at weeky intervals brings significant reduction in fruit damage by *Helicoverpa armigera* (Krishnamoorti and Mani 1988). Two releases of *Trichogramma chilonis* 2.5 lakh ha^{-1}caused maximum egg reduction of (55.79%) *H. armigera* (Kulkarani and Shekarappa 2001). Based on fruit damage and marketable fruit yield, five weekly releases of *Trichogramma brasiliense* or *T. pretiosum* @ 50000 eggs ha^{-1} coupled with three sprays of HaNPV @ 1.5 x

10^{12} POBs ha^{-1} at 10 days interval and three sprays of endosulfan @ 700 g a.i. ha^{-1}at fornightly interval proved most effective against *H. armigera* in open field conditions (Singh *et al.* 2001).

Natural enemies like *Encarsia haitiensis* and *E. guadeloupae* can satisfactorily control exotic spiraling whitefly, *Aleurodicus disperses* in chillies. *E. guadeloupae* caused 20.54 per cent parasitism in Andhra Pradesh and 70.26 per cent in Kerala (Mani *et al.* 2002).

Kim *et al.* (2005) noticed that the parasitoid *Aphidius colemani* developed normally (approximately 90% adult emergence) when its cotton aphid (*A. gossypii*) host was treated with *Verticillum lecanii* conidia 5 or 7 days after parasitization. The results suggest that the parasitoid and fungus may be used together for aphid biocontrol as long as fungus applications are timed to allow late-instar development of the parasitoid.

Zamani *et al.* (2007) evaluated the performance of two aphid parasitoids, *Aphidius colemani* Viereck and *Aphidius matricariae* (Haliday), against *A. gossypii* Glover on greenhouse cucumber (*Cucumis sativus* L.) and *M. persicae* (Sulzer) on sweet pepper (*Capsicum annuum* L.) at various constant temperatures (5, 10, 15, 20, 25, 30, and 35°C). The optimal temperatures for development of *A. colemani* and *A. matricariae* were approximately 30 and 25°C, respectively, and high mortality occurred at higher temperatures.

Pathogens. The spraying of microbial pesticide *B. bassiana* @ 1 x 10^4 cfu/g two times with a week interval satisfactorily managed whitefly nymphs and adults as reviewed by Dhandapani *et al.*, 2003. *Metarhizium anisopilae var. anisopilae* and *Spinosyn,* Spinosad 48 SC were found effective against *P. xylostella, S. littiralis* and *H. undalis* (Gopalakrishanan *et al.* 2002; Day and Somchaudhary 2001). Diamondback moth was effectively controlled by weekly sprays of Dipel 8L Biolep / Biobit @ 0.750 l/kgha^{-1} and this formulation gave superior control of pest over endosulfan and fenvelrate. This formulation was also found moderately effective against other lepidopterous pests: Head borer, *Hellula undalis* (Fabricius), tobacco caterpillar, *Spodoptera litura* and *H. armigera* attacking cabbage and cauliflower crops in the state. Further, weekly sprays of SINPV @ 375 LEha^{-1} gave effective control of *S. litura* on cauliflower. Total pest complex of cruciferous vegetables could be managed by use of biopesticides (Battu *et al.*, 1998; Arora *et al.*, 2000). Monobrullah and Shankar (2008) reported that the application of different doses of SpltMNPV on final instars of *Spodoptera litura* showed dose-related mortality. A significant increase in the time for pupation and progeny mortality as well as reduced fecundity was observed in surviving larvae. Talekar *et al.* (2003) reported that *Bacillus thurigiensis* is effective against *S. litura.*

Ugine *et al.* (2007) found statistically significant reductions in western flower thrips, *F. occidentalis* population when three to four sprays of clay-based wettable powder formulation of *Beauveria bassiana* strain GHA conidia were applied at the highest label rate (2 x10^{14} conidia) in the highest volume (3740 L carrier ha^{-1}) at < 7-day intervals over the control treatment. The most effective treatment programs reduced pest populations by 30-40% compared to untreated controls. This slowed, but did not stop the growth of pest populations. Results indicate that use of fungi for thrips management will require integration with other control agents.

The influence of *Lecanicillium attenuatum* CS625 (=*Verticillium lecanii* CS625) on development and reproduction of cotton aphid (*A. gossypii*) was studied by Kim (2007). A significant dose effect on reduction of life span, reproductive period and fecundity was observed in 1st and 3rd instars after spore application. Reproduction period was also significantly shortened with increasing spore concentration. It was concluded that the isolate of *L. attenuatum* is able to affect populations of cotton aphid by reducing life span and total fecundity as well as by killing the aphids directly.

Wraight *et al.* (2000) conducted Collaborative research at the USDA-ARS Subtropical Agricultural Research Center in southern Texas to assess the microbial control potential of *B. bassiana* and *Paecilomyces fumosoroseus* against whitefly, *Bemisia argentifolii*. The results indicated that both *B. bassiana* and *P. fumosoroseus* (@ 5 x 10p#pd conidia in 180 liters water/ha at 4-5 days interval have strong potential for microbial control of nymphal whiteflies infesting cucurbit crops.

Kaur 2005 reported that *V. lecani*, *Azadirachta indica* 0.03% AzEC and Sulfur 80 WP produced delayed but constant effect on aphid, *A. gossypii* population in polyhlouse. These formulations being safe, more eco friendly and having no residue problems can be fit in to the IPM programme of aphid on chilli in polyhouse.

Fiedler e*t al.* (2007) found the nematophagous fungus *Paecilomyces lilacinus* (Thom) Samson effective against the winged and wingless forms of the cotton aphid, *A. gossypii* and completely eliminated the aphid population from green house. It was also effective in controlling greenhouse whitefly, *T. vaporariorum* specially against nymphal growth stages and the western flower thrips against their prepupal and pupal stages when the fungus was applied as a water spore suspension to the soil. When the fungus was applied at temperatures below 10 pC, it was able to reduce a glasshouse red spider mite population by 60%.

Many workers also tried the combinations of biocontrol agents and biopathogens for the effective management of insect pests.

Stansly and Liu (1997) found that neem extract, insecticidal soap and two sugar esters tested had little or no effect on *Encarsia pergandiella* Howard. These later materials could be considered selective in respect to

E. pergandiella and might be used to suppress *B. argentifolii* without decimating parasitoid populations in south Florida vegetable fields.

Senthil *et al.* (2005) evaluated the effects of *Azadirachtin* (AZA) and nucleopolyhedrovirus (NPV) on midgut enzyme activity in *S. litura* Fabricius (Lepidoptera: Noctuidae) (Tobacco cutworm). Gut enzyme activities were decreased by AZA and NPV individually and in combination. When *S. litura* larvae were fed a diet of castor leaves treated with AZA and NPV in bioassays, gut enzyme acid phosphatases, alkaline phosphatases, adenosine triphosphatases, and lactate dehydrogenase activities were decreased. There were statistically significant differences (P < or = 0.05) in enzyme activities between combined and individual treatment. A synergistic effect of botanical insecticides and virus was found when combined in low doses. These effects are most pronounced in early instars. Maximum weight loss (59-72%) occurred, when AZA and NPV were combined.

Ebssa *et al.* (2006) recommended weekly applications of entomopathogenic nematodes alone (*Heterorhabditis bacteriophora* Poinar or *H. indica* Poinar, Karunakar and David), alone at 200 Infective juviniles and a predatory mite species alone [*Amblyseius cucumeris* (Oudemans) (Acarina: Phytoseiidae)] alone at three adults per plant, or the combined application of the natural enemies at these rates significantly reduced western flower thrips populations on beans in green house. Thus, this study clearly shows a great potential of combined applications of the mites and the nematodes to control foliage-feeding and soil-dwelling life stages of thrips.

24.6. BOTANICAL CONTROL

Immaraju (1998) discussed the broad-spectrum activity of *Azadirachtin* at low use rates (12.5 to 40 g AI ha^{-1}) coupled with the insect growth regulator activity (in all larval nymphal instars including the pupal stage) and unique mode of action (ecdysone disruptor), make *azadirachtin* an ideal candidate for insecticide resistance, integrated pest control and organic pest control programmes. *Azadirachtin* has been exempted from residue tolerance requirements by the US Environmental Protection Agency for food crop applications. Azadirachtin exhibits good efficacy against key pests such as whiteflies, leaf miners, fungus gnats, thrips, aphids and many leaf-eating caterpillars. *Azadirachtin* has minimal to no impact on non-target organisms, is compatible with other biological control agents and has a good fit into classical Integrated Pest Management programmes. The world's largest *Azadirachtin* extraction facility has been fully commissioned in India to process over 10,000 tonnes neem seeds per annum. This will ensure the wide availability of *azadirachtin* technical grade material in the future.

Tobacco caterpillar, *S. litura* (Fabricius) on solenaceous crops is currently being managed with neem extracts in South India. The extract is not only cheeper, but also effective as synthetic insecticides (Joshi *et al.,* 1993). The neem seed kernel extract 5 per cent concentration found promising in control of leaf miner and recorded highest yield of bittergourd (Rosaiah, 2001). The use of neem and pongamia, *Pongamia pinnata* (L.) cakes for the management of tomato pests was successfully demonstrated (Krishna and Krishnamoothy, 2002). The use of neem and pongamia soap is also recommended against by IIHR (www.iihr.res.in/technologies%20developed.HTM). Application of neem cake four times during the crop growth, decresed the incidence of brinjal shoot and fruit borer to 8 per cent as against 40 per cent in control. The yield was increased by 68 per cent as compared to control (Sreenivasamurthy *et al.,* 2002).

Brito *et al.* (2006) found Neemseto, among tested neem formulations, performed better against the two spotted spider mite and exhibited relatively low impact against the predatory mites studied.

Venzon *et al.* (2008) found that all broad mites, *Polyphagotarsonemus latus* on plants treated with abamectin died without ovipositing before the end of the experiment. The rate of increase for *P. latus* decreased linearly with increasing neem concentration, with the population growth rate becoming negative when mites were exposed to plants treated with neem at concentrations higher than 0.13g a.i/l.

Eileen and Sydney (2009) discussed the general traits of the natural products for insect pest management like Botanicals: Neem, Pyrethins/Pyrethrum, Rotenone, Ryania, Sabadilla, Soaps and Oils which are as under:

Eileen and Sydney (2009) mentioned Horticultural oils (from plants) and Insecticide soaps act as contact insecticides thus, affects the surfaces of insect bodies. They also mentioned some safe insecticides: Diatomaceous Earth, Sulfur for the control of vegetable pests.

REFERENCES

Anonymous (2008) Net house cultivation. Packages of practices for vegetable crops. Punjab Agricultutral University, Ludhiana, Punjab, India. Pp: 123.

Arno, J., Roig, J. and Riudavets, J. (2008) Evaluation of *Orius majusculus* and *O. laevigatus* as predators of *Bemisa tabaci* and estimation of their prey preference. *Biol. Control Theor. Appl. Pest Mngt* **44**(1): 1-6.

Arora, R., Battu, G.S. and Bath, D.S. (2000) Management of insect pests of cauliflower with biopesticides. *Indian J. Ecol.* **27**(2):156-162.

Battu, G.S., Arora R. and Bath, D.S. (1998) Evaluation of native nuclear polyhedrosis virus for ecosafe management of *Spodoptera litura* (Fabricius). In Dhaliwal GS, Randhawa NS.

Arora R and Dhawan AK (eds.). Ecological Agriculture and Sustainable Development Vol. 2. Indian Ecological Society and CRRID, Chandigarh. Pp: 315-322.

Brito, H.M., Gondim, G.C. Jr, Oliveira, JV-de, Camara, CAG-da (2006) Toxicity of neem (*Azadirachta indica* A. Juss) formulations for two spotted spider mite and *alatus Euseius* De Leon and *Phytoseiulus macropilis* (Banks) (Acari: Phytoseiidae). *Neotropic. Ent.* **35**(4): 500-505.

Chen, T.Y,. Chu, C.C., Fitzgerald, G., Natwick, E.T. and Henneberry, T.J. (2004) Trap evaluations for thrips (Thysanoptera: Thripidae) and hoverflies (Diptera: Syrphidae). *Envir. Ent.* **33**(5): 1416-1420.

Chen, T.Y., Chu, C.C., Henneberry, T.J. and Natwick, E.T. (2004) Catching thrips (Thysanoptera: Thripidae) with various traps. *Proc. Beltwide Cott. Conf.*: 1703-1705.

Chiranjeevi, C., Reddy, I.P., Naryanamma, M. and Neeraja, G. (2002) Effect of shoot clipping and insecticides on the incidence of fruit borer in brinjal. *Proc. Int. Conf. Veg. held on Nov. 11-14, Bangalore, India.* Pp: 271.

Costa, H.S., Robb, K.L. and Wilen, C.A. (2002) Field trials measuring the effects of ultraviolet absorbing greenhouse plastic films on insect populations. *J. Econ. Ent.* **95**(1): 113-120.

Dey, P.K. and Somchoudhary, A.K. (200) Evaluation of spinosyn A+D (spinosad 48 SC) against lepidopterous pest complex of cabbage and its effect on natural enemies of diamondback moth under field conditions of W.B. *Pestology* **25** (1):54-58.

Dhandapani, N., Umeshchandra, R.S. and Murugan, M. (2003) Biointensive pest management (BIPM) in major vegetable crops — An Indian Perspective. *Fd. Agric. Envir.***1** (2): 333-339.

Dreistadt, S.H., Phillios, P.A., O'Dennell, C.A. and Davis, U.C. (2007) Pest Notes: Thrips. University of California, Agricultural and Natural Resources UC IPM Statewide Integrated Pest Management Programme. Publ. 7429.

Ebssa, L., Borgemeister, C. and Poehling, H.M. (2006) Simultaneous application of entomopathogenic nematodes and predatory mites to control western flower thrips *Frankliniella occidentalis. Bio. Control Theor. Appl. Pest Mngt* **39**(1): 66-74.

Eileen AB and Sydney GPB 2009. Natural products for insect pest management. http://edis.ifas.ufl.edu/IN197

Fiedler, Sosnowska and Danuta (2007) Nematophagous fungus *Paecilomyces lilacinus* (Thom) Samson is also a biological agent for control of greenhouse insects and mite pests. *Biocontrol* **52**(4): 547-558.

Frank, D.L. and Liburd, O.E. (2005) Effects of living and synthetic mulch on the population dynamics of whiteflies and aphids, their associated natural enemies and insect transmitted plant diseases in zucchini. *Envir. Ent.* 34(4): 857-865.

Gopalakrishanan, G., Krishanamoorthy, A. and Gangavisalakhsi, P.N. (2002) *Metarhizium anisopliae* var. *anisopliae*: A potential entomopathogenic fungus for the management of *P. xylostella* on cabbage. *Proc. Int. Conf. Veg. held on Nov. 11-14, Bangalore, India*. Pp: 262.

Gu, X.S., Bu, W.J., Xu, W.H., Bai, Y.C., Liu, B.M. and Liu, T.X. (2008) Population suppression of *Bemisia tabaci* (Hemiptera: Aleyrodidae) using yellow sticky traps and *Eretmocerus nr. rajasthanicus* (Hymenoptera: Aphelinidae) on tomato plants in greenhouses. *Insect Sci.* **15**(3): 263-270.

Harmanto, Tantau H.J. and Salokhe, V.M. (2006) Microclimate and air exchange rates in greenhouses covered with different nets in the humid Tropics. *Biosystems Engng.* **94**(2): 239-253.

Held, D.W., Potter, D.A., Gates, R.S. and Anderson, R.G. (2001) Modified atmosphere treatments as a potential disinfestation technique for arthropod pests in greenhouses. *J. Econ. Ent.* **94**(2): 430-438.

Hoelmer, K.A. and Simmons, A.M. (2008) Yellow sticky trap catches of parasitoids of *Bemisia tabaci* (Hemiptera: Aleyrodidae) in vegetable crops and their relationship to in-field populations. *Envir. Ent.* **37**(2): 391-399.

Hogendorp, B.K. and Cloyd, R.A. (2006) Insect management in floriculture: how important is sanitation in avoiding insect problems? *Hort. Technol.***16**(4): 633-636.

Immaraju, J.A. (1998) The commercial use of *azadirachtin* and its integration into viable pest contol programmes. *Pestic. Sci.* **54**(3): 285-289.

Jitendar, K. (2009) Management of tomato fruit borer, *Helicoverpa armigera* (Hubner) on tomato. Punjab Agricultural University, Ludhiana, Punjab, India.

John, L.C. (2005) Featured Creatures. Publ. No. EENY-222. Department of Entomology and Nematology, Unversity of Florida Institute of Food and Agricultural Sciences and Division of Plant Industry, Florida Depatment of Agriculture and Consumer Services. http://entnem.ufl.edu/creatures/veg/aphid/ green peach aphid.htm

Joshi, B.G., Sriprakash, K. and Murthy, P.N.S. (1993) Efficacy of neem seed kernel suspension, neem seed bitter, *Pongamia* cake water extract, *Bacillus thuringiensis* Berliner, nuclearpolyhedrosis virus alone and in combination against tobacco caterpillar, *Spodoptera litura* F. in nursery. In: Chari MS and Ramprasad G (eds.) Botanical Pesticides in Integrated Pest management. *Proc. Symp. Bot. Pestic. IPM, Rajmundry, India.* Pp:154-158.

Kaur, S. (2005) Management of aphids, *Aphis gossypii* Glover on chilli in polyhouse. *Indian J. Ecol.* **32** (2): 241-242.

Kaur, S., Bal, S.S., Singh, G., Sidhu, A.S. and Dhillon, T.S. (2004) Management of brinjal shoot and fruit borer, *Leucinodes orbonalis* Guenee through net house cultivation. *Acta Hort.* **659**:345-350.

Kaur, S. and Singh, D. (1997) Management of tomato fruit borer, *Helicoverpa armigera* (Hubner) through planting periods and staking of tomato crop. *J. Insect Sci.* **10**(2):160-162.

Kim, J.J. (2007) Influence of *Lecanicillium attenuatum* on the development and reproduction of the cotton aphid, *Aphis gossypii*. *Biocontrol.* **52**(6): 789-799.

Kim, J.J., Kim, K.C. and Roberts, D.W. (2005) Impact of the entomopathogenic fungus *Verticillium lecanii* on development of an aphid parasitoid, *Aphidius colemani. J. Invert. Path.* **88**(3): 254-256.

Kim, J.K., Park, J.J., Park, H. and Cho, K. (2001) Unbiased estimation of greenhouse whitefly, *Trialeurodes vaporariorum*, mean density using yellow sticky trap in cherry tomato greenhouses. *Entomologia Exp. Appl.* **100**(2): 235-243.

Krishan, K.N.K. and Krishnamoorthy, P.N. (2002) Advances in the use of botanicals for the IPM of major vegetable pests. *Proc. Int. Conf. Veg. held on Nov. 11-14, Bangalore, India.* Pp: 233.

Krishanamoortty, A. and Mani, M. (1988) Feasibility of managing *H. armigera* on tomato through parasitoid. *Proc. Natn. Wkshop Heliothis Mngt. held on Feb. 18-19, TNAU, Coimbatore.*

Kulkarni, K.A. and Shekarappa. (2001). Integrated management of chilli fruit borer, H. armigera. Proc. Second Natn. Symp. Integ. Pest Mngt Hort. Crops: New Mole. Biopestic. Envir. held on October 17-19, Bangalore, India. PP: 59-60.

Mani, M., Dinesh, M.S. and Krishnamoorthy, A. (2002) Survey and evaluation of natural enemiese of the exotic spiraling whitefly, *Aleurodicus disperses* on vegetables in Peninsular India. *Proc. Int. Conf. on Veg. held on Nov. 11-14, Bangalore, India.* Pp:238.

McDonald, S.A., Halbert, S.E., Tolin, S.A. and Nault, B.A. (2003) Seasonal abundance and diversity of aphids (Homoptera: Aphididae) in a pepper production region in Jamaica. *Envir. Ent.* **32**(3): 499-509.

Monobrullah, M. and Shankar, U. (2008) Sub-lethal effects of SpltMNPV infection on developmental stages of *Spodoptera litura* (Lepidoptera: Noctuidae). *Biocontrol Sci. Technol.* **18**(3-4): 431-437.

Omkar, and Bind, R.B. (2004) Prey quality dependent growth, development and reproduction of a biocontrol agent, *Cheilomenes sexmaculata* (Fabricius) (Coleoptera: Coccinellidae). *Biocontrol Sci. Technol.* **14**(7): 665-673.

Omkar, Pervez, A and Gupta, A.K. (2006) Why do neonates of aphidophagous ladybird beetles preferentially consume conspecific eggs in presence of aphids? *Biocontrol Sci. Technol.* **16**(3-4): 233-243.

Pervez, A. and Omkar (2004[a]) Prey-dependent life attributes of an aphidophagous ladybird beetle, *Propylea dissecta* (Coleoptera: Coccinellidae). *Biocontrol Sci. Technol.* **14**(4): 385-396.

Pervez, A. and Omkar (2004[b]) Temperature-dependent life attributes of an aphidophagous ladybird, *Propylea dissecta. Biocontrol Sci. Technol.* 14(6): 587-594.

Perdikis, D.C. and Lykouressis, D.P. (2002) Life table and biological characteristics of *Macrolophus pygmaeus* when feeding on *Myzus persicae* and *Trialeurodes vaporariorum. Entomologia Exp. Appl.* **102**(3): 261-272.

Prabhat, K. and Poehling, H.M. (2006) UV blocking plastic films and nets influence vectors and transmission on greenhouse tomato in humid Tropics. *Envir. Ent.* 35(4):1069-1082.

Rosaiah, B. (2001) Effect of different plant products against leaf miner, *Liriomyza trifolii* on ridge gourd *Luffa accutangula. Pestology* **25**(4):7-9.

Senthil, N.S., Kalaivani, K. and Chung, P.G. (2005) The effects of *azadirachtin* and nucleopolyhedrovirus on midgut enzymatic profile of *Spodoptera litura* Fab. (Lepidoptera: Noctuidae). *Pestic. Biochem. Physiol.* **83**(1): 46-57.

Shipp, J.L., Binns, M.R., Hao, X. and Wang, K. (1998) Economic injury levels for western flower thrips (Thysanoptera: Thripidae) on greenhouse sweet pepper. *J. Econ. Ent.***91**(3): 671-677.

Singh, D., Singh, N.S. and Kaur, S. (2002) Insect pests. In Field Problems of Vegetable Crops. Punjab Agricultural University, Ludhiana, Punjab, India. Pp: 88-155.

Singh, J., Brar, K.S., Shenhmar, M., Kaur, S. and Kaur, S. (2001) Biocontrol based management of tomsto fruit borer, *Helicoverpa armigera* (Hubner). *Proc. Symp. Biocontrol Pest Mngt : Qual. Crop Prot. Curr. Millen. held on July 18-19, PAU, Ludhiana.*Pp:165-166.

Srinivasamurthy, D., Krishnamoorthy, P.N., Raja, M.E. and Shivanna (2002) Economic assessment of pest management in brinjal and cabbage using neem products. *Proc. Int.Conf. Veg. held on. Nov. 11-14, Bangalore, India.* Pp: 243.

Stansly, P.A. and Liu, T.X. (1997) Selectivity of insecticides to *Encarsia pergandiella* (Hymenoptera: Aphelinidae), an endoparasitoid of *Bemisia argentifolii* (Hemiptera: Aleyrodidae). *Bull. Ent. Res.* **87**(5): 525-531.

Talekar, N.S., Su, F.C. and Lin, M.Y. (2003) How to grow leafy vegetables in nethouses and net tunnels. AVRDC International Copperators' Guide. AVRDC Publ # 03-558. htt://www.avrdc.org

Thiruvudainambi S, Doraisamy D, Veeraraghavanatham D and Srinivasan K 2002. Chemical and non chemical methods of managing mosaic disease of chilli. *Proc. Int. Conf. Veg. held on Nov. 11-14, Bangalore, India.* Pp: 258.

Ugine, T.A., Wraight, S.P. and Sanderson, J.P. (2007) Effects of manipulating spray-application parameters on efficacy of the entomopathogenic fungus *Beauveria bassiana* against western flower thrips, *Frankliniella occidentalis*, infesting greenhouse impatiens crops. *Biocontrol Sci. Technol.* **17**(1-2): 193-219.

Venzon, M., Rosado, M.C., Molina-Rugama, A.J., Duarte, V.S., Dias, R. and Pallini, A. (2008) Acaricidal efficacy of neem against *Polyphagotarsonemus latus* (Banks) (Acari: Tarsonemidae). *Crop Prot.* **27**(3-5): 869-872.

Xu, X., Borgemeister, C. and Poehling, H.M. (2006) Interactions in the biological control of western flower thrips, *Frankliniella occidentalis* (Pergande) and two-spotted spider mite, *Tetranychus urticae* Koch by the predatory bug, *Orius insidiosus* Say on beans. *Biol. Control Theory. Appl. Pest Mngt.* **36** (1): 57-64.

Weintraub, P.G., Kleitman, S., Mori, R., Shapira, N. and Palevsky, E. (2003) Control of the broad mite (*Polyphagotarsonemus latus* (Banks) on organic greenhouse sweet peppers (*Capsicum annuum* L.) with the predatory mite, *Neoseiulus cucumeris* (Oudemans). *Biol. Control Theor. Appl. Pest Mngt.* **27**(3): 300-309.

Wraight, S.P., Carruthers, R.I., Jaronski, S.I., Bradley, C.A., Garza, C.J., Galaini-Wraight, S. (2000) Evaluation of entomopathogenic fungi *Beauvaria bassiana* and *Paecilomyces fumosoroseus* for microbial control of silver whitetfly, *Bemisia argentifolii*. *Biol. Control Theor. Appl. Pest Mngt.* **17**(3): 203-217.

Zamani, A.A., Talebi, A., Fathipour, Y. and Baniameri, V. (2007) Effect of temperature on life history of *Aphidius colemani* and *Aphidius matricariae* (Hymenoptera: Braconidae), two parasitoids of *Aphis gossypii* and *Myzus persicae* (Homoptera: Aphididae). *Envir.Ent.* **36**(2): 263-271.

Theory and Practice of Integrated Pest Management
R. Arora, B. Singh and A.K. Dhawan (Eds)
Scientific Publishers (India), Jodhpur 2011, pp. 443 - 453.

25

MANAGEMENT OF INSECT PESTS OF OILSEED CROPS

Sarwan Kumar

Department of Plant Breeding and Genetics
Punjab Agricultural University, Ludhiana - 141 004, India

'Yellow Revolution' in India during early 90's made country self sufficient in oilseeds production, but this self sufficiency lasted for a short period. At present the demand for edible oils has outstripped the supply. About 40 per cent of country's edible oils requirements is met by imports. Oilseed crops, unlike wheat and rice, are grown under energy deprived conditions with low inputs. For the past 4-5 years area under oilseeds is more or less stagnant and there is very little scope for increasing the area further. Thus, the major challenge before oilseed workers is to increase production in limited area. One of the strategies to increase production is by reducing the losses caused by insect-pests. According to one estimate, insect-pests are known to cause 15-30 per cent loss in yield in different oilseed crops (Dhaliwal *et al.*, 2004), thus devouring upto one third of the agricultural produce. A timely and wise adoption of pest management strategies can help reduce these losses.

The oilseed *Brassicas* are known to be attacked by about 40 species of insect-pests out of which a few are serious pests. Among these, the mustard aphid (*Lipaphis erysimi*) is the key pest. Besides, cabbage caterpillar, (*Pieris brassicae*), sawfly (*Athalia lugens proxima*), painted bug (*Bagrada hilaris*), leaf miner (*Chromatomyza horticola*) and diamond back moth *Plutella xylostella* have been traditionally associated with rapeseed and mustard crops. A number of new insect pests including

Dasineura hisarensis, Melanagromyza clemi, Japanagromyza sp., *Helicoverpa armigera, Bagrada hirta* and *Myllocerus setilifer* have been reported attacking these crops during the last two decades (Arora, 1999).

Sunflower is a host for wide array of insect-pests, among which, head borer (*Helicoverpa armigera*) causes direct loss in seed yield as it feeds on developing seeds. Six insect pests namely head borer, Bihar hairy caterpillar, semilooper (*Trichoplusia ni*), tobacco caterpillar (*Spodoptera litura*), jassid (*Amrasca biguttula biguttula*) and cutworm (*Agrotis* sp.) are of common occurrence in one or other part of the country. Likewise groundnut, sesame and linseed are attacked by one or the other pest.

These insect-pest problems warrant economically and farm worthy pest management strategies particularly in rapeseed-mustard, sunflower and groundnut.

25.1. RAPESEED-MUSTARD

A number of insect pests are known to attack rapeseed-mustard in India. Of these, the mustard aphid, *Lipaphis erysimi*, is the key pest and causes colossal yield losses year after year. If not managed timely it may lead to complete crop failure. Continuous feeding by large aphid colonies drains out plant sap, thereby making the plant weak and produce less seed. Due to high fecundity and short generation time it multiplies very fast. Prolific breeding and short generation time makes this pest a major constraint in successful cultivation of rapeseed-mustard. Yield losses due to mustard aphid varied from 4 to 81 per cent during different years. The losses also varied in different *Brassica* species. Higher losses were reported in *B. campestris* and *B. mapus* while losses were lower in *B. carinola*. In B. juncea the losses were highly variable (Arora, 1999). The chemical insecticides, although, bring the population below economic injury level, but population again shoots up after a few days of insecticide application. This necessitates the need to develop a holistic pest management strategy under the umbrella of 'Integrated Pest Management'.

On rapeseed-mustard in India, it breeds from December to March with peak activity in end February to early March (Table 25.1). Population build up of this pest is largely governed by weather parameters with cold and cloudy conditions most favourable for its development.

Host Plant Resistance

An insect resistant cultivar is the most preferred option in pest management as it provides the farmers with economical, effective and environmental friendly option for pest management. Further, it can be well integrated with other pest management methods and does not require any elaborate efforts in technology transfer to farmers. Their effect is relatively stable and adverse weather does not alter the action of resistant varieties as is often experience with other control methods such as chemical and biological control. The first step in the development of insect resistant cultivar is the precise knowledge of sources of resistance. Variable level of resistance has been reported by different workers in different genotypes of oilseed *Brassica*. In general, a few strains of *B. juncea* are considered to possess tolerance to *L. erysimi* (Bakhetia and Sandhu, 1973; Singh *et al.,* 1984). Different genotypes of *B. nigra, B. carinata* and *Eruca sativa* were reported to possess antibiosis (Kundu and Pant, 1967; Dilawari and Dhaliwal, 1988). However, antixenosis and antibiosis were found to be responsible for resistance in some lines of *E. sativa* and *B. juncea* (Abraham, 1991; Abraham and Bhatia, 1992) while *B. napus* strains were known to possess antibiosis resistance (Gill and Bakhetia, 1985; Kalra *et al.,* 1987). A very large number of attempts have been made in the past to identify sources of resistance in primary gene pool of crop *Brassica* species (Brar and Sandhu, 1978; Amjad and Peters, 1992; Sekhon and Ahman, 1993; Bhandoria *et al.,* 1995; Saxena *et al.*, 1995). Despite these investigations and resultant publications claiming resistance, no germplasm that carries genetically characterized resistance factor(s) seems to be available as source of resistance that can be used in breeding programmes. The inability to identify suitable sources of resistance has precipitated attempts at developing *Brassica* transgenics carrying genes for lectin production that offer appreciable levels of resistance against *L. erysimi* (Kanrar *et al.*, 2002; Hossain *et al.*, 2006). However, field testing of these transgenics is still awaited.

Table 25.1. Peak activity period of *L. erysimi* on rapeseed-mustard in different regions of India

State	Peak period	Crop	Reference
Punjab	Mid February	Mustard	Bakhetia and Sidhu, 1983
	January-February	*B. rapa*	Bakhetia, 1986
	January-March	*B. juncea*	
	March	*B. napus*	
		B. carinata	
Rajasthan	End January	Mustard	Ahuja, 1990
Haryana	January-February	*Brassicas*	Rana *et al.*, 1993
Delhi	February	*B. rapa*	Phadke, 1986
Bihar	January-February	Rapeseed-Mustard	Sinha *et al.*, 1990
Uttar Pradesh	January	Mustard	Srivastava and Srivastava, 1972

Among the other pests of economic significance, very limited information is available on leaf miner, mustard saw-fly and painted bug. According to Singh *et al.* (1986) the thinner and non-hairy leaves along with yellow colour were more apparent to leaf miner flies.

Biological Control

The mustard aphid is known to be attacked by a number of natural enemies such as coccinellids, syrphids, green lace wing and *Aphidius* sp. etc. The predators of mustard aphid compriesed 8 species of coccinellids, 15 syrphids, 2 chrysopids, 1 chamaemyid and 1 hemerbid* (Arora, 1999). Kumar *et al.* (1987) enlisted six species of syrphids predating on aphids on *B. juncea* in Punjab. Kalra (1988) noticed the association of 3 coccinellids, 2 chrysopids, one syrphid and one chamaemyiid with aphid colonies. Among these six species, *Leucopis* sp., *Coccinella septempunctata* and *Episyrphus* sp. were recorded in significantly high numbers. Yadava and Singh (1992) conducted survey on natural enemy complex of aphids on *B. campastris* & *B. juncea* and identified seven predatory species such as *C. septempunctata, C. septempunctata* var. *divarieata, C. transvarsalis, Coccinella* sp., *Micrapsis discolour* and *Menochilus* sp. Their population increased from 0.33 to 19.5 individuals/10 plants during first to third week of February. More than 50 per cent of the coccinellid population in February was constituted by *C. septempunctata* alone. These coccinellids and syrphids could serve as biological control agents for suppressing the aphid population (Raj and Dass, 1992). Fungal diseases caused by *Entomophthora coronata* and *Cephalosporium aphidicola* were found inflict heavy mortality in population of mustard aphid (Arora, 1999). Though there have been reports on the predation of natural enemies on mustard aphid, there have been virtually no consistent study on the field efficacy of these natural control agents.

Cultural Control

(i) Sowing time. Alterations in sowing time employs the phenomenon of phonological asynchrony between the susceptible stage of host plant and peak pest population. It has been clearly shown that early sown crop of rapeseed-mustard suffers less damage by mustard aphid compared to the one sown late (after end October) (Singh *et al.*, 1984; Singh and Bakhetia, 1987). Another possible strategy to minimize loss is by identify the gene for earliness can be identified and incorporated in crop *Brassica* to make maximum asynchrony to manage insect-pests.

(ii) Fertilizer Application. Increase nitrogen application favours the multiplication by insect-pests. High nitrogen application resulted in high fecundity, faster development and heavy aphid attack in rapeseed-mustard (Kalra *et al.*, 1983). Kalra (1988) further observed that levels of K higher than 50 ppm adversely affected the multiplication of mustard aphid while there was rapid multiplication of aphids under increased application of sulfur irrespective of K levels.

(iii) Irrigation. Generally aphid fecundity and subsequent multiplication was reduced on water stressed plants. Sidhu and Kaur (1976) observed a significant decrease in the mean fecundity of mustard aphid as the soil water regimes decreased from 8 to 2%. Kalra (1988) made similar observations and found that aphid multiplication declined under water stress conditions.

Mechanical Control

Hand picking and destruction of mustard sawfly, tobacco caterpillar on small scale in kitchen garden could be feasible and subsequently reduce the pesticide load. A regular survey of field is necessary for this method to be effective. Similarly, on large scale mechanical removal of twigs and twigs harbouring hairy caterpillars can be very effective. Under low population level, this method can be of great value in insecticide free management of insect-pests.

Chemical Control

The rationale of insecticide use on oilseed *Brassicas* is based on the economic threshold of mustard aphid. The insecticides are and will continue to be an important component of IPM. They are the only control options available in hand when other methods fail to give satisfactory control or when the pest outbreak occurs. However, insecticides should not be used as the only strategy for pest management and they should be used as last alternative. A number of insecticides are available for rapeseed-mustard pests, but they should be judiciously and only when they are actually needed.

25.2. SUNFLOWER

Growing sunflower on ridges reduces the damage by cutworm significantly. Increased spacing and N doses favour multiplication by jassids (Mohammad *et al.,* 1980).

Sunflower harbours a diverse array of natural enemies such as lady bird beetles, green lace wing and spiders which prey on eggs and small larvae of head borer and tobacco caterpillar. On the basis of extensive study in Northern districts of Karnatka, Bhat *et al.* (1994) suggested that among various treatments NPV, *C. carnea* and endosulfan alone or in various combinations with NPV, the chrysopids hold a great promise for effective management of head borer on sunflower.

H. armigera is highly susceptible to NPV and Bt. In a field study, Muthuswami *et al.* (1993) observed an effective control of this pest. There is a good scope for combining NPV formulation with insecticides (Jayaraj *et al.*, 1989).

Plant products like neem, sweet flag etc. have been found effective against *S. litura* and *H. armigera* but with variable levels of toxicity.

Pheromone traps can be used in the field for monitoring, mass trapping and mating disruption of *H. armigera* and *S. litura*, of which first two are important as a component of IPM. Monitoring by pheromone traps has been successfully used in *H. armigera* management (Srivastava *et al.*, 1990). The pheromones have been successfully used in the prediction of the population trends of different insect pests for better control with chemicals (Singh and Sachan, 1991, 1993).

25.3. GROUNDNUT

Indian farmers often grow rows of castor around groundnut to attract *Spodoptera*. Groundnut covered with synthetic shining plastic mulches had lower thrips incidence and consequently lesser injury than the no mulch areas (Ranga Rao and Wightman, 1993). Summer ploughing is known to reduce the white grubs population significantly (>70%). The ploughing exposes the grubs to predatory birds. In addition to summer ploughing, early sowing helps to allow the establishment of better root system to enable plants to tolerate grub damage to a considerable extent. (Yadava, 1993). Adult beetle control is reported to be more rewarding than grub control as it brings about overall reduction in pest population.

S. litura and hairy catepillars occur in gregarious phases at some stage of their life when they can be easily picked and destroyed. Similarly, shaking of threes during evening hours helps to collect the adult beetles of white grub which can then be destroyed. Such a simple practice is highly rewarding in situations where field population is low.

For the control of aphids, the coccinellids are of great promise and it has been found that intercropping of pearl millet with groundnut increases the population of these predatory coccinellids (Kennedy *et al.*, 1990). A good control of *Aphis craccivora* by Coccinellids, Syrphids and Chrysopans has been reported from Gujarat (Patel *et al.*, 1976). Many predatory vertebrates like black drongo, house sparrow etc. are known to prey on grubs and beetles of white grubs. Thus, ploughing of fields in summer in the presence of these becomes very effective in curbing the population of grubs up to 70 % (Yadava, 1991).

Since the first report on use of *Cyprus rotundus* and *Cynodon dactylon* against the red hairy caterpillars (Dandapani *et al.*, 1984), many more including neem and garlic have been shown to reduce the leaflet damage (Mani *et al.*, 1990; Patel *et al.*, 1990a, b). Neem seed extract is found as effective against *A. craccivora* and *Aproaerema modicella* as phosphamidon (Chandramohan and Sivasubramanian, 1987).

The chemical control of the insect pests is studied quite extensively and more than one insecticide against each of the major/important pests are identified. However, use of toxic insecticides is advisable only judiciously based on monitoring of pest population in the field.

25.4. SESAMUM

Sesame leaf webber and capsule borer, *Antigastra catalaunalis* causes heavy losses in yield in years with normal rainfall. However, jassid, *Orosius albicinctus* also causes considerable damage in sesame especially as a vector of phyllody.

There is scanty information on important resistant sources in sesame against capsule borer and hence, there is urgent need for large scale screening of germplasm. B-67, RAUSS-17-6, TC 289 and C1036 were reported to posses some level of resistance against capsule borer (Yadav, 1985; Chemma *et al.,* 1982; Singh and Dhamdhere, 1986).

Early sowing of sesame in June helps to escape from phyllody damage than in July (Sandhu *et al.*, 2009). Similary Murli Baskaran *et al.* (1990) reported less damage by capsule borer on early sown crop.

Ahuja and Bakhetia (1993) reported the efficacy of various insecticides against important pests particularly *A. catalaunalis*. Though insecticides are highly effective, yet there is an urgent need to develop ETL for need based spraying of insecticides.

25.5. LINSEED

Bud fly is an important pest in linseed. Most of the high yielding varieties are susceptible to this pest. The released varieties found resistant to linseed bud fly include Jawahar, LS2, Neela, RL-43-5, RLC27, RLC37, RLC45, MP55, Bengal 6, Bengal 15, LCK8528 and LCK8062 (Malik, 1993).

Phosphamidon has been reported as quite effective against bud fly (Sood and Pathak, 1984). Jakhmola (1973) observed that number of insecticides are more important than choice of insecticides.

Sowing the crop early suffers less damage from bud fly (Singh *et al.,* 1985). Intercropping with gram also results in less damage than the sole crop (Malik, 1993).

25.6. TECHNOLOGY TRANSFER IN OILSEEDS

Though a number of pest control tactics are available pest management of oilseed pests, farmers are still largely dependent on the use of chemical insecticides as they find it most easily available, effective and economical method of pest management. Hence at present there is a need to educate the farmers about the adverse effects of pesticides and environmental pollution. Equally important is the training of extension agencies and field functionaries about the available IPM technologies and their adoption at farmers' level.

REFERENCES

Abraham, V. (1991) Rooting of excised leaves in oilseed crops. *Cruciferae Newsl.* **14**: 112-113.

Abraham, V. and Bhatia, C.R. (1992) Multiple screening for aphid resistance in Indian mustard, *Brassica juncea* (Linn.) Czern & Coss. *Cruciferae Newsl.* **16**: 1-11.

Ahuja, D.B. (1990) Population dynamics of mustard aphid, *Lipaphis erysimi* on Indian mustard, *Brassica juncea. Indian J. Pl. Prot.* **18**: 233-235.

Ahuja, D.B. and Bakhetia, D.R.C. (1993) Bio-ecology and management of insect pests of sesame- A review. Group Discussion on IPM strategies in oilseeds in India, December 23-24, 1993, Punjab Agricultural University, Ludhiana.

Amjad, M.D. and Peters, C. (1992) Survival, development and reproduction of turnip aphids (Homoptera: Aphididae) on oilseeds *Brassica. J. econ. Ent.* **85**: 2003-2007.

Arora, R. (1999) Major insect pests of rapeseed mustard and their management. In: R.K. Upadhyay, K.G. Mckeji and R.C. Rajak (eds). *IPM system in Agriculture.* Vol. V. *Oilseeds.* Aditya Books Pvt. Ltd., New Delhi, pp. 35-75.

Bakhetia, D.R.C. (1986) Pest management in rapeseed-mustard. *Pesticides* **5**: 32-39.

Bakhetia, D.R.C. and Sandhu, R.S. (1973) Differential response of Brassica species/varieties to the aphid Lipaphis erysimi Kalt infestation. *J. Res. Punjab Agric. Univ.* **10**(3): 272-279.

Bakhetia, D.R.C. and Sekhon, B.S. (1989) Insect-pests and their management in rapeseed-mustard. *J. Oilseeds Res.* **6**: 269-299.

Bakhetia, D.R.C. and Sidhu, S.S. (1983) Effect of rainfall and temperature on the mustard aphid *Lipaphis erysimi* Kalt. *Indian J. Ent.* **45**:203.

Bhadoria, N.S., Jakhmola, S.S. and Dhamdhere, S.V. (1995) Relative susceptibility of mustard cultivars to *Lipaphis erysimi* in North West Madhya Pradesh (India). *J. entomol. Res.* **19**: 143-146.

Bhat, N.S., Virupakshappa, K. and Chakravarty, A.K. (1994) Effect of biocontrol agents in management of *Helicoverpa armigera* in sunflower. pp. 312-314. In: Prasad, M.V.R. 1994. (ed.) *Sustainability in Oilseeds.* Indian Society of Oilseeds Research, Hyderabad.

Brar, K.S. and Sandhu, G.S. (1978) Comparative resistance of different *Brassica* species/varieties to the mustard aphid (*Lipaphis erysimi* Kalt.) under natural and artificial conditions. *Indian J. agric. Res.* **12**: 198-200.

Chandramohan, N. and Sivasubramanian, P. (1987) Evaluation of neem against leaf miner, *Aproaerema modicella. Neem Newsl.* **4**: 44.

Cheema, J.S., Singh, G., Grewal, G.S. and Labana, K.S. (1982) Comparative incidence of the sesame shoot and leaf webber, *Antigastra catalaunalis* in different cultivars of sesame. *J. Res. Punjab agric. Univ.* **19**: 170-171.

Dandapani, N., Rajmohan, N.S. and Kareem, A.Z. (1984) Plant products as antifeedants to control insect pests. *Pesticides* **19**: 53-60.

Dhaliwal, G.S., Arora, R. and Dhawan, A.K. (2004) Crop losses due to insect pests in Indian Agriculture: An update. *Indian J. Ecol.* **31**: 1-7.

Dilawari, V.K. and Dhaliwal, G.S. (1988) Population build up of mustard aphid on cruciferous cultivars. *J. Insect Sci.* **1**: 149-153.

Gill, R.S. and Bakhetia, D.R.C. (1985) Resistance of some *Brassica napus* and *B. campestris* strains to *Lipaphis erysimi*. *J. Oilseeds Res.* **2**: 227-239.

Hossain, M.A., Maiti, M.K., Basu, A., Sen, S., Ghosh, A.K. and Sen S.K. (2006) Transgenic expression of onion leaf lectin in Indian mustard offers protection against aphid colonization. *Crop Sci.* **46**: 2022-2032.

Jakhmola, S.S. (1973) Chemical control of linseed bud fly, *Dasineura lini*. *Indian J. agric. Sci.* **43**: 1078-1080.

Jayaraj. S., Rabindra, R.J. and Narayanan, K. (1989) Development and use of microbial agents for the control of *Helicoverpa armigera* in India. In: King, E.G. and Jackson, R.D. (eds.) *Proc. National Workshop Biol Control of Helicoverpa armigera: increasing the effectiveness of natural enemies,* TNAU, Coimbatore. pp. 483-503.

Kalra, V.K. (1988) Multiplication of mustard aphid, *L. erysimi* on *Brassica campestris*. *J. Aphidology* **2**: 69-71.

Kalra, V.K., Gupta, D.S. and Yadava, T.P. (1983) Effects of cultural practices and aphid infestation on seed yield and component traits in *Brassica juncea*. *Haryana agric. Univ. Res. J.* **13**: 115-120.

Kalra, V.K., Singh, H. and Rohilla, H.R. (1987) Influence of various genotypes of *Brassica juncea* on biology of mustard aphid, *Lipaphis erysimi* (Kalt.). *Indian J. agric. Sci.* **57**: 277-279.

Kanrar, S., Venkateswari, J., Kirti, P.B. and Chopra, V.L. (2002) Transgenic India Mustard (*Brassica juncea*) with resistance to the mustard aphid (*Lipaphis erysimi* Kaltenbach). *Pl. Cell Rep.* **20:** 976-981.

Kennedy, F.J.S., Rajamanickam, K. and Raveendran, T.S. (1990) Effect of intercropping on insect pests of groundnut and their natural enemies. *J. Biol. Cont.* **4**: 63-64.

Kumar, A., Kapoor, V.C. and Laska, P. (1987) Immature stages of aphidophagous syrphid flies of India. *Zool. Scr.* **16**: 83-88.

Kundu, G.G. and Pant, N.C. (1967) Studies on the *Lipaphis erysimi* with special reference to insect plant relationship I Susceptibility of different varieties of *Brassica* and *Eruca sativa* to the mustard aphid infestation. *Indian J. Ent.* **29**: 241-251.

Malik, Y.P. (1993) Management of linseed pests. Paper presented at group discussion on IPM strategies in oilseeds in India, December 23-24, 1993, Punjab Agricultural University, Ludhiana.

Mani, A., Kumudanathan, K. and Jagadish, C.A. 1990. Antifeedant properties of neem against the red hairy caterpillar, *Amsacta albistriga. Neem Newsl.* **7**: 16-17.

Mohammand, S., Hussain, M. and Sagar, G.C. (1980) Preferential infestation of the leaf hopper on sunflower in relation to nitrogen nutrition and plant density. *Indian J. agric. Sci.* **50**: 955-957.

Murli Baskaran, R.K., Ganesh, S.K. and Thangavelu, S. (1990) Effect of different time of sowing on shoot webber and phyllody incidence and yield of sesame. *Sesame and Safflower Newsl.* **5**: 40-45.

Muthuswami, M., Rabindra, R.J. and Jayaraj, S. (1993) Use of a baculovirus mixture for the control of *Helicoverpa armigera* and *Spodoptera litura. J. Biol. Cont.* **7**: 105-108.

Patel, J.R., Jyani, B.B., Patel, C.C. and Board, P.K. (1990b) Insecticidal properties of some plants and their available formulations against *Amsacta moorei. Proc. Symp. Botanical Pesticides in IPM.* Rajamundry, pp. 332-337.

Patel, J.R., Patel, J.I., Mehta, D.N.M. and Shah, B.R. (1990a) Integrated management of *Amsacta moorei* with botanical insecticides. *Proc. Symp. Botanical Pesticides in IPM.* Rajamundry, pp. 327-331.

Patel, R.C., Yadav, D.N. and Patel, J.R. (1976) Natural control of the groundnut aphid, *Aphis craccivora* in central Gujarat. *Curr. Sci.* **45**: 34-35.

Phadke, K.G. (1986) Ecological factors influencing aphid, *Lipaphis erysimi* (Kaltenbach) incidence on mustard crop. In: Agarwal, B.K. (Ed.) *Aphidology in India*, AR Printers, Calcutta, pp. 37-42.

Raj, D. and Dass, G. (1992) Population build up of aphids and their predators in mid-hill zone of HP. *Proc. Nat. Symp. Recent Adv. IPM*, October 12-15, 1992, PAU, Ludhiana. pp. 35-36.

Rana, J.S., Khokhar, K.S., Singh, H. and Sucheta, K.S. (1993) Influence of abiotic environment on the population dynamics of mustard aphid, *Lipaphis erysimi. Crop Res.* **6**: 116-119.

Ranga Rao, G.V. and Wightman, J.A. (1993) Sucking pests as vectors and pests of groundnut. Group Discussion on IPM strategies in oilseeds in India, December 23-24, 1993, Punjab Agricultural University, Ludhiana.

Sandhu, P.S., Kaur, S., Sheoran, P. and Kumar, S. (2009) Influence of sowing dates on development of phyllody on sesame, *Sesamum indicum* L. *J. Oilseeds Res.* **26** (Special Issue): 495-496.

Saxena, A.K., Bhadoria, S.S., Gadewadikar, P.N., Barteria, A.M., Tomar, S.S. and Dixit, S.C. (1995) Yield losses in some improved varieties of mustard by aphid, *Lipaphis erysimi* Kalt. *Agric. Sci. Digest* **15**: 235-237.

Sekhon, B.S. and Ahman, I. (1993) Insect resistance with special reference to mustard aphid. In: *Monographs on Theoretical and Applied Genetics: Breeding oilseed brassicas*, Springer Verlag, New York Inc. pp. 206-221.

Sidhu, H.S. and Kaur, P. (1976) The influence of water stress in the host plant on the reproduction of the mustard aphid, Lipaphis erysimi (Kaltenbach). *Indian J. Ecol.* **3**:163-166.

Singh, B. and Bakhetia, D.R.C. (1987) Screening and breeding techniques for aphid resistance in oleiferous *brassicae*: A review. The *Oilcrops Network,* International Development Res. Centre, Canada, 50p.

Singh, D.S., Sircar, P., Srivastava, V.S. and Dhingra, S. (1985) Biological efficacy of botanical products against some important insect pests. *Indian J. Ent.* **47**: 445-451.

Singh, H., Kalra, V.K. and Rohilla, H.R. (1986) Studies on the efficacy and economics of sprays of synthetic pyrethroids against mustard aphid, *Lipaphis erysimi* (Kalt.) on rapeseed. *Indian J. Pl. Prot.* **10**: 1-5.

Singh, H., Rohilla, H.R., Kalra, V.K. and Yadava, T.P. (1984) Response of *Brassica* varieties sown on different dates to the attack of mustard aphid, *Lipaphis erysimi. J. Oilseeds Res.* **1**: 49-56.

Singh, K.N. and Sachan, G.C. (1991) Monitoring of *Spodoptera litura* population pheromone baited traps at Pantnagar, India. *Indian J. Pl. Prot.* **8**: 97-101.

Singh, K.N. and Sachan, G.C. (1993) *Spodoptera litura* male moth catches in pheromone traps and their relationships with ovipositional inhibitors in groundnut fields at Pantnagar, India. *Insect Sci. Appl.* **14**: 11-14.

Singh, U.C. and Damdhere, S.V. (1986) Response of sesame varieties to *Antigastra catalaunalis. N.D. Univ. J. agric. Res.* **1**: 109-111.

Sinha, R.P., Yazdani, S.S. and Verma, S.D. (1990) Population dynamics of mustard aphid, *Lipaphis erysimi* in relation to ecological parameters. *Indian J. Ent.* **52**: 387-392.

Srivastava, A.S. and Srivastava, J.L. (1972) An induced aphid resistant mutant in turnip, *Brassica rapa. Curr. Sci.* **41**: 820-821.

Srivastava, C.P., Pimbert, T. and Reed, W. (1990) Spatial and temporal changes in abundance of *Helicoverpa armigera* in India. *Insect Sci. Appl.* **11**: 869-876.

Yadav, S.K. (1985) Studies on the susceptibility/ resistance of sesame genotypes to the infestation of the til leaf roller and pod borer, *Antigastra catalaunalis.* M.Sc. Thesis, Department of Entomology, Haryana Agricultural University, Hisar, 77pp.

Yadava, C.P.S. (1991) White grub management in groundnut. Technical bull. no. 1, 14 p, AICRP on whitegubs, Agricultural Research Station, Durgapura, Jaipur.

Yadava, C.P.S. (1993) Management of white grubs with particular reference to groundnut crop. Group Discussion on IPM strategies in oilseeds in India, December 23-24, 1993, Punjab Agricultural University, Ludhiana.

Yadava, R.P. and Singh, V.S. (1992) Species complexity and dominance structure of predatory coccinellids associated with mustard aphid, *Lipaphis erysimi* in rapeseed-mustard crop ecosystem. *Proc. Nat. Symp. Recent Adv. IPM,* October 12-15, 1992, PAU, Ludhiana. pp. 49-50.

Theory and Practice of Integrated Pest Management
R. Arora, B. Singh and A.K. Dhawan (Eds)
Scientific Publishers (India), Jodhpur 2011, pp. 454 - 472.

26

INTEGRATED MANAGEMENT OF INSECT PESTS OF *RABI* PULSES

G.K. Taggar and Ravinder Singh
Department of Plant Breeding & Genetics
Punjab Agricultural University, Ludhiana - 141 004, India

Pulses are the major sources of dietary protein in the vegetarian diet in our country. Besides being a rich source of protein, they maintain soil fertility through biological nitrogen fixation in soil and thus play a vital role in furthering sustainable agriculture. Pulses are ideal crop for mixed and intercropping and also serve as nutritious vegetables and fodders. Chickpea, lentil and field pea are the major *rabi* pulses grown in Punjab. However, lathyrus (*Lathyrus sativus*) and rajmash (*Phaseolus vulgaris* L.) are also grown as *rabi* pulses in other parts of India. Among the various constraints, insect pests are the major and important one affecting the productivity of pulses apart from ecological and biological constraints.

26.1. INSECT PEST PROBLEMS IN CHICKPEA

Chickpea or gram (*Cicer arietinum* Linn.) is the one of the most important legume crops in the world. It is a premier *rabi* pulse crop in the Indian subcontinent. Nearly 80% of the world chickpea crop is grown in India, with an average yield of around 800 kg/ha in north India. It is cultivated in 42 countries in south Asia, north and central America, the Mediterranean region, West Asia and north and east Africa. In recent years, it has become an important crop in Australia, Canada and the USA. Relatively few insect pests attack chickpea compared to other legumes. This is probably because it is a cool-season crop and also because of the dense glandular trichomes found on all of its green tissues. The glandular trichomes excrete an acidic substance comprised mainly of malic and oxalic acids, which deter many potential herbivores. Nearly 60 insect species are known to feed on chickpea (Reed *et al.*, 1987). The important insect pests damaging chickpea in different regions are discussed hereby:

Insects attacking vegetative parts

i. Aphids: *Aphis craccivora* Koch and *Acyrthosiphon pisum* (Harris)

ii. Cutworm: Black cutworm, *Agrotis ipsilon* (Hfn.) and *Agrotis flammatra* Schiff.

iii. Leaf miners: *Liriomyza cicerina* (Rodani) and *L. congesta* (Becker)

iv. Leaf-feeding caterpillars: Cabbage looper, *Trichoplusia ni* (Hub.), leaf caterpillar, *Spodoptera exigua* (Hub.) and hairy caterpillar, *Spilosoma oblique* Walker

v. Semilooper: *Autographa nigrisignia* Walker

vi. Termite: *Microtermes obesi* (Holm.) and *Odontotermes* sp.

Insect attacking reproductive structures

i. Pod borers: Gram pod borer, *Helicoverpa armigera* (Hub.), native budworm, *H. punctigera* (Wallengren) and corn earworm, *H. zea* (Boddie).

Insect attacking stored seeds

I. Bruchids: Chinese bruchid, *Callosobruchus chinensis* L., pulse weevil, *C. analis* F. and pulse bruchid, *C. phaseoli* (Gylh.)

The gram pod borer, *H. armigera* and the aphid, *A. craccivora* are the major pests of chickpea in the Indian Subcontinent. The black aphid, *A. craccivora* is important as a vector of the chickpea stunt disease, while, *C. chinensis* is the most dominant species in storage (Sharma *et al.*, 2007). Among the two chickpea types, *kabuli* genotypes are generally more susceptible to insect pests than *desi*-types.

26.1.1. Gram pod borer *H. armigera*

H. armigera is known to be the 'key' pest of chickpea in north India. Being of polyphagous nature, *H. armigera* has been reported in damaging proportions on more than 182 plant species, including chickpea, pigeon-pea, cowpea, cotton, sorghum, peas, tomato, tobacco, maize, field beans, millet, groundnut, safflower, vegetable crops, wild grasses and many ornamental plants.

Helicoverpa females lay eggs singly on leaves, flowers, and young pods. The young larvae feed on the tender portion of the leaves and shoots by making scratches. The young seedlings of chickpea may be destroyed completely. After second instar, larvae consume whole of the opened leaves, leaf buds, flower buds and flowers. Under severe pest infestation, the whole crop may get defoliated. With the availability of the pods, 3rd instar larva makes a hole in the pods and moves inside to feed inside the grains, whereas full-grown larva feeds after cutting a hole in the pods and thrusts its head therein, while keeping hind part of the body outside. A single larva is capable of damaging 30-40 pods in its life time.

The global problem

Intensification of agriculture on a global level is the root cause for the enhanced damage caused by *H. armigera*. Farmers around the world use chemical pesticides indiscriminately to control *H. armigera*. Pesticide manufacturers have produced more potent and toxic pesticides whenever *H. armigera* developed resistance to the prevailing insecticides. The vicious cycle has led to a stage where *H. armigera* has developed resistance to most of the available insecticides, especially to the synthetic pyrethroids. It has, therefore, become necessary to devise a suite of pest management tactics to control this pest. Environmentally safe technologies are not yet ready in form to be delivered to the farmers, although the use of integrated pest management (IPM) strategies can reduce the negative environmental effects of the existing chemical pesticides (Sharma, 2001). Chickpea production is severely threatened by increasing difficulties in controlling *H. armigera* (Matthews, 1999).

The extent of losses due to *H. armigera* in chickpea have been estimated to be over $328 million in the semi-arid tropics (ICRISAT, 1992). Worldwide losses due to *Helicoverpa* in cotton, legumes, vegetables, cereals etc. may exceed $2 billion, and the cost of insecticides used to control these pests may be over $1 billion annually (Sharma, 2005). In Punjab, the loss in yield due to *H. armigera* on chickpea has been reported to be about 35%.

Components of *Helicoverpa* management. The major pest status of *H. armigera* is because of its high mobility, polyphagous nature, high reproductive rate and ability to undergo diapause. All these traits make it well adapted to exploit the transient habitats created by intensification of agriculture. Management strategies for *H. armigera* require different control tactics based on the relationship between population density and economic loss (Gowda, 2005). Pest management strategies for *H. armigera* include cultural management of the crop, biological control using predators, parasites and microbial pesticides, sex pheromones for population monitoring or mating disruptions, host-plant resistance, and chemical control.

Population monitoring and forecasting. Field monitoring and determining the economic threshold levels (ETLs) for different crops and cropping systems is essential for a rational pest management effort, especially in case of chemical pesticide application. Monitoring of *Helicoverpa* populations is necessary to determine if threshold has been exceeded and control measures are required. Action thresholds based on egg numbers have been used to make control decisions. One larva per meter row in chickpea causes economic loss. Efforts have been made to develop a forewarning system for *H. armigera* on cotton, chickpea and pigeonpea in India. A thumb rule has been developed to predict *H. armigera* population using surplus/deficit rainfall in different months in south India (Das *et al.*, 2001). A combination of surplus rains during the monsoon and deficit rainfall during November indicated low incidence, while deficit rains during the monsoon and surplus rains during November indicated severe attack. Models for long-range forecasts of *H. armigera* populations in Australia have also been developed. These population-forecasting models may be incorporated into crop production models for pest management.

The data accumulated on the moth catch through pheromone traps in Punjab reveal that *H. armigera* attains peaks twice a year. The population of moth starts swelling abruptly during the second half of March and first half of April. This is the first peak period. It, then, gradually starts declining during May. Again in September, it starts ascending and attains the second peak during October (Chhabra and Kooner, 1990). Thus, the first peak of *H. armigera* population coincides with the peak fruiting stage of chickpea crop, a stage more conducive for breeding of the pest and damage. In Australia, population monitoring with sex pheromone-baited traps is used to detect the onset of immigration or emergence from local diapause. Timing of control is determined by field monitoring of larval densities in crops through the period of crop susceptibility. Control is only recommended when larval population in post-flowering crops exceeds the threshold of 2-4 larvae per metre row (Lucy and Slatter, 2004).

Host-plant resistance. Host-plant resistance is one of the cheapest and most effective management tools for reducing damage by *H. armigera*. The development of crop cultivars resistant or tolerant to *H. armigera* has considerable potential for use in integrated pest management, particularly under subsistence farming conditions in developing countries. More than 14,000 chickpea accessions have been screened for resistance towards *H. armigera* in India under field conditions. No line with immunity has so far, been identified in chickpea against *H. armigera*, therefore, the emphasis must be on reduced susceptibility. Several germplasm accessions (ICC 506EB, ICC 10667, ICC 10619, ICC 4935, ICC 10243, ICCV 95992 and ICC 10817) with resistance to *H. armigera* have been identified, and varieties such as ICCV7, ICCV10 and ICCL 86103 with moderate levels of resistance have been released for cultivation (Gowda *et al.*, 1983, Lateef, 1985, Lateef and Pimbert, 1990). Higher concentration of malic acid in the leaf exudates is reported to repel the larvae of *H. armigera* or results into poor oviposition and low larval preferences (Lateef, 1985; Rembold, 1981).

Most of the characters associated with resistance to *H. armigera* are polygenic in nature and are associated with undesirable traits such as susceptibility to major diseases, other insect pests, poor plant type and poor yield (Table 26.1). However, several wild relatives of chickpea have shown high levels of resistance to *H. armigera,* and efforts are underway to transfer resistance from the wild relatives into high yielding varieties of chickpea. Wild relatives of chickpea with accessions belonging to *C. bijugum, C. judaicum, C. pinnatifidum* and *C. reticulatum* showed resistance to *H. armigera* (Sharma *et al.*, 2005). With the use of interspecific hybridi-zation, it is possible to transfer resistance genes from the wild relatives to cultivated chickpea.

Table 26.1. Characters associated with resistance to *H. armigera* in chickpea

S.No.	Mechanism	Character(s)
1.	Non-preference	Pod shape, pod wall thickness, foliage color and glabousness
2.	Antibiosis	Malic acid, oxalic acid, crude fibre, non-reducing sugars, low starch, cellulose, hemicellulose, lignin in pod wall, trypsin inhibitors and proteinase inhibitors
3.	Escape	Earliness and cold tolerance

(Source: Dua *et al.*, 2005)

Molecular marker-assisted selection (MAS) can be used to accelerate the introgression of desirable genes into improved cultivars. Preliminary results on development of molecular markers for resistance to *H. armigera* have been reported in chickpea (Lawlor *et al.*, 1998). Kar *et al.* (1997) developed transgenic chickpea plants with *cry1AC* gene. Efforts are underway at ICRISAT and elsewhere to develop transgenic plants of chickpea with *Bacillus thuringiensis* (Bt) and soybean trypsin inhibitor (SBTI) genes for resistance to *H. armigera.* Chickpea cultivars ICCV 1 and ICCV 6 have been transformed with *cry IAc* gene. Insect feeding assays indicated that the expression level of the *cry IAc* gene was inhibitory to the development and feeding by *H. armigera.*

For long-term effectiveness of the host-plant resistance, it is essential to deploy a combination of strategies (Gowda, 2005):

- Identify lines with diverse resistance mechanisms and pyramid resistance genes into the adapted genotypes.
- Tap resistance genes from wild relatives.
- Combine HPR with novel genes (*Bt*, SBTI and lectins) in order to make plant resistance a viable option in pest management.
- Deploy more than one gene in transgenics to increase the levels and durability of resistance.

Cultural manipulation of the crop. Chickpea is a crop of low return per hectare and is sensitive to adverse weather. In view of high cost of pesticide and other associated problems with chemical control, the role of cultural practices has been recognized as a main component of integrated pest management.

A number of cultural practices such as time of sowing, spacing, fertilizer application, deep ploughing, and interculture have been reported to reduce the survival of and damage by *H. armigera*. Early sown crop shows minimum pod damage due to *H. armigera* and maximum grain yield. However, delay in planting causes gradual increase in the per cent pod damage and reduction in grain yield (Srivastava and Singh, 1979). It is therefore possible to escape peak infestation period of *H. armigera* by growing early maturing cultivars or by early sowing of crop in north India. Last week of October has been reported to be the best period for sowing in Punjab conditions, both for keeping down the pod borer's incidence and getting maximum yield (Chhabra *et al.,* 1978). Delayed sowing by middle of November or first week of December, usually shifts crop maturity to March or middle of April, which coincides with the peak period of *H. armigera* activity.

The plant population also significantly influences pod damage. Closer spacing in chickpea leads to greater pod damage than wider spacing. Phosphatic fertilizers tend to reduce pod damage. Assured irrigation leads to better development of foliage which provides favorable microclimate for *H. armigera* to lay maximum eggs and food for larval development. Irrigation also decreases malic content on the plant surface, thereby increasing the susceptibility of the crop grown under irrigated conditions.

Intercropping or strip-cropping with wheat, barley, sorghum, marigold, sunflower, linseed, mustard and coriander can minimize the extent of damage to the main crop. Yadava *et al* (1989) found intercropping of chickpea with mustard, linseed and wheat having significantly low pod damage than recorded on sole chickpea. Strip-cropping also increases the efficiency of chemical control. Hand-picking of large-sized larvae can also be practiced to reduce *H. armigera* damage. Crop rotations do not help manage this polyphagous and very mobile pest, although it has been noted that some crops (lucerne) are more attractive to the moths, and susceptible crops should not be planted too close to the main crop.

Biological Control. The importance of biotic and abiotic factors on the seasonal abundance of *H. armigera* is poorly understood. Low activity of parasitoids, particularly egg parasitoids *Trichogramma* spp., has been reported from chickpea because of dense later of trichomes and their acidic exudates (Jalali *et al.*, Romeis *et al.*, 1999). The ichneumonid, *Campoletis chloridae* (Uchida), is probably the most important larval parasitoid of *H. armigera* in chickpea, causing 20-76% parasitization under field conditions in India. *Carcelia illota* (Curran), *Goniophthalmus halli* Mesnil and *Palexorista laxa* (Curran) have also been reported to parasitize *H. armigera* larvae up to 54% on chickpea. Predators such as *Chrysoperla* spp., *Chrysopa* spp., *Nabis* spp., *Geocoris* spp., *Orius* spp. and *Polistes* spp. are the most common in India. Of late, a nematode parasite, *Ovomermis albicans* has been recorded on *H. armigera* especially in the rainy season. *O. albicans* grows to a length of 10 cm or more and as it grows, it coils up inside *H. armigera* larvae, killing them before they can pupate by consuming and disrupting their internal organs. Provision of bird perches or planting of tall crops that serve as resting sites for various insectivorous birds such as *mynah* and drongo helps reduce the number of caterpillars.

Biopesticides are an emerging area of promise. The use of microbial pathogens including *H. armigera* nuclear polyhedrosis virus (*Ha*NPV), entomopathogenic fungi (*Beauveria bassiana* and *Metarrhizium anisopliae*) Bt, some strains of *Pseudomonas* spp. and nematodes have shown some potential to control *H. armigera* (Sharma, 2001). Two sprays of *Ha*NPV @ 500 LE/ha are as effective as two sprays of endosulfan (0.05%) in reducing the infestation of *H. armigera* and pod damage, resulting in increase in the seed yield (Pawar *et al.*, 1987). NPV either alone or mixed with endosulfan gave good yield under moderate levels of pest infestation (Ujagir *et al.*, 1997). Incidence of cytoplasmic polyhdrosis virus (CPV) and granulosis virus has also been reported on the larvae of *H. armigera*. Spraying *Bt* formulations in the evening results in better control than spraying at other times of the day. *Bt* formulations such as Dipel 8L and Delfin WG have been found to be effective for the control of *H. armigera*. Entomopathogenic fungus, *Nomuraea rileyi* (10^6 spores per ml), results in 90-100% larval mortality, while *B. bassiana* (2.68×10^7 spores per ml) resulted in 6% damage in chickpea compared to 16.3% damage in the untreated control plots (Saxena and Ahmad, 1997). Liquid formulation of EPN (*Steinernema masoodi* in water) applied @ 30×10^5 IJs/ha recorded lower pod damage (21.8%) as compared to 59.6% in control (Anonymous, 2010). A number of natural plant products such as neem leaf extract, neem seed kernel extract (NSKE), neem oil, custard apple and karanj (*Pongamia*) kernel extracts have been found showing good insecticidal activity against low to moderate levels of infestation of *H. armigera* in chickpea. NSKE (5%) is highly effective against early instars of *H. armigera* in chickpea.

Chemical control. Management of *H. armigera* in India and other countries in chickpea and other high-value crops relies heavily on insecticides. The economic losses in chickpea occur at the podding stage. Endosulfan, cypermethrin, fenvalerate, thiodicarb, profenophos, spinosad and indoxacarb have been found to be effective for *H. armigera* control on chickpea. The pod borer can be controlled by spraying any one of the following insecticides in 200 litres of water per hectare at the pod initiation stage. Repeat after two weeks, if necessary: 250 ml of Sumicidin/ Fenlik/Agrofen 20 EC (fenvalerate) or 400 ml of Decis 2.8 EC (deltamethrin) or 200 ml of Cymbush 25 EC (cypermethrin) or 2.5 litres of Thiodan 35 EC (endosulfan) or dust 25 kg Malathion 5% dust (malathion) per hectare. For consuming any part of raw gram plant, observe a waiting period of 20 days after the application of the insecticide.

The possibility of using chitin inhibitors such as diflubenzuron with chlorpyriphos or endosulfan for controlling *H. armigera* more advantageously than conventional insecticides has also been shown. It was also found to be safe to the larval parasitoid, *C. chloridae* (Ravi and Verma, 1997). Insecticide use in the form of sprays should be undertaken at 50% flowering stage, based on the pheromone monitoring (when there are 5-6 moths at least for 2-8 days or action threshold of 1 larva per meter row length or per 10 plants). *H. armigera* populations in northern Australia are largely resistant to pyrethroids, carbamates and organphosphates. Introduction of new chemistry, notably indoxacarb and spinosad, is being managed to minimize the development of resistance in *H. armigera* through a strategy that takes into account all crops throughout the year. Many new insecticides are being tested for the management of *H. armigera* on chickpea in various AICRP trials throughout the country. These include: flubendiamide, rynaxipr, emamectin benzoate, indoxacarb and spinosad.

Integrated Pest Management (IPM). Considering the complexity of managing *H. armigera* effectively, it is imperative to follow an integrated approach, which involves agronomic and cultural management, host plant resistance, biological control, natural pesticides and judicious use of chemical pesticides. Several management tactics have been investigated, which provide a framework for improved management of pod borer in chickpea cropping system worldwide. For example, crop cultivars with resistance to *Helicoverpa* (derived through conventional plant breeding or biotechnological approaches) can play an important role. Cultural practices such as deep ploughing, interculture, and intercropping could potentially reduce the intensity of *Helicoverpa*. Although the role of natural enemies as biological control agents is unclear, their impact could be improved by reducing pesticide applications, and adopting cropping practices that encourage their activity. Most studies have shown that insecticide applications are more effective than neem kernel extracts, *Bt*,

HaNPV, or augmentative releases of natural enemies. However, biopesticides and synthetic insecticides, applied alone, together, or in rotation, are effective for *Helicoverpa* control in chickpea. Maximum yield of 1.17 t ha^{-1} was obtained with IPM, followed by endosulfan (1.05 t ha^{-1}) compared to 0.74 t ha^{-1} in untreated control plots. IPM also showed the best cost-benefit ration (1:6.3), followed by endosulfan (Table 26.2).

Moreover, scouting for eggs and young larvae is critical for initiating timely control measures. Insecticides with ovicidal action, and/or systemic action are effective against *Helicoverpa* during the flowering stage. Finally, the development of transgenic plants with different insecticidal genes, molecular marker assisted selection, and exploitation of the wild relatives of *Cicer* species should be pursued to develop comprehensive programs for *Helicoverpa* management on chickpea.

Table 26.2. Cost-benefit ratio of different treatments for *H. armigera* mana-gement in chickpea

Treatments	Grain yield (t ha^{-1})	Total cost of plant protection (Rs ha^{-1})	Total income (Rs ha^{-1})	Cost:benefit ratio
Azadirachtin 0.006%	0.96	1750	9600	1:5.5
HaNPV 250 LE ha^{-1}	0.96	2000	9600	1:4.8
Bird perches	0.86	-	8600	-
Endosulfan 0.07%	1.05	1725	1050	1:6.1
IPM	1.17	1845	1170	1:6.3
Control	0.74	-	7400	-

Source: Ranga Rao *et al.*, 2005

In the ICAR-ICRISAT collaborative program during 1996-97 at Nanded, the following module was followed on farmer's fields:

- Adoption of field sanitation
- Growing of linseed or coriander crop to attract natural enemies
- Installation of pheromone traps (5 ha^{-1}) and bird perches (5 ha^{-1})
- Spraying of 5% neem seed kernel extract (NSKE)
- Mechanical collection of larvae
- Spraying of HaNPV @ 250 LE ha^{-1}
- Intercropping with linseed (one row after every 10th row)
- Spraying of endosulfan @0.07%

A thorough analysis of multi-trophic interactions in the context of benefits versus crop damage and yield loss will help in deciding management options. At the same time, the possibilities of maximizing the efficacy of insecticide use should be explored, while minimizing the harmful effects on the environment.

26.1.2. Black Aphid, *Aphis craccivora* Koch

The black aphid, *Aphis craccivora* (Hemiptera: Aphidiidae) is an important pest of chickpea in the Indian subcontinent. It is also important as a vector of the chickpea stunt disease. It is commonly known as cowpea aphid, black aphid, black legume aphid, groundnut aphid, bean aphid, lucerne aphid, black lucerne aphid, and African bean aphid. It has an extensive host range, and damages several grain legumes including chickpea, lentil, cowpea, groundnut, alfalfa, and other legumes.

Both nymphs and adults suck the plant sap from the tender leaves, stems and pods, and mostly colonize the young leaves and growing points, which become characteristically deformed. Young leaves of seedlings become twisted under heavy infestation. Yield can be drastically reduced, and if infestations are early and severe, plants can be killed completely. Infestations during the bloom and early pod stages reduce yield and crop quality by removing plant sap, impairing pod appearance, reducing seed fill, and by the presence of aphid honeydew.

Management

- *Monitoring*: Aphid infestation should be monitored regularly, particularly when the crops begin to flower.
- *Cultural Control*: Early sowing leads to early canopy closure, which also helps in reducing the spread of viruses in chickpea. Aphid infestation is greater under wider spacing.
- *Biological control*: *Aphis craccivora* is attacked by several parasites and predators. Ladybirds, *Coccinella septempunctata, C. transversalis, C. nigritis, Menochilus sexmaculatus*, and *Brumus suturalis*; and chrysopids, *Chrysoperla* spp. are common aphid predators. The parasitoids such as *Trixys indicus* and *Lipolexix scutellaris* are important natural enemies of *A. craccivora* in India. Although parasitism as high as 95% has been documented, aphid population levels can be so high that enough non-parasitized individuals remain to cause significant injury.
- *Biopesticides:* Plant products such as *Azadirachtin*, neem oil, and pyrethrins have shown some suppression of *A. craccivora* populations, but do not provide effective control of the aphids.
- *Host plant resistance:* A number of genotypes have been reported to be less susceptible to aphid damage. Chickpea varieties with low trichome density or those devoid of trichomes, and low acid exudates are highly susceptible to aphid damage. The winged

adults that land on chickpea may be deterred and fly off, or be killed by the acid exudates on leaves and pods. Chickpea genotypes with little or no exudates maintain the largest colonies.

- *Chemical control:* Generally, there is no need for aphid control on chickpea in India, but chemical control may become necessary to prevent secondary spread of the chickpea viruses. It can be controlled by spraying 625 ml of Malathion 50 EC (malathion) or 375 ml of Rogor 30 EC (dimethoate) or 375 ml of Metasystox 25 EC (oxydemeton methyl) in 200 litres of water per hectare as soon as the pest appears.

26.1.3. Termite, *Odontotermes obesus*

The termite, *Odontotermes obesus* is a serious pest of chickpea throughout India, particularly in Rajasthan, Haryana, Punjab and UP. Termites damage the plants soon after sowing and continue till late growing stage. The pest can be generally observed feeding on roots or near the root zone of the damaged plants. The leaves of the damaged plant droop down, which later wither and dry. Such a plant, if pulled, is easily uprooted and termite cuts or galleries are seen. The incidence of pest is more in light soils.

Management

- Seed treatment with chlorpyrifos 20 EC or endosulfan 35 EC @ 10 ml/kg seed before sowing is effective in controlling termites in chickpea.
- The use of oil cakes of *karanja, mahua*, neem or castor in the field or soil application of chlorpyriphos is also reported to be effective.

26.1.4. Cutworm, *Agrotis ipsilon* (Hufnagel) and *A. flammatra* Schiff. (Lepidoptera: Noctuidae)

In India, both species of *Agrotis* occur throughout the Northern belt from Punjab in the north-west to West Bengal in the east and MP in the south. Cutworms are polyphagous feeders with a wide host range. Among the grain legumes, chickpea, lentil, kidney bean, Lathyrus (*khesri*) and peas are the common host plants. However, it is most serious on chickpea.

Larvae come out of their hiding places at night and damage the plants under the cover of darkness. Cutworms attack chickpea crop in two stages of its growth. In the seedling stage, the stem is generally cut at an average height of 5 cm from the surface of the soil. They consume only a little of the stem, then nibble and move to attack other stems. Consequently, a few cutworms in the field damage many small-growing plants in one night. During the later stage of growth, the shoots are damaged by cutting them at an average height of 24 cm from the soil surface, where the stem is soft, thus leading to the retardation in growth and ultimately causing reduction in the yield.

Agrotis ipsilon is widespread and feeds on many host plants throughout Asia. Caterpillars cut seedling stems at ground level, killing the plant. Cutworm larvae are found in the soil or in leaf litter near damaged plants. Larvae feed on the foliage. Larvae are nocturnal, residing in the soil during the day.

Management. Cutworms appear sporadically in chickpea crops, though they can reach damaging levels in some years or locations. Their populations may be overrun with parasites, prompting an awareness of natural enemies and their impact before implementation of other control strategies.

26.1.5. Leafminers *Liriomyza cicerina* (Rondani) and *Chromatomyia horticola* (Goureau) (Diptera: Agromyzidae):

Liriomyza cicerina is an important insect pest of chickpea in West Asia. In India, *Chromatomyia horticola* reportedly attacks chickpea leaflets. Females puncture the upper surface of the foliage to lay eggs and feed on plant juices exuding from the wound. Larvae feed on the chlorophyllous tissue between the epidermal layers. The feeding damage produces plainly discernible pale, serpentine mines on the upper surface of the leaflet. Heavy attacks can cause leaflets to drop.

Management. Heavy attacks by leafminers, which are more common in West Asia, can result in significant defoliation and yield loss. In South Asia, leaf miners generally do not cause any economic impact and hence need no control.

26.1.6. Semilooper, *Autographa nigrisigna* (Walker) (Lepidoptera: Noctuidae)

Several semiloopers have been reported to feed on chickpea in Asia, but the most important, appears to be *Autographa nigrisigna.* Larvae feed on leaflets and pods. Semiloopers produce ragged, irregular damage to the chickpea pod walls in contrast to the neat, round hole that *H. armigera* chews out.

Management. This species is only occasionally a pest in chickpea. The general management principles and tactics described for *H. armigera* may be effectively applied for *A. nigrisigna.*

26.2. INSECT PEST PROBLEMS IN LENTIL

Lentil, *Lens culinaris* Medikus is one of the important *rabi* pulse crop grown in northern and central parts of India including MP, UP, West

Bengal, Orissa and Punjab. About 36% world coverage of lentil (1.15 m ha) is in India. Lentil is also an important pulse crop grown in southern Europe, North Africa, Asia (Turkey, Syria, Lebanon, Jordan, and India), the former USSR, and North America. Poor crop management and abiotic and biotic stresses reduce grain yields. Among the biotic constraints, insect pests play a major role in yield reduction.

About three-dozen insect pests have been reported to infest lentil under field and storage conditions, out of which 21 species have been reported from India. However, only a few of these are economically important, and require control measures. The important insect pests feeding on lentil under field conditions include:

i. Lentil pod borer (*Etiella zinckenella* Treit.)

ii. Black aphid (*Aphis craccivora* Koch)

iii. Thrips (*Megalurothrips distalis*) (Karny)

iv. Cut worm (*Agrotis ipsilon*) (Hufnagel)

However, under Punjab conditions, the most important insect pests attacking lentil include lentil pod borer or lima bean pod borer, *Etiella zinckenella* and *Aphis craccivora.*

26.2.1. Lentil pod borer, *Etiella zinckenella*

Lentil pod borer, *Etiella zinckenella* (Lepidoptera: Pyralidae) is an important insect pest of several pulse crops, including lentil. It is commonly known as lima bean pod borer, spiny pod borer, or pea pod borer. The lentil pod borer is widely distributed in Asia, Africa, and Europe. It feeds on several leguminous species, especially lima bean, cowpea, pigeonpea, lentil, horse gram, green gram, field pea, and sunhemp.

The presence of a hole on the pod surface, dry light-colored frass, and webbing in the pod are indications of infestation by the lima bean pod borer, *E. zinckenella*. As a result of insect damage, the pods are poorly developed. Individual seeds have holes and internal portions are gutted out. The pods are partially or completely consumed inside. Externally, the pods give shrunken appearance, with small surface punctures. Larvae generally feed on maturing pods. Lentil pod borer population builds up by the end of the season, when the temperature is high. The caterpillar is greenish or pinkish red, with a yellow head. It moves violently when disturbed. The adults are small, brown, and active at night. Its infestation can be detected by the presence of small punctures on the surface of the pods, and the larvae can be observed by splitting the pods.

Management

- Host plant resistance can play a major role in reducing the losses due to *E. zinckenella* in lentil. Chhabra and Kooner (1980) screened over 800 cultivars of lentil from 25 countries and as many as 84 were found promising. Short-duration genotypes are more susceptible than the medium- and long-duration genotypes. The line LH 90-39 is resistant, while LL 147 is tolerant to *E. zinckenella* damage.

- Several natural enemies have been reported to control *E. zinckenella*. These include *Bracon etiellae, B. pectoralis, Phanerotoma planifrons, Pigeria piger, P. hendecasisella, Exorista roborator*, and *Tetrastichus* spp.
- It can be controlled by spraying the crop with 2.25 kg Sevin/ Hexavin 50 WP (carbaryl) in 200 litres of water per hectare at the flower initiation stage. Repeat the spray after three weeks, if necessary. Sprays of methidathion (0.5 kg ai ha^{-1}), deltamethrin (38 g ai ha^{-1}), or endosulfan (6 ml L^{-1}) at flowering and early pod setting have been reported to provide effective control of this pest. Abamectin 1.8% EC (1 ml 20 L^{-1}), α-cyper-methrin 10% (1.5 ml 2 L^{-1}), and cyhalothrin 2.5% (2 ml 2 L^{-1}) have also been found to provide effective control.

26.3. INSECT PEST PROBLEMS IN FIELD PEAS

Field pea, *Pisum sativum* L. is an important *rabi* pulse crop grown in Punjab (mainly in Hoshiarpur district under rainfed conditions). Field peas are cultivated around the world in countries such as Ethiopia, China, Canada and Russia and historically have been an important cash crop in certain regions of the United States like the Midwest. However, field peas are attacked by a number of insect pests including:

1. Pea leaf miner (*Chromatomyia horticola*) (Goureua)
2. Pea pod borers (*Etiella zinckenella* Treit. and *Helicoverpa armigera* (Hub.) and pea blue butterfly, *Lampides boeticus* (Linn.)
3. Pea stem fly (*Ophiomyia phaseoli* Tryor)
4. Pea aphid (*Acyrthosiphon pisum*) (Harris)
5. Pea thrips (*Thrips indicus*)

26.3.1. Pea leaf miner, *Chromatomyia horticola* (Goureua)

Leaf miner occurs throughout the world (Russia, UK, Holland, Italy, India, New Zealand and North America). In India, damage due to this insect is moderate to serious in the states of UP, MP, Bihar, Orissa, Punjab, West Bengal and Delhi. It is a major pest of peas. Besides this, its hosts are cabbage, cauliflower, berseem, lentil, mustard, brinjal, carrot etc.

Larvae are the damaging stage of the insect. They make serpentine mines in leaves of the host plants. Larvae feed below the epidermis, which results in the construction of a linear leaf-mine on the dorsal or ventral surface of the leaf. Usually, the mine starts from the periphery and ends towards the midrib. As the larva grows in size, the mine widens and becomes distinct, being marked by dark excretal pellets. Generally the channels are much larger when they are in the underside of the leaf, than when they are on the upper side in the same food plant. The direct effect of tunneling is the loss of chlorophyll bearing tissues. Adults with their sucking and sponging type mouth parts feed on the cell sap. Normally, the female makes triangular punctures on the leaf with the help of ovipositor and then feeds on the exuding sap. The male is unable to make such puncture and sucks the sap from the punctures made by the female. Flowering and fruiting of the infected crop is much reduced. The adult also feeds on the pollen of the flowers.

Management. Leaf miner can be controlled by spraying one litre of Rogor 30 EC (dimethoate) in 200-250 litres of water per hectare when attack starts. Repeat the spray after 15 days, if necessary.

26.3.2. Pea pod borers

***Etiella zinckenella* Treit.** It is a serious pest of green peas, field peas and lentil in the northern India and is also found on a variety of other pulses in various states of the country. The newly emerged tiny greenish caterpillars feed on the floral parts and subsequently, they bore into the pods to feed on the seeds.

Management. The pest may be controlled by spraying 750 ml of Thiodan 35 EC (endosulfan) or 2.25 Kg of Sevin/Hexavin 50 WP (carbaryl) in 200-250 litres of water per hectare. Repeat the application after 15 days, if necessary.

***Helicoverpa armigera* (Hub.).** Larvae feed on the tender leaves, floral parts, penetrates the pods and usually destroys one or two seeds in each pod. One larva attacks 7-10 pods before reaching pupation. Under severe attack, most pods on the plant have a number of injury holes.

Management. Spray 2.0 litres Durmet 20 EC (chlorpyriphos) or 800g Asataf 75 SP (acephate) in 100 litres of water per acre. Repeat the application after 15 days, if necessary. Pluck the harvestable green pods before spray and observe a waiting period of 7 days after spray for re-plucking the pods.

Pea blue butterfly (*Lampides boeticus*) (Linn.). It has been reported from the states of UP, MP, Punjab, Haryana, Andhra Pradesh, Tamil Nadu, Maharashtra and Gujarat. The pest causes serious damage to green peas and field peas. Besides, it damages mungbean, urdbean, ricebean, cowpea and pigeonpea. The larvae which are not very active, start feeding on tender leaves and then bore into the buds, flowers, green pods and feed inside the pod on the developing grains.

Management. Same as Lentil pod borer

26.3.3. Pea stemfly, *Ophiomyia phaseoli* Tryor

The maggot is creamy in color and damages the crop by tunneling through the stem of young seedlings and later on, the growing points. As a result of severe infestation, the leaves turn yellow, giving the plants a dry appearance. The stem turns brown, becomes swollen and breaks down. The attacked plants bear less number of pods, which are mostly empty or else their seeds may be very small. It causes serious damage in the early sown crop at the seedling stage.

Management. Sow the crop in the second fortnight of October to escape the damage of the pest. Avoid sowing of the crop earlier than mid-October to check its attack.

Pea aphid (*Acyrthosiphon pisum*) (Harris). It is widely distributed throughout the country. The main host plants include peas, clovers, vetch, Lucerne and beans. The growing part of the plant is most severely infested and development is retarded. It is especially serious on peas and can cause complete loss of the crop. Plants become stunted and yellow. Leaves and pods are malformed and grains are undersized.

Management. Spray 1 litre of Rogor 30 EC (dimethoate) in 200-250 litres of water per hectare when the attack starts and repeat after 15 days, if necessary.

Pea thrips (*Thrips indicus*). It causes severe damage to the young crop by sucking the cell sap. It can be managed by spraying one litre of Rogor 30 EC (dimethoate) in 200-250 litres of water per hectare and repeat the spray after 15 days, if necessary.

26.4. CONCLUSIONS

Based on the fact that *rabi* pulses are mostly grown under rainfed cropping system as a part of subsistence farming by a majority of small and medium farmers, the integrated pest management system with low-input technology seems to be the answer for handling insect pest problems in the near future. In overall context, the presently available plant protection technology can be suitably put to use for evolving a system of managing insect pests in *rabi* pulses. However, the pest management system in *rabi* pulses needs to be evaluated in the light of general socio-economic conditions of the farmers, if it is to be of any practical benefit. Effective implementation of IPM is a gradual and time taking process and needs concerted efforts on the part of IPM specialists, policy makers, government agencies as well as the farmers. Consequently, future research in various methods of insect pest control will emphasize on their mutual compatibility.

REFERENCES

Anonymous (2010). *Annual Report 2009-10*. Indian Institute of Pulses Research, Kanpur.

Chhabra, K.S. and Kooner, B.S. (1980). Field resistance in some lentil cultivars against pod borer, *Etiella zinckinella* Treit. *LENS Newsletter* **7**: 46-49.

Chhabra, K.S. and Kooner, B.S. (1990). *Rabi Pulses Research Progress Report, 1976-77 to 1989-90*. Department of Plant Breeding, Punjab Agricultural University, Ludhiana.

Chhabra, K.S., Kokner, B.S., Dhingra, K.K. and Tripathi, H.P. (1978). Preliminary studies on the influence of sowing dates on the incidence of gram pod borer, *Heliothis armigera* (Hub). *Proceedings of All India Rabi Pulses Workshop,* Bhubaneshwar, 3-6 October, 1978.

Das, D.K., Trivedi, T.P. and Sricastava, C.P. (2001). Simple rule to predict attack of *Helicoverpa armigera* on crops growing in Andhra Pradesh. *Indian J. Agric. Sci.* **71**: 421-423.

Dua, R.P., Gowda, C.L.L., Kumar, S, Saxena, K.B., Govil J.N., Singh B.B., Singh A.K., Singh, R.P., Singh, V.P. and Kranthi, S. (2005). Breeding for resistance to *Heliothis/Helicoverpa*: effectiveness and limitations. *In*: Sharma HC (ed.) (2005). *Heliothis/Helicoverpa Management: Emerging Trends and Strategies for Future Research*. Oxford and IBH, New Delhi, India, pp. 223-242.

Gowda, C.L.L. (2005). *Helicoverpa-* the global problem. *In*: Sharma HC (ed.) (2005). *Heliothis/Helicoverpa Management: Emerging Trends and Strategies for Future Research*. Oxford and IBH, New Delhi, India, pp. 1-6.

Gowda, C.L.L., Lateef, S.S., Smithson, J.B. and Reed, W. (1983). Breeding for resistance to *Heliothis armigera* in chickpea. *In*: *Proceedings of the National Seminar on Breeding Crop Plants for Resistance to Pests and Diseases*, 25-27 May, 1983. School of Genetics, Tamil Nadu Agricultural University, Coimbatore, Tamil Nadu, India, pp. 36-39.

ICRISAT (International Crops Research Institute for the Semi-Arid Tropics) (1992). *Mid Term Plan*. ICRISAT, Hydrabad, India.

Jalali, S.K., Singh, S.P., Kumar, P. and Ballal, C.R. (1988). Influence of the food plants on the degree of parasitism of larvae of *Heliothis armigera* by *Cotesia kazak*. *Entomophaga* **33**: 65-71.

Kar, S., Basu, D., Das, S., Ramkrishnan, N.A., Mukherjee, P., Nayak, P. and Sen, S.K. (1997). Expression of *cry1A(c)* gene of *Bacillus thuringiensis* in transgenic chickpea plants inhibits development of pod borer (*Heliothis armigera*) larvae. *Transgenic Res.* **6**: 177-185.

Lateef, S.S. (1985). Gram pod borer *Heliothis armigera* (Hub.) resistance in chickpea. *Agril. Ecosys. Environ.* 14: 95-102.

Lateef, S.S. and Pimbert M.P. (1990). The search for host plant resistance of *Helicoverpa armigera* in chickpea and pigeonpea at ICRISAT. *In*: *Proceedings of the Consultative Group Meeting on the Host Selection Behavior of Helicoverpa armigera,* 5-7 March, 1990. ICRISAT, Hydrabad, India, pp. 185-192.

Lawlor, H.J., Siddique, K.H.M., Sedgley, R.H. and Thurling, N. (1998). Improving cold tolerance and insect resistance in chickpea and the use of AFLPs for the identification of molecular markers for these traits. *Acta Hort.* **461**: 185-192.

Lucy, M. and Slatter, J. (2004). Chickpeas. *In*: Lucy M, French V and Bullen K (eds). *Crop Management Notes*, Winter Edition. Queensland Department of Primary Industries, Brisbane, Queensland, Australia.

Matthews, M. (1999). *Heliothine Moths of Australia. A Guide to Pest Bollworms and Related Noctuid Groups.* Monograph on Australian Lepidoptera, Vol. 7. CSIRO Publishing, Callingford, Victoria, Australia.

Pawar, V.M., Aleemuddin, Mohd. and Bhosle, B.B. (1987). Bioefficacy of *Ha*NPV in comparison with endosulfan against pod borer in chickpea. *Int. Chickpea Newsl.* 16: 4-5.

Ranga Rao, G.V., Chari, M.S., Pawar, C.S., Sharma, O.P. and Rameshwar Rao, V. (2005). *Helicoverpa* Management: Successes and Failures-Lessons for the Future. *In*: Sharma HC (ed.) (2005). *Heliothis/Helicoverpa Management: Emerging Trends and Strategies for Future Research.* Oxford and IBH, New Delhi, India, pp. 431-452.

Ravi, G. and Verma, S. (1997). Persistence and dissipation of insecticides against *Heliothis armigera* on chickpea. *Indian J. Ent.* **59**: 62-68.

Reed, W., Cardona, C., Sithanantham, S. and Lateef, S.S. (1987). The chickpea insect pests and their control. *In*: Saxena MC and Singh KB (eds). *The Chickpea.* CAB International, Wallington, UK, pp. 283-318.

Rembold, H. (1981). Malic acid in chickpea exudates- a marker for *Heliothis* resistance. *Int. Chickpea Newsl.* **40**: 18-19.

Romeis, J., Shanower, T.G. and Zebitz, C.P.W. (1999). Why *Trichogramma* (Hymenoptera: Trichogrammatidae) egg parasitoids of *Helicoverpa armigera* (Lepidoptera: Noctuidae) fail on chickpea. *Bull. Ent. Res.* **89**: 89-95.

Saxena, H. and Ahmad, R. (1997). Field evaluation of *Beauveria bassiana* (Balsamo) Vuillemin against *Helicoverpa armigera* (Hubner) infecting chickpea. *J. Biol. Cont.* **11**: 93-96.

Sharma, H.C. (2001). Cotton Bollworm/Legume Pod Borer, Helicoverpa armigera (Hubner) (Noctuidae: Lepidoptera): Biology and Management. Crop Protection Compendium, Wallington, UK: CAB International, pp. 72.

Sharma, H.C. (2001). Crop Protection Compendium: *Helicoverpa armigera. Electronic Compendium for Crop Protection.* CAB International, Wallingford, UK.

Sharma, H.C. (ed.) (2005). *Heliothis/Helicoverpa Management: Emerging Trends and Strategies for Future Research.* Oxford and IBH, New Delhi, India.

Sharma, H.C., Gowda, C.L.L., Stevenson, P.C., Ridsdill-Smith, T.J., Clement, S.L., Ranga Rao, G.V., Romeis, J., Miles, M. and Bouhssini, M.E.L. (2007). Host plant resistance and insect pest management in chickpea. *In*: Yadav SS, Redden R, Cen W and Sharma B (eds). *Chickpea Breeding and Management.* CAB International, Wallington, UK, pp. 638.

Sharma H.C., Pampapathy G. and Ridsdill-Smith T.J. (2005). Exploitation of wild *Cicer reticulatum* germplasm for resistance to *Helicoverpa armigera. J. econ. Ent.* **98**: 2246-2253.

Srivastava K.M. and Singh R.P. (1979). Forecasting of yield losses and infestation level of pod borer, *Heliothis armigera* Hub. in chickpea. *Indian J. Ent.* **41**: 375-378.

Ujagir R., Chaubey A.K., Sehgal V.K., Saini G.C. and Singh J.P. (1997). Evaluation of some insecticides against *Helicoverpa armigera* (Hubner) on chickpea at Badaun, Uttar Pradesh, India. *Int. Chickpea Pigeonpea Newsl.* **4**: 22-24.

Yadava C.O., Lal S.S. and Sachan J.N. (1989). Effect of mixed/inter cropping on gram pod borer damaging chickpea. *In*: *National Symposium on New Frontiers in Pulses Research and Development*, Nov. 10-12, 1989, ISPRD, Directorate of Pulses Research, Kanpur.

Theory and Practice of Integrated Pest Management
R. Arora, B. Singh and A.K. Dhawan (Eds)
Scientific Publishers (India), Jodhpur 2011, pp. 473 - 500.

27

INTEGRATED PEST MANAGEMENT IN *KHARIF* PULSES

Ravinder Singh, G.K. Taggar and H.K. Cheema
Department of Plant Breeding and Genetics
Punjab Agricultural University, Ludhiana - 141 004, India

Pulses occupy an unique position in Indian economy and society. In India and other developing countries people get 80% of food protein form plant sources, mainly pulses. Yield of pulses vary considerably among locations, cultivars, seasons and cropping periods. The yield of pulses remained low against the proposed yields of 1.5 to 3.5 t/ha. The main reasons for the low yields of pulses are various biotic and abiotic conditions. Among biotic factors insect pests are the most important yield reducing factor.

The *kharif* pulse crops mainly mungbean, urdbean, soybean, pigeon-pea and ricebean are attacked by a wide range of insect pests both in the fields and storage. The insect pests which are economically important and attack the reproductive stages are the pod borers. The development and use of varieties resistant to insect pest has a long way to minimize crop losses in the farmers' fields. Although a number of plant protection schedules, based on the use of insecticides and microbial have been recommended but management practices employed to contain these at this stage rely mostly on insecticides or a combination of different groups of insecticides. This has led to the problems of residues, resurgence and resistance in insects. There is an urgent need that an efficient integrated pest management module based on the judicious use of insecticides, cultural control and other methods of control and appropriate cropping systems is developed for wider adoption of these pulses in the country.

27.1. LOSSES CAUSED BY KEY INSECT PESTS

On an average of 30-80% crop losses occur in pulses due to ravages of insect pests valued at 4000-5000 crores (Asthana *et al.*, 1997). In case of pigeonpea, pod borer, *Helicoverpa armigera* (Hubner) has been reported to cause yield losses varying from 40 to 50% in major pigeonpea growing states. The average loss due to insect pests in mungbean and urdbean crops was estimated to be 34.7 and 28.7% in different states of India. Lal and Ahmad (2002) reported nearly 300 insect species on pigeonpea, 60 on mungbean and urdbean. Among these, 34 insects were serious pests on one or more of these crops. Kooner *et al.* (2006) had reviewed the various insect pests of *rabi* and *kharif* pulses in India.

27.2. WHITEFLY *Bemisia tabaci* (Gennadius) (Hemiptera: Aleyrodidae)

It occurs throughout the northern and western regions of India. It is a serious pest of cotton, mung and mash. It also feeds on new crops like ricebean, soybean, arhar, cabbage, cauliflower, rapeseed and mustard, melon, potato, brinjal, okra, ornamental plants and some weeds, etc.

Nature of damage. It is a sucking insect pest and generally causes damage by sucking the sap from leaves and tender parts of the host plants. In case of grain legumes like mung, mash and soybean, it damages the plants in three ways: (i) sucking of cell sap from leaves, thus lowering the vitality of the plants. Under high population pressure of the pest, the plant becomes very weak, giving a sickly look and may die. (ii) The insect excretes honey dew, on which growth of sooty mould takes place which result in blackening of leaves, resulting in decrease in photosynthesis. In severe attacks there is total blackening of the crop, drying of leaves and total failure of the crop. (iii) The whitefly is a vector for a number of viral diseases especially mungbean yellow mosaic virus (MYMV).

Biology. The insect breeds throughout the year and all the developmental stages are noticed. During winter season generally the adults are predominant. The female lay eggs singly on lower side of the leaves, averaging 120 eggs per female. The eggs are light yellow first, turning brown later on. The eggs hatch in 3-5, 5-17 and 33 days in April-September, October-November and December-January, respectively. The nymphs soon after emergence fix their mouth parts into the plant tissues and feed on the cell sap. They grow into three stages to form the pupae within a period of 9-14 and 17-81 days in April and October-March, respectively. The pupal period generally lasts for 2-8 days. The total life cycle is completed in 14-122 days and there are 11-12 generations of the insect in a year.

Management. Chhabra *et al.* (1993) reported that whitefly can be controlled by spraying the mung bean, urd bean, soybean and rice bean with any of the following insecticides using 250 litres of water per hectare with manually operated sprayer: 950 ml of Malathion 50 EC (malathion), 625 ml of Rogor 30 EC (dimethoate) or Anthio 25 EC (formothion) or Metasystox 25 EC (oxydemeton methyl), and 225 ml of Dimecron 85 WSC (phosphamidon). Disulfoton at the rate of 1.5 kg ha^{-1} as basal application at sowing gave significant reduction of population of whitefly (Chaudhary *et al.,* 1981). Combination of phorate l0 G and monocrotophos (Vadodaria and Vyas, 1987), endosulfan, monocrotophos, chlorpyriphos and dimethoate (Yein, 1983) were most effective for the control of the pest. Cheema *et al.* (2009) reported that pest can be controlled by spraying the crop with 1.25 litre triazophos 40 EC or 100 g thiamethoxam 25 WG /ha by using 250 litre water. Shirale and Bidgire (2009) reported that spraying triazophos 40 EC @ 800 ml/ha twice, recorded less number of whiteflies at three, seven and ten days after spraying. The treatment *B. bassiana* @ 1 l/ha followed by Triazophos 40 EC @ 800 ml/ha was at par with treatment triazophos 40 EC @ 800 ml/ha followed by *B. bassiana* @ 1 l/ha at three days after spraying. Nene (1973) reported nuclear polyhedrosis virus (NPV) and a fungal pathogen, *Poecilomyces fairnosus,* as biocontrol agents of adult whitefly. Lal (1987) reported *Prospaltella flava* and *Geocaris tricolor* parasitising nymphal, pupal and immature stages. He further reported *Brumus* sp. as a predator of immature stages of the pest.

Many genotypes of mung bean and urd bean were identified resistant to whitefly and jassids by Kooner (1998), Kooner and Cheema (2007a, b), where genotypes ML 508, ML 537, ML 1265, ML 1229, ML 1260, ML 1256 and ML 1191 of mungbean and were identified as resistant. Chhabra *et al.* (1981) while studying the mechanism of resistance in mung bean and black gram against whitefly investigated that bio-chemicals like phenols, amino acids and non-reducing sugars were responsible for resistance to the insect. Kooner *et al.* (1994) studied the phytochemical basis of resistance in urd bean and concluded that high contents of phenols, free amino acids, total sugars and low content of non-reducing sugars governed the resistance.

Ravindrababu (1987) reported that maize, sorghum and pearl millet when grown as barrier crops for mung bean, significantly reduced the population of *B. tabaci* and incidence of MYMV and resulted in higher yield. Intercropping rice bean with sorghum had a suppressing effect on the incidence of whitefly on rice bean (Satyanarayana *et al.,* 1997).

27.3. GREEN JASSID, *Empoasca* spp. (Hemiptera : Cicadellidae)

It is widely distributed in India and is a destructive pest of mung. Besides, it also feeds on mash, ricebean, soybean, arhar, cotton, okra, potato, brinjal and some wild plants.

Nature of damage. Damage to the crop is caused by the adults as well as the nymphs, both being very agile and move briskly forward and side ways. Injury to plants is due to the loss of sap and probably due to the injection of toxins. The attacked leaves turn pale, the rust-red and ultimately cup downwards, dry up and fall.

Biology. The pest breeds practically throughout the year except during winter. Adults are about 3 mm long and greenish yellow during the summer, acquiring a reddish tinge in the winter. The winged adults jump or fly away at the slightest disturbance and are attracted to light at night. The female lays about 15 yellowish eggs on the underside of the leaf, embedding them into the leaf veins. The eggs hatch in 4-11 days and give rise to nymphs. They suck cell-sap from the underside of the leaves and pass through six stages of growth in 7-21 days. On transformation into winged adults, they live for 5-7 weeks, feeding constantly on the plant juice. The pest complete seven generations in a year.

Management. Jassid can be controlled by spraying any of the following insecticides, using 200-250 litre of water per hectare with manually operated knap sack sprayer.

> 937.5 ml Malathion 50 EC (malathion), 625 ml Rogor 30 EC (dimethoate), 625 ml of Metasystox 25 EC (oxydemeton methyl)

Nayar *et al.* (1976) reported *Chrysopa symbela* and *Camponotus* sp. predating upon nymphs of hoppers. Dhuri and Singh (1983) reported the predation of nymphs of *E. kerri* by *Acantholepis simplex* and *Coccinella septumpunctata.*

The resistant genotypes possess a higher percentage of reducing and non-reducing sugars, total phenols and free amino acids as compared to the check varieties. Similar results have been reported on urd bean (Kooner *et al.,* 1994). Kooner (1998) screened 48 genotypes of mung bean and isolated ML 508 and ML 537 possessing resistance against *Empoasca* spp.

Singh and Singh (1977) have reported that intercropping of mung bean with non-host plants like sorghum and pearl millet decreased incidence of jassid at Delhi. Mung bean variety Co 4 intercropped with MCU 10 variety of cotton recorded a minimum incidence of leaf hopper (Janagarajan, 1987).

27.4. TOBACCO CATERPILLAR, *Spodoptera litura* Fabricius (Lepidoptera : Noctcidae)

Tobacco caterpillar is found throughout the tropical and sub-tropical parts of the world. It is wide spread in India and feeds on more than 115 species of cultivated plants. The important host plants include tobacco, castor, groundnut, tomato, sunflower, cabbage, mung, mash, arhar, gram, cowpea, linseed, cotton, sugarcane, okra, onion, chillies and soybean.

Nature of Damage. The damage is done by larval stage and is often serious. They feed on the foliage gregariously in their early stages while disperse in their later stages. These are voracious feeders of the foliage, remaining hidden under soil during the day time. The maximum damage is caused to the young plants, which in severe cases of attack are totally destroyed. Large areas of the leaves of the grown up plants may also be eaten up by the caterpillar. Besides leaves they also damage buds, flowers, pods and grains inside pods.

Biology. The pest breeds throughout the year, although its development is retarded during winter. The moths are active at night when they mate and the female lays about 300 eggs in clusters on lower side of leaves. The clusters are covered over by brown hair and they hatch in about 3-5 days. The small larvae are black whereas grown up larvae are dark green with black triangular spot on their body. The larvae measure 35-40 mm in length at maturity. They feed gregariously for the first few days and then disperse to feed individually. They pass through 6 stages and are full fed in 15-30 days. The full-grown larvae enter the soil where they pupate. The pupal stage lasts for 7-15 days. The adult is a stout built moth with the forewings greyish-brown, bearing white markings and the hind wings are white. The moths are about 22 mm long and measure 40 mm across the spread wings. The forewings have beautiful golden and greyish-brown patterns. The life cycle is competed in 32-60 days and the pest completes eight generations in a year.

Control measures. The pest can be controlled by spraying one litre of Malathion 50 EC or 925 ml of Thiodan 35 EC (endosulfan) ha^{-1}. Foliar application of diflubenzuron chitin synthesis inhibitor belonging to the acyl urea group has shown excellent residual activity against eggs of this pest and has been used commercially for its control. It can be controlled by spraying 375 ml novaluron 10 EC, 2.5 kg acephate 75 SP and 3.75 litre chlorpyriphos 20 EC using 200-250 litres of water per hectare with manually operated knap sack sprayer (Annonymous, 2009). The plant extracts of custard apple act as feeding deterrent against *S. litura*. However, it has been reported that this polyphagous pest has become resistant to nearly all the available groups of pesticides in India (Atwal and Dhaliwal, 1997).

The eggs are parasitised by *Trichogramma australicum*, larvae by *Chelonus formosanus, Campoletis* sp. and pupae by *Blepherella setigera* and *Sarcophaga dux*. The larvae are predated upon by the pentatomids *Canthecona furcellata* and *Harpactor costalis* and infected by *Nomuraea rileyi*. Larva of the elaterid *Agrypnus fuscipes* and the adult of the carbid *Brascus punctatus* feed on the caterpillars (Nair, 1986; Atwal and Dhaliwal, 1997).

Atwal and Dhaliwal (1997) reported that hand picking and mechanical destruction of larvae during early stage of attack can reduce infestation. They further reported that application of potash and sometimes phosphorus either singly or in combination results in lower incidence of *S. litura* on cowpea.

27.5. BIHAR HAIRY CATERPILLAR, *Spilosoma obliqua* Walker (Lepidoptera: Arctiidae)

It is a sporadic pest and is very serious in Bihar, Madhya Pradesh, Uttar Pradesh and Punjab as a polyphagus pest, particularly on mung, mash, sesame, linseed, mustard and some vegetables.

Nature of damage. Damage is caused by the caterpillars. When young, they feed gregariously in scattered spots and eat the green matter of leaves, leaving behind the mid-rib. The mesh like leaves give whitish appearance and the infested plants can be spotted from a distance. The grown up caterpillars feed voraciously on leaves, soft stems and branches and totally denude the crop in a few days (Sinha *et al.,* 1985; Chhabra and Kooner, 1987).

Biology. The pest breeds from March to April and again from July to November. It passes the summer (May to June) and the winter (December to February) in the pupal stage in plant debris. The adults emerge from the over wintering pupae in March. The moths are nocturnal in habit. The female lays 400-1200 light green, spherical eggs in clusters on the underside of the leaves. The eggs hatch in 8-13 days and caterpillars grow to maturity through seven stages, within 4-8 weeks. Body of the larvae is covered with numerous long hairs and when full grown (4 to 4.5 cm), the larva spins a loose silken cocoon. The pupal stage lasts 1-2 weeks in the active period and moths live for about a week. The life cycle is completed in 6-12 weeks and the pest passes through 3 to 4 broods in a year (Chhabra *et al.,* 1993).

Management. Chhabra and Kooner (1987) reported that the insect can be controlled by spraying 1.25 litres of Thiodan 35 EC (endosulfan) or Ekalux 25 EC (quinalphos) or 500 ml of Nuvan 100 (dichlorvos) or 375 ml of Fenval 20 EC (fenvalerate) in 250 litres of water ha^{-1} Bhattacharya and Pramanik (2002) reported that azadirachtin gave higher rate of larval mortality than other botanicals & imidacloprid and cartap hydrochloride.

Biswas *et al.* (1994) reported that *Bacillus thuringiensis* formulation Dipel was effective in controlling this pest.

Young larvae are gregarious and are on only few plants in a field. They can be destroyed by pulling out the infested plants and burrying them in the ground. The grown up caterpillars can be destroyed by crushing them under feet or by picking and putting them into kerosenised water.

27.6. GREEN SEMILOOPER, *Anomis flava* (Fabricius) (Lepidoptera: Noctuiidae)

It is a sporadic pest of mung. It also attacks mash, ricebean, soybean, arhar, cowpea, cotton, okra and many other plants.

Nature of damage. Damage is done by caterpillar which eats the entire leaf tissue leaving only the veins. Some times they also damage the tender shoots and buds resulting in complete defoliation. It occurs in an epidemic form in years of heavy rainfall.

Biology. The moth is medium-sized (about 2 cm) with light yellow to orange coloured body. They are active fliers, nocturnal and are attracted to light. Their life span is 8-12 days. A single female moth can lay up to 600 eggs in its life time. Eggs are laid singly on upper surface of leaves and may be scattered at irregular distances. Freshly laid eggs are leaf-green and later turn pale-green. Eggs hatch in 3-4 days. The full grown larva is 3 to 3.5 cm and pale-yellowish green with five longitudinal white lines on the dorsal surface. The larval stage lasts for 15-20 days. Pupation occurs in folded leaf and in a loosely spun cocoon on the plant or in the soil debris. Pupal period lasts for 9 days. Total life cycle is completed in 4-6 weeks. It overwinters in pupal form.

Management. The pest can be controlled by spraying 1250 ml of Thiodan 35 EC (endosulfan) or 500 ml of Nuvan 100 (dichlorvos) in 200 litres of water per hectare. Spray should be done as soon as the pest appears. One spray is sufficient if the pest is controlled at the initial stage (Annonymous, 2010a).

Bacillus thuringiensis and *Baculovirus* microbial pesticides produced 85 to 90 per cent control (Chen *et al.,* 1991). *Apanteles anomidis* and *Rhynocoris kumarii* are reported to be important natural enemies (Xiong *et al.,* 1994; Ambrose, 2000).

27.7. BEAN MITE, *Polyphagotarsonemus latus* (Banks) (Acarina: Tarsonemidae)

The bean mite (*Polyphagotarsonemus latus*) has assumed a pest status on a number of crops and fruit trees throughout India. The pest has been recorded on mung bean, urd bean, common bean, horse gram, cowpea, brinjal, chilli, potato, cotton, sesame, citrus, hollyhock, marigold, tea, weeds and ornamental plants (Gupta and Dhooria, 1973; Dhooria, 1984; David, 1991; Singh, 1996).

Nature of damage. Both the adults and nymphs suck the cell-sap from the leaves. In case of serious infestation, it causes downward curling of leaves and necrosis of young leaves and flowers. Initially the infested leaves appear pale to light brown. Severely infested leaves first turn light brown to brown and then to dark reddish brown.

Biology. Adult mites are hyaline in colour, oval in shape and have four pairs of legs. The females are larger than the males and are broader at abdominal end. The eggs are laid on the ventral surface of the new leaves in the depressions and are firmly stick to the plant substrate. The eggs are oval, elongate, covered with whitish longitudinal rows of tubercles with flat base and are transparent shiny. Dhooria (1984) reported that Egg stage lasts for 1-3 days. Hatched out larvae have 3 pairs of legs and resemble the adults. The larvae are pear-shaped, milky-white with a dorsal white stripe in the middle region. Larval period ranges from 3 to 11 days. Under optimal conditions a complete life cycle can be completed in about one week and 20-30 generations may actually be raised under tropical and subtropical regions. Various climatic factors such as temperature, relative humidity and precipitation affect field populations of mite. Mite population is low during winter due to a combination of low temperature and heavy rains (Gerson, 1992).

Management. The mite can be controlled by spraying 375 ml of Rogor 30 EC (dimethoate) in 200-250 liters of water per hectare (Annonymous, 2010a).

About 11 natural enemies have been reported to predate on mite. *Amblyseius laryoensis, Typhlodromus annectens, T. porresi, Euseius victoriensis, T. rickeri* and *T. stipulatus* are some of the major phytoseid mites which help to control mite in different parts of the world on different hosts.

27.8. GRAM POD BORER, *Helicoverpa armigera* (Hubner) (Lepidoptera: Noctuidae)

Helicoverpa armigera is widely distributed throughout the country and recorded feeding on 181 cultivated and uncultivated plant species. On pulse crops it is more serious in the states of Karnataka, Maharashtra, Andhra Pradesh, Tamil Nadu, Orissa, Punjab, Gujarat and Uttar Pradesh. In Punjab, the most preferred hosts are chickpea, pigeonpea, mung bean, cotton, maize, cowpea, lentil, fieldpea, urd bean, soybean, sunflower, tomato, brinjal, *berseem*, and lucerne (Singh and Singh, 1975; Chhabra and Kooner, 1980).

Nature of damage. The young larvae feeds on tender leaves, buds, flowers and subsequently it bores into the pods and feeds on the seeds with its head and part of the body inside the pod and the rest of the body remaining outside. A single larva may destroy 30-40 pods before it reaches maturity. All the seeds within a pod are not damaged. Although *H. armigera* is the most important pest of pulses in India but it is the key pest of pigeonpea and chickpea. Reed and Pawar (1982) estimated the combined loss due to this pest on pigeonpea and chickpea in India to exceed US$ 300 million in some years. 30-40% in chickpea and 10-35% in pigeonpea in Haryana (Chauhan, 1992), 17-30% in pigeonpea and 45-75% in pigeonpea in Andhra Pradesh (Rao *et al.,* 1992). On the basis of survey, Chaudhary *et al.* (1982) reported average loss of 31.9% by pod borer in chickpea from different districts of Haryana. The loss in yield of chickpea by *H. armigera* has been reported to be 40-50%, 5.32% and 36.0% in Madhya Pradesh, Haryana and Punjab, respectively (Chaudhary and Sharma, 1982).

Biology. The adult is a medium-sized light brown moth with a 'V' shaped mark on the upper side and a kidney shaped mark on the underside of the fore-wings. Hind-wings are dull grey with a black border on the distal end. The female moth is bigger than the male with tuft of hair on the tip of the abdomen. The adult female lay eggs late in the evening mostly singly. Freshly laid eggs are dome shaped, yellowish-white and change to dark brown before hatching. Incubation period is 3-4 days. The newly hatched caterpillars are whitish-green. There are normally six larval instars. The full-grown larvae are pale green and 3.5 to 4.0 cm in length. However, the colour varies according to the food intake. Dorsal surface bears dark brown stripes. The larval period lasts from 18-25 days. Larva pupates in the soil in earthen cell. The pupal period lasts from 7-14 days.

Forecasting through traps. For monitoring adult population of *H. armigera* the pheromone trap and light trap may be used. One of the most important criterian for any monitoring system used for short-term pest forecasting is that the relationships between the trap catch and the corresponding field population must be consistent across time and space. Srivastava and Srivastava (1995) reported that the peaks in pheromone trap catches of moths were invariably followed by the peaks in egg and larval counts (early instars) of this insect in chickpea. Different workers worked on the monitoring of *H. armigera* moth population through pheromone traps or light traps. Chhabra and Kooner (1992) monitored the population of male moths through pheromone trap (1983-90) and reported that peak population coincided with the flowering/ podding stage of chickpea/pigeonpea. These informations can be effectively used for short-term pest forecasting purposes (Balasubramanian *et al.,* 1985; Sachan, 1987; Pawar *et al.,* 1988).

Management. In India, the chemical control is the only practical method so far, which is available to the farmer for effective control of *H. armigera* on pigeonpea.It has been reported that only small segment of farmers, i.e., up to 5-10% use insecticides on pigeonpea (Lal and Yadava 1987). There are some problems regarding application of insecticides. At flowering and podding stage, the crop attains a height of 2-3 m and becomes a dense *jungle* and spraying of such crop becomes a difficult task. In India, about 90% area under pigeonpea is either of medium- or long-duration varieties. As pigeonpea is generally grown in rainfed areas, as intercrop or in mixed cropping system, so arrangement of water for spraying is again a big problem. To solve this problem, varieties with short stature and compact nature like ICPL 87, ICPL 151 (early maturity group) and Bahar, C 11 (medium/long maturity group) should be grown. A pathway of 1.5-2 m wide after a strip of 10-12 rows of pigeonpea can be left, which can be utilised for growing suitable intercrops before flowering stage of pigeonpea; this will provide enough space for the movement of sprayers and operators (Sachan *et al.*1992; Srivastava and Sachan, 1998). In order to solve the problem of arrangement of water for spray, the use of ultra low volume (ULV) like controlled droplet applicator (CDA) has been recommended (Sachan, 1992). Many groups of chemicals, viz., organochlorines, organophosphates, carbamates and synthetic pyrethroids have been evaluated for the control of *H. armigera* on pigeonpea. Besides these, use of neem leaf extract and NSKE have also been tested and recommended against this pest (Table 27.1). Among the botanicals, Nimbecidine, Achook and Niconeem tested in combination with Sevin 50 WP against *H. armigera* on pigeonpea, Achook @ 1250 g ha^{-1} + Sevin 50 WP @ 1750 g ha^{-1} gave better control of the borer (Kooner and Malhi, 2001a).

The gram pod borer can also be controlled by spraying 2.5 litre Thiodan 35 EC (endosulfan) in 250-300 litres of water per hectare depending upon variety, with manually operated knapsack sprayer at the time of 50 per cent flowering. Repeat the spray after 10 days, if necessary (Annonymous, 2010b).

Campoletis chlorideae Uchida is the most important and most potential parasite of *H. armigera* larvae before they are large enough to damage the pods (Sachan and Bhaumik, 1998). The eggs of *H. armigera* collected from tomato, potato and lucerne were parasitised by *Trichogramma chilonis* up to 98.2%; however, no parasitism was recorded in eggs collected from chickpea. The acid exudate of chickpea plant was thought to be a deterrent to the parasitoid (Yadav, 1980; Yadav and Patel, 1981). Similarly 25.4% and 0.1% parasitism has been reported from the eggs collected from sorghum and pigeonpea, respectively. The almost negligible parasitism on pigeonpea and no parasitism on chickpea is considered to be one of the major factors in the high population built up of the pest on these crops. *Chrysopa* sp., a general predator, also predates upon eggs and small larvae of the pest (Remeis and Shanower, 1996).

Table 27.1. Insecticides commonly recommended against pod borer complex of pigeonpea

Insecticide	Formulation	Dose a.i. ha^{-1}	Reference
Organochlorines			
Endosulfan	35 EC	700	Upadhyay *et al.* (1998)
Endosulfan	2 & 4% dust	500-1000	Upadhyay *et al.* (1998)
Organophosphates			
Dimethoate	30 EC	300	Upadhyay *et al.* (1998)
Malathion	5% dust	1250	Chhabra and Kooner (1979)
Methyl parathion	50 EC	500	Upadhyay *et al.* (1998)
Methyl parathion	2% dust	500	Upadhyay *et al.* (1998)
Monocrotophos	40 EC	400	Upadhyay *et al.* (1998)
Monocrotophos	36 SL	400	Upadhyay *et al.* (1998)
Quinalphos	25 EC	500	Chhabra and Kooner (1979)
Phosphamidon	85 SL	500	Upadhyay *et al.* (1998)
Carbamates			
Carbaryl	50 WP	875	Upadhyay *et al.* (1998)
Carbaryl	10% dust	2500	Chhabra and Kooner (1980)
Carbaryl	50 WP	1000	Chhabra and Kooner (1980)
Cypermethrin	25 EC	100	Upadhyay *et al.* (1998)
Fenvalerate	20 EC	100	Upadhyay *et al.* (1998)
Fenvalerate	0.4% dust	100	Upadhyay *et al.* (1998)
Synthetic pyrethroids			
Delta methrin	2.8 EC	30	Upadhyay *et al.* (1998)

For the control of *H. armigera,* nuclear polyhedrosis virus (NPV), *Bacillus thuringiensis (Bt),* some fungi, nematodes and protozoa have been studied in some detail. However, from the commercial point of view, the NPV and Bt are currently considered the most potential pathogens for the management of this pest.

Genotypes ICPL 6, PDR 54-2, ICP1903 and MA 1 have been identified as promising in the All India Coordinated Pulses Improvement Project (AICPIP) in multi-location testing. Resistant genotypes ICPX 7730, ICPL 87088, ICPL 84060 and ICPL 87089 have been developed through crosses between resistant cultivars (Reed and Lateef, 1990;

Singh *et al.,* 1990). Genotypes AL 1340 and AL 587 (Kooner *et al.,* 2002) AL 1122 (Malhi *et al.,* 2001) and AL 323 and AL 607 (Kooner and Malhi, 2001b) have been reported to be tolerant to *H. armigera.* Recently, genotypes AL 1340, AL 1498, AL 1502, AL 1495, AL 1488 and AL 1493 were also found promising against the pod borer complex (Kooner and Cheema, 2006; Kooner *et al.*, 2008).

Field sanitation, soil tillage and hand picking are the other cultural methods, which provide preventive and limited control of borer in pigeonpea. Removal of alternate hosts and avoiding ratooning are useful in reduction of damage. Timely and proper tillage of soil may help in exposure of hibernating pupae to natural enemies. Hand picking of larvae could prevent damage in a limited way. Though cultural control alone is rarely sufficient to maintain pest populations below the economic threshold level (ETL), however, it is the most logical initial step in pest management on which other control measures can be structured within the frame work of IPM.

Sharma and Choudhary (1984) reported that early sown (16 June) crop had less pod damage (7.9%) as compared to late sown (15 July; 18.6%) at Hisar. Similarly early sown crop had shown less pod infestation than late planting at other places also (Parsad *et al.,* 1986; Kushwaha and Malik, 1987)

Pigeonpea is generally intercropped with sorghum, millet, cowpea, mung bean, urd bean, sesame and groundnut. Incidence of *H. armigera* was reduced when pigeonpea was intercropped with cowpea (Singh and Singh, 1978), groundnut, mung bean, urd bean and finger millet (Patnaik *et al.,*1989). Early maturing varieties of pigeonpea have been reported to suffer greater than late varieties in southern India (Rajagopalan and Devakumar, 1965) while in northern India, late varieties have been shown to exhibit more damage (Yadava *et al.,* 1988).

In Punjab, it has been observed that the determinate type of varieties of pigeonpea is more susceptible to *H. armigera* than the indeterminate varieties.

IPM in pigeonpea. Different researchers have listed different possibilities to manage *H. armigera* on pigeonpea. The various components of IPM are listed as following:

i. Prefer cultivation of varieties which may escape attack of pod borer. Variety Bahar in and around Kanpur matures early and escapes incidence of the pest. Similarly in southern India tolerant cultivars like ICPL 332 (Abhaya) should be preferred.

ii. Crops like sorghum, maize, millet, etc. should be intercropped in medium/late maturing pigeonpea because these crops are harvested early and thus provide the required gangway for undertaking spray in effective manner to control the pest.

iii. Avoid cultivation of varieties of pigeonpea with different maturity groups in the same area.

iv. Collection and destruction of fully grown larvae in smaller plots

v. To monitor population of male moths use sex pheromone traps @ 10 ha^{-1}. A threshold level of 6-7 moths $night^{-1}$ $trap^{-1}$ continuously for 3 nights should be an indicator to start pest control operations.

vi. For chemical control of *H. armigera* the economic threshold level of 8-10 eggs or 3-5 small larvae $plant^{-1}$ at flowering stage of crop should be considered as signal to start the control.

vii. Prefer use of relatively environmentally safe (RES) chemicals like endosulfan (0.07%), quinalphos (0.05%), monocrotophos and NSKE (5.0%). In case there is problem of both *H. armigera* and tur pod fly prefer monocrotophos.

viii. At initial egg laying, spray NSKE (5.0%) or profenophos (0.15%).

ix. Alternate row spraying of insecticides have been found equally effective in the control of borer as well as to minimise the cost of spray operations.

x. Bird perches should be arranged @ 50 ha^{-1}. This can be done easily by growing traditional type of sorghum plants here and there in the field.

27.9. TUR POD FLY, *Melanagromyza obtusa* (Malloch) (Diptera: Agromyzidae)

Tur pod fly is a serious pest of pigeonpea and widely distributed throughout the Asian countries like India, Sri Lanka, Myanmar, Bangladesh, Indonesia, Japan, Malaysia and Thailand.

Nature of damage. The damage is done by maggot of pod fly which upon hatching mines into soft seed and starts feeding on the contents therein. Such infested seeds show presence of a dirty white or a creamy-band. The full-grown larva before pupation comes out of seed and cut partial holes into the pod wall leaving thin circular layer of epidermis which is forced upon by the emerging fly. The damaged pods have holes of the pinhead size. When the damaged pods are split open, the immature stages or their exuviae can be seen attached to the pod wall or seed.

Biology. The adult fly is small, shining and metallic blue or black and its emergence takes place in the morning hours. Mating occurs during day time and majority copulate within 2-5 days of emergence and survives for 3-5 days. The fecundity rate is 30 eggs per female. Large-sized pods were preferred over small-sized pods for oviposition. The adult female inserts the eggs singly into the lumen of 10-35-day-old pod. The freshly laid egg is white, broad and rounded at the posterior end which is embedded in the tissue of the pod. The average incubation period is 5-8 days. Freshly hatched larvae are white with dark brown mouth. Young larva attaches itself on the immature seed inside the pod and feed on it. One seed is enough for the development of a larva. The larva never leaves the pod in which once it enters and completes its 3 instars of larval period in 6-19 days in the same pod. Pupation takes place in hard chitinous puparium, which is usually found sticking to the side of the pod or the groove eaten into the grain by the larva. Fresh pupa is yellowish white, but becomes darker subsequently. The pupal stage lasts in between 8 and 31 days.

Control measures. Same as in gram pod borer

Management. Lal and Katti (1997) reported that many insecticides have been tested and found effective against pod borer complex including pod fly in pigeonpea. List of these insecticides is given in Table 27.1 under chemical control of *H. armigera* on pigeonpea. Among these insecticides endosulfan (0.07%), monocrotophos (0.04%), dimethoate (0.03%), fenvalerate (0.02%) and cypermethrin (0.01%) have been reported to be superior. In addition, plant products such as neem seed kernel extract are recommended for application. Usually 2-3 applications beginning with the pod initiation stage are recommended for effective control.

Despite the fact that *M. obtusa* completes its entire life cycle within pigeonpea pods, ten hymenopteran parasitoids have been recorded to parasitise its immature stage. Of these, *Euderus lividus, E. agromyzae, Ormyrus orientalis, Eurytoma* sp. *Antistropholex* spp., *Eupelmus urozonus, Trichopria* spp., *Bracon fletcheri* and *Bracon* spp. are commonly found throughout India. In a survey conducted on the extent of parasitism in 11 different states of India. *Euderus* sp. was found to be prominent, however, the overall parasitism was generally low, i.e., 3.7-13.8% (Talekar 1990).

It has been reported that resistant donors like PDA 88-2E and PDA 89-2E provided 60-70% crop protection against pod fly damage in comparison to local checks (Lal and Sachan 1992).

The pod fly is known to thrive during off-season (May-August) on pigeonpea ratoons and alternate wild hosts like *Rhycosia minima, Flemingia congesta, Atylosia* spp. and *Tephrosia purpurea* and, therefore, removal or destruction of these alternate host plants will help to reduce carryover population of the pest (Lal and Katti, 1997). Multiplication rate of pod fly is low during winter months, therefore early-maturing crop (February-March) has low damage than late-maturing (April-May). Similarly early-sown crop has low damage than late-sown crop. Prefer to not to sow indeterminate varieties of pigeonpea as these are more susceptible to *M. obtusa* than determinate varieties.

27.10. LEGUME POD BORER, *Maruca vitrata (testulalis)* (Geyer) (Lepidoptera: Pyralidae)

It is an important pest of grain legumes such as cowpea, pigeonpea, mung bean, common bean, soybean, adzuki bean, groundnut, fieldpea, broad bean, and lima bean.

Nature of Damage. The larvae web the leaves and inflorescence and feed inside on flowers, flower buds and pods. This typical feeding habit protects the larvae from natural enemies and other adverse factors, including insecticides. The flower bud stage is preferred most for oviposition, and it is at this stage that the young larvae cause substantial damage, and reduce the crop potential for flowering and fruit setting. The young larvae bore into the flower buds, and cause flower shedding by destroying the young flower parts enclosed in the sepals. The larval movement to other flowers and flower buds is facilitated by the silken threads, which are used as bridges between flowers. The larvae are photo-negative, emerge early in the evening and feed on the plant throughout the night. The third instar larvae prefer pods rather than flowers or young leaves. First instar larvae prefer flowers over pods and leaves.

Biology. Eggs are normally deposited on flower buds and flowers. A female lays 6-189 eggs. Eggs are light yellow, translucent and have faint reticulate sculpturing on the delicate chorion. Eggs are usually deposited singly or in batches of 2-16. Females live for 4-8 days. Eggs hatch in about 5 days. There are five larval instars. Total larval development is completed in 8-14 days. The larvae are translucent and shining, and have 6 rows of black spots running from thorax to abdomen. Due to these prominent black spots it is also called a spotted caterpillar. Pupation occurs in a silken cocoon amongst webbed leaves/pods or in soil. The life cycle is completed in 18-35 days. Adults are brown to black with a white patch on the wings. In normal resting posture, the moths hold the wings in a horizontal position, unlike other moths which rest with folded wings. There is no diapause in this insect. Alghali (1993) reported that the peak population, in general, coincided with peaks in rainfall. Significant relationships have been observed between borer counts and cumulative rainfall, and number of rainy days between crop emergences to 50% flowering.

Management. It has been reported that 10 larvae per 100 flowers can be considered as a tentative threshold for *M. vitrata* on pigeonpea (Dharmasena *et al.* 1992).

Early-maturing pigeonpea varieties suffer greater pod borer damage than the late-maturing varieties such as CC 11 and Berhampur Local (Sahoo and Patnaik, 1993). Different workers have reported a number of varieties/entries least susceptible to *M. vitrata* (Patnaik *et al.,* 1986; Parsad *et al.,* 1989a, 1989b). Saxena *et al.* (2000) screened 152 determinate and 119 non-determinate short-duration germplasm accessions and observed that the damage was higher in determinate than in non-determinate genotypes. A determinate selection MPG 537-M1-2-1B and a non-determinate selection MPG 664-M1-2-M13 were identified as resistant to the pest.

Foliar application of cypermethrin (0.008%) or dimethoate (0.07%) at flowering of pigeonpea and repeating after 10-15 days provided effective protection against the pest as observed by Rahman (1991). Four sprays, i.e., at flower initiation, 50% flowering, 100% flowering and 100% pod setting, in pigeonpea were effective to control the borer (Rahman and Rahman, 1988) Sundara Babu and Rajasekaran (1984) reported that sprays of 0.07% triazophos or endosulfan and 0.04% monocrotophos gave maximum reduction in borer damage. Three sprays of cypermethrin (75 g a.i. ha^{-1}) have been found to be effective against pod borer on pigeonpea (Sontakke and Misra, 1991).

The list of parasitoids/predators of *M. vitrata* includes *Phanerotoma hendec asisella* Cameron, *Braunsia* spp., *Microdus* spp., *Anthrocephalus subelongatus* Kohl., *Telenomus* spp., *Pseudoperichaeta laevis* Villeneure and *Dinothrombium* spp. (Lal and Katti, 1998). A summary of natural enemies has been prepared by Sharma (1998).

27.11. PEA BLUE BUTTERFLY, *Lampides boeticus* Linnaeus (Lepidoptera: Lycaenidae)

The pest causes serious damage to pigeonpea, cowpea, common bean, mung bean, urd bean, field bean, dolichus bean and peas.

Nature of Damage. The larva bores inside the buds, flowers and green pods just within couples of hours after hatching and feeds inside the developing grains. The holes can be seen even after maturity of the pods.

Biology. The adult is medium-sized butterfly. The colour of the wings is violet metallic blue to dusky blue. The tail of the hind wings is black and tipped with white. The female is slightly bigger than the male. Copulated female laid eggs on the buds, flowers and green pods and to a lesser extent on shoots and leaves. The eggs are greenish-white and round in shape with a slight depression at the top. The incubation period varies from 4 to 7 days. The larval period varies from 9 to27 days. The pupation takes place on weeds, leaves, twigs or on pods, on ground or in soil cracks. The insect over winters as pupa and partly as larva. The pupal colour is grayish-pink to reddish-brown and pupal period is completed in 7 to 19 days. The longevity of adult butterflies ranges from 2 to 6 days. The entire life cycle is completed in 20-55 days. The pest is active and found in all stages from the last week of November to the first week of March. The insect has 8-9 generations in a year. The duration of one generation is 30-50 days (Verma, 1986; Qingzhou *et al.,* 1995; Uphadhyay *et al.,* 1998).

Management. The pest can be controlled by spraying 1.75 kg of Sevin 50 WP (carbaryl) (Chhabra *et al.,* 1993). Irulandi and Balasubramanian (2000) tested some botanical insecticides against this pest and found neem seed kernal extract to be effective than monocrotophos in terms of cost benefit ratio. In mung bean, urd bean and rice bean pest can be controlled by spraying 500 ml of Nuvan 100 (dichlorvos) or 1000 ml of Thiodan 35 EC (endosulfan) in 200 litres of water per hectare.

27.12. PLUME MOTH, *Exelastis atomosa* (Walschingham) (Lepidoptera: Pterophoridae)

Plume moth is a specific pest of pigeonpea but it has also been reported on horse gram and lablab and mung bean.

Nature of Damage. The young larvae bore into the unopened flower buds for consuming the developing anthers. Grown up larvae first scrap the surface of the pods and then bore into pods. The larva never enters the pod completely. The damage is more in determinate varieties as compared to indeterminate ones (Reddy *et al.,* 2001). The pest infests the pigeonpea crop during seedling, vegetative, flowering, pod-maturing and pod-filling stage (Singh, 2001). Bindra and Jakhmola (1967) reported an average loss of 80.22 per cent in grain weight by this pest.

Biology. The adult moth is light brown, with deeply fissured wings. The legs are long and slender. The abdomen is dark brown. Females lay minute eggs at night singly on tender leaves, flower, buds and young pods. Eggs hatch in 3-5 days and the tiny larvae feed on the pods and become full-grown in 11-16 days. The full-grown caterpillar is about 1.3 cm in length, greenish-brown fringed with hairs and spines. It pupates on the pod surface or in the entrance hole itself. The pupa also looks like larva. The pupal period lasts 4-7 days. The total life cycle is completed in 18-28 days.

Management. Fenvalerate 20 EC (0.02%), deltamethrin 2.8 EC and methomyl 24 L (0.1%) are the effective insecticides against this pest (Singh *et al.,* 2001). Singh *et al.* (2001) reported fenvalerate 0.02% to possess better efficiency to check the pest as compared to endosulfan 0.07%, monocrotophos 0.03%, dimethoate 0.03%, quinalphos 0.05% and 5% neem seed kernal extract. Treatment of pigeonpea with HaNPV recorded the minimum pod damage and was comparatively better than endosulfan 0.07% and acephate 0.07% (Neharkar *et al.,* 1999).

Early sowing of pigeonpea varieties exhibit less *E. atomosa* damage as compared to late sown varieties. In an experiment conducted at Coimbatore early sown pigeonpea crop, i.e., on 15th May recorded 6.4 per cent pod damage due to *E. atomosa* as compared to 15.9 per cent on late sown crop, i.e., on 15th June

27.13. LEAF BINDER, *Grapholita (Cydia) critica* Meyrick (Lepidoptera: Tortricidae)

Leaf binder is very common pest of arhar in Punjab. It is also serious pest in the state of Bihar, Uttar Pradesh, West Bengal, Tamil Nadu, Andhra Pradesh, Maharashtra, Madhya Pradesh, Gujarat, Haryana and Delhi. It has also been reported on mung, mash, cowpea, gram, lentil and pea.

Nature of damage. During vegetative stage of the crop, the caterpillars cause damage by binding leaves and growing tips of tender shoots. The larva get itself concealed into the froth produced during the course of scratching. It remains there for 2-4 days. The grown up larva then draws the two leaves together and spins a thread between them. The spun up threads readily dry up and bind the adjacent leaves and shoots. At the reproductive stage they enter the buds, flowers and pods and feed on the immature seeds.

Biology. Adult moth is dusky brown with fore-wings having four black dots and a silvery transparent mark. Females are bigger than males. Freshly laid eggs are oval and creamy white. The average incubation period is five days. Young hatched out larvae are pale. Full grown larvae are shining yellow, moderately stout, smooth except for a few short scattered hairs. Average larval period is 19 days. Full grown larvae in the folded parts of the leaves develop white silken covering around themselves and pupate. The average pupal period is seven days.

Chhabra *et al.* (1993) reported that various species of *Grapholita* cause damage to different leguminous crops in different parts of the world. Though *G. critica* is a regular and serious pest of pigeonpea, it has also been reported on mung bean, urd bean, cowpea, chickpea, lentil, bread bean, peas and sunhemp.

Management. The leaf binder can be controlled by spraying Decis 2.8 EC (deltamethrin) @ 200 ml ha^{-1} or Hostathion 40 EC (triazophos) @ 125 ml ha^{-1} or Monocil 36 WSC (monocrotophos) @ 250 ml ha^{-1}. Chaudhary and Bhattacharya (1974) recommended the spraying of Thiodan 35 EC @ 1.25 l ha^{-1}. Srivastava (1974) recommended the spraying of dimethoate 30 EC (0.02%), endosulfan 35 EC (0.02%) or carbaryl 50 WP (0.02%) for the effective control of this pest. Chhabra and Kooner (1979) observed quinalphos 25 EC (0.07%) and monocrotophos 40 EC (0.04%) to be effective against the pest.

As the knots of leaves are readily seen and could be rapidly and easily collected by hand so the very simple remedy is pulling off the affected leaves, which is sufficient to check the pest damage.

27.14. POD BUG OR GREEN BEAN BUG, *Clavigralla gibbosa* Spinola (Hemiptera: Coreidae)

Green bean bug has been reported to feed on pigeonpea, mung bean, urd bean, chickpea, cluster bean, common bean, lablab, etc.

Nature of Damage. Both nymphs and adults cause damage to leaves, flower buds, stem and pods by sucking juice. Major damage is caused to the pods before the maturity of the crop. The attacked pods show pale yellow patches. When the pest population is high there is severe damage to pods which become shrivelled and the grains in the attacked pods also become shrivelled and small in size resulting in great loss in yield.

Biology. The adult bugs are greenish-brown measuring about 2 cm in length. The female bug is bigger and has a round and swollen abdomen as compared to male. The female lays fertile eggs generally on green pods. The eggs are laid in clusters of 3-15, which are dull white and gradually the colour changes to brown to dark brown. The incubation period is about 8 days. The newly hatched nymphs are red which become greenish-brown with age. Nymphal period varies from 8 to17 days.

Management. Shaking the infested plants over the vessels of oil and water or oily cloth gives most effective control.

27.15. BLISTER BEETLE, *Mylabris pustulata* (Thunberg) (Coleoptera: Meloidea)

Blister beetle attacks arhar, mung, mash and many other cultivated plants.

Nature of damage. Beetles are diurnal and general feeders. The major damage is caused at the flowering stage. The adult beetles feed on flowers (mainly pollen), leaves and tender panicles, preventing grain formation.

Biology. Beetles are robust with bright metallic blue, green, black, yellow or red stripes on the forewings and length ranging from 1.2 to 2.5 cm. Head is joined to the prothorax by a distinct narrow neck. The legs are long. When disturbed the beetles emit a fluid containing canthardine that has irritant effect on human body. The female lays a large number of eggs, often 2000-10000 on the ground or in the soil. The larvae prey upon the eggs of Orthoptera and Hymenoptera. It pupates after passing through different moulting stages and finally becomes adult.

Management. Blister beetle can be controlled by spraying 0.1% endosulfan or 0.03% dimethoate or the mixture of dimethoate 0.03% and endosulfan 0.1% or permethrin ULV @ 150 ml ha^{-1}.

Long-eared Indian hedgehog, *Hemiechinus auritus* (Gray), feed on the blister beetle (Maheshwari 1986).

The pest can be controlled through hand collection and destruction in early morning hours.

27.16. FUTURE RESEARCH NEEDS

Pulses are severely ravaged by a complex of insect pests and are a major constraint to their production in India. However, researches have been concentrated almost exclusively on *H. armigera* and *M. obtuse* with little or negligible attention to other major yield reducing species such as *M. vitrata, L. boeticus, E. atomosa, C. gibbosa,* etc. Therefore, the needs is to work on pest complexes for each agro-ecological niche rather than concentrating only on one or two key species, as pulses are damaged by a group of pest complex and rarely by an individual insect pest.

- The use of insecticides has shown a great promise for controlling pest complexes in pulses. However, past experience has shown that pesticide use alone is inappropriate and unsafe and that farmers frequently fall into a cycle of increasing the amounts and/or frequency of pesticide applications. To avoid this pesticide treadmill farmers need effective alternate pest management practices. This could become possible by focusing on ways to enhance natural control processes i.e. biological control and host plant resistance. Till such date insecticides which are selective and safe to non-target organism has to be recommended in pulses.
- Various control options including cultural, biological, host plant resistance and chemicals are available for minimizing the losses due to pest complex and therefore, deserve to be tested under farmers fields to identify and evolve most effective and economical control strategy for agro-ecological niche to strengthen the development of IPM.
- Biotechnological tools for improving pest resistance level by transferring resistance genes from wild sources need to be exploited. The genetic engineering to develop more efficient strains of microbial agents and their mass multiplication may prove a great success.

27.17. CONSTRAINTS AND POLICIES IN ADOPTION OF IPM IN PULSES

IPM adoption studies revealed that farmers in general are aware of biopesticides and IPM tools such as pheromones and microbial pesticides but due to non-availability they are not able to use them. Much of the research, development, technology transfer, education and training undertaken by the crop protection industry supports implementation of IPM. However, more can be done to research and develop IPM technologies and promote their transfer into practice at field level. The future strategies for insect pests control may be oriented in the following directions:

- Pulses being grown by marginal farmers has never received adequate attention either in terms of human power or infrastructure support.
- Education and dissemination of technology over larger areas should be taken up with the support of NGO and private pesticide industries and their network. IPM being knowledge based more attention to self help groups, extension workers and village level workers may be given.
- Need for transgenic pulses can see light of the day. Likelihood of socio-economic effect of these GM pulses need to be worked out alongwith its effect on local niche.
- Develop IPM compatible crop protection products including chemicals, semi-chemicals to modify pest behavior and biological control products.

REFERENCES

Alghali, A.M. (1993) The effects of some agrometeorological factors on fluctuation of the legume pod borer, *Maruca testulalis* (Geyer) (Lepidoptera : Pyralidae), on two cowpea varieties in Nigeria. *Insect Sci. Appl.* **14:** 55-59.

Ambrose, D.P. (2000) Suppression of *Helicoverpa armigera* (Hubner) and *Anomis flava* (Fabricius) infesting okra by the reduviid predator Kumarii Ambrose and Livingstone in field cages. *Pest Mngmt. Horticu. Ecosyst.* **6:** 32-35.

Annonymous (2009) *Package of Practices for Kharif crops*, pp 44-60, Punjab Agricultural University, Ludhiana.

Annonymous (2010a) *Package of Practices for Kharif crops*, pp 58-63, Punjab Agricultural University, Ludhiana.

Annonymous (2010b) *Package of Practices for Kharif crops*, pp 64-66, Punjab Agricultural University, Ludhiana.

Asthana, A.N. and Ali, M. (Eds.) (1997) *Recent Advances in Pulses Research.* Kanpur : Indian Society of Pulses Research and Development.

Asthana, A.N., Lal, S.S. and Dhar, V. (1997) Current problems in pulse crops and future needs. In *National Seminar on Plant Protection towards Sustainability,* p. L-3. Hyderabad : Plant Protection Association of India.

Atwal, A.S. and Dhaliwal, G.S. (1997) *Agricultural Pests of South Asia and their Management.* Ludhiana: Kalyani Publishers.

Balasubramanian, P. (1992) Food plants of blister beetle *Mylabris pustulata* Thunb. (Coleoptera: Conthasidae) from Pt. Colinero Wildlife Sanctuary, Tamil Nadu. *J. Bomb. Nat. Hist. Soc.* **89:** 262-263.

Bhattacharya, S. and Pramnik, A. (2002) Comparative efficacy of botanicals alone and in combination with synthetic insecticides against *Diacrisia obliqua* (Walker) under laboratory conditions. *Environ. Ecol.* **20:** 727-731.

Bindra, O.S. and Jakhmola, S.S. (1967) Incidence of and losses caused by some pod infesting insects in different varieties of pigeonpea (*Cajanus cajan* (L) Millsp.) *Indian J. Agricult. Sci.* **37:** 177-188.

Biswas, S., Upadhyay, K.D. and Kumar, A. (1994) Bio-efficacy of various *Bacillus thuringiensis* formulations and dosage against hairy caterpillar, *Spilosoma (Diacrisia) oblique. J. Ecotoxic. Environ. Monng.* **4:** 185-188.

Chaudhary, J.P. and Sharma, S.K. (1982) Biology of gram pod borer, *Heliothis armigera* (Hubner) in the Haryana State. *Bull. Ent.* **22:** 101-112.

Chaudhary, J.P., Rustagi, K.P. and Yadav, L.S. (1982) Intensity of attack of *Heliothis armigera* (Hubner) on gram in Haryana. *Indian J. Ent.* **44:** 191-192.

Chaudhary, J.P., Singh, A.K. and Yadav, L.S. (1981) Effect of disulfoton granules on pest complex, crop growth and yield of green-gram. *Indian J. Ent.* **43:** 369-372.

Chaudhary, R.R.P. and Bhattacharya, A.K. (1974) Keeping pests of arhar at bay. *Indian Farmers' Dig.* **7:** 23-33.

Chauhan, R. (1992) Present status of *Heliothis armigera* in pulses and strategies for its management in Haryana. In *Helicoverpa Management : Current Status and Future Strategies. Proceedings of First National Workshop* (Ed J.N. Sachan), pp. 49-54. Kanpur : Directorate of Pulses Research.

Cheema, H.K., Taggar G.K., Singh R. and Kooner B.S. (2009) Evaluation of insectides against *B. tabaci* (Gennadius) on urdbean [*Vigna mungo* (L.) Hepper]. *J. Insect Sci* **22**(4): 388-392.

Chen, H.F., Zhang, H.F., Xue, S.D., Deng, P.J. and Li, G.Q. (1991) Occurrence and control of *Anomis flava* Fabs. *China's Fiber Crops* **3:** 41-44.

Chhabra, K.S. and Kooner, B.S. (1978) Control of gram pod borer. *Prog. Fmg.* **14(10):** 13.

Chhabra, K.S. and Kooner, B.S. (1979) Field evaluation of certain insecticides for the control of Tur-pod fly, *Melanagromyza obtusa* Mall. and pod borer complex. *Pestic. Inf.* **4:** 89-92.

Chhabra, K.S. and Kooner, B.S. (1980) Pests of moong, mash and soybean and their control. In *Proceedings of Discussion-cum-Training Seminar on Pest Disease Managment in Pulses,* pp. 41-46. Ludhiana: Punjab Agricultural University.

Chhabra, K.S. and Kooner, B.S. (1987) Susceptibility of different stages of Hairy caterpillar *Diarcrisia obliqua* (Walker) to some insecticides. *Pesticides* **21:** 22-24.

Chhabra, K.S. and Kooner, B.S. (1992) Sources of resistance in mungbean against major insect-pests and yellow mosaic virus. *Legume Res.* **14:** 175-184.

Chhabra, K.S., Kooner, B.S. and Brar, J.S. (1980) Resistance behaviour of mungbean, *Vigna radiata* (L.) Wilczek cultivars against insect-pest complex and yellow mosaic virus. *Indian J. Ecology* **7:** 276-280.

Chhabra, K.S., Kooner, B.S., Saxena, A.K. and Sharma, A.K. (1981) Effect of biochemical components on the incidence of insect pest complex and yellow mosaic virus in mungbean. *Crop Imprmnt.* **8:** 56-59.

Chhabra, K.S., Lal, S., Kooner, B.S. and Verma, M.M. (1993) *Insect Pests of Pulses — Identification and Control Manual.* Ludhiana: Punjab Agricultural University, and Kanpur : Directorate of Pulses Research.

David, P.M. (1991) Resurgence of yellow mite, *Polyphagotarsonemus latus* (Acari : Tarsonemidae) on chilli following application of insecticides. *Madras Agricul. J.* **78:** 20-25.

Dharmasena, C.M.D., Subasinghe, S.M.C., Lateef, S.S., Menike, S., Saxena, K.B. and Ariyaratne, H.P. (1992) Entomology research. In *Pigeonpea Varietal Adaptation and Production Studies in Sri Lanka. Report of Work,* pp. 104-108. Sri Lanka : Department of Agriculture, and Patancheru : International Crops Research Institute for the Semi-Arid Tropics.

Dhooria, M.S. (1984) Preliminary observation on the biology and host range of the tarsonemid mite, *Polyphagotarsonemus latus* (Acari : Tarsonemidae) *Acaricide Newsl.* **4:** 7-9.

Dhuri, A.V. and Singh, K.M. (1983) Pest complex and succession of insect pests in blackgram, *Vigna mungo* (L.) Hepper. *Indian J. Entomology* **45:** 396-401.

Gerson, U. (1992) Biology and control of the broad mite, *Polyphagotarsonemus latus* (Banks) (Acari : Tarsonemidae) *Exper. Appli. Acaric.* **13:** 163-178.

Gopali, J.B., Teggelli, R., Mannur, D.M. and Yelshetty, S. (2010) Web-forming lepidopteran, *Maruca vitrata* (Geyer): an emerging and destructive pest in pigeonpea. *Karnataka J. Agricult. Sci.* **23(1):** 35-38.

Gupta, S.K. and Dhooria, M.S. (1973) Occurrence of *Hemitatorsonemus latus* (Banks) on blackgram and its control in Punjab. *Sci. Cult.* **37:** 484-485.

Irulandi, S. and Balasubramanian, G. (2000) Efficacy of insecticides against pod borer, *Lampides boeticus* L. (Lycaenidae: Lepidoptera) on green gram. *Insect Environ.* **6:** 108-109.

Janagarajan, A. (1987) Effect of intercropping on the incidence of pest in cotton and sorghum. In *Proceedings of Regional Research Council Southern Zone,* p 13 Aruppukkottai : Tamil Nadu Agricultural, University.

Kooner, B.S. (1998) Identification of sources of resistance to whitefly, jassids and MYMV in *kharif* mungbean. In *Proceedings of National Symposium on Management of Biotic & Abiotic Stresses in Pulse Crops,* p. 40. Kanpur: Indian Society of Pulses Research and Development.

Kooner, B.S. and Cheema, H.K. (2006) Evaluation of pigeonpea genotypes for resistance to pod borer complex. *Indian J. Crop Sci.* **1:** 194-96.

Kooner, B.S. and Cheema, H.K. (2007a) Screening of mungbean germplasm against whitefly, *Bemisia tabaci* Genn. and MYMV. *J. Food Legumes* **20:** 100-102.

Kooner, B.S. and Cheema, H.K. (2007b) Screening of mungbean germplasm against whitefly, *Bemisia tabaci* and MYMV. *Acta Hort.* **752:** 307-310.

Kooner, B.S. and Malhi, B.S. (2001a) Identification of sources of resistance against *Helicoverpa armigera* in pigeonpea. In: Proceedings of Diamond Jubilee International Symposium on Hundred Years of Post-Mendelian Genetics and Plant Breeding - Retrospect and Prospects, p. 184. New Delhi : Indian Agricultural Resrearch Institute.

Kooner, B.S. and Malhi, B.S. (2001b) Management of *Helicoverpa armigera* (Hubner) by neem based insecticides on pigeonpea. In: *Proceedings of Biopesticide Conference on Biopesticides: Emerging Trends.* p. 117. Chandigarh : Society of Biopesticide Science.

Kooner, B. S., Cheema, H. K. and Kaur Rubaljot. (2006) Insect pests and their management. In: Masood Ali and Shiv Kumar (Eds.) *Advances in Mungbean and Urdbean.* Indian Institute of Pulses Research, Kanpur. India. pp. 335-401.

Kooner, B. S., Cheema, H.K. and Singh, Pritpal. (2008) Reaction of some pigeon-pea genotypes to pod borer complex in Punjab. *J. Insect Sci.* **21:** 389-93.

Kooner, B.S., Chhabra, K.S., Sharma, A.K. and Saxena, A.K. (1994) Phytochemical resistance in urdbean genotypes against major insect-pests. In *Proceedings of International Symposium on Pulses Research,* p. 162. Kanpur : Indian Society of Pulses Research and Development.

Kooner, B.S., Malhi, B.S. and Singh, I. (2002) Screening pigeonpea material to isolate least susceptible genotypes to *Helicoverpa armigera.* In: *Proceedings of National Conference on Agricultgural Heritage of India,* pp. 66-67. New Delhi : ICAR and Udaipur : Rajasthan College of Agriculture and Rajasthan Chapter of Asian Agricultural History Foundation.

Kushwaha, K.S. and Malik, B.P.S. (1987) Effect of sowing time and plant type on pod borer incidence and grain yield in some pigeonpea genotypes. *Internat. Pigeonpea Newsl.* **6:** 65-67.

Lal, S.S. (1987) Insect pests of mung, urd, cowpea and pea and their management. In *Plant Protection in Field Crops* (Eds M. Veerabhadra Rao and S. Sithanantham), pp. 185-201. Hyderabad: Plant Protection Association of India.

Lal, S.S. and Ahmad, R. (2001) Final report, 1993-98. ICAR scheme on development of IPM package under selective crop conditions against *Helicoverpa armigera* (Hubner) IIPR, Kanpur.

Lal, S.S. and Ahmad, R. (2002) Integrated insect pest management present status and future strategies in pulses. In: *Pulses for SustainableAgriculture and Nutritional Security* (Eds M. Ali, S.K. Chaturvedi and S.N. Gurha), pp. 101-110. Kanpur : IIPR.

Lal, S.S. and Katti, G. (1997) Podfly, *Melanagromyza obtusa* Malloch - A key pest of pigeonpea. Kanpur : IIPR.

Lal, S.S. and Katti, G. (1998) IPM of pod borer complex infesting pigeonpea. In *IPM System in Agriculture. Vol. IV. Pulses* (Eds R.K. Upadhyey, K.G. Mukerji and R.L. Rajak), pp. 79-128. New Delhi : Aditya books Pvt. Ltd.

Lal, S.S. and Sachan, J.N. (1992) Controlling pod fly in pigeonpea through host plant resistance. *Internat. Pigeonpea Newsl.* **15:** 28-30.

Lal, S.S. and Yadava, C.P. (1987) Estimate of crop losses in pigeonpea caused by the pod borer complex. *FAO Pl. Prot. Bull.* **35:** 93-98.

Maheshwari, U.K. (1986) Biological control of major agricultural pulse pest, *Mylabris pustulata* - A new approach. *Indian J. Ent.* **48:** 381-387.

Malhi, B.S., Kooner, B.S. and Chhabra, K.S. (2001) Isolation of resistant sources of *Helicoverpa armigera* in pigeonpea: A basic component of IPM for future research. In: *Abstracts of Fourth Punjab Science Congress,* p. 6. Ludhiana: Punjab Agricultural University.

Nair, M.R.G.K. (1986) *Insects and Mites of Crops in India.* New Delhi : Publications and Information Division, Indian Council of Agricultural Research.

Nayar, K.K., Ananthakrishnan, T.N. and David, B.V. (Eds) (1976) *General and Applied Entomology.* New Delhi : Tata McGraw Hill Publishing Company Ltd.

Neharkar, P.S., Patang, N.R., Kendre, M.S. and Munde, A.T. (1999) Comparative efficacy of insecticides against pod borers complex of pigeonpea. *J. Soils Crops* **9:** 178-181.

Nene, Y.L. (1973) Control of *Bemisia tabaci* Genn.: A vector of several plant viruses. *Indian J. Agricul. Sci.* **43:** 433-436.

Patnaik, H.P., Samolo, A.P. and Samolo, B.N. (1986) Susceptibility of some early varieties of pigeonpea for pod borers under protected conditions. *Legume Res.* **9:** 7-10.

Patnaik, N.C., Dash, A.N. and Mishra, B.K. (1989) Effect of intercropping on the incidence of pigeonpea pests in Orissa. *Internat. Pigeonpea Newsl.* **9:** 24-25.

Pawar, C.S., Sithanantham, S., Bhatnagar, V.S., Srivastavaa, C.P. and Reed, W. (1988) The development of sex pheromone trapping of *Heliothis armigera* at ICRISAT, India. *Tropi. Pest Mngmnt.* **34:** 39-43.

Prasad, D., Premchand and Haque, M.F (1989a) Relative susceptibility of different cultivars of red gram. *J. Res. Birsa Agricul. Unive.* **1:** 179-181.

Prasad, D., Premchand and Haque, M.F (1989b) Incidence of pod boring insects in different cultivars of pigeonpea. *J. Res. Birsa Agricul. Univ.* **1:** 79-80.

Prasad, D., Premchand and Srivastava, G.P. (1986) Effect of sowing dates on the infestation of insect pests and grain yield of pigeonpea. *Indian J. Ent.* **48:** 230-231.

Quinzhou, L., Yangsheng, C., Long, Q.Z. and Cai, Y.S. (1995) Biological characteristics of *Lampides boeticus* L. and its control. *Pl. Prot.* **21:** 15-18.

Rahman, M.M. (1991) Control measurers for important insect pests of major pulses. In *Advances in Pulses Research in Bangladesh : Proceedings of the Second National Workshop on Pulses,* pp. 139-146. Patancheru : International Crops Research Institute for the Semi-Arid Tropics.

Rahman, M.M. and Rahman, M.S. (1988) Timing and frequency of insecticide application against *Maruca testulalis* Geyer infesting short duration pigeonpea in Bangladesh. *Legume Res.* **11:** 173-179.

Rajagopalan, C.K. and Devakumar, J.P. (1965) Prelimnary studies on the infestation of *Agromyza obtusa* Mall. in red gram (*Cajanus cajan* Linn.) Millsp. *Madras Agricul. J.* **52:** 345-346.

Rao, N.V., Rao, K.T. and Rao, A.S. (1992) Present status of *Helicoverpa armigera* in pulses and strategies for its management in Andhra Pradesh. In *Helicoverpa Management: Current Status and Future Strategies. Proceedings of First National Workshop* (Ed J.N. Sachan), pp. 68-74. Kanpur: Directorate of Pulses Reseach.

Ravindrababu, R. (1987) Studies on mungbean yellow mosaic virus disease. Ph.D. thesis, Tamil Nadu Agricultural University, Coimbatore.

Reddy, C.N., Singh, Y., and Singh, V.S. (2001) Effect of sowing time and plant type on pod borer incidence and grain yield in some pigeonpea genotypes. *Indian J. Ent.* **63:** 215-220.

Reed, W. and Lateef, S.S. (1990) Pigeonpea : Pest management. In: *The Pigeonpea* (Eds Y.L. Nene, S.D. Hall and V.K. Sheila), pp. 349-374. Wallingford : CAB International.

Reed, W. and Pawar, C.S. (1982) *Heliothis :* A global problem. In *Proceedings of International Workshop on* Heliothis *Management,* pp. 9-14. Patancheru : International Crops Research Institute for the Semi-Arid Tropics.

Romeis, J. and Shanower, T.G. (1996) Arthropod natural enemies of *Helicoverpa armigera* (Hubner) (Lepidoptera : Noctuidae) in India. *Biocontrol Sci. Tech.* **6:** 481-508.

Sachan, J.N. (1987) Status of the gram pod borer, *Heliothis armigera* (Hubner) in India and its management. In *Recent Advances in Entomology* (Eds Y.K. Mathur, A.K. Bhattacharya, N.D. Pandey, K.D. Upadhyay and J.P. Srivastava), pp. 91-110. Kanpur: New Gopal Printing Press.

Sachan, J.N. (1992) Present status of *Helicoverpa armigera* in pulses and strategies for its management. In Helicoverpa *Management : Current Status and Future Strategies. Proceedings of First National Workshop* (Ed J.N. Sachan), pp. 7-23. Kanpur: Directorate of Pulses Research.

Sachan, J.N. and Bhaumik, R. (1998) Extent of parasitisation of *Campoletis chlorideae* on *Helicoverpa armigera* damaging chickpea. *Indian J. Pulses Res.* **11:** 65-69.

Sahoo, B.K. and Patnaik, N.C. (1993) Susceptibility of pigeonpea cultivars to pod borers in Orissa. *Internat. Pigeonpea Newsl.* **18:** 31-33.

Satyanarayana, J., Singh, K.M., Singh, R.N. and Singh, S. (1997) Influence of intercropping and insecticide. I. Incidence of major pests in ricebean, *Vigna umbellata. Indian J. Ent.* **59:** 316-325.

Sharma, H.C. (1998) Bionomics, host plant resistance, and management of the legume pod borer, *Maruca vitrata* — A review. *Crop Prot.* **17:** 373-386.

Sharma, S.K. and Chaudhary, J.P. (1984) Effect of some cultural practices on incidence of *Heliothis armigera* (Hubner) in pigeonpea. *Haryana Agricul. Univ. J. Res.* **14:** 361-365.

Shirale D. and Bidgire U. (2009) Integration of bioagents and synthetic insecticide in the management of whitefly and its effect on yield in soybean. *Karnataka J. Agricul. Sci.* **22(3):** 631-632.

Singh, D.P., Singh, R.P., Singh, S.K. and Kumar, A. (2001) Evaluation of certain insecticides against major insect pests of pigeonpea. *Ann. Pl. Prot. Sci.* **9:** 313-314.

Singh, H. and Singh, G. (1975) Biological studies on *Heliothis armigera* (Hubner) in the Punjab. *Indian J. Ent.* **37:** 154-165.

Singh, J., Sandhu, S.S. and Singla, M.M. (1990) Ecology of *H. armigera* (Hub) on chickpea in Punjab. *J. Insect Sci.* **3:** 47-52.

Singh, K.M. and Singh, R.N. (1977) Succession of insect-pests in mungbean and blackgram under dryland conditions at Delhi. *Indian J. Ent.* **39:** 365-370.

Singh, M. (1996) Studies on the biology of leaf curl mite *Polyphagotarsonemus latus* (Banks) (Acari : Tarsonemidae) on chilli (*Capsicum annum*) M.Sc. thesis, Punjab Agricultural Univerity, Ludhiana.

Singh, R.N. and Singh, K.M. (1978) Influence of intercropping on succession and population build-up of insect pests in early varieties of red gram, *Cajanus cajan* (L.) Millsp. *Indian J. Ent.* **40:** 361-375.

Singh, S.K. (2001) Pest faunal complex of pigeonpea in western UP. *Insect Environ.* **7:** 3-4.

Sinha, M.M., Yazdani, S.S., Kumar, A. and Singh, R. (1985) Observation on the seasonal occurrence of insect pests of greengram in northern Bihar. *Bull. Ent.* **26:** 210-211.

Sontakke, B.K. and Mishra, P.R. (1991) Comparative efficacy and economics of synthetic pyrethroids in the management of pod borer complex of pigeonpea. *Indian J. Pl. Prot.* **19:** 167-170.

Srivastava, C.P. and Sachan, J.N. (1998) Management of Gram Pod Borer, *Helicoverpa armigera* in India. Kanpur : IIPR.

Srivastava, C.P. and Srivastava, R.P. (1995) Monitoring of *Helicoverpa armigera* (Hubner) in chickpea (*Cicer arietinum* L.) *J. Appl. Ent.* **119:** 607-609.

Srivastava, O.S. (1974) Chemical control of the tur leaf caterpillar (*Eucelis critica*) *Indian J. Ent.* **36:** 359-360.

Sundara Babu, P.C. and Rajasekaran, B. (1984) Studies on the control of pod borers and pod fly on redgram (*Cajanus cajan* L.) *Pesticides* **18:** 24-26.

Talekar, N.S. (1990) *Agromyzid Flies of Food Legumes in the Tropics.* New Delhi : Wiley Eastern Limited.

Upadhyay, R.K., Mukerji, K.G. and Rajak, R.L. (Eds) (1998) *IPM System in Agriculture. Vol. IV. Pulses.* New Delhi : Aditya Books.

Vadodaria, M.P. and Vyas, H.N. (1987) Control of whitefly, *Bemisia tabaci* (Gennadius) and its impact on yellow mosaic virus (YMV) in greengram, (*Vigna radiata* (L.) Wilczek) and the grain yield. *Indian J. Agric. Res.* **21:** 21-26.

Verma, B.R. (1986) First record of the bean pod borer, *Cosmolyce* (*Lampides*) *boeticus* Linn. (Lepidoptera : Lycaenidae) on jainti. *Indian J. Ent.* **48:** 120-121.

Xiong, S.L., You, L.S., Zhou, Z.H. and Chen, Z.G. (1994) Studies on braconid wasp *Apanteles anomidis* Watanabe, an endoparasitic of *Anomis flava* Fabricius. *J. Hunan Agricul. College* **20:** 57-65.

Yadav, D.N. (1980) Studies on the natural enemies of *Heliothis armigera* (Hubner) and its biological control using an egg parasite, *Trichogramma australicum* Germault (Hymenoptera : Trichogrammatidae) *Gujarat Agric. Univ. Res. J.* **6:** 62-63.

Yadav, D.N. and Patel, R.C. (1981) Egg parasitism of *Heliothis armigera* (Hubner) (Lepidoptera : Noctuidae) *Gujarat Agric. Univ. Res. J.* **7:** 19-22.

Yadava, C.P. and Lal, S.S. (1988) Relationship between certain abiotic and biotic factors and the occurrence of gram pod borer, *Heliothis armigera* (Hubner) on chickpea. *Entomon.* **13:** 269-273.

Yein, B.R. (1983) Efficacy of certain insecticides against *B. tabaci* (Genn.) and Pagria signata (Motsch) on blackgram. *J. Res. Assam Agric. Univ.* **4:** 45-48.

Theory and Practice of Integrated Pest Management
R. Arora, B. Singh and A.K. Dhawan (Eds)
Scientific Publishers (India), Jodhpur 2011, pp. 501 - 517.

28

PEST MANAGEMENT IN JUTE AND MESTA

T. Ramasubramanian and S. Satpathy

Division of Crop Protection, Central Research Institute for Jute and Allied Fibres (Indian Council of Agricultural Research), Barrackpore, Kolkata - 700 120, India

28.1. INTRODUCTION

Jute, "the golden fibre of India" though occupies only 0.42% of the gross cropped area, provides livelihood to more than 4 million farm families. The productivity of jute has become doubled from nearly 10 q/ha during 1950-51 to more than 22 q/ha during 2008-09. However, the area under jute has been dwindling around one million hectares over the six decades. It may be difficult to break this plateau, but concerted efforts may be taken to enhance the productivity. There are several constraints in increasing the productivity of jute, of which the yield loss due to insect pests is of major concern and hence, the "Jute Entomology" deserves more attention for economic production of the crop. The allied fibre crop mesta, which includes two spp. namely *Hibiscus cannabinus* (kenaf) and *Hibiscus safdariffa* (roselle) cannot be ignored to meet the increasing demand for natural fibres. They also suffer due to the ravages caused by many insect pests and hence, the entomology of bast fibre crops requires much attention for economic production of the crop.

28.2. PEST MANAGEMENT IN JUTE

Intensive cultivation of high yielding, fertilizer responsive cultivars of jute brought forth the problem of insect pests and mites. A few relatively less important pests turned to be major ones. Bihar hairy caterpillar, *Spilosoma obliqua* Walker that was once considered as a sporadic pest on jute is now a major threat to jute crop during later part of the season every year. Semilooper, *Anomis sabulifera* Guenee, stem weevil, *Apion corchori* Marshall, yellow mite, *Polyphagotarsonemus latus* (Banks), and the indigo caterpillar, *Spodoptera exigua* Hübner are the major pests of jute. The jute girdler, *Nupserha bicolor postbrunnea* Dutt, scale insect, *Pinnaspis* sp., thrips, *Ayyaria chaetophora* Karny, mealy bugs, *Ferrisia virgata* Ckll. and *Pseudococcus filamentosus* var. *corymbatus* are the minor pests of jute (Ramasubramanian and Ghosh, 2009). Grey weevil *Myllocerus discolor* Boheman was first time recorded as a pest on *tossa* (*Corchorus olitorius*) jute in 1973 (Das and Ghosh, 1973). *Tossa* jute occupies 80% of the jute growing area as against 20% by the *white* jute, *Corchorus capsularis*, but unfortunately the incidence of major pests is more on *C. olitorius* than on *C. capsularis.*

28.2.1. Semilooper, *Anomis sabulifera* Guenee (Noctuidae: Lepidoptera)

Biology. Semilooper is one of the most important foliage pests of the jute and occurs in all the jute growing tracts of India (Das *et al.*, 1995). Crop loss due to this pest has been estimated at up to 50%. Eggs are laid on both the surfaces of leaf and preferably on the lower surface. About 83% of the eggs are laid between the growing point and the eighth leaf from the top. Larval period ranged from 9 to 16 days and the third instar larvae is the most destructive stage of the pest (Tripathi and Ram, 1971).

Nature of damage. Damage starts in all cases from unopened leaves, which represent the most susceptible portion of the plant. It was reported that up to 95% of the damage is restricted to nine fully opened leaves from the top. Repeated damage by this pest checks crop growth and induces profuse branching resulting in reduction of fibre yield. Pods and unripe seeds of seed crop are also damaged by *A. sabulifera.* The *C. olitorius* pods are more susceptible than the *C. capsularis* pods. Although the damage caused by *A. sabulifera* to the jute crop was recognized as early as 1954 in Hooghly district of West Bengal (Dutt, 1958), farmers are realizing the magnitude of loss due to such ravages only recently. The incidence of semilooper is increasing over the years irrespective of the locations across the jute growing belts of the country. The level of incidence was reported to be higher across the centres of AINP on Jute and Allied Fibres viz., Nagaon, Assam; Coochbehar, West Bengal and Bahraich, Uttarpradesh during the 2009 cropping season (Ramasubramanian *et al.*, 2009). Hence, prevention of huge crop loss as often occurs due to semilooper attack deserves the utmost consideration.

Conducive climate. High forenoon relative humidity for 15 days followed by drizzling during night hours and bright sunshine during daytime are more conducive for the outbreak of the semilooper.

Management

- All the notified varieties of jute (*tossa* and *white*) are equally susceptible to semilooper.
- Both *white* and *tossa* jute varieties showed increasing trend in infestation of semilooper with the increase in fertilizer dose (Das and Singh, 1985, 1986b). Hence, higher dose of nitrogenous fertilizers may be discouraged.
- *Sisiropa formosa, Trichologa sorbilans, Litomastix sp., Mesochorus sp.* were reported as parasitoids on semilooper (Das *et al.*, 1999). Since the mass multiplication technique has not yet been perfected, these parasitoids are not being utilized for applied biological control programmes.
- *Bacillus thuringiensis* Berliner was reported to be effective against the semilooper, *A. sabulifera.* Complete dissolution of midgut epithelial cells was observed 12 hours after the ingestion of *B. thuringiensis* spores (Das and Singh, 1976). However, published reports are not available with respect to bioefficacy of *Bt* commercial formulations in the field against the semilooper. Hence *Bt* commercial formulations may be incorporated as one of the components of jute IPM after detailed investigations in the field. This may also promote the organic farming of jute.
- Row length of 1.5 - 2.0 m has been determined as the suitable size for getting the most reliable estimates of semilooper incidence (Das and Basak, 1995). The most suitable direction for sampling has been determined to be south (Das, 1995b).
- Insecticidal interference may be avoided up to 15% of infestation by the semilooper on JRO 524, the variety that occupies more than 80% of the area under jute (Chatterjee *et al.*, 1988). However, control measures need to be initiated as soon as the infestation reaches beyond 15% level. Cypermethrin (0.03%) or carbaryl (0.1%) may be recommended against the semilooper (Das *et al.*, 1995; Das and Pradhan, 1995).
- The insecticidal sprays need to be targeted towards the apical portion of the plant rather than covering the whole plant as the incidence of the pest is confined to the top leaves.

28.2.2. Stem weevil, *Apion corchori* Marshall (Curculianidae: Coleoptera)

Biology. Jute stem weevil, *Apion corchori* Marshall occurs in all the jute growing tracts of India. The adult female lays eggs singly in holes gnawed by it on the stem at the bases of apical petioles. Up to 675 eggs are laid by a weevil during the oviposition period of around 124 days. Egg, larval and pupal periods last for 3, 8 - 18 and 4 days, respectively. The grub makes tunnel within the stem, feeds on the internal tissues and pupates within. Adults survive up to 209 days (Das, 1944).

Nature of damage. Though the stem weevil is a pest of both the species of *Corchorus, C. capsularis* suffers more than *C. olitorius*. Early sown crop is more susceptible and the infestation starts from seedling stage and continues till harvest. The yield loss was estimated at about 18% on JRC 212. There may be one or more punctures by the female at the nodes, and there are corresponding numbers of grubs in the plant. Nodal region below leaf base is the most susceptible spot. Feeding of tissues by grubs affects the growth of the plant, and where the growing point is lost, branching is induced. Loss of apical meristem checks the vertical growth and thus affects the fibre both in terms of quality and quantity. In older plants mucilaginous substances produced around the tissues damaged by the larva bind the fibres together; the fibres break at these points during processing and results in 'knotty fibre' (Tripathi and Ram, 1971).

Management

- The *tossa* jute variety JRO 878 is less susceptible to stem weevil attack (Das and Singh, 1985, 1986b).
- Both *tossa* and *white* jute varieties sown in late-April had shown remarkably less incidence of stem weevil as compared to those sown in late-March or early-April (Das and Singh, 1985).
- Application of phosphatic and potassic fertilizers reduces the attack, but nitrogenous fertilizers increase the incidence of the pest. Hence, inorganic fertilizers need to be applied in a judicious manner.
- It is recommended to remove and destroy the stubbles and self-sown plants in order to avoid the carryover of the pest.
- *Entedon manilensis* Crawford was reported to be the most efficient among the parasitoids. Hence insecticidal interference may be avoided if the population of *E. manilensis* is high in the ecosystem.
- Adults feed very little on the apical leaves and rest of the life stages are passed well hidden inside the stem. Therefore *A. corchori* always escapes from the direct contact of the insecticides. Cypermethrin (0.03%) or carbofuran (1 kg a.i ha^{-1}) are effective in managing the stem weevil (Das and Singh, 1986a; Das *et al*., 1995).
- Generally, the infestation is more in the periphery of the field than in the core of the plot and hence, proper care should be taken while sampling as well as spraying the insecticides.

28.2.3. Yellow mite, *Polyphagotarsonemus latus* Banks (Tarsonemidae: Acari)

Biology. Yellow mite, *Polyphagotarsonemus latus* Banks is one of the important destructive pests of jute. It is widely distributed in all the jute growing tracts of India. The newly hatched adults of both the sexes are white in colour and gradually turn to yellow after 24 hours. The colour becomes deeper with the ageing of the mites. Females are larger than males and their average life span is about one week. The average life span of males is 6 days. After hatching, male yellow mite moves very slowly for 2-3 hours and sucks juice from the ventral surface of leaf. After that the male starts moving over the entire leaf lamina in search of quiescent female. As soon as male gets a quiescent female, it catches that female with the sucker and lifts the female vertically up and moves very quickly for safer places. One male can carry one quiescent female at a time. When the adult female comes out from the quiescent stage, the male mates with the young female and the mating period on an average lasts for 20 seconds. After mating the female rests for sometime before it lays eggs in a suitable place preferably the young leaves that are like buds. The female lays eggs singly on the ventral surface of leaves between vein lets. Approximately 23 eggs are laid throughout the life period. The egg stage lasts for 30 hours. Active nymph of yellow mite is crystal white in colour and three pairs of legs mark the nymphal stage, the fourth pair of legs developed during adult stage. The average nymphal period in case of female is 24 hours, whereas in case of male it is 18 hours. In the inactive quiescent stage, the nymphs are boat-shaped and crystal white in colour. This quiescent stage has also been termed as 'pupa'. The mite has 14 generations in a season on jute crop and sometimes one overlaps the other. Due to large number of generations the population build-up is faster (Das and Singh, 1988).

Nature of damage. Both nymphs and adults suck the sap from ventral surface of young leaves, which curls ventrally in due course of time. The leaves do not grow to their full size; they turn coppery-brown and drop prematurely. The vertical vegetative growth of the crop is arrested, internodes become shortened and significant yield loss occurs regularly. Under severe damage the internodal length is reduced by two folds. *Tossa* jute suffers more due to yellow mite infestation than the *white* jute. The quality of the fibre was also found to be deteriorated due to the attack of yellow mite (Pandit *et al.*, 2002). The yield loss due to this pest has been estimated between 10% and 42% depending upon the level of infestation between 20% and 90%.

Conducive climate. The weather condition prevails during the 25th and 26th standard meteorological weeks were reported to be more conducive for faster multiplication of yellow mite in CRIJAF, Barrackpore. Though it attacks the crop during mid-June to mid-August, cloudy weather with high humidity and morning temperature were found congenial for multiplication of yellow mite in the field (Pandit *et al.*, 2002).

Management

- JRO 524, JRO 7835 and JRC 212 are least susceptible (Das *et al.*, 1999), whereas JRO 878 is highly susceptible to yellow mite (Das *et al.*, 1987).
- Crops sown during March and April are liable to more damage than the late sown crops.
- Whenever the population crosses the threshold of 110 mites per leaf (preferably in the second unfolded leaf) (Das *et al.*, 2002), insecticidal interference is required.
- Fenazaquin 10 EC (0.02%) followed by fenpyroximate 5 EC (0.01%) and propargite 57 EC (0.14%) are quite effective in reducing the yellow mite population in jute.

28.2.4. Bihar hairy caterpillar, *Spilosoma obliqua* Walker (Arctiidae: Lepidoptera)

Biology and Nature of damage. *Spilosoma obliqua* Walker is a polyphagous pest attacking several crops including jute. It was once considered as a sporadic and irregular pest on jute, but nowadays it is a regular and major pest on jute. The female moth lays eggs in groups and a single female may lay up to 1000 eggs. Upon hatching the larvae are gregarious and scrap the chlorophyll content of the leaves. Later they disperse to the entire field and prefer to defoliate the older leaves. The larval period ranges from two to three weeks and pupation occurs in the soil.

Management

- Ploughing the field after the harvest of the crop may expose the pupae to the hot sun and predatory birds.
- In the early stage, when the caterpillars remain gregarious on leaf, it is easy to destroy them after plucking such infested leaves and then dipping them in insecticides.
- When caterpillars dispersed, their control is achieved by insecticidal spraying. Cypermethrin has been proved to be most toxic to the hairy caterpillar (Das, 1989b). Preliminary studies at CRIJAF revealed that lambda cyhalothrin 5 EC (0.003%) and indoxacarb 14.5 SC (0.02%) were very effective in managing the hairy caterpillar on jute.
- *Apanteles obliquae, Protapanteles obliquae* and *Eurytoma* sp. were recorded to parasitize the larvae of *S. obliqua*. If more activity of parasitoids is noticed, insecticidal interference may be avoided.

28.2.5. Indigo caterpillar, *Spodoptera exigua* Hübner (Noctuidae: Lepidoptera)

Biology. It is an important pest of jute at seedlings stage and occurs in all the jute growing tracts of India. The eggs are laid on the under surface of the leaves and covered with buff-coloured hairs. A single female lays more than 1000 eggs in 2 - 11 days. The first instar caterpillars are gregarious in habit. Egg, larval and pupal stages last for 2, 13 - 16 and 5 - 7 days, respectively (Dutt, 1964).

Nature of damage and management. *S. exigua* was once considered as a minor pest on jute with a maximum yield loss of 20%. It has become a major pest responsible for entire crop failure. Re-sowing is hence, warranted very often to maintain the desired plant population in the field, failing which lot of gaps may be there in the field. Nowadays, *S. exigua* causes serious damage to grown up jute plants also. The larvae cut the growing points and hence, the apical vegetative growth is arrested. The *C. capsularis* jute sown early is more liable to damage by this pest than the late sown *C. olitorius* varieties. This pest can be managed with insecticides like quinalphos or any synthetic pyrethroids such as cypermethrin or fenvalerate or lambda cyhalothrin.

28.2.6. Grey weevil, *Myllocerus discolor* Boheman (Curculianidae: Coleoptera)

Nature of damage. *Myllocerus discolor* Boheman causes considerable damage to *tossa* jute by defoliating the crop. Fecundity of *M. discolor* is around 200 eggs during the oviposition period of 20-25 days. Egg, larval and pupal periods ranges from 3-4, 20-55 and 6-8 days, respectively. Adult's longevity is about two months. The weevil hibernates in the larval stage during winter (Mallick, 1985). Unopened tender apical leaves are the first choice of attack by *M. discolor* (Mallick, 1986). Irregular holes on the apical leaves are the characteristic symptom of damage due to grey weevil. *White* jute varieties are immune to the attack of grey weevil. One or two pre-monsoon showers, temperature range of 35-39°C and relative humidity of 85 - 94% are conducive for the maximum incidence of the grey weevil (Mallick, 1989a).

Management

- There is no resistant variety to grey weevil, and JRO 7835 is the most susceptible variety (Mallick, 1995a).
- A larger sampling size of 25-30% plant population has been estimated to provide reliable estimate of grey weevil incidence (Das, 1995a).

- With respect to directional effect, south was found to give most reliable estimate for all the pests except grey weevil for which, east was found to be the most reliable (Das, 1995a, 1995b).
- Two rounds of cypermethrin (0.03%) spray at fortnightly intervals is the most effective followed by oxydemeton methyl (0.05%) and carbaryl (0.1%) (Mallick, 1989a, 1995b).

28.2.7. Red mite, *Oligonychus coffeae* Nietner (Tetranychidae: Acarina)

Nature of damage. Red mite is more common on *white* jute. This mite resides under web and sucks the sap from upper surface of the leaves. Its incidence is sporadic in nature, but occasionally it causes serious damage to the crop. The affected leaves become leathery, turn yellow and drop off prematurely. Intermittent shower followed by dry spell with high humidity are quite favourable for faster multiplication of the mite (Tripathi and Ram, 1971).

Management

- Dicofol (0.04%) is effective in managing the red spider mite on jute.
- *Physalis minima* is serving as an alternate host to red mite in the jute ecosystem and hence, weed free condition needs to be maintained to avoid the carryover of the pest.

28.2.8. Jute girdler, *Nupserha bicolor postbrunnea* Dutt (Cerambycidae : Coleoptera)

Nature of damage. This is a pest on *tossa* jute only. Among the *tossa* jute varieties, 'Chinsurah green' showed highest degree of susceptibility. Sufficient rainfall during early crop season is conducive for early emergence of the pest and spread. Unlike most other Cerambycidae, the larva practically causes no damage to the host plant. On hatching, the larva moves down along the central hollow and feeds upon the remnant of the pith tissue. Such feeding has no apparent adverse effect on the growth of the plant. The larva neither travels to the side branches, if any, nor makes any tunnels to cause damage to the fibre. In *olitorius* jute, pith disintegrates early, resulting in the formation of hollow central region in the stem. The female cuts the cortex of the stem at two points by its mandibles, making two circular rings (girdles) approximately one cm apart around the stem and thrust the egg in the pith of the stem between the girdles. The girdled plant withers, droops and finally dies above the level of lower girdle (Dutt, 1961; 1964). The loss of stem varies from 12.51 - 17.11 cm on an average in different height groups, while 101 - 200 cm height group seems to suffer most by this pest.

Management

- Among the several parasitoids reported, *Neocatolaccus nupserhae* and *Norbanus acuminatus* are the most important.
- Nowadays the incidence and the extent of damage are insignificant. Collection and destruction of drooping shoots and casings containing the diapausing larvae may reduce the further spread of the pest.
- Any contact insecticide may take care of the pest population.

28.2.9. Mealy bugs *Phenacoccus solenopsis* (Pseudococcidae: Hemiptera)

The cotton mealybug *Phenacoccus solenopsis* (Pseudococcidae: Hemiptera) has become a new threat to jute farming. The infestation of *P. solenopsis* on jute was to the extent of 60-80% on JRO 524 with average intensity of 2-3 in 0-4 scale during summer 2009 in some parts of south Bengal. This species is capable of quick multiplication and spread under favourable climatic condition. This has become the first report of *P. solenopsis* on jute The infestation of such proportion in the early crop growth (40-65 days old) was unprecedented. An insight into the weather parameters persisting during the past six years indicated that there was increase in the maximum and minimum temperature compared to the previous years, besides low rainfall and less number of rainy days between January and May. The warm and dry condition during the summer months might be the pre-disposing factor for mealy bug outbreak on jute (Satpathy *et al.*, 2009).

***Ferrisia virgata* Ckll (Pseudococcidae: Hemiptera).** It is widely distributed in tropical countries including most parts of India. The damage is caused mostly by the nymphs, which suck the sap from leaf, petiole, stem and pod. Repeated attacks on the stem cause the development of crust due to which fibre bundles resist separation at the time of retting, resulting in the formation of "barky fibre". More damage is caused to seed crop, which does not put forth more number of pods.

***Pseudococcus filamentosus* var. *corymbatus* (Pseudococcidae: Hemiptera).** It is another species of mealy bug causing serious damage to *white* and *tossa* jute. The infestation by this pest arrests the vertical growth of the stem and the attacked region is characterized by its deep green colour, swelling and shortened internodes. Leaves arising from the closely opposed nodes give the attacked region a bushy appearance. The first instar and early second instar nymphs mainly induce this kind of malformation. Apical meristem is the most susceptible portion of the crop.

28.2.10. Thrips, *Ayyaria chaetophora* Karny (Thripidae: Thysanoptera)

Tossa jute particularly in seedling stage is infested by thrips, whereas *white* jute is resistant to thrips attack. This pest generally avoids light and remains within the folds of apical buds. They macerate the leaf tissues in between veins and when an infested bud begins to unfold, it moves up to the next bud and thus macerates all the buds in a regular succession. Soon the macerated tissues die and become prominent as white streaks in unfolded leaves (Tripathi and Ram, 1971).

28.2.11. Integrated pest management

The following strategies may be considered for inclusion in the IPM programme after detailed investigations across the locations

- Tolerant varieties such as JRO 524 (Navin) and JRC 212 (Basudev) may be recommended for cultivation in the yellow mite endemic areas. Cultivation of JRO 7835 may be discouraged in grey weevil endemic areas.
- The intensity of pest attack in jute is greatly influenced by the date of sowing. The level of incidence by stem weevil, grey weevil and yellow mite is more on early sown crops, but the reverse is true with respect to semilooper incidence. The optimal date of sowing is the 4th week of March for JRO 524, whereas 2nd week of April is optimum for other varieties in order to minimize the pest incidence.
- Self-sown plants and weeds should be removed properly to avoid carryover of the pests.
- Bt commercial formulations may be incorporated as a component of jute IPM after detailed investigations in the field. This may also promote the organic farming of jute.
- Neem oil @ 4ml l^{-1} was reported to be effective against the jute semilooper, *A. sabulifera*. This may be incorporated as one of the important components in the IPM practices
- It is suggested to pluck the leaves with early stage gregarious larvae of Bihar hairy caterpillar prior to spray insecticides. It is one of the important cultural practices needs to be followed in the management of jute pests.
- The incidence gradually declines from border to inner rows in case of yellow mite and grey weevil with reverse trend in case of semilooper and no such trend with respect to jute stem weevil. Hence, care should be taken accordingly to get reliable population estimates as well as spraying insecticides.

- Need based application of insecticides such as cypermethrin (0.03%) or carbaryl (0.1) may be recommended to reduce the pest population below ETL.

28.3. PEST MANAGEMENT IN MESTA

Mesta suffers due to ravages caused by a number of insect pests. Spiral borer, *Agrilus acutus* Thunb, mealy bugs, *Maconellicoccus hirsutus* (Green) and *Phenacoccus solenopsis,* hairy caterpillar, *Spilosoma obliqua,* flea beetle, *Nisotra orbiculata* (Motsch) and the green leafhopper, *Amrasca bigutulla bigutulla* Ishida are the major pests of both kenaf (*Hibiscus cannabinus*) and roselle (*Hibiscus sabdariffa*) (Pandit and Pathak, 2000). The major pests however, have limited geographical distribution and do not occur in all the mesta growing states of India. West Bengal is the only state, wherein all the major pests cause economic damage to mesta crop.

28.3.1. Spiral borer, *Agrilus acutus* Thunb. (Buprestidae: Coleoptera)

Biology: *A. acutus* is a pest on both kenaf and roselle. The adult female mates within a day of emergence, but does not oviposit for at least seven days. The eggs are laid singly on the stem, most frequently at the nodal region below the leaf base. The incubation period was observed tobe around 12 days under laboratory condition. There are generally three moults and exceptionally four besides a pre-pupal moult. Larval and pupal periods last for about 26 and 12 days, respectively. The entire life cycle of *A. acutus* was observed to be around two and half months (Dutt, 1969).

Nature of damage. Larva makes tunnel spirally throughout the entire length of the stem beneath the cambium layer. The course of the spiral inside the stem is generally recognized by elongated gall (9-15 cm). The portion of the stem above the gall dries up and ultimately dies.

Management. Crop sown during mid-April to mid-May suffer more than what was sown on later dates. Carbofuran @ 1kg ha^{-1} was observed to be very effective in managing the spiral borer. Swabbing the stem with methyl demeton (2.5%@ 20 ml plant^{-1}) at the height of 45 cm above the soil surface can also take care of this pest.

28.3.2. Mealybugs *Phenacoccus solenopsis* and *Maconellicoccus hirsutus* (Green) (Pseudococcidae: Hemiptera)

The cotton mealybug *Phenacoccus solenopsis* (Pseudococcidae: Hemiptera) has become a new threat to mesta. The infestation of *P. solenopsis* on mesta was to the extent of 60% with average intensity of 4 in 0-4 scale during summer 2009 in some parts of south Bengal. This has become the first report of *P. solenopsis* on mesta (Satpathy *et al.*, 2009).

The hibiscus mealybug *Maconellicoccus hirsutus* was first observed as a pest of roselle in 1948 and at present it is also a major pest on both the species of *Hibiscus*.

Nature of damage. Nymphs and adults suck the sap from apical portion of the plant and eventually arrest the vertical growth of the inter-nodes. The attacked region becomes deep green in colour with swellings. Secondary branches are put forth with the appearance of "bunchy tops". The fibre of attacked plants snaps during retting at the infested points.

Management. Preventive management through seed treatment with thiamethoxam 70 WS @ 5 g/kg seed provides protection against *P. solenopsis* at least for 30-40 days after sowing. Later on foliar spray of profenofos 50 EC (0.1%) is very effective in the management of *P. solenopsis* (Satpathy *et al.*, 2010). *M. hirsutus* may be managed easily by spraying dimethoate or methyl demeton

28.3.3. Flea beetle, *Nisotra orbiculata* (Motsch) [= *Podagrica orbiculata* (Mots.); *Podagrica bowringi* Baly] (Curculianidae: Coleoptera)

Biology: Flea beetle was recorded as a pest of kenaf as early as 1961 (Chowdhuri, 1961). The egg, larval, pre-pupal and pupal stages last for 7-9, 10-14, 1-2 and 7-9 days, respectively. The adult longevity varies between 30 and 32 days in kenaf as compared with 40-45 days in roselle. The adult beetle is shiny black having the habit of quick jumping rather than flying (Pandit and Chatterji, 1978).

Nature of damage. Its attack commences with seedling stage and ceases just before harvest. Week-old seedlings are highly susceptible to this pest and hence re-sowing is warranted frequently. The eggs are laid in clusters either under dry leaves or under the cracks and crevices of the soil. The larvae on hatching feed on roots. Adults are more injurious to the crop as they feed on the tender portion of the stem besides damaging the leaves. Intermittent showers followed by dry spell with high humidity observed to be conducive for its multiplication (Pandit and Pathak, 2000).

Management. Neem oil @ 5ml l^{-1} was reported to be effective against the flea beetle. Carbaryl was also found to be highly toxic to *N. orbiculata* (Pandit and Chakravorty, 1986). Preventive management through seed treatment with thiamethoxam 70 WS @ 5 g/kg seed provides protection against *N. orbiculata* at least for 60 days after sowing. Later on foliar spray of profenofos 50 EC (0.1%) is very effective in the management of *N. orbiculata*.

28.3.4. Minor pests

The yellow vein mosaic disease of mesta is spreading across the country in an alarming rate. Whitefly *Bemisia tabaci* Gennadius is the vector of the virus. The neonicotinoid insecticides imidacloprid and thiamethoxam were found very effective in managing the vector population.

Green leaf leafhopper, *Amrasca biguttula biguttula* Ishida, Semilooper, *Cosmophila erosa* Hub., large black weevil, *Desmidophorus hebes* Fab., grey weevil *Myllocerus discolor*, spittle bug, *Ptylus nebulosa* Fab., defoliator, *Trabala visnou* L. thrips, *Haplothrips ganglbaueri* Schmutz, hairy caterpillar, *Porthesia scintillans* Walk., pod sucking bug, *Oxycaraenus hyalinipennis* Costa, *Spermophagus tesselatus* Mot., termite *Microtermes obesi* H. and the wasp, *Psen orientalis* Cameron were also recorded as pests on mesta (Pandit and Pathak, 2000), but their occurrence is highly erratic and often less important.

28.3.5. Integrated pest management

The following points may be considered as components of IPM after detailed investigation in the research institutes:

- The flea beetle *N. orbiculata* thrives well on the alternate hosts *Sida* and *Hibiscus vitifolius* in the absence of mesta crop (Pandit and Pathak, 2000). Hence the field should be maintained weed free.
- Preventive management through seed treatment with thiamethoxam 70 WS @ 5 g/kg seed is effective to manage the infestation of *P. solenopsis* and flea beetle during the early crop growth stage.
- The possibilities of exploring the entomogenous fungi *Beauveria bassiana* (Balls) Vuill and *Metarhizium anisopliae* (Metsch) need to be studied on priority basis as the natural mortality of *N. orbiculata* was recorded up to 87 percent (Pandit *et al.,* 1979).
- Neem oil @ 5ml l^{-1} was observed to exhibit excellent antifeedant effect on *N. orbiculata* and hence, be included as one of the components of IPM.
- Regular monitoring is essential for effective management of mealybug as natural enemies are abundant and at the initial stage the distribution of the pest is clumpy. Spraying of profenofos 50 EC (0.1%) on the basis of pest and natural enemy monitoring will ensure and enhance the performance of insecticides. Carbofuran may be advocated against the coleopterans (spiral borer and flea beetle) and methyl demeton may be recommended against the hemipteran pests (mealy bugs and jassids) of mesta.

28.4. FUTURE STRATEGIES

28.4.1. Transgenic jute

Lack of resistant varieties as well as hidden nature of the stem weevil infestatio, which precludes the contact from insecticides, necessitates genetic engineering of jute with coleopteran specific *cry* genes. Moreover, the emerging threat from *A. sabulifera* and *S. exigua* also warrants genetic engineering with lepidopteran specific *cry* genes. Hence, it is suggested to go for pyramiding both lepidopteran and coleopteran specific *cry* genes to contain the colossal loss from major pests of jute.

28.4.2. Resistance risk assessment

The exposure of the jute pests to the new chemistry in jute growing belts of West Bengal as well as other parts of the Indian subcontinent is nil or at least minimal, and hence, it is the ideal time to get the susceptible population for generating baseline susceptibility to insecticides with novel modes of action. For effective pest management, detection and continuous monitoring of insecticide resistance are essential. This will also help to phase out the insecticides for which resistance has been built up, and thus the expenditure on ineffective insecticides by the resource poor jute farmers may be minimized. Endosulfan has been in use for more than four decades in the jute growing tracts of India. The possibility of development of resistance to endosulfan due to several decades of exposure cannot be ruled out (Ramasubramanian, 2007). A detailed investigation on status of susceptibility of the jute pests to insecticides with different modes of action is required to prevent / delay the development of resistance, if any, and effective management.

28.4.3. Traps for stem weevil

Stem weevil adults feeds very little on the apical leaves and rest of the life stages are passed well hidden inside the stem; as a result the *A. corchori* always escapes from the direct contact of the insecticides. Therefore, identification of chemical components and subsequent synthesis of commercial sex pheromones to attract the adults may pay the way for better management of stem weevil without compromising the environmental health. Similarly, research activities may also be focused to design other kinds of traps with poison baits besides pheromone traps.

28.4.4. Need to promote biocontrol agents in mesta eco system

In the Caribbean sub-region, the introduction of *C. montrouzieri* from India resulted in very good control of hibiscus mealy bug, *M. hirsutus*. It has been a tremendous success in all crops with the exception of *Annona muricata* and *Hibiscus sadariffa* (Keiro *et al.,* 2001). It is a challenge rather than opportunity for the entomologists working on mesta mealy bug to rectify the lacunae in release technology and make it successful in roselle also. It is unfortunate that the potential of indigenous *C. montrouzieri* has remained untapped or under-utilized against the mesta mealy bug *M. hirsutus* in India. Moreover, the white fluffy wax, which covers the entire body of the mealy bug often, prevents the entry of insecticides, and hence it is necessary though it may be risky to promote entomophagous insects in the mesta ecosystem.

REFERENCES

Chatterji, S.M., Das, L.K., and Singh, B. (1988) Effect of different degrees of infestation by jute semilooper, *Anomis sabulifera* Guen. on fibre yield of jute crop. *Sci. Cult.* 52, 136-137.

Chowdhuri, A.N. (1961) *Podagrica bowringi* Baly as a major pest of *Hibiscus cannabinus* commonly known as "Mesta". *Indian J. Entomol.* 22, 152.

Das, G.M. (1944) Studies on the jute stem weevil *Apion corchori* Marshall I: Bionomics and life history. *Indian J. Agric. Sci.* 14, 295-303.

Das, L.K. (1989b) Relative toxicity of newer insecticides against jute pests under laboratory condition. *In* "Annual Report: 1987-1988 and 1988-1989", Central Research Institute for Jute and Allied Fibres, Barrackpore, p.95.

Das, B.B. (1995a) Studies on the seasonal incidence and population dynamics of major insect pests of jute. *In* "Annual Report: 1990-1991 and 1991-1992", Central Research Institute for Jute and Allied Fibres, Barrackpore, p.66.

Das, B.B. (1995b) Studies on the seasonal incidence and population dynamics of major insect pests of jute. *In* "Annual Report: 1993-1994 and 1994-1995", Central Research Institute for Jute and Allied Fibres, Barrackpore, p.47.

Das, B.B. and Basak, S.L. (1995) Size of sampling units for the estimation of jute semilooper incidence. *Indian Biol.* 27: 59-61.

Das, L.K. and Ghosh, T. (1973) *Myllocerus discolor* Boheman on *Corchorus olitorius* —a new pest. *Jute Bull.* 36, 199.

Das, L.K. and Pradhan, S.K. (1995) Studies on dissipation of important effective pesticides in/on jute crop. *In* "Annual Report: 1990-1991 and 1991-1992", Central Research Institute for Jute and Allied Fibres, Barrackpore, p.66-67.

Das, L.K. and Singh, B. (1976) The effect of *Bacillus thuringiensis* Berliner on the gut of the jute semilooper, *Anomis sabulifera* Guen. *Sci. Cult.* 42, 567-569.

Das, L.K. and Singh, B. (1985) Effect of varieties, sowing dates and fertilizer levels on the incidence of pest complex of *capsularis* and *olitorius* jute. *In* "Annual Report: 1985", Central Research Institute for Jute and Allied Fibres, Barrackpore, pp.68-69.

Das, L.K. and Singh, B. (1986a) Effective control measures against the pest complex of capsularis jute *Corchorus capsularis* L. *Jute Dev. J.* 25-27.

Das, L.K. and Singh, B. (1986b) Effect of varieties, sowing dates and fertilizer levels on the incidence of pest complex of *capsularis* and *olitorius* jute. *In* "Annual Report: 1986", Central Research Institute for Jute and Allied Fibres, Barrackpore, pp.59-60.

Das, L.K. and Singh, B. (1988) Life history of yellow mite, *Polyphagotarsonemus latus* (Banks) on jute crop. *Ann. Agric. Res.* 9, 20-25.

Das, L.K., Laha, S.K., and Pandit, N.C. (1999) Entomology. *In* "Fifty years of research on jute and allied fibres agriculture", pp. 142-164. Central Research Institute for Jute and Allied Fibres, Barrackpore.

Das, L.K., Mitra, G.C., and Roychaudhuri, D.N. (1987) Anatomical changes of jute stem by yellow mite. *Sci. Cult.* 53, 252-253.

Das, L.K., Rao, P.V., Pradhan, S.K., and Chakraborty, A.K. (2002) Studies on determination of economic threshold level and economic injury level of major pests of jute. *In* "Annual Report: 2000-2001 and 2001-2002", Central Research Institute for Jute and Allied Fibres, Barrackpore, p.69.

Das, L.K., Singh, B., and Pradhan, S.K. (1995) Efficacy of different synthetic pyrethroids insecticides against pest complex of jute. *Sci. Cult.* 61, 203-204.

Dutt, N. (1958) *Anomis sabulifera* Guen and *Apion corchori* Marsh —Incidences and control. *Jute Bull.* 121-128.

Dutt, D.K. (1969) Bionomics of *Agrilus acutus* (Thunb.) on mesta. *Bull. Entomol. Res.* 58, 421-430.

Dutt, N. (1961) Studies on bionomics of jute girdler, *Nupserha bicolor postbrunnea* Dutt. *Sci. Cult.* 29, 302.

Dutt, N. (1964) Jute pests. In: *"Entomology in India"*, pp. 59-71. Entomological Society of India, New Delhi.

Kairo, M. T.K., Pollard, G.V., Peterkin, D.D., and Lopez, V.F. (2001) Biological control of the hibiscus mealybug, *Maconellicoccus hirsutus* Green (Hemiptera: Pseudococcidae) in the Caribbean. *Integrated Pest Manage. Rev.* 5, 241-254.

Mallick, R.N. (1985) Studies on the biology, bionomics and control of grey weevil (*Myllocerus discolor* Boheman) on *olitorius* jute. *In* "Annual Report: 1985", Central Research Institute for Jute and Allied Fibres, Barrackpore, p. 65.

Mallick, R.N. (1986) Studies on the biology, bionomics and control of grey weevil (*Myllocerus discolor* Boheman) on *olitorius* jute. *In* "Annual Report: 1986", Central Research Institute for Jute and Allied Fibres, Barrackpore, p.56.

Mallick, R.N. (1989a) Studies on the biology, bionomics and control of grey weevil (*Myllocerus discolor* Boheman) on *olitorius* jute. *In* "Annual Report: 1987-1988 and 1988-1989", Central Research Institute for Jute and Allied Fibres, Barrackpore, p. 94.

Mallick, R.N. (1995a) Studies on the biology, bionomics and control of grey weevil (*Myllocerus discolor* Boheman) on *olitorius* jute. *In* "Annual Report: 1990-1991 and 1991-1992", Central Research Institute for Jute and Allied Fibres, Barrackpore, p.66.

Mallick, R.N. (1995b) Studies on the biology, bionomics and control of grey weevil (*Myllocerus discolor* Boheman) on *olitorius* jute. *In* "Annual Report: 1993-1994 and 1994-1995", Central Research Institute for Jute and Allied Fibres, Barrackpore, p.48.

Pandit, N.C., and Chakravorty, S. (1986) Relative toxicity of some insecticides to adults of mesta flea-beetle *Nisotra orbiculata* (Mots.) (Coleoptera: Chrysomelidae) *Indian J. Agric. Sci.* 56, 136-138.

Pandit, N. C., and Chatterji, S.M. (1978) Bionomics of *Nisotra orbiculata* (Motsch) (Coleoptera: Chrysomelidae) *J. Entomol. Res.* 2, 49-54.

Pandit, N.C., and Pathak, S. (2000) "Management of insect pests in mesta". Central Research Institute for Jute and Allied Fibres, Barrackpore, 39p.

Pandit, N.C., Som, D. and Chatterjee, S.M. (1979) *Beauveria bassiana* (Bals) kill as a pathogen of *Nisotra orbiculata* (Mots.) infesting mesta (*Hibiscus cannabinus* L.) *J. Entomol. Res.* **3**, 111-113.

Pandit, N.C., Rao, P.V., and Chakraborty, A.K. (2002) Studies on the biotic and abiotic factors on the incidence of yellow mite of jute. *In* "Annual Report: 2000-2001 and 2001-2002", Central Research Institute for Jute and Allied Fibres, Barrackpore, p.71.

Ramasubramanian, T. (2007) Need for resistance risk assessment in major pests of jute to insecticides. *Resist. Pest Manage. Newsl.* 17, 2-4.

Ramasubramanian, T. and S.K. Ghosh (2008) Pest management in jute and allied fibres: present status and future strategies. *In*: Karmakar, P.G., S.K. Hazra, T. Ramasubramanian, R.K. Mandal, M.K. Sinha and H.S. Sen (Eds.) 2008. Jute and Allied Fibre Updates: Production and Technology *Jute and Allied Fibre Updates: Production and Technology*, pp. 194-216.

Ramasubramanian, T., Chakraborty, G. Sarma, K.K., Samal, T., Anjana Arun, Sharma, I.N., Srivastava, R.K., Laha, S.K., Satpathy, S. and Sinha, M.K. (2009) Bioefficacy of insecticides against major pests of jute. Paper presented in the "International Conference on Emerging Trends in Production, Processing and Utilization of Natural Fibres" held in Mumbai between 16.04.2009 and 18.04.2009.

Satpathy, S. Gotyal, B.S., Ramasubramanian, T., Bhattacharya, S.K. and Laha, S.K. (2009) Mealybug infestation in jute and mesta- a case study. National symposium on climate change, Plant Protection and Food security interface. 17-19, December, 2009. Association for Advancement in Plant Protection, Bidhan Chandra Krishi Viswa Vidyalaya, Kalyani, West Bengal, pp 108-109.

Satpathy, S., Ramasubramanian, T., Gotyal, B.S., Bhattacharya, S.K. and Laha, S.K. (2010) Evaluation of seed-treating and foliar insecticides against kenaf mealybug, *Phenacoccus solenopsis* Tinsely (Hemiptera : Pseudococcidae) National Symposium on Emerging Trends in Pest Management Strategies under Changing Climatic Scenario 20- 21 December, 2010 Department of Entomology, OUAT, Bhubaneswar, Odisha pp 175-176.

Tripathi, R. L., and Ram, S. (1971) *A Review of Entomological Researches on Jute, Mesta, Sunnhemp and allied fibres.* ICAR Technical Bulletin No. 36.

Theory and Practice of Integrated Pest Management
R. Arora, B. Singh and A.K. Dhawan (Eds)
Scientific Publishers (India), Jodhpur 2011, pp. 518 - 529.

29

MANAGEMENT OF INSECT PESTS OF TOBACCO

A.D. Patel, D.M. Mehta and L.V. Ghetiya
Bidi Tobacco Research Station
Anand Agricultural University, Anand - 388 110, India

29.1. INTRODUCTION

Tobacco is the most important cash crop in the world's agriculture and is cultivated in more than 100 countries on approximately 4.2 million hectares of crop land. Tobacco belongs to the order tubiflorae and family solanaceae, is one of the important commercial crop of India. The genus *Nicotiana* embraces over 66 species out of which only two species viz., *Nicotiana tabacum* L. and *Nicotiana rustica* L. are grown commercially in the world and which are naturally amphidiploids (2n=48) (Goodspeed, 1954).

In India, tobacco is grown mainly in Gujarat, Andhra Pradesh, Karnataka, Uttar Pradesh, West Bengal, Tamil Nadu, Orissa, Bihar and Maharashtra. Gujarat occupies first place from productivity view point followed by Andhra Pradesh. During 2007-08, tobacco production in the state was 78.7 M kg from about 0.45 lakh ha area under cultivation and productivity was 1722 kg/ha. Ninety per cent of tobacco grown in the state is accounted by bidi tobacco. Anand, Kheda, Vadodara, Mehesana, Banaskantha, Panchmahals, Dahod, Ahmedabad, Gandhinagar and Sabarkantha are important tobacco growing districts of Gujarat. Among all, Anand district stands first in production as well as hectrage.

N. tabacum is grown almost all throughout the country, whereas hardly 5 to 6% of total area under tobacco cultivation is accounted by *N. rustica,* confined mainly to the northern and north eastern areas of the country. India is the only country where different types of tobacco, viz. Flue Cured Virginia (FCV), burley, natu, cigar filler, cigar wrapper, cheroot, hookah, bidi and chewing are grown under different agro-climatic conditions.

Tobacco at present is a leading commercial crop valued for its leaf containing nicotine. In India, tobacco is grown on 0.45 million ha of area (accounting for only 0.31% of net cultivated area in the country) with 750 million kg production. The world tobacco production is ~7 billion kg, China occupying the first place with 2.35 billion kg. There are ten distinct tobacco types grown in 15 states of the country which include cigarette (FCV, burley, Oriental) and non-cigarette types (Bidi, chewing, hookah, natu, cheroot, cigar and HDBRG). India stands second in tobacco production and exports in the world. Tobacco earns annually 4,402 crores as foreign exchange and 13,853 crores as excise revenue. Its total contribution to the national economy is 18,255 crores. Tobacco cultivation in many countries is a family business. In India, the tobacco crop directly or indirectly supports ~36 million people engaged in production, processing, marketing and exports which include 6 million farmers and 5 million people involved in bidi rolling and tendu leaf-plucking. Thus, the crop is a lifeline for sizeable chunk of population, particularly rural women, tribal's and other weaker sections of the society. Conventionally, tobacco is used in the manufacture of cigarettes, bidis, scented chewing mixtures, cigars, cheroots, zarda, hookah, hookah tobacco paste, snuff, gutka and quiwam. However, the worldwide anti-tobacco campaign due to the perceived health hazards associated with the traditional form of tobacco consumption is posing a serious challenge to tobacco production, commerce and industry. Since the economic life line of millions of people world over depends on tobacco, the crop needs to be sustained by taking advantage of its potential for alternative uses. To convert this threat into an opportunity and to sustain the crop, it is imperative to intensify the research efforts towards channelizing tobacco into non-conventional and economically viable alternative uses. Owing to its easy amenability to genetic transformation and *in vitro* regeneration, tobacco is considered as a model plant for biotechnological research (Krishnamurthy, 2011).

Tobacco has potential to provide pesticides (nicotine sulphate and tobacco dust), progress in current research on use of nicotine compounds as pharmaceutical products for treatment of Alzheimer's disease, Parkinson's disease, Attention Deficit Disorder, Tourette's Syndrome and Ulcerative Colitis (Rao, 2005). Solenosol and its derivatives, nicotinamide and vitamin K, tobacco leaf proteins for food, organic acids for industries and pharmaceutical use, tobacco based cosmetics and toiletries like deodorants.

Tobacco could be developed as an important food crop in combination with its traditional use for smoking and chewing. As a food crop, tobacco gown in dense spacing could produce about four times more protein per acre than soybean. Green tobacco leaf contains about 2.3 to 3.5 per cent of crude protein. The biochemical analysis showed that tobacco leaf proteins (crude) contains 9.67% moisture, 6.50% ash, 0.90% crude fiber, 16.50% starch and 62.80% pure protein, which is superior to FAO provisional patterns of essential amino acids in protein with 4.2% isoleucine, 8.8% leucine, 5.8% lysine, 4.4% phenylalanine, 4.9% tyrosine, 1.6% methionine, 5.2% threonine, 1.5% tryptophan and 7.2% valine (Raju and Rao, 2005). Furthermore, biochemical profile insures that it doesn't contain any antinutritional factors like nicotine, polyphenol and nitrate (Patel and Parmar, 2005). In addition to this, we can get edible oil from tobacco seeds. In world, refined tobacco seed oil is being used as edible oil in Bulgaria, Tunisia, Turkey and Greece (Singh and Rao, 2005). The oil content in tobacco seed varies from 35-39% and is comparable to mustard (33%), sunflower (35%), safflower (30%), higher than soybean (17%) and cotton seed (12%). Fatty acid profile of tobacco seed oil revealed that it contained about 72% linoleic acid, which makes the oil more nutritive as it provides more essential fatty acids PUFA (Polyunsaturated fatty acids) value of tobacco seed oil is better than other edible oil and is almost comparable to safflower oil. Moreover, chemical analysis of crude oil didn't show presence of nicotinoids.

It is paradoxical that tobacco, which yields nicotine- a very effective insecticide, suffers from damage by many insect pests right from sowing to the leaf kept in storage. This is because of either selective feeding in phloem as in aphids or efficient excretory mechanism as in tobacco hornworm or capacity to metabolize nicotine to nontoxic cotinine and other alkaloids (Self *et al.*, 1964). The ants carry away the seeds from the nursery beds. Earthworms damage tiny seedlings by uprooting and covering them with the soil. Rove beetles (*Bledius latiusculus* Kraatz) making burrows which covers tender seedlings and also disturb the root system. Mole cricket (*Gryllotalpa africana* Palisot de Beauvois) feed on roots and near ground plant parts and damage to seed beds. Caterpillars (*Spodoptera litura* Fabricius *Laphygma exigua* Hubner) of various types including borers (stem borer, *Scrobipalapa heliopa* Lower) and grass hoppers invade the nursery as well as fields. Cutworm (*Agrotis ipsilon* Hufnagel) cuts the stalk of plants both in nurseries and fields. The ground beetles (*Mesomorphus villager* Blanch.; *Seleron latipes* Guer. and *Opatroides frater* Farum) cut the young transplants causing gaps in the fields. Aphids (*Myzus persicae* Sulzer) and mealy bug (*Phenococcus solenopsis* Tinsley) infest plants in the growth stage and suck the juice from the leaves and by secreting honey-dew, on which black sooty mould develops, makes the leaves unfit for curing, whiteflies (*Bemasia tabaci* Gennadius) transmits leaf curl disease. Capsule borers (*Helicoverpa armigera* Hubner) attack capsules and sometimes also climb down to feed on leaves. Apart from these, there are other pests such as green semilooper (*Plusia signata* Fabricius), thrips (*Thrips tabaci* Linderman) and stink bugs (*Nezara viridula* Linneus) which cause damage. On the stored tobacco, cigarette beetle commonly known as cigarette weevil (*Lasiodrma serricorne* Fabricius) is a major pest. The works carried out by various scientists at different places on management tactics to control them are as under.

29.2. SOIL PESTS

29.2.1. Rove beetle, *Bledius latiusculus*

The beetle is a soil inhabiting pest and very active in sandy and sandy loam soils. The beetles cause damage during germination of seeds in the nursery. They create characteristic burrows in soil, excavate soil particles, which cover tender seedlings and also disturb the root system. The damage becomes more serious at each watering or when there is rains. The population of seedlings reduced drastically up to 80 per cent.

Application of beaurdex mixture (6-3-100) @ 4 l/1.2 m^2 bed, chlorpyriphos 0.05%, fenitrothion 0.05% and quinalphos 0.05% reduced rove beetle burrows at Anand, Gujarat. Moreover, the pest can be managed effectively by one drenching of either chlorpyriphos 0.04% or endosulfan 0.07% @ 2 l / m^2 one week after sowing (Patel, 2002). Recently, the drenching of 10% neem leaf extract @ 2 l/m^2 seven days after seeding recomme nded against the rove beetle damaging to tobacco seedlings (Anon., 2010a).

29.2.2. Tobacco cutworm, *Agrotis ipsilon*

Caterpillar cuts the stalk of seedlings as well as transplanted plants at ground level. They feed only at night and may be found in the day time either just below soil level or under leaves.

Carbaryl 5% @ 25 kg/ha was found superior and gave good control of cutworms under artificial infestation conditions in tobacco field at Anand, Gujarat (Patel, 2002). Whereas, the application of parathion 2% dust @ 30 kg/ha and endosulfan 4% dust @ 25 kg/ha on soil were found to be effective against cutworm in tobacco crop (Amin, 1977). Similarly, endosulfan dust @ 1.25 kg a.i./ha as soil application around the plant base at 15 and 35 days after transplanting or two application of chlorpyriphos @ 0.16 kg a.i./ha, endosulfan @ 0.28 kg a.i./ha and monocrotophos 0.24 kg a.i./ha as foliar spray were found effective against cutworm damage in cigar wrapper tobacco field (Chavan, 1985).

29.2.3. Mole cricket, *Gryllotalpa africana*

The mole cricket found damaging to Indian bidi-growing areas is *G. africana*. They feed on roots and near ground plant parts, but the main harm to tobacco is through physical damage to seedbeds. In the process of coming to the surface at night to feed, burrows are created just under the surface, which uproot germinating seedlings. Their preference for damp conditions tends to make seedbed areas attractive. Drenching of chlorpyriphos 0.04% and endosulfan 0.07% insecticides to control rove beetles also control this pest.

29.3. ABOVE-GROUND PESTS

29.3.1. Tobacco leaf eating caterpillar, *Spodoptera litura*

It is a worst enemy of tobacco nurseries. When the seedlings are small they cut the seedlings while they voraciously feed on the leaves and entire seedbeds will be defoliated within a week. The pest causes damage to an extent of 80-100% in the nurseries under favourable conditions (Chari *et al.*, 1994) and 10-25% in the field crop (Sitaramaih *et al.*, 2001).

Cultural practices such as ploughing in summer, collection and destruction of egg masses and clipping of leaves having swarms of tiny larvae help in checking the pest. Fenvalerate 0.02% gave quick mortality of *S. litura* in the tobacco field at Anand, Gujarat (Patel, 1984). (Chari *et al.,* 1985) reported that the spray application of diflubenzuron and nuclear polyhedrosis virus (NPV) and neem seed endosperm suspension as antifeedent gave healthy transplantable seedlings, whereas the use of NPV and antifeedent with castor as trap crop on all sides of the nursery is ecologically sound and economically beneficial to the farmers. Similarly, diflubenzuron 0.009% and neem seed kernel suspension (NSKS) treatments were found effective for reduction in larval population of *S. litura* and obtained high number of healthy tobacco seedlings for transplanting at Rajahmundry, Andhra Pradesh (Sitaramaiah *et al.*, 1986). Whereas, the application of NSKS 2% alone showed effective in protecting the leaf from *S. litura* with its antifeeding activity (Murthy *et al.*, 1989).

Prasad *et al.* (1991) found methomyl @ 240 and 300 g a.i./ha and endosulfan 265 g a.i./ha effective against *S. litura* on FCV tobacco. However, the *S. litura* in FCV tobacco crop effectively managed by integrating castor as a ring crop (trap crop), quinalphos 0.05%, NPV 250 LU/ha and NSKS 1% at Shimoga, Karnataka (Yogeshwara, 1991). Besides these, Koshiya and Patel (2000) at Anand, Gujarat reported that nicotine sulphate found superior against *S. litura* infesting tobacco seedlings under nursery conditions. Whereas Ramaprasad *et al.* (2000), advocated Bioasp (*Bacillus thuringiensis* var. *kurstaki asporogenic*) and Biolep (*B.t.* var. *kurstaki sporogenic*) @ 2 kg/ha against *S. litura* in tobacco nurseries. The SNPV caused 96% mortality of *S. litura* within a period of 10 days at a dosage of 6 x 10^8 PIBs/larva at Malaysia (Sajap *et al.,* 2000). Similarly, at Rajahmundry, Andhra Pradesh the sequential spray of biopesticides *viz., Sl*NPV 250 LE (1.5 x 10^{12} PIB/ha, *Btk* @ 1.0 kg/ha were found effective against *S. litura* in tobacco nurseries at Rajahmundry, Andhra Pradesh (Rao and Sitaramaiah, 2001).

The integration of hand collection and destruction of egg masses and grown up larvae, application of poison bait and need based spray of endosulfan @ 2 ml/l was found effective against *S. litura* in cigarette *natu* tobacco at Nandyal (Rao *et al.*, 2004). Indoxacarb, methomyl and chlorpyriphos @ 100, 400 and 500 ml per 100 l of water were found effective against *S. litura* on tobacco at Faisalabad, Pakistan (Ahmed *et al.,* 2004). A synthetic gene coding for the hybrid δ-endotoxin Cry1EC was designed for high level expression (0.1- 0.7%) of soluble leaf protein in tobacco plants were completely resistant to *S. litura* and caused 100% mortality in all stages of larval development (Singh *et al.,* 2004). The methofenozide, a moult accelerating compound @ 0.01% recorded cent per cent mortality at 2 days after treatment which was at par with its med-ium concentration, 0.0075% and chlorpyriphos 0.05% and it was superior to diflubenzuron 0.01% at 8 days after treatment under laboratory conditions. Methofenozide 0.01% was found to be the most persistent (up to 18 days) than that of methofenozide 0.0075%, chlorpyriphos 0.05%, methofenozide 0.005% and diflubenzuron 0.01% on tobacco nurseries against *S. litura* (Sreedhar *et al.*, 2004).

The entomopathogenic nematodes (*Steinrnema carpocapse*) @ 10 infective juveniles/larva was found effective against tobacco leaf eating caterpillar, *S. litura* and gave cent per cent mortality of 3rd, 4th and 5th instars larvae within 4 days under laboratory conditions (Gupta *et al.*, 2008). Whereas, Krishnamurthy (2009) reported the emamectin benzoate @ 11g a.i./ha effective against *S. litura* in tobacco seedbeds. The IPM module having castor as a trap crop and one spray of NSKS 2% and application of acephate 0.075% at economic threshold level ETL (one egg mass/20 plants) and chlorpyriphos 0.04% at ETL found effective to manage *S. litura* in transplanted bidi tobacco at Anand, Gujarat (Anon., 2010 b).

29.3.2. Tobacco stem borer, *Scrobipalpa heliopa*

The tiny caterpillars of the stem borer bore inside the stem and mid ribs and feed on internal tissues. As a result, swelling appears outside where the borer stays. It causes stunting and unusual branching of the seedlings. The transplanted crop is also affected by this pest and caterpillars bore their way through leaf veins into the stems and feed on inner tissues, thereby giving a swelling appearance outside, which hinders the growth of plant and ultimately resulted into heavy economical loss up to the extent of 25 per cent.

Application of carbaryl 0.15% at an interval of 15 days found to be effective against tobacco stem borer (Patel, 1973). According to Sorathia

(1980), quinalphos 0.05% and cypermethrin 0.0125% were toxic to eggs and quinalphos 0.05% to the first instar larvae on rustica tobacco. Similarly, cypermethrin 0.0125% and quinalphos 0.05% gave cent per cent mortality of eggs of *S. heliopa*, whereas phenthoate 0.05%, fenvalerate 0.01% and triazophos 0.06% caused more than 80% egg mortality at Anand, Gujarat (Anon., 1980). Whereas, methomyl @ 240 g a.i./ha and 300 g a.i./ha as well as endosulfan 265 g a.i./ha were found effective against stem borer on FCV tobacco and harvested higher cured leaf yield with good ICBR compared to unsprayed control (Prasad *et al.*, 1991). Moreover to these, removal and destruction of stubbles after harvesting of the crop helps in avoiding carry over of infestation in next season (Patel, 2002). *Chelonus heliopae* parasitizes the stem borer larvae to an extent of 3.72 to 49.6% (Reddy *et al.*, 2006).

29.3.3. Whitefly, *Bemasia tabaci*

Whiteflies are small fly like insects seen on the underside of leaves. They will fly immediately when the plant is disturbed. The adult fly sucks the sap from leaves and transmits the leaf curl virus diseases from infected to healthy seedlings.

Sanitary measures such as removal and destruction of curled plants and alternate hosts reduces the incidence. Moreover to this, application of dimethoate 0.045% or methyl-s-demeton 0.0375% or monocrotophos 0.06% and flucythrinate 0.005% gives effective protection at 15 days interval in nurseries starting from 30 days after seeding and in field crop, 2 to 3 sprays of the same insecticides at 10 days interval commencing from 15 days after planting (Patel, 2002). When tobacco field surrounded by bajara as border crop, which is cheap food for farming community the low population of whitefly was recorded on tobacco plants (Anon., 2010 a).

29.3.4. Aphids, *Myzus persicae*

Aphids sucks the sap and make the plants pale, sickly and thereby retard the growth. Secondly, the growth of black sooty mould which develops on the sugary secretions of aphids on leaves, makes them unfit for curing and if cured the quality of leaf gets reduced considerably.

Methyl-s-demeton 0.025%, dimethoate 0.03% and phosphamidon 0.03% were found superior to control aphid infestation on tobacco at Hebbal, Mahya and Mayakonda in Karnataka (Jagadish, 1978). Amongst various species of tobacco, *Nicotiana gossei* possessed immunity against aphids (Murthy and Ramaprasad, 2000). The sorghum reported as barrier crop and obstructed the movement of aphid into the tobacco field at Andhra Pradesh (Anon., 2010 b).

29.3.5. Mealy bug, *Phenococcus solenopsis*

The nymph and adult feed on phloem and excrete honey dew, which encourages the development of sooty mould that reduces plant photosynthetic abilities that ruins the plant's appearance and makes the leaves unfit for curing.

Bhatt *et al.* (2009) at Anand, Gujarat reported the methomyl 0.08%, profenophos 0.1% and carbaryl 0.2% were found effective against the mealy bug on tobacco.

29.3.6. Tobacco budworm/capsule borer, *Helicoverpa armigera*

Caterpillars bore the bud as well as capsule and destroy the seeds. Under the severe infestation they feed on the leaves also.

Chemosterilant tepa 0.05% and thiotepa 0.05% induce cent per cent sterility, whereas hempa was able to induce 68 per cent sterility in male moths of tobacco capsule borer (Mule, 1976). Whereas, decamethrin 0.025% was found effective to control capsule borer, *H. armigera* on bidi tobacco at Anand, Gujarat (Rao, 1983). However, the *Bt* dust formulation on tobacco seed capsules @ 7.5 x 10^{12} kg a.i./ha was at par with fenvalerate 0.4% dust @ 1 kg a.i./ha in protecting the capsule damage caused by *H. armigera* in tobacco at Rajahmundry, Andhra Pradesh (Rao and Sitaramaiah, 2001). Whereas, Ahmed *et al.* (2004) at Faisalabad, Pakistan reported that the *H. armigera* on tobacco effectively controlled with the application of indoxacarb, methomyl and chlorpyriphos @ 100, 400 and 500 ml per 100 l of water.

29.4. STORAGE PEST

Cigarette beetle, *Lasioderma serricorone* Fabricius. The larvae feed on the leaves making large galleries and when the infestation is heavy, turn the leaf into powder that deteriorate its quality, which affects price of the product.

Even small quantity of tobacco dust is enough for breeding of this insect and hence it is absolutely essential to keep the premises very clean by removing and destroying waste tobacco pieces and dust (Philip and Parkhe, 1957).

The godown should be free of cracks and crevices, properly white washed, well lighted, with double doors, windows and ventilators screened with wire mesh (20 meshes to 2.54 cm) to prevent entry from beetle from outside (Gopalachari, 1984). Moreover to these, the tobacco can be kept free of infestation by storing it at low temperatures (5 to 18°C) as cigarette beetles are very susceptible to low temperature. The twenty tablets of 3g of aluminium phosphide per 1000 cubic feet with 96 hours exposure kills all the stages of the beetle (Childs *et al.*, 1969).

29.5. TOBACCO IN PEST MANAGEMENT

Nicotine is the principal tobacco constituent having insecticidal property. No insect has ever developed immunity to this chemical and it biodegrades rapidly and has no residual effect. Chari *et al.* (1992) carried out biological evaluation of nicotine sulphate against various insect pests attacking different crops and reported the results against cotton pests. The study revealed that it caused quick knockdown effect against brown plant hopper (*Nilaparvata lugens* Stal.). The mortality of this pest was up to 97 to 100% within 30 and 60 minutes, respectively and the same level of mortality was observed in paddy green leaf hoppers (*Nephotettix* spp.). Significant efficacy of nicotine sulphate at 0.05% against safflower aphid (*Euroleucon compositae* Theeh) was observed. Among the vegetable pests, grubs of *Epilachna* beetle (*Henosepilachna vigintioctopunctata* F.) on egg plant, diamond back moth (*Plutella xylostella* L.) on cauliflower and *Aphis gossypii* Glover on tomato was controlled effectively with application of nicotine sulphate at 0.2%. Tobacco decoction at 2.5% against *B. tabaci* recorded 100% mortality in 96 hr. Tobacco has been also used as a bird repellent against rose ringed parakeet, *Psittacula krameri* through application of tobacco leaf powder decoction (10%) on sorghum ear head at Anand, Gujarat (Mehta *et al.*, 1994).

REFERENCES

Ahmed, S., Rasool, M.R., Irfanullah and Rauf, I. (2004) Comparative efficacy of some insecticides against *Helicoverpa armigera* Hub. and *Spodoptera spp.* on tobacco. *Internat. J. Agri. Biol.* **6**(1): 93-95.

Amin, P.R. (1977) Life history and control measures of the tobacco cutworm *Agrotis ipsilon* (*Hufnagel*) (Lepidoptera : Noctuidae) M.Sc. (Agri.) Thesis submitted to GAU, Sardar Krishi Nagar.

Anonymous (1980) Annual Report (1979-80) BTRS, GAU, Anand pp. 104-105.

Anonymous (2010a) Annual Report (2009-10), 6th Plant Protection Sub-Committee Meeting, BTRS, AAU, Anand. pp. 18-34.

Anonymous (2010b) Annual Report (2009-10), AINRP on Tobacco, CTRI, Rajahmundry. pp. 268-279.

Bhatt, N.A., Jayani, D.B. and Patel, A.D. (2009) Proc. of the National Symposium IPM Strategies to Combat Emerging Pest in the Current Scenario of Climate Change held at CAU, Pasighat on January 28-30, 2009, p. 38.

Chari, M.S., Bharpoda, T.M. and Patel, S.N. (1985) Studies on integrated management of *Spodoptera litura* F. in tobacco nursery. *Tob. Res.* **11**(2): 93-98.

Chari, M.S., Rao, R.S.N. and Prabhu, S.R. (1992) Bioefficacy of nicotine sulphate against pests of different crops. *Tob. Res.* **18**: 113-116.

Chari, M.S., Rao, R.S.N. and Sreedhar, U. (1994) Integrated management of *Spodoptera litura* F. in India. *Bull. Spec. CORESTA. 1994 Congress*, Harare, pp. 99.

Chavan, V.M. (1985) Field evaluation of some insecticides against *Agrotis ipsilon* Hafnagel on cigar wrapper tobacco. *Tob. Res.*, **11**(2): 167-171.

Childs, D.O., Overby, J.E. and Niffengger, D. (1969) Phosphine fumigation of flue-cured tobacco warehouses for control of the cigarette beetle. *Tob. Sci.* **13**: 64-69.

Goodspeed, T.H. (1954) *The genus Nicotiana.* Chronica Bot Wallham, Mass, USA.

Gopalachari, N.C. (1984) *Tobacco.* ICAR, New Delhi. 327 pp.

Gupta S., Kaul, V., Srivastava, K. and Monbrullah, M.D. (2008) Pathogenicity and in vitro culturing of a local isolate of *Steinernema carpocapsae* against *Spodoptera litura* (Fab.) *Indian J. Ent.* **70** (4) : 346-349.

Jagadish, A. (1978) Evaluation of insecticides against aphids (*Myzus persicae* Sulzer) and caterpillars (*Spodoptera litura* F.) on tobacco in Karnataka state. *Phytoparasitica* **6**(2): 45-50.

Koshiya, D.J. and Patel, A.R. (2000) Management of leaf eating caterpillar (*Spodoptera litura* F.) through neem based formulations in tobacco nursery. A paper presented in the National Symposium on Tobacco held at Rajahmundry, December 13-16, 2000. 32 pp.

Krishnamurthy, V. (2009) Tobacco research in India : Opportunities and challenges. *Souvenir : National Symposium on Tobacco-2009; Challenges and Opportunities*, November 24-26, 2009. CTRI, Rajahmundry. 1-7 pp.

Krishnamurthy, V. (2011) "Tobacco-New Dimensions" at the 219th meeting of the governing body of the ICAR society on 4.1.2011 at the NASC complex, New Delhi. (http://www.ctri.org.in)

Mehta, D.M., Parasharya, B.M. and Sukhadia, A.G. (1994) Evaluation of tobacco leaf powder decoction as a bird repellent against Rose ringed Parakeet *Psittacula krameri*. A paper presented in *"National Symposium on Emerging Trends in Pest Management"* held at Solan, Himachal Pradesh, from 28-30, 1994.

Mule, N.M. (1976) Chemosterilization of adult male moth of tobacco capsule borer, *Heliothis armigera* (Hub.) with tepa, hempa and thiotepa. M.Sc. (Agri.) Thesis submitted to GAU, Sardar Krishi Nagar.

Murthy, P. S.N., Ramaprasad, G., Sitaramaiah, S., Rao, B.V.K., Prabhu, S. R. and Rao, S. N. (1989) Studies on the enrichment of neem fractions and their relative efficacy against tobacco caterpillar (*Spodoptera litura* F.) on Lanka and FCV tobacco. *Tob. Res.* **15**(1): 65-70.

Murthy, T.G.K. and Ramaprasad, G. (2000) Aphid resistance in some *Nicotiana* species and interspecific hybrids. *Tob. Res.* **26**(2): 52-56.

Patel, A.D. (2002) Fifty Years of Tobacco Research. Bidi Tobacco Research Station, Gujarat Agricultural University, Anand. p. 82.

Patel, B.H. (1973) The bionomics and control measures of the tobacco stem borer (*Gnorimoschema heliopa* Low) M.Sc. (Agri.) Thesis submitted to GAU, Sardar Krishi Nagar.

Patel, B.K. and Parmar, N.B. (2005) Bidi tobacco for protein and seed oil. pp. 64-73. (Source: Alternative uses of Tobacco by Singh, K.D. and Jain, K.C., CTRI, Rajahmundry)

Patel, D.N. (1984) Bioefficacy of fenvalerate on tobacco leaf eating caterpillar, *Spodoptera litura* F. M.Sc. (Agri.) Thesis submitted to GAU, Sardar Krishi Nagar.

Philip, A. and Parkhe, D.P. (1957) Take care of your storage premises, cigarette beetle can live part of its life in the dust outside the tobacco storage. *Indian Tob.* **7**: 267.

Prasad, N.K.P., Mallikharjunaih, H., Yogeshwara, B.S. and Shetty, M.V.N. (1991) Evaluation of Methomyl in the management of FCV tobacco pests. *Tob. Res.* **17**(1): 41-43.

Raju, K.S. and Rao, C.V.N. (2005) Tobacco proteins. pp. 44-55. (Source: Alternative uses of Tobacco by Singh, K.D. and Jain, K.C., CTRI, Rajahmundry)

Rao, B.S. (1983) Population studies of *Heliothis armigera* (Hub.) and its insecticidal control on bidi tobacco. M.Sc. (Agri.) Thesis submitted to GAU, Sardar Krishi Nagar.

Rao, K.L. (2005) Value added products from tobacco. pp. 89-119. (Source: Alternative uses of Tobacco by Singh, K.D. and Jain, K.C., CTRI, Rajahmundry)

Rao, R.G.M.V., Saralamma, S. and Sudhakar, K. (2004) Management of the tobacco leaf eating caterpillar *Spodoptera litura* F. in cigarette *natu* tobacco. *Tob. Res.* **30**(1): 61-64.

Rao, S.G. and Sitaramaiah, S. (2001) Application strategies for microbial pesticides against *Spodoptera litura* F. and *Helicoverpa armigera* Hb. in Tobacco. *Tob. Res.* **27**(1): 12-18.

Reddy, P. R.S., Singh, K.D. and Jain, K.C. (2006) *Achievements in Tobacco Research.* CTRI, Rajahmundry. 71 pp.

Sajap, A.S., Kotuli, J.R., Bakir, M.A., Hong, L.W., Samad, N.A. and Kadir, H.A. (2000) Pathogenicity and characteristics of *Spodoptera litura* Nucleopolyhedrovirus from Peninsular Malaysia. *Pertanika J. Trop. Agric. Sci.* **23**(1): 23-28.

Ramaprasad, G., Sitaramaiah, S. and Sreedhar, U. (2000) Efficacy of *Bacillus thuringiensis* Berliner formulations against tobacco caterpillar, *Spodoptera litura* F. in tobacco nurseries. *Tob. Res.* **26**(2): 1-6.

Self, L.S., Guthire, F.E. and Hodgson, E. (1964) Metabolism of nicotine by tobacco feeding insects. *Nature* London. **204**: 300-301.

Singh, K.D. and Rao, C.V.N. (2005) Tobacco seed oil. pp. 56-63. In: *Alternative uses of Tobacco*, Singh, K.D. and Jain, K.C. (eds.) CTRI, Rajahmundry.

Singh, P.K., Kumar, M., Chaturvedi, C.P., Yadav, D. and Tuli, R. (2004) Development of a hybrid δ-endotoxin and its expression in tobacco and cotton for control of a polyphagous pest *Spodoptera litura*. *Transgenic Res.* **13:** 397-410.

Sitasramaiah, S., Ramaprasad, G. and Joshi, B.G. (1986) Efficacy of diflubenzuron and neem seed kernel suspension against the tobacco caterpillar, *Spodoptera litura*, in tobacco nurseries. *Phytoparasitica* **14**(4): 265-271.

Sitaramaih, S., Sreedhar, U. and Satyanarayana, S.V.V. (2001) Management of tobacco leaf eating caterpillar, *Spodoptera litura* F. with insecticide baits in NLS tobacco. *Tob. Res.* **27**: 7-11.

Sorathia, B.K. (1980) Laboratory and field evaluation of different insecticides against tobacco stem borer, *Scrobipolpa heliopa* Low. and its general biology on rustica tobacco (*Nicotiana rustica* L.) M.Sc. (Agri.) Thesis submitted to GAU, S.K. Nagar.

Sreedhar, U., Ramaprasad, G. and Sitaramaiah, S. (2004) Bio-efficasy of methofenozide, a novel moult accelerating compound against tobacco caterpillar, *Spodoptera litura* F. in tobacco nurseries. *Tob. Res.* **30**(1) : 72-77.

Yogeshwara, B.S., Mallikharjunaiah, H., Krishnaprasad, N.K. and Shetty, M.V.N (1991) Integrated management of *Spodoptera litura* Fabricius (Noctuidae : Lepidoptera) in FCV tobacco crop. *Tob. Res.* **17**(2): 59-61.